Purines in Cellular Signaling

Awards Presented at the Conference on Purines Nucleosides
and Nucleotides in Cell Signalling,
Sept. 17, 1989, Bethesda, Maryland to:

Dr. John W. Daly
for "Pioneering Research on
the Biology and Chemistry
of Adenosine",
sponsored by CIBA-Geigy
Corp., Summit, NJ

Prof. Geoffrey Burnstock
for "The Concept of
Purinergic Transmission"
sponsored by Nova
Pharmaceutical Corp.,
Baltimore, MD

K.A. Jacobson   J.W. Daly   V. Manganiello
Editors

Purines in Cellular Signaling

Targets for New Drugs

With 95 Illustrations

Springer-Verlag
New York Berlin Heidelberg
London Paris Tokyo Hong Kong

Kenneth A. Jacobson, PhD, Laboratory of Chemistry, National Institute of Diabetes and Digestive and Kidney Diseases, NIH, Bethesda, Maryland 20892, USA

John W. Daly, PhD, Laboratory of Bioorganic Chemistry, National Institute of Diabetes and Digestive and Kidney Diseases, NIH, Bethesda, Maryland 20892, USA

Vincent Manganiello, MD, PhD, Laboratory of Cellular Metabolism, National Heart, Lung, and Blood Institute, NIH, Bethesda, Maryland 20892, USA

On the front cover: Fluorescence histochemical localization of neurons in the rat intestine showing positive staining for quinacrine, a fluorescent compound known to bind strongly to ATP (prepared by Dr. Abi Belai, Anatomy and Developmental Biology, University College, London). Graphs: A comparison of the responses to exogenous adenosine triphosphate (ATP) and intramural nerve stimulation (NS) in the presence of atropine and guanethidine, inhibitory response of guinea pig taenia coli. *Upper left*: Ns, 1 Hz, 0.5 msec pulse duration, supramaximal voltage; *lower right*: ATP, conc. 2 μM. Courtesy of Prof. G. Burnstock, University College London.

Purines in cellular signaling: targets for new drugs/edited by Kenneth A. Jacobson, John W. Daly, Vincent Manganiello.
p. cm.
   Includes bibliographical references.
   ISBN 0-387-97244-7 (alk. paper)
   1. Purine nucleotides. 2. Adenosine--Receptors. 3. Second messenger (Biochemistry) I. Jacobson, Kenneth Alan, 1953- .
II. Daly, John W. III. Manganiello, Vincent.
QP625.P87P87 1990
615'.0157--dc20   90-30962

Printed on acid-free paper.

Camera-ready copy provided by the editors.
Printed and bound by Edwards Brothers, Ann Arbor, Michigan.
Printed in the United States of America.

9 8 7 6 5 4 3 2 1

ISBN 0-387-97244-7 Springer-Verlag New York Berlin Heidelberg
ISBN 3-540-97244-7 Springer-Verlag Berlin Heidelberg New York

Preface

The meeting "Purine Nucleosides and Nucleotides in Cell Signaling: Targets for New Drugs," held on September 17-20, 1989, in Rockville, Maryland, sparked renewed interest in purines as biological regulators and as potential targets for drug development. The purine molecules which are the main focus of this volume are adenosine, adenosine 5'-triphosphate (ATP), cyclic 3',5'-adenosine monophosphate (cAMP), and cyclic 3',5'-guanosine monophosphate (cGMP). These purines act at different recognition sites, to produce a wide range of biological effects. Adenosine acts at its specific receptors on the cell surface, which occur in many organs and tissues and have different regulatory functions. Adenosine receptors will be the major, single topic considered in this volume. ATP, related structurally and metabolically to adenosine, acts at its own distinct receptors on the cell surface. The cyclic purine nucleotides, cyclic AMP and cyclic GMP, act as intracellular signals. Cyclic nucleotide phosphodiesterases represent a group of distinct and highly regulated enzymes which control cellular concentrations of cyclic nucleotide second messengers by regulation of their destruction. This volume encompasses research on the biology and chemistry of all of these purines and the proteins with which they interact.

The subject matter of this book has been assembled to have a broad appeal, to purely basic researchers and those with a clinical orientation, alike, from medicinal chemists to physiologists. Specifically, internationally recognized experts have contributed papers on: i) various aspects of formation, action, and deposition of endogenous adenosine and adenine nucleotides as related to the physiological functions of these regulatory substances; ii) the definition of the nature of the receptors and effector systems subserving these substances; iii) the development of synthetic agents as agonists and antagonists for such receptors and as inhibitors of phosphodiesterases that serve to limit the extent and duration of action of effector systems acting through cyclic AMP and cyclic GMP; and iv) the definition of the nature and roles of phosphodiesterases.

The presentations are organized so as to focus on the potential development of the selective and potent agents that activate or block purinergic receptors or inhibit phosphodiesterases as clinically useful drugs. Adenosine is useful clinically in the treatment of cardiac arrhythmias, and selective adenosine agonists and antagonists are being evaluated as potential drugs for the cardiovascular system, the central nervous system, kidneys, etc. In the future, agents acting selectively at ATP receptors, present notably in vasculature and muscle, may provide a means of treating hypertension.

Different phosphodiesterases have been characterized and their roles defined in the regulation of various physiological processes, including visual transduction, platelet aggregation, myocardial contractility, and lipolysis. In some cases the recent development of specific and selective inhibitory drugs has provided indispensable probes for examination of intracellular function of specific phosphodiesterases. In the near future, molecular cloning and cellular biological approaches should further elucidate structure-function characteristics of the different phosphodiesterases, as well as organ- and cell-specific regulatory mechanisms. The appropriate probes will be used to define functions of phosphodiesterases in normal and pathological states and to devise appropriate therapeutic agents.

The abstracts that appear at the end of this volume were presented in poster format at the Rockville meeting and are numbered according to their original sequence in the program. Abstracts are included at the option of the authors: any gap in the sequence of abstracts reflects a desire of the participant not to submit material for inclusion in the book. Approximately 80 contributions included here complete a comprehensive treatment of purine derivatives and their receptors.

K.A. Jacobson
J.W. Daly
V. Manganiello

Contents

Contributors

H . Ali
Laboratory of Chemical Pharmacology
National Heart, Lung,
and Blood Institute
Building 10, Room 8N114
National Institute of Health
Bethesda, MD 20892, USA

J.A. Beavo
Department of Pharmacology
SJ-30, University of Washington
Seattle, WA 98195, USA

L. Belardinelli
Department of Medicine
Division of Cardiology
University of Florida
Gainesville, FL 32610, USA

R. Berne
Department of Physiology
University of Virginia
School of Medicine
Box 449, Jordan Hall
Charlottesville, VA 22908, USA

M.L. Billingsley
Department of Pharmacology
Cell and Moleuclar Biology Center
Pennsylvania State University
School of Medicine
Box 850
Hershey, PA 17033, USA

J.L. Boyer
Departmento de Endocrinologia
Instituto Nacional de Cardiología
"Ignacio Chávez"
Juán Badiano # 1
Tlalpan, México D.F. 14080

R.F. Bruns
Eli Lilly and Company
Lilly Corporate Center
Indianapolis, IN 46285, USA

G. Burnstock
University College London
Gower Street
London WC1 E6BE
United Kingdom

A.S. Clanachan
Department of Pharmacology
Faculty of Medicine
University of Alberta
Edmonton, Canada T6G 2H7

M. Collis
Bioscience II
ICI Pharmaceuticals
Alderley Park
Macclesfield, Cheshire
United Kingdom

R.W. Colman
Thrombosis Research Center
Temple University
School of Medicine
3400 North Broad Street
Philadelphia, PA 19140, USA

M. Conti
CB#7500, MacNider Building
The Laboratories for Reproductive Biology
School of Medicine
The University of North Carolina
Chapel Hill, NC 27599-7500, USA

B.N. Cronstein
Department of Medicine
NYU Medical Center
550 First Avenue
New York, NY 10016, USA

N.J. Cusack
Director, Pharmacology
Nelson Research
1001 Health Sciences Road West
Irvine, CA 92715, USA

J.W. Daly
Laboratory of Bioorganic Chemistry
National Institute of Diabetes
and Digestive and Kidney Diseases
Building 8A, Room B1A-17
National Institute of Health
Bethesda, MD 20892, USA

R.L. Davis
Department of Cell Biology
Baylor College of Medicine
Houston, TX 77030, USA

T.W. Dunwiddie
Department of Pharmacology
C236 University of Colorado
4200 East 9th Avenue
Denver, CO 80262, USA

I.H. Fox
Department of Internal Medicine
University of Michigan
University Hospital
Ann Arbor, MI 48109-0108, USA

S.H. Francis
Howard Hughes Medical Institute
Vanderbilt University
School of Medicine
Nashville, TN 38232, USA

B.B. Fredholm
Department of Pharmacology
Karolinska Institut
Box 60400
S-104 01 Stockholm
Sweden

J. Geiger
Neuropharmacology Section
Department of Pharmacology
University of Manitoba
770 Bannatyne Avenue
Winnipeg, Manitoba R3E OW3
Canada

J.G. Gerber
Division Clinical Pharmacology
Box C-237
University of Colorado
4200 East 9th Avenue
Denver, CO 80262, USA

L.E. Gustafsson
Department of Physiology
Karolinska Institute
Box 60 400
104 01 Stockholm
Sweden

R.E. Howell
Nova Pharmaceutical Corporation
6200 Freeport Centre
Baltimore, MD 21224, USA

A.J. Hutchison
Neurogen Corporation
35 NE Industrial Road
Branford, CT 06495, USA

K.A. Jacobson
Laboratory of Chemistry
National Institute of Diabetes
and Digestive and Kidney Diseases
Building 8A, Room B1A-17
National Institute of Health
Bethesda, MD 20892, USA

R.A. Johnson
Department of Physiology and Biophysics
School of Medicine
Health Sciences Center
State University of New York
Stony Brook, NY 11794-8661, USA

H. Kather
Institut für Herzinfarktforschung
Meizinische Universität-Klinik
Bergheimerstrasse 58
6900 Heidelberg
Federal Republic of Germany

J. Linden
Department of Physiology and Medicine
University of Virginia
School of Medicine
Box 449
Charlottesville, VA 22908, USA

M.J. Lohse
Duke University Medical Center
HHH1, Box 3821
Durham, NC 27710, USA

J. Lowenstein
Department of Biochemistry
Brandeis University
415 South Street
Waltham, MA 02254, USA

V. Manganiello
Laboratory of Cellular Metabolism
National Heart, Lung, and Blood
Institute
Bldg. 10, Room 5N323
National Institute of Health
Bethesda, MD 20892, USA

P.J. Marangos
Research Director Gensia
Pharmaceuticals, Inc.
11075 Roselle Street
San Diego, CA 92121-1207, USA

T.F. Murray
Oregon State University
College of Pharmacy
Corvallis, OR 97331, USA

H. Nakata
Laboratory of Clinical Science
National Institute of Mental Health
National Institutes of Health
Building 10, Room 3D48
Bethesda, MD 20892, USA

D. Ortwine
Parke-Davis Pharmaceutical Research
Division
2800 Plymouth Road
Ann Arbor, MI 48105-2430, USA

J.D. Pearson
Vascular Biology Section
Medical Research Council
Clinical Research Center
Watford Road Harrow, HA1 3UJ
United Kingdom

J.W. Phillis
Department of Physiology
Wayne State University
540 East Canfield
Detroit, MI 48201, USA

K.G. Proctor
Department of Physiology
University of Tennessee
Health Science Center
894 Union Avenue
Memphis, TN 38163, USA

A. Rascon
Department of Medicinal and
Physiological Chemistry
University of Lund
P.O. Box 94
S-221 00 Lund
Sweden

J.A. Ribeiro
Laboratory of Pharmacology
Gulbenkian Institute of Science
2781 Oeiras
Portugal

D.W. Robertson
Lilly Research Laboratories
Lilly Corporate Center
Indianapolis, IN 46285, USA

H. Schneider
Schering AG
Department of
Neuropsychopharmacology
Postfach 65 03 11
1000 Berlin 65
Federal Republic of Germany

J. Schrader
Department of Physiology
University of Düsseldorf
Moorenstrasse 5
4000 Düsseldorf
Federal Republic of Germany

U. Schwabe
Pharmakologisches Institut
der Universitat Heidelberg
Im Neuenheimer Feld 366
6900 Heidelberg 1
Federal Republic of Germany

P. Silver
Section Director
Cardiopulmonary Pharmacology
Sterling Drug Inc.
81 Columbia Turnpike
Rensselaer, NY 12144, USA

C.J. Smith
Laboratory of Cerebral Metabolism
National Institute of Mental Health
Building 36, Room 1A27
National Institutes of Health
Bethesda, MD 20892, USA

W.S. Spielman
Department of Biochemistry
Michigan State University
East Lansing, MI 48824, USA

G. Stiles
Department of Medicine
Duke University
School of Medicine
Box 3444
Durham, NC 27710, USA

B.K. Trivedi
Parke-Davis Warner Lambert Company
2800 Plymouth Road
Ann Arbor, MI 48105, USA

R.E. Weishaar
Parke-Davis Warner Lambert Company
2800 Plymouth Road
Ann Arbor, MI 48105, USA

J.N. Wells
Department of Pharmacology
School of Medicine
Vanderbilt University
Nashville, TN 37232, USA

D. Westfall
Department of Pharmacology
School of Medicine
University of Nevada
Howard Medical Science Building
Reno, NV 89557, USA

M. Whalin
Department of Pharmacology
University of South Alabama
College of Medicine
Mobile, AL 36688, USA

✓ M. Williams
D-47W Abbott Laboratories
Abbott Park, IL 60064-3500, USA

Section 1
Adenosine Receptors and Effector Systems

1
Adenosine Agonists and Antagonists

J.W. Daly

Introduction

The modern era of research on the nature of adenosine receptors began in the early seventies. At that time the stimulatory effect of adenosine on formation of cyclic AMP in brain slices and blockade by xanthines, such as caffeine and theophylline, was discovered (Sattin and Rall, 1979). The class of adenosine receptors that are coupled to adenylate cyclase in an inhibitory manner have become known as A_1-receptors, while the class of adenosine receptors that are coupled to adenylate cyclase in a stimulatory manner have become known as A_2-receptors. The rank order of potencies for certain agonists differed at A_1- and A_2-receptors and it was suggested that rank order of potencies at 5'-N'ethylcarboxamidoadenosine (NECA), 2-chloro-adenosine, N^6-R-phenylisopropyladenosine and N_6-S-phenylisopropyladenosine might be useful for definition of the class of adenosine receptor involved in physiological responses (see Daly, 1982). This series of agonists has not proven to be entirely satisfactory and more definitive series of selective agonists and antagonists have been and are being sought. The lack of complete correspondence of rank orders of potencies of agonists in certain physiological systems to that expected of A_1 receptors coupled to adenylate cyclase has led to the proposal of an A_3-receptor, which is localized in presynaptic nerve terminals and which serves to inhibit neurotransmitter release (Ribeiro and Sebastião, 1986). In the synaptic terminals a variety of evidence suggests that the receptor is coupled not to adenylate cyclase, but instead in an inhibitory manner to calcium channels or calcium function. There also is evidence for adenosine receptors that are coupled in a stimulatory manner to potassium channels in brain and heart (see Fredholm and Dunwiddie, 1988). At present the adenosine receptors coupled to calcium and potassium channels are perhaps best still treated as subclasses of A_1 receptors, since unambiguous paradigms for distinguishing these receptors from A_1 receptors coupled to adenylate cyclase have not been developed.

A variety of data have provided evidence for various "subclasses" of A_1- and A_2-adenosine receptors. Certainly, a unique "high affinity" subclass of A_2-receptor exists in the limbic system, but not in other brain regions (Daly et al., 1983). Two subclasses of A_2 receptors have been proposed with A_{2a} being the "high affinity" subclass present in striatum and other limbic structures, while A_{2b} is the "low affinity" subclass present throughout brain (Bruns et al., 1986). 2-Phenylaminoadenosine appears a useful tool for differentiating A_{2a} and A_{2b} subclasses, since it is relatively potent at the former, but not at the latter.

In smooth muscle preparations the nature of the effector system, to which adenosine receptors are coupled, has been unclear. However, the profile of potencies of agonists suggested that relaxation of smooth muscle involved an A_2 receptor. Recent evidence (Munshi et al., 1988) suggests either that this response is mediated by a subclass of A_2 receptors different in their response to 5'-methylthioadenosine than A_2-adenosine receptors coupled to adenylate cyclase in other cell types, or that the response is mediated by a further class of adenosine receptors not coupled to adenylate cyclase.

Adenosine receptors appear to be able in some systems to mediate an inhibition of phospholipase C-catalyzed phosphoinositide breakdown (see Linden and Delahunty, 1989) and of phospholipase A_2-catalyzed arachidonate formation (Schimmel and Elliott, 1988). Whether such effects are direct or indirect at present is unknown. In contrast, activation of an adenosine receptor enhances phosphoinositide breakdown in cultured RBL2H3 cells (H. Ali et al, this volume). Adenosine has been reported to stimulate cyclic AMP formation in brain slices (Ohga and Daly, 1977) and, recently, guanylate cyclase in vascular smooth muscle cells (Kurtz, 1989).

Virtually all adenosine receptor-mediated responses are blocked by caffeine/theophylline or xanthine analogs. Indeed, blockade by xanthines remains an important criteria for definition of an adenosine receptor-mediated response. However, there also is an intracellular regulatory site on adenylate cyclase, the "P-site" (Londos et al., 1980) through which adenosine and certain analogs inhibit adenylate cyclase activity. This site is not blocked by xanthines.

Activation of central receptors for adenosine leads to sedation, indeed a hypnotic-like state, while other effects of adenosine receptor activation include muscle relaxation, hypotension, hypothermia, cardiac depression and analgesia (see Williams, 1987). Certain adenosine analogs have been purported to have antipsychotic activity (Bridges et al., 1987). The anticonvulsant properties of adenosine and analogs are well known and adenosine has been referred to as a "neuroprotective" agent (see Dragunow and Faull, 1988). Xanthines can antagonize such effects and indeed caffeine and other xanthines can cause behavioral excitation and anxiety, muscle contractures, hypertension, hyperthermia, cardiac stimulation and can reverse morphine-induced analgesia. Furthermore, caffeine has been thought to be linked to relapses in schizophrenic patients (Mikkelsen, 1978), and caffeine and theophylline are well known to have convulsant activity at high dosages.

While the effects of agonists (adenosine and analogs) and antagonists (xanthines) are consonant with a major involvement of blockade of adenosine receptors as the basis for pharmacological effects of theophylline, caffeine, and other xanthines, further research is needed to define the class of adenosine receptor involved in different effects of caffeine/theophylline and to ascertain to what extent other mechanisms may play a role. To this end, the development of further "molecular probes" for adenosine receptors clearly had a high priority, since such probes if selective and potent as agonists or antagonists would provide tools for detailed investigation of the properties and roles of adenosine receptors, phosphodiesterases and other sites involved in the pharmacological profile of action of caffeine and theophylline.

Development of Molecular Probes

Two approaches have been employed for the development of "molecular probes" for adenosine receptors. One is the classical approach in which new compounds are developed, based on modification at one or more centers of the parent pharmacophore, in this case adenosine as the parent agonist,

4

and theophylline (1,3-dimethylxanthine) or caffeine (1,3,7-trimethylxanthine) as the parent agonist. The other is the functional congener approach in which a spacer chain attached at one end to the parent pharmacophore and terminating at the other end in a functional group, such as an amine, alcohol or carboxylic acid is developed. Such a functional group in this congener then serves as a reactive site for the covalent attachment to a variety of carrier entities such as amino acids, peptides, lipids, fluorophores, spin-labels, biotin, photo-affinity labels, etc. (see K.A. Jacobson, this volume)

The Classical Approach: Adenosines

The adenosine molecule proved to have a limited number of centers at which the structure could be modified with retention of activity.

The 1-position: N^6-substituted 1-deazaadenosines retain both A_1- and A_2-receptor activity (Cristalli et al., 1988). 1-Methylisoguanosine is active at both A_1 and A_2 receptors (Ukena et al., 1987a).

The 2-position: A variety of groups including 2-fluoro, 2-chloro, 2-hydroxy, or 2-phenylamino can be introduced with retention of activity at both A_1- and A_2-receptors (see Bruns, 1980; Ukena et al., 1987a). The 2-phenylamino group increases A_2-selectivity by markedly reducing affinity for the A_1 receptor. In contrast, 2-p-methoxyphenyladenosine is inactive at an A_2 receptor coupled to adenylate cyclase (Bruns, 1980), but is active at A_1 receptors (Paton, 1980). In binding assays it is highly A_1 selective (Bruns et al., 1980, 1987a). Recently, certain 2-alkynyladenosines have appeared to be highly selective for A_2-receptors (Matsuda and Ueda, 1987). Certain 2-alkoxyadenosines are also A_2 selective (R.A. Olsson, personal communication). Thus, the 2-position may prove to be an important entry point in the development of A_2-selective agonists. This has proven to be true and a 2-p-(2-carboxyethyl)phenethylamino-NECA has been recently introduced as a potent, highly selective A_2 receptor agonist (Hutchison et al., 1989).

The 6-position: An amine at the 6-position of purine ribosides appears to be required for agonist activity at adenosine receptors. However, a large variety of substituents can be accepted on the N^6-amino group of adenosine and structure activity analyses have allowed some delineation of the so-called N^6-binding domain of both A_1 and A_2 receptors (see Daly et al., 1986b; Ukena et al., 1987a). Most N^6-substituted adenosines are relatively selective for A_1 receptors and, indeed, some can be used as nearly specific A_1 receptor agonists. These include the N^6-cycloalkyl-adenosines (Daly et al., 1986b; Moos et al., 1985; Ukena et al., 1987a), 2-chloro-N^6-cyclopentyladenosine (Lohse et al., 1988b) and certain N^6-bicycloalkyladenosines (Trivedi et al., 1989). However, there are certain N^6-fluorenylmethyl- and N^6-2,2'-diphenylethyl- adenosine and NECA derivatives that are very potent at both A_1 and A_2 receptors (Bridges et al., 1988; Trivedi et al., 1988). The ability of the binding domain for N^6-substituents to accept, while still being influenced by such a wide range of structural entities led to the choice of N^6-(p-carboxylmethyloxyphenyl)adenosine as a starting point for a "functionalized congener" approach to adenosine receptor antagonists (see K.A. Jacobson, this volume). An N^6-cyclohexyl-2',3'-dideoxyadenosine has activity as an antagonist at both A_1 and A_2-receptors (Lohse et al., 1988a). It did not appear to have "P-site" activity.

The 5-position: Other positions containing hydroxyl groups (2',3') in the ribose ring appear essential to activity at adenosine receptors. The oxygen ring function can be replaced with sulfur or a methylene group, albeit with loss of potency at A_2 receptors (Bruns, 1980). Replacement of

5

the 5'-CH$_2$OH moiety of adenosine with 5'-CONHC$_2$H$_5$ yields NECA, an extremely potent A$_2$ agonist, which, however, retains high potency at A$_1$ receptors. An N-ethyl or N-cyclopropyl substituent appears optimal for activity at adenosine receptors; the primary amide and methylamide are less active, while larger substituents result in complete loss of activity (see Bruns, 1980; Olsson et al., 1986 and ref. therein). It appears likely that the acidic hydrogen of the amide group serves the same hydrogen donor function as the alcohol hydrogen in adenosine. Consonant with this interpretation, the 5'-dimethylamide is inactive. A comparison of activities of N^6-substituted adenosines with activities of corresponding N^6-substituted NECAs indicates that interaction of the N^6-substituent with its binding subdomain tends to negate the effect of binding of the 5'-N-ethylcarboxamide group with its own binding subdomain (Olsson et al., 1986).

Although a hydrogen donor, as in the 5'-CH$_2$OH of adenosine and the 5'-CNHC$_2$H$_5$ of NECA, appears important to agonist activity at adenosine receptors, 5'-deoxyadenosine does have partial agonist activity (Bruns, 1980). 5'-Methylthioadenosine is, however, a competitive antagonist at A$_2$ receptors linked to adenylate cyclase (Bruns, 1980; Munshi et al., 1988). The 5'-ethylthio analog is less potent, while the 5'-isobutylthio analog is inactive. In marked contrast to clear <u>antagonist</u> activity at A$_2$ receptors linked to adenylate cyclase, 5'-methylthioadenosine is an <u>agonist</u> at A$_1$-receptors linked to adenylate cyclase, and has <u>agonist</u> activity at A$_2$-receptors mediating smooth muscle relaxation (Munshi et al., 1988). 5-Methylthioadenosine has no activity at "P-sites" (Nimit et al., 1982), but was reported in 1980 to bind to A$_1$ receptors (Bruns et al., 1980). 5'-Methylthioadenosine, 2-fluoro-5'-methylthioadenosine, and 5'-chloroadenosine are relatively potent at A$_1$ receptor binding sites in brain and all are agonists at A$_1$ receptors coupled to adenylate cyclase in fat cell membranes (Padgett, W. and Daly, J.W., unpublished results). The 5'-chloro analog is the most potent.

The Classical Approach: Xanthines

Efforts to develop more potent and highly selective antagonists for different subtypes of adenosine receptors have been extensive and to survey the structure activity relationships would be a monumental task best deferred at present. The classical approach, based on caffeine and theophylline as prototypic pharmacophores, has been to alter substituents at the 1-, 3-, and 7-position, in combination with addition of aryl or other ring systems at the 8-position. Indeed, the discovery that an 8-phenyl substituent markedly enhanced the potency of theophylline and other 1,3-disubstituted xanthines at adenosine receptors (Bruns, 1981; Smellie et al., 1979) has dominated further research on xanthines. Early studies also emphasized the importance of activity as phosphodiesterase inhibitors to any development of xanthines as highly selective adenosine antagonists: 9-Methylisobutylmethylxanthine, 8-bromoisobutylmethylxanthine and 1-isoamyl-3-isobutylxanthine were among the potent phosphodiesterase inhibitors with little or no activity as A$_2$-adenosine antagonists (Smellie et al., 1979). Later work has shown that 1-isoamyl-3-isobutylxanthine does have antagonist activity at A$_1$-adenosine receptors (Daly et al., 1985) and at certain A$_2$-adenosine receptors (Shamim et al., 1989). In contrast, 8-phenyltheophylline was nearly inactive as a phosphodiesterase inhibitor, while being a potent adenosine receptor antagonist (Smellie et al., 1979).

Remarkably, the effects of 8-aryl substituents on antagonist activity of xanthines is strongly influenced by the nature of substituents at the 1-,3- and 7-position. For example, unlike 8-phenyltheophylline, which is much more potent as an adenosine antagonist than theophylline, 8-phenylcaffeine is only slightly more potent than caffeine (Daly et al., 1985; Shamim et al., 1989). A wide range of aryl, heteroaryl and cycloalkyl substituents have been introduced in the 8-position of theophylline, caffeine and analogs thereof (Bruns et al., 1983; Daly et al., 1985, 1986c; Hamilton et al., 1985; Jacobson et al., 1985a, 1988, 1989; Martinson et al., 1987; Shamim et al., 1988, 1989). Most of the 8-arylxanthines and 8-cycloalkylxanthines are selective for A_1-receptors. The high potency of substituted 8-arylxanthines for adenosine receptors led to the choice of 8-(p-carboxylmethyloxyphenyl)-1,3-dipropylxanthines as the starting point for a "functionalized congener" approach to adenosine receptor antagonists (see K.A. Jacobson, this volume).

Recently, the effects of 2-thio and/or 6-thio substitutions on activity of theophyllines and 1,3-dipropylxanthines has been explored (Jacobson et al., 1989). Certain of the 2-thio analogs were nearly specific for antagonism of A_1 receptors; these contained an 8-cyclopentyl or an 8-aryl substituent.

A systematic evaluation of the effects of alterations in the 1-,3- and 7-substituents on activity of theophylline and caffeine has been initiated (Choi et al., 1988; Daly et al., 1986a; Shamim et al., 1989; Ukena et al., 1986). Certain analogs of caffeine, such as 3,7-dimethyl-1-propargyl-xanthine, 1,3-dimethyl-7-propylxanthine, and 1,3-dipropyl-7-methylxanthine were selective for A_2 receptors (Daly et al., 1986a, Ukena et al., 1986). Further studies have indicated that apparent selectivity can be greatly dependent on the A_1- and A_2-receptor systems that are being compared (see Shamim et al., 1989). Such differences strongly suggest the existence of a number of subtypes of A_1- and A_2-adenosine receptors.

The Classical Approach: Other Heterocycles

Both agonists (adenosines) and the classical antagonists (xanthines) for adenosine receptors contain a planar heterocyclic ring, suggesting that both bind to a common site, on adenosine receptors, which represents a subdomain that interacts strongly with the planar heterocyclic ring. Thus, one approach to developing new classes of agonists and antagonists would be to replace the purine ring of adenosines/xanthines with other heterocyclic rings. As yet no agents other than nucleosides have been found to be potent agonists at adenosine receptors. Indeed, only a few purine ring alterations to adenosine yield nucleosides that retain activity; i.e. 1-deazaadenosines (Cristalli et al., 1988) and 2-azaadenosine (Bruns, 1980). A variety of heterocycles other than xanthines have been discovered to have antagonist activity at adenosine receptors (see Daly et al., 1988a and ref. therein). Most are nonselective or slightly selective towards A_1 receptors and few have been investigated except in biochemical model systems. The 9-methyladenines (Ukena et al., 1987b) and related 7-deaza-9-phenyladenines and 7-deaza-9-phenylhypoxanthines (Daly et al., 1988b) are deserving of further study, since some are somewhat selective towards A_2 receptors, while others such as N^6-cyclohexyl-9-methyladenine are quite selective for A_1 receptors. None of the 9-methyladenines or 7-deaza-9-phenyl compounds have been evaluated in physiological systems.

Selective Agonists

The past decade has seen intensive efforts to develop selective or even specific agonists for different classes of adenosine receptors. The initial agonists proposed for characterization of A_1 and A_2 adenosine receptors, namely the diastereomeric R- and S-N^6-phenylisopropyladenosine, N^6-cyclohexyladenosine, 2-chloroadenosine, and NECA have not proven entirely satisfactory. Evaluation of some 63 adenosine analogs at A_1-binding sites and A_1 and A_2 receptors coupled to adenylate cyclase (Ukena et al., 1987a) led to suggestion of two additional agonists, namely N^6-cyclohexyl-NECA and 2-phenylaminoadenosine. Of this set of seven analogs only 2-phenylaminoadenosine is A_2-selective and only about 10-fold. Recently other classes of A_2 selective agonists have been developed. These include N^6-fluorenyl- and N^6-2,2-diphenylethyl-adenosines, some of which in brain A_1 and A_2 binding show 10 to 40-fold selectivity for A_2 receptors (Trivedi et al., 1988; Bridges et al., 1988). Even more selective is the 2-substituted NECA derivative CGS 21680, which would appear to represent the long-sought highly A_2 selective agonist (Hutchison et al., 1989).

A series of highly A_1 selective agonists have been developed. These include ADAC and certain other functionalized congeners (see K.A. Jacobson, this volume) N^6-cyclopentyladenosine (Moos et al., 1985); 2-chloro-N^6-cyclopentyladenosine (Lohse et al., 1988b); N^6-cyclopentyl-1-deazaadenosine (Cristalli et al., 1988) and certain N^6-norbornyladenosines (Trivedi et al,. 1989). The N^6-cyclopentyladenosine has already proven invaluable as a research tool.

Certain adenosine analogs recently were found to have either antagonist or agonist activity depending on the type of adenosine receptor: Thus, 5'-deoxy-5'-methylthioadenosine is an agonist at A_1-receptors inhibitory to adenylate cyclase, while being an antagonist at A_2-receptors stimulatory to adenylate cyclase (Munshi et al., 1988). Remarkably, this analog is an agonist at what have been considered to be the A_2 receptors that elicit smooth muscle relaxation.

Selective Antagonists

The search for selective adenosine receptor antagonists has in recent years yielded several selective A_1 antagonists. These include the functionalized congeners XAC and XCC (Jacobson et al., 1985a, 1986) and 8-cyclopentyl-1,3-dipropylxanthine (Bruns et al., 1987b; Shamim et al., 1988). These xanthines, because of solubility and selectivity, are proving useful as research tools. XAC was selective in vivo versus cardiovascular A_1 receptors (Evoniuk et al., 1987a; Fredholm et al., 1987; Jacobson et al., 1985b). An earlier A_1-selective xanthine, namely 8-(o-amino-p-chlorophenyl)-1,3-dipropylxanthine (Bruns et al., 1983) has not proven generally useful because of low water solubility resulting in pseudo-irreversible partitioning into lipids. Conversely, highly water soluble xanthines, such as 8-p-sulfophenyltheophylline and 8-p-sulfophenyl-1,3-dipropylxanthine (Daly et al., 1985), have in spite of lack of selectivity proven useful research tools, since they do not penetrate into cells and therefore are specific to extracellular sites of action; i.e., no effect on phosphodiesterases (Evoniuk et al., 1987b; Heller and Olsson, 1985; Seale et al., 1988; Wiklund et al., 1985). Further efforts to develop more selective A_1 antagonists are needed and, indeed, 2-thio-XCC ethylester appears based on brain A_1 and A_2 binding assays to be virtually specific for A_1 receptors (Jacobson et al., 1989).

Although antagonists with selectivity of several hundred to a thousand-fold for A_1 receptors have been developed, no antagonist with a selectivity towards A_2 receptors of more than about ten-fold has been forthcoming. A systematic evaluation of structural alterations at the 1-, 3-, and 7-

positions of caffeine did afford several caffeine analogs with selectivities for A_2 receptors ranging from 3- to 20-fold, depending both on the compound and on the A_1- and A_2-receptor assays that were compared (Daly et al., 1986a; Ukena et al., 1986). Among these caffeine analogs was 3,7-dimethyl-1-propargylxanthine, which has now been shown to selectively antagonize the hypothermic and behavioral depressant effects of NECA, compared to blockade of the hypothermic and depressant effects of the A_1-selective agonist, N^6-cyclohexyladenosine (Seale et al., 1988). 3,7-Dimethyl-1-propargylxanthine is a behavior stimulant with only very weak activity as a phosphodiesterase inhibitor (Choi et al., 1988). Another A_2-selective caffeine analog, 1,3-dipropyl-7-methylxanthine was slightly A_2 selective in vivo versus cardiovascular effects of adenosine analogs (Evoniuk et al,. 1987a). Recently, 8-cyclohexylcaffeine was found to be somewhat selective for A_2 receptors although the selectivity was dependent on the A_2 receptor assays being compared (Shamim et al., 1989). Further evaluation of antagonists in physiological systems rather than biochemical models always must be conducted before any conclusions are warranted as to their usefulness in delineating sites of action relevant to the pharmacology of caffeine and theophylline.

Functional Studies with Adenosine Agonists and Antagonists

The wide range of adenosine agonists now available, some A_1 selective and others A_2 selective, exhibiting a wide range of potencies, has led to many studies attempting to define functional roles for adenosine receptors. These studies have met with some success, particularly in peripheral systems. Thus, for a wide range of adenosine analogs the ability to enhance coronary blood flow correlates well with the A_2 potency of the analogs (Hamilton et al., 1987; Ukena et al., 1987a), while the ability of various analogs to decrease heart rate correlates well with the A_1 potency of the analogs (Hamilton et al., 1987). More recently both agonist and antagonist profiles for such cardiovascular effects were shown to correlate with those predicted from biochemical data on A_1 and A_2 receptors (Oei et al, 1988). For the central affects of adenosine analogs, in particular behavioral depression, the rank order of potency of adenosine analogs does not correspond satisfactorily with potencies expected for either A_1 or A_2 receptor. Thus, NECA is more potent than any of the N^6-substituted analogs, suggestive of an involvement of A_2 receptors. But certain highly selective A_1 agonists, such as N^6-cyclopentyladenosine cause behavioral depression, providing strong evidence for involvement of A_1 receptors. Of course, such studies are complicated by pharmacokinetics. It remains quite possible that the behavioral depression elicited by adenosine analogs is a complex vectorial response resulting from activation of A_1 and A_2 receptors, perhaps both central and peripheral in location. However, 8-p-sulfophenyltheophylline, which blocks only peripheral adenosine receptors, does not affect the behavioral depression elicited by N^6-cyclohexyladenosine or that elicited by NECA in mice (Seale et al., 1988).

One approach to determining whether the potent adenosine agonist NECA elicits behavioral depression through interaction with central A_2 receptors, while selective A_1 agonists, such as N^6-cyclohexyladenosine, elicit behavioral depression through interaction with central A_1 receptors, would be to use a series of A_1 and A_2 selective xanthine antagonists. Caffeine and theophylline are relatively nonselective and reverse NECA and N^6-cyclohexyladenosine-elicited depression with similar potencies (Seale et al., 1988). However, an A_2-selective analog of caffeine, namely 3,7-dimethyl-1-propargylxanthine, is 11-fold more potent versus NECA than versus N^6 cyclohexyladenosine (Seale et al., 1988) providing evidence that

NECA and N^6-cyclohexyladenosine, indeed, may elicit behavioral depression through interactions at A_2 and A_1 receptors, respectively. Further studies of this type are warranted. However, consideration must be given to the role of phosphodiesterase inhibition in evaluating xanthines and other heterocycles as adenosine antagonists. A recent study demonstrated that xanthines that are potent inhibitors of brain calcium-independent phosphodiesterases will be behavioral depressants and will not reverse N^6-cyclohexyladenosine-elicited depression (Choi et al, 1988). Earlier studies (Snyder et al., 1981; see also Katims et al, 1983) suggested a strong correlation between behavioral stimulant effects of a set of ten xanthines and their potencies as antagonists at the central A_1 receptor. The only exception was isobutylmethylxanthine, which was a depressant and is now known to be a potent inhibitor for the calcium-independent phosphodiesterases (Choi et al, 1988). At present it is not clear to what extent blockade of A_1 and/or A_2 receptors subserves the behavioral stimulant effects of caffeine/theophylline and various analogs. It is noteworthy that the ED_{50} of 3 mg/kg for central stimulant effects of 3,7-dimethyl-1-propargylxanthine corresponds fairly well with the IC_{50} of 1.3 mg/kg for reversing the depression elicited by the A_1 agonist N^6-cyclohexyladenosine and is far greater than the IC_{50} of 0.11 mg/kg for the reversal of the depression elicited by NECA, a potent A_1 and A_2 agonist (Seale et al., 1988).

Summary

There are now available an extensive array of "molecular probes" for adenosine receptors, ranging from radioligands, photoaffinity and irreversible ligands, fluorescent and spin-label probes, and biotin-containing probes to highly A_1 selective agonists and antagonists. As yet only one highly selective A_2 agonist is available. It is now clear that the two monolithic classes of A_1 and A_2 adenosine receptors are insufficient to describe both the variation by species and tissue of A_1 and A_2 receptors and the apparent existence of receptors that do not fit well the A_1 and A_2 classification. The existence of effector systems, in particular ion channels, other than adenylate cyclase for adenosine receptors introduces an additional complication. With the extensive range of molecular probes available for further investigation of adenosine receptors, it appears likely that the physiological functions, mechanisms, and molecular factors affecting protein-agonist-antagonist interactions for adenosine receptors are amenable to solution in the near future. One of the major challenges that remains is the delineation of the diverse pharmacological effects of caffeine/theophylline and exploitation of such insights in the development of xanthine-based therapeutic agents.

References

Bridges, A.J., Moss, W.H., Szotek, D.L., Trivedi, B.K., Bristol, J.A., Hoffner, T.G., Bruns, R.F. and Downs, D.A. (1987) J. Med. Chem. 30, 1709-1711.
Bridges, A.J., Bruns, R.F., Ortwine, D.F., Priebe, S.R., Szotek, D.L. and Trivedi, B.K. (1988) J. Med. Chem. 31, 1282-1285.
Bruns, R.F. (1980) Can. J. Physiol. Pharmacol. 58, 673-691.
Bruns, R.F., Daly, J.W. and Snyder, S.H. (1980) Proc. Natl. Acad. Sci. USA 77, 5547-5551.

Bruns, R.F. (1981) <u>Biochemical Pharmacol.</u> **30**, 325-333.
Bruns, R.F., Daly, J.W. and Snyder, S.H. (1983) <u>Proc. Natl. Acad. Sci. USA</u> **80**, 2077-2080.
Bruns, R.F., Lu, G.H. and Pugsley, T.A. (1986) <u>Mol. Pharmacol.</u> **29**, 331-346.
Bruns, R.F., Lu, G.H. and Pugsley, T.A. (1987a) In "Topics and Perspectives in Adenosine Research. Eds., E. Gerlach and B.F. Becker. Springer-Verlag Berlin Heidelberg, pp. 59-73.
Bruns, R.F., Fergus, J.H., Badger, E.W., Bristol, J.A., Santay, L.A., Hartman, J.D., Hays, S.J. and Huang, C.C. (1987b) <u>Naunyn-Schmiedeberg's Arch. Pharmacol.</u> **335**, 59-63.
Choi, O.H., Shamim, M.T., Padgett, W.L. and Daly, J.W. (1988) <u>Life Sci.</u> **43**, 387-398.
Cristalli, G., Franchetti, P., Grifantini, M., Vittori, S., Klotz, K.-N. and Lohse, M.J. (1988) <u>J. Med. Chem.</u> **31**, 1179-1183.
Daly, J.W. (1982) <u>J. Med. Chem.</u> **25**, 197-207.
Daly, J.W., Butts-Lamb, P. and Padgett, W. (1983) <u>Cellular Mol. Neurobiology</u> **3**, 69-80.
Daly, J.W., Padgett, W., Shamim, M.T., Butts-Lamb, P. and Waters, J. (1985) <u>J. Med. Chem.</u> **28**, 487-492.
Daly, J.W., Padgett, W. and Shamim, M.T. (1986a) <u>J. Med. Chem.</u> **29**, 1305-1308.
Daly, J.W., Padgett, W., Thompson, R.D., Kusachi, S., Bugni, W.J. and Olsson, R.A. (1986b) <u>Biochem. Pharmacol.</u> **35**, 2467-2481.
Daly, J.W., Padgett, W. and Shamim, M.T. (1986c) <u>J. Med. Chem.</u> **29**, 1520-1524.
Daly, J.W., Hong, O., Padgett, W.L., Shamim, M.T., Jacobson, K.A. and Ukena, D. (1988a) <u>Biochemical Pharmacol.</u> **37**, 655-664.
Daly, J.W., Padgett, W.L. and Eger, K. (1988b) <u>Biochemical Pharmacol.</u> **37**, 3749-3753.
Dragunow, M. and Faull, R.L.M. (1988) <u>Trends Pharmacological Sci.</u> **9**, 193-194.
Evoniuk, G., Jacobson, K.A., Shamim, M.T., Daly, J.W. and Wurtman, R.J. (1987a) <u>J. Pharmacol. Exper. Therap.</u> **242**, 882-887.
Evoniuk, G., Von Borstel, R.W. and Wurtman, R.J. (1987b) <u>J. Pharmacol. Exp. Therap.</u> **240**, 428-432.
Fredholm, B.B., Jacobson, K.A., Jonzon, B., Kirk, K.L., Li, Y.D. and Daly, J.W. (1987) <u>J. Cardiovascular Res.</u> **9**, 396-400.
Fredholm, B.B. and Dunwiddie, T.V. (1988) <u>Trends Pharmacological Sci.</u> **9**, 130-134.
Hamilton, H.W., Ortwine, D.E., Worth, D.F., Badger, E.W., Bristol, J.A., Bruns, R.E., Haleen, S.J. and Steffen, R.P. (1985) <u>J. Med. Chem.</u> **28**, 1071-1079.
Hamilton, H.W., Taylor, M.D., Steffen, R.P., Haleen, S.J. and Bruns, R.F. (1987) <u>Life Sci.</u> **41**, 2295-2302.
Heller, L.J. and Olsson, R.A. (1985) <u>Am. J. Physiol.</u> **248**, H907-H913.
Hutchison, A.J., Oei, H.H., Ghai, G.R. and Williams, M. (1989) CGS 21680, an A2 selective adenosine (ADO) receptor agonist with preferential hypotensive activity. <u>FASEB J.</u> **3**, A281.
Jacobson, K.A., Kirk, K.L., Padgett, W.L. and Daly, J.W. (1985a) <u>J. Med. Chem.</u> **28**, 1334-1340.
Jacobson, K.A., Kirk, K.L., Daly, J.W., Jonzon, B., Li, Y.-O. and Fredholm, B.B. (1985b) <u>Acta Physiol. Scand.</u> **125**, 341-342.
Jacobson, K.A., Ukena, D., Kirk, K.L. and Daly, J.W. (1986) <u>Proc. Natl. Acad. Sci. USA</u> **83**, 4089-4093.
Jacobson, K.A., de la Cruz, R., Schulick, R., Kiriasis, L., Padgett, W., Pfleiderer, W., Kirk, K.L., Neumeyer, J.L. and Daly, J.W. (1988) <u>Biochemical Pharmacol.</u> **37**, 3653-3661.

Jacobson, K.A., Kiriasis, L., Barone, S., Bradbury, B.J., Kammula, U., Campagne, J.M., Secunda, S., Daly, J.W., Neumeyer, J.L. and Pfleiderer, W. (1989) Sulfur-containing xanthine derivatives as selective antagonists at A_1-adenosine receptors. J. Med. Chem. 32, 1873-1879.

Katims, J.J., Annau, Z. and Snyder, S.H. (1983) J. Pharmacol. Exp. Therap. 227, 167-173.

Kurtz, A. (1987) Adenosine stimulates guanylate cyclase activity in vascular smooth muscle cells. J. Biol. Chem. 262, 6296-6300.

Linden, J. and Delahunty, T.M. (1989) Trends Pharmacological Sci. 10, 114-120.

Lohse, M.J., Klotz, K.-N., Diekmann, E., Friedrich, K. and Schwabe, U. (1988a) Eur. J. Pharmacol. 156, 157-160.

Lohse, M.J., Klotz, K.-N., Schwabe, U., Cristalli, G., Vittori, S. and Grifantini, M. (1988b) Naunyn-Schmiedeberg's Arch. Pharmacol. 337, 687-689.

Londos, C., Cooper, D.M.F. and Wolff, J. (1980) Proc. Natl. Acad. Sci. USA 77, 2551-2554.

Martinson, E.A., Johnson, R.A. and Wells, J.N. (1987) Mol. Pharmacol. 31, 247-252.

Matsuda, A. and Ueda, T. (1987) Nucleosides Nucleotides 6, 85-94.

Mikkelsen, E.J. (1978) J. Clin. Psychiatry 39, 732-736.

Moos, W.H., Szotek, D.S. and Bruns, R.F. (1985) J. Med. Chem. 28, 1383-1384.

Munshi, R., Clanachan, A.S. and Baer, H.P. (1988) Biochemical Pharmacol. 37, 2085-2089.

Nimit, Y., Law., J. and Daly, J.W. (1982) Biochem. Pharmacol. 31, 3279-3287.

Oei, H.H., Ghai, G.R., Zoganas, H.C., Stone, G.A., Zimmerman, M.B., Field, F.P. and Williams, M. (1988) J. Pharmacol. Exp. Therap. 247, 882-888.

Ohga, Y. and Daly, J.W. (1977) Biochim. Biophys. Acta 498, 46-60.

Olsson, R.A., Kusachi, S., Thompson, R.D., Ukena, D., Padgett, W. and Daly, J.W. (1986) J. Med. Chem. 29, 1683-1689.

Paton, D.M. (1980) J. Pharm. Pharmacol. 32, 133-135.

Ribeiro, J.A. and Sebastião, A.M. (1986) Prog. Neurobiology 26, 179-209.

Sattin, A. and Rall, T.W. (1970) Mol. Pharmacology 6, 13-23.

Schimmel, R.J. and Elliott, M.E. (1988) Biochem. Biophys. Res. Commun. 152, 886-892.

Seale, T.W., Abla, K.A., Shamim, M.T., Carney, J.M. and Daly, J.W. (1988) Life Sci. 43, 1671-1684.

Shamim, M.T., Ukena, D., Padgett, W.L., Hong, O. and Daly, J.W. (1988) J. Med. Chem. 31, 613-617.

Shamim, M.T., Ukena, D., Padgett, W.L. and Daly, J.W. (1989) J. Med. Chem. 32, 1231-1237.

Smellie, F.W., Davis, C.W., Daly, J.W. and Wells, J.N. (1979) Life Sci. 24, 2475-2482.

Snyder, S.H., Katims, J.J., Annau, Z., Bruns, R.F. and Daly, J.W. (1981) Proc. Natl. Acad. Sci. USA 78, 3260-3264.

Trivedi, B.F., Bristol, J.A., Bruns, R.F., Haleen, S.J. and Steffen, R.P. (1988) J. Med. Chem. 31, 271-273.

Trivedi, B.K., Bridges, A.J., Patt, W.C., Priebe, S.R. and Bruns, R.F. (1989) J. Med. Chem. 32, 8-11.

Ukena, D., Shamim, M.T., Padgett, W. and Daly, J.W. (1986) Life Sci. 39, 743-750.

Ukena, D., Olsson, R.A. and Daly, J.W. (1987a) Can. J. Physiol. Pharmacol. 65, 365-376.

Ukena, D., Padgett, W.L., Hong, O., Daly, J.W., Daly, D.T. and Olsson, R.A. (1987b) FEBS Lett. 215, 203-208.

Wiklund, N.P., Gustafsson, L.E. and Lundin, J. (1985) Acta Physiol. Scand. 125, 681-691.

Williams, M. (1987) Ann. Rev. Pharmacol. Toxicol. 27, 315-345.

2
Intracellular and Extracellular Metabolism of Adenosine and Adenine Nucleotides

D. Pearson and L.L. Slakey

Introduction

ATP is not solely the major intracellular biochemical energy currency, directly regulating the activities of a wide variety of proteins and other molecules which are substrates for phosphorylation. It also serves as the precursor for an important intracellular second messenger molecule, cyclic AMP, which in turn regulates the activities of a subset of proteins via cyclic AMP-dependent protein kinases, as discussed elsewhere in this volume.

The recognition that hydrolysis of ATP can in addition lead to the production of an extracellular messenger, adenosine, which binds to specific receptors (P_1 purinoceptors) at the cell surface, is relatively recent despite the pioneering studies of Drury & Szent-Gyorgi in the 1920s and Green & Stoner in the 1940s (cited in Pearson, 1987). The existence of a second class of purinoceptor, which is activated by ATP or ADP but not by adenosine, was first proposed by Burnstock in 1978 (for review see Burnstock et al., 1986). Since then pharmacological evidence has accrued demonstrating that ATP itself acts extracellularly as an autocrine or paracrine signal, in a manner different from adenosine, at P_2 purinoceptors.

P_2 purinoceptors are present on a wide variety of cell types, and fall into at least two main subtypes (P_{2X} and P_{2Y}) in terms of responsiveness to analogs of ATP. The transduction pathways stimulated by one of these (P_{2Y}) have been elucidated in some detail, and, unlike responses to adenosine, involve the rapid metabolism of phosphoinositides and the liberation of intracellular calcium from membrane-bound stores (reviewed in Olsson et al., 1989).

This review examines briefly, with particular reference to the cardiovascular system where most study has so far taken place, what is known in answer to two major questions posed by the existence of P_2 purinoceptors: what are the sources of extracellular adenine nucleotides, by comparison with those of adenosine; and what is the mechanism by which their concentrations are regulated? The cardiovascular system has been investigated in most detail because of the distinct and potent biological actions exhibited by adenosine and adenine nucleotides in this system. Thus while ADP stimulates platelets by acting at a unique subclass of the P_2 receptor, adenosine antagonises this via action at A_2 receptors. ATP and ADP are both powerful indirect vasodilators in most vascular beds due to their action at P_{2Y} receptors on vascular endothelium, causing the generation of the labile vasodilator nitric oxide (and also the release of the anti-aggregatory and dilator eicosanoid, prostacyclin), whereas adenosine is a dilator by direct action at smooth muscle A_2 receptors (reviewed in Pearson et al., 1989).

Sources of extracellular adenosine and adenine nucleotides

Understanding of how adenosine is generated extracellularly has been important since the initial hypothesis by Berne (see Berne, 1980) and the subsequent general agreement that adenosine is produced in response to a fall in the energy state within a cell, notably the coronary myocyte in response to hypoxia and/or ischemia, and then acts extracellularly as a retaliatory metabolite to cause local vasodilation (Newby, 1984). The precursor for adenosine formation by this route is AMP, generated by dismutation from ADP by the disturbance of the myokinase reaction steady state when ATP is dephosphorylated. Subsequently AMP is dephosphorylated by a 5'-nucleotidase to yield adenosine, which is then transported by the adenosine carrier out of the cell. A cytosolic 5'-nucleotidase with the appropriate characteristics (preference for AMP over IMP, activated by ATP) and sufficient capacity for adenosine formation has only recently been described (Truong et al., 1988; Newby, 1988).

It is difficult, however, to ascertain whether extracellular adenosine is generated by this pathway in sufficient quantities and at the right time, e.g. by hypoxic cardiac myocytes to account for the coronary dilatation observed. It is impractical to determine directly the appropriate interstitial adenosine concentration, and all measurements from effluent are underestimates due to the rapid re-uptake of adenosine by other cellular compartments and/or its extracellular catabolism by surface-bound adenosine deaminase. The extent to which hydrolysis of cytosolic AMP can lead to the production of extracellular adenosine is also limited by competitive intracellular processes. AMP is deaminated to IMP, a process which in several cell types other than the cardiomyocyte is preferred to dephosphorylation. Adenosine once formed can be rephosphorylated via adenosine kinase, or deaminated to form inosine (and subsequently hypoxanthine), in addition to being transported out of the cell by a symmetrical carrier protein that will only act down a concentration gradient (see Clanachan, this volume). For a more detailed discussion of these points see Olsson et al. (1989) and Schrader (this volume).

Since the description of P_2 purinoceptors, and with the knowledge that in the coronary bed exogenous intravascular ATP and ADP are more potent vasodilators than adenosine (e.g Fleetwood et al., 1987), it is pertinent to consider two possibilities. First, that at least a proportion of the response attributed to the generation of extracellular adenosine is due to extracellular adenine nucleotides. Second, that a proportion of the adenosine found extracellularly is formed there from nucleotides, rather than deriving from intracellular catabolism of AMP. In the particular case of the ischemic heart there is evidence suggestive that this process occurs in response to hypoxia (e.g. Van Belle et al., 1987), though it remains controversial. Nonetheless it is clear that under certain pathophysiological conditions sufficient ATP and ADP are generated extracellularly to have biological activity. Stimulation of blood platelets causes secretion of ATP and ADP from granules, leading transiently to local concentrations well in excess of those required to activate endothelial P_{2y} receptors. Similarly, blood vessel injury leading to hemolysis or increased permeability of cells in the vessel wall produces biologically active levels of ATP (Born et al., 1984). Studies with endothelial cells cultured *in vitro* have demonstrated that brief exposure to proteases such as thrombin, or an abrupt change of shear forces by changing the rate of flow of extracellular medium, is capable of producing transient release of cellular ATP in the absence of cytosolic enzyme release or any irreversible cell damage (Pearson et al., 1979; Milner et al., 1989). Also, ATP released as a co-transmitter in perivascular nerves can modulate vascular tone (reviewed in White, 1988). The mechanism by which the local concentrations of extracellular ATP or ADP and hence their ability to act at P_2 purinoceptors (whether in the cardiovascular system or elsewhere) are regulated is thus at least as important as that leading to the inactivation of adenosine. The following section describes the characteristics of the ectonucleotidase enzymes responsible for this process.

Catabolism of extracellular adenine nucleotides by ectonucleotidases

The presence of endothelial cell ectoenzymes that dephosphorylate adenine nucleotides was first shown by the Ryans (reviewed in Ryan et al., 1984). It is clear, from comparison of the half lives of nucleotides in the circulation and in whole blood or plasma *ex vivo*, that these enzymes and not their counterparts on leukocytes or any soluble enzymes are responsible for the substantial catabolism of nucleotides in a single passage through a microvascular bed (see Coade et al., 1989 and references therein).

The biochemical and pharmacological characterisation of these enzymes on cultured aortic endothelial cells demonstrated that ATP was sequentially dephosphorylated by distinct ecto-ATPase, ecto-ADPase and 5'-nucleotidase enzymes, with little or no contribution from non-specific phosphatases (Pearson, 1987). Subsequent studies in isolated perfused lungs and heart indicate the presence of a similar ectonucleotidase system with comparable kinetic properties in these microvascular beds (Hellewell et al., 1987; Fleetwood et al., 1989).

A detailed examination of the kinetic features of the ectonucleotidases on endothelial cells was carried out by recirculating a small volume of perfusate, initially containing known concentrations of an added adenine nucleotide, through columns of packed microcarrier beads bearing monolayers of cultured endothelium. The pattern of extracellular nucleotides generated was studied by sequential subsampling of the

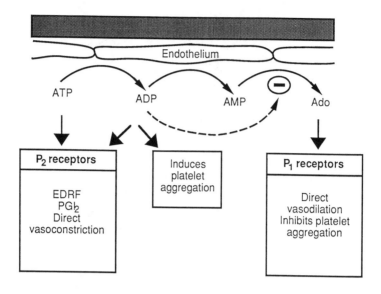

Figure 1. Ectonucleotidases at the endothelial cell surface control the conversion of ATP and ADP (acting at endothelial P_{2Y} receptors to cause EDRF (nitric oxide) and prostacyclin release, and at P_{2X} receptors on smooth muscle cells in some beds to cause vasoconstriction) to adenosine. The rate of adenosine formation is limited by feed-forward inhibition of 5'-nucleotidase by ADP.

perfusate (Gordon et al., 1986). The most striking feature was a lag in the production of adenosine from AMP evident when ADP or ATP was used as the initial substrate: indeed, the rate of adenosine production was inversely related to the initial ATP concentration. Computer modelling of the complete reaction sequence was achieved by numerical integration of Michaelis-Menten kinetic equations, which demonstrated that the observed time courses for ATP catabolism and ADP formation and dephosphorylation could be adequately fitted to simple Michaelis-Menten equations describing the ATPase and ADPase activities, with kinetic parameters (K_m, V_{max}) similar to those previously found by initial velocity experiments. Satisfactory fitting of the AMP and adenosine data, however, was only obtained when a model in which AMP hydrolysis was competitively inhibited by ADP was used, a property of 5'-nucleotidase anticipated by studies with the purified enzyme from other sources.

The ectonucleotidases on endothelial cells thus act concertedly to maximise the separation in time between the presence of pools of P_2 agonists (ATP and ADP) and of adenosine, and therefore regulate the balance between P_2- and P_1-mediated effects (figure 1). *In vivo*, where blood is flowing past the endothelium, this time separation will be reflected by a spatial separation.

A similar set of 3 ectonucleotidases is present on the surface of

16

vascular smooth muscle cells (Pearson et al., 1980). Recent kinetic analysis based on examination of the whole time course of the reactions with an analogous system of perfusion of substrates over cells in microcarrier bead columns revealed two properties of the enzymes that differ significantly from those on endothelium (Gordon et al., 1989). First, despite evidence for inhibition of 5'-nucleotidase by ADP, the production of adenosine proceeds rapidly. This was due in part to the higher V_{max} for 5'-nucleotidase (relative to the V_{max} for ATPase and ADPase) in smooth muscle cells. Additionally, however, there was kinetic evidence for the preferential delivery to the ectoADPase and 5'-nucleotidase of nucleotides newly-formed at the cell surface, by comparison with nucleotides in the bulk phase, which did not occur in endothelium, and this may indicate clustering of enzymes on the smooth muscle cell surface. In combination with the less rapid uptake of adenosine by smooth muscle cells than by endothelial cells (Pearson et al., 1978) these properties suggest that the ectonucleotidases on smooth muscle cells, unlike on endothelial cells, are organized to generate pericellular adenosine rapidly, for example to modify neurotransmission prejunctionally.

In current experiments we are studying the properties of the 3 ectonucleotidases on cardiac myocytes, where they are also present, in similar detail. Although it is easy to hypothesize a regulatory role for the ectonucleotidases on cardiovascular cells, where P_2 and P_1 agonists have obvious and powerful effects, it should be noted that these enzymes are found in a wide variety of cell types (reviewed in Pearson, 1987) - the first description of an ectoATPase in mammalian cells was in the central nervous system over 35 years ago - and their functions are still far from understood.

Conclusions

This review has highlighted one aspect of the pathways of intracellular and extracellular metabolism of adenine nucleotides and adenosine, reflecting the authors' bias, which is derived from the need to understand how the concentration of extracellular P_2 agonists (ATP and ADP) are regulated. The kinetic studies outlined here have revealed subtle, but perhaps functionally important, differences in the ectonucleotidase cascade on different cells of the cardiovascular system. Only one of these enzymes, 5'-nucleotidase, has been extensively purified (from liver): it belongs to the growing family of ectoenzymes anchored through phosphatidylinositol to the plasma membrane (Low et al., 1978). Very recently, cDNA coding for an ectoATPase from liver has been sequenced and cloned, and it appears to be a conventional transmembrane protein (Lin et al., 1989), though whether there are significant structural or evolutionary homologies between the three enzymes remains to be determined.

It is relevant to point out that in addition to the small differences found in these studies, there is considerable evidence of heterogeneity of the metabolism of adenine derivatives on a wider scale, with certain pathways being preferentially located in some cell types rather than others. Histochemical or immunological localization, together with perfusion studies, show for example that adenosine deaminase (a proportion of which is an exoenzyme), purine nucleoside phosphorylase and xanthine dehydrogenase are predominantly within endothelium, thus compartmentalizing the catabolic pathway leading to hypoxanthine (a substrate for the purine salvage pathway) and uric acid to the vasculature within an organ (see references in Olsson et al., 1989). The functional consequences of this intercellular specialization in the metabolism of adenosine have also not been elucidated, but its existence emphasizes the complexity of the task facing those who seek to provide an integrated picture of the generation and removal of extracellular adenosine compounds that induce physiological responses via purinoceptors.

References

Berne, R.M., 1980, The role of adenosine in the regulation of coronary blood flow, *Circ. Res.* **47**:807-813.

Born, G.V.R., and Kratzer. M.A.A., 1984, Source and concentration of extracellular adenosine triphosphate during haemostasis in rats, rabbits and man, *J. Physiol. (Lond.)*, **354**:419-429.

Burnstock, G., and Kennedy, C., 1986, Is there a basis for distinguishing two types of P_2-purinoceptor?, *Gen. Pharmacol.* **16**: 433-440.

Coade, S.B., and Pearson, J.D., 1989, Metabolism of adenine nucleotides in human blood, *Circ. Res.* **65**: (in press).

Fleetwood, G., and Gordon, J.L., 1987, Purinoceptors in the rat heart, *Br. J. Pharmacol.* **90**:219-227.

Fleetwood, G., et al., 1989, Kinetics of adenine nucleotide catabolism in the coronary circulation, *Am. J. Physiol.* **256**:H1565-H1572.

Gordon, E.L., Pearson, J.D., and Slakey, L.L., 1986, The hydrolysis of extracellular adenine nucleotides by cultured endothelial cells from pig aorta, *J. Biol. Chem.* **261**:15496-15504.

Gordon, E.L., et al., 1989, The hydrolysis of extracellular adenine nucleotides by arterial smooth muscle cells, *J. Biol. Chem.* **264**: in press.

Hellewell, P.G., and Pearson, J.D., 1987, Adenine nucleotides and pulmonary endothelium, In: *Pulmonary Endothelium in Health and Disease*, ed. Ryan U; New York, Dekker, pp.327-348.

Lin, S.-H., and Guidotti, G., 1989, Cloning and expression of a cDNA coding for a rat liver plasma membrane ecto-ATPase, *J. Biol. Chem.* **264**:14408-14414.

Low, M.G., and Finean, J.B., 1978, Specific release of plasma membrane enzymes by a phosphatidylinositol-specific phospholipase C, *Biochim. Biophys. Acta* **508**:565-570.

Milner, P., et al., 1989, Endothelial cells cultured from human umbilical vein can release ATP, Substance P and acetylcholine, *Experientia* (submitted).

Newby, A.C., 1984, Adenosine and the concept of 'retaliatory metabolites', *Trends Biochem. Sci.* **9**:42-44.

Newby, A.C., 1988, The pigeon heart 5'-nucleotidase responsible for ischaemia-induced adenosine formation, *Biochem. J.* **253**:123-130.

Olsson, R.A., and Pearson, J.D., 1989, Cardiovascular purinoceptors, *Physiol. Rev.* (in press).

Pearson, J.D., 1987, Nucleotide metabolism, In: *Mammalian Ectoenzymes*, eds. Kenny AJ & Turner AJ; Amsterdam, Elsevier, pp.139-167.

Pearson, J.D., and Gordon, J.L., 1979, Vascular endothelial and smooth muscle cells in culture selectively release adenine nucleotides, *Nature* **281**:384-386.

Pearson, J.D., and Gordon, J.L., 1989, P$_2$ purinoceptors in the blood vessel wall, *Biochem. Pharmacol.* (in press).

Pearson, J.D., et al., 1978, Uptake and metabolism of adenosine in pig aortic endothelial and smooth muscle cells in culture. *Biochem. J.* **170**:265-271.

Pearson, J.D., Carleton, J.S., and Gordon, J.L., 1980, Metabolism of adenine nucleotides by ectoenzymes of vascular endothelial and smooth muscle cells in culture, *Biochem. J.* **190**:421-429.

Ryan, J.W., and Ryan, U.S., 1984, Endothelial surface enzymes and the dynamic proessing of plasma substrates, *Int. Rev. Exp. Path.* **26**:1-43.

Truong, V.L., Collinson, A.R., and Lowenstein, J.M., 1988, 5'-Nucleotidases in rat heart. Evidence for the occurrence of two soluble enzymes with different substrate specificities, *Biochem. J.* **253**: 117-121.

Van Belle, H., Goossens, F., and Wynants, J., 1987, Formation and release of purine catabolites during hypoperfusion, anoxia, and ischemia, *Am. J. Physiol.* **252**:H886-H893.

White, T.D., 1988, Role of adenine compounds in autonomic neurotransmission, *Pharmacol. Ther.* **38**:129-168.

3

Adenosine Deaminase and Adenosine Transport Systems in the CNS

J.D. Geiger, R.A. Padua, and J.I. Nagy

Introduction

New strategies for drug development have evolved following the introduction of radioligand binding methods that are now commonly used for the pharmacological characterization of agents which selectively activate or block hormone and neurotransmitter receptors. Such stratagies are now being applied to develop compounds for possible therapeutic use that mimic or interfere with the widespread regulatory and neuromodulatory actions of adenosine. An alternate strategy under consideration is to design drugs that influence adenosine transport, metabolism or release processes and thereby may alter intracellular or extracellular adenosine levels and subsequently adenosine receptor-mediated events. Some inhibitors of adenosine transport and adenosine deaminase (ADA) are already available for both experimental and clinical use and these have been shown to increase adenosine levels under certain circumstances (German et al., 1989; Phillis et al., 1988; Caciagli et al., 1988). However, very little is known about adenosine transport and metabolism, especially in the CNS, and this may have hindered the drug discovery process. In order to gain a better understanding of the relationship between adenosine transport, metabolism and the neuroregulatory actions of this nucleoside (Geiger and Nagy, 1989; Nagy et al., 1990), we and others have investigated the pharmacological characteristics and anatomical localization of adenosine transport sites labelled by [^{3}H]nitrobenzylthioinosine ([^{3}H]NBI) and [^{3}H]dipyridamole ([^{3}H]DPR), the functional significance of these sites as assessed through studies of adenosine transport, and the possible contribution of ADA to adenosine catabolism by documentation of its distribution in the CNS and the effects of its inhibition by 2'-deoxycoformycin (DCF).

Adenosine Transport

Transport of adenosine across cell membranes in the CNS was described about two decades ago (Shimizu and Daly, 1970). Based on evidence that adenosine uptake inhibitors enhance various effects of adenosine in both in vitro and in vivo preparations, it has been proposed that the actions of adenosine may be regulated in part by mechanisms involving its removal from extracellar spaces by transport into cells (Wu and Phillis, 1984; Geiger and Nagy, 1989). As with peripheral tissues and cultured cells, adenosine transport systems in the CNS are more complex than previously appreciated.

20

Kinetics: Until recently and with the exception of a single report in 1981 (Bender et al., 1981), adenosine transport systems in, for example, rat and guinea pig synaptosomes were thought to consist of a single high affinity process with Km values ranging from 0.9 to 21 μM (Bender et al., 1981; Barberis et al., 1981). It is now clear that both high and low affinity adenosine transporters exist in CNS tissues (Geiger, 1987; Geiger et al., 1988; Johnston and Geiger, 1989; Baldy and Shank, 1988; Shank et al., 1989). A combination of two methodological factors appear to have contributed to the differences noted between these studies. First, prolonged incubation intervals led to extensive metabolism of the substrate adenosine and this likely interfered with measurement of transport as the rate limiting step. Second, minor deviations of data from linearity may have been missed and those may now be reliably determined through sophisticated computer analyses.

Facilitated Diffusion and Sodium Dependent Transport: In the past, nucleosides were thought to be transported at physiological levels by facilitated diffusion and at supra-physiological levels by passive diffusion. Sodium gradient-dependent systems have now been described in various tissues including choroid plexus, hepatocytes, renal brush border vesicles, cultured epithelial cells, and dissociated brain cells (see Johnston and Geiger, 1989). For cell bodies isolated from adult rat brain, transport levels were consistently and significantly 20% lower in the absence than in the presence of physiological concentrations of sodium. The K_T values for the high affinity portion were significantly (two-fold) higher when sodium was substituted with choline or mannitol. The selective reversal of the sodium effect by ouabain and 2,4-dinitrophenol suggested that the process was dependent on a maintained sodium gradient. Unlike peripheral tissues, however, the sodium dependent component was inhibited by NBI (Johnston and Geiger, 1989).

Substrate Specificity: Nucleoside transport systems in the CNS, as in many other tissues, can apparently transport a variety of purine and pyrimidine nucleosides. However, studies with synaptosomes (Bender et al., 1981), primary cultures of chick neurons (Thampy and Barnes, 1983a), and dissociated brain cells (Geiger et al., 1988) have shown that adenosine is the preferred nucleoside substrate for the transporter, whereas glial cells lack this specificity (Thampy and Barnes, 1983b). Additional evidence, albeit indirect, has shown that adenosine competed more potently than other nucleosides for [3H]nitrobenzylthioinosine binding sites in rat brain (Geiger et al., 1988).

Pharmacology: In studies of nucleoside transport systems with the radioligands [3H]NBI and [3H]DPR, it appears that [3H]DPR labels more sites in the CNS than [3H]NBI. Nevertheless, the usefulness of [3H]DPR may be somewhat limited because these sites have little preference for adenosine as the nucleoside substrate and because of an inability to detect specific [3H]DPR binding in rat tissues (Marangos and Deckert, 1987). As useful as both ligands have been, the exact molecular nature of their attachment sites is unknown. Furthermore, the conceptualization of mechanisms underlying nucleoside transport is made difficult by the presence of inhibitor-sensitive and-insensitive transport sites in the CNS, the proportion of which varies not only among species but also among brain regions within a given species (Morgan and Marangos, 1987; Lee and Jarvis, 1988). Finally, as in some peripheral tissues, adenosine may be 'released' through bidirectional transporters. Very little is known about the physiological relevence of this bidirectionality. Nevertheless this represents a mechanism that might be amenable to pharmacological manipulation.

21

Immunohistochemical and biochemical studies have shown that ADA, a soluble cytoplasmic enzyme that catabolizes adenosine to the physiologically less active metabolite inosine, has a highly restricted, heterogeneous distribution in the rat CNS not unlike patterns observed for neurotransmitter synthetic and degradative enzymes (Nagy et al., 1984; Geiger and Nagy, 1989; Nagy et al., 1990). The ADA-positive neurons were found to have discrete as well as widespread projections to clearly defined terminal fields. Of particular importance with respect to the central actions of ADA inhibitors and adenosine uptake inhibitors was the finding that distribution patterns of adenosine transport sites autoradiographically labelled by [^{3}H]NBI coincided with ADA-immunoreactivity (Nagy et al., 1985).

The role of ADA in the CNS is open to speculation. Apart from the possibility that ADA regulates the levels of adenosine available for release (Geiger and Nagy, 1989; Nagy et al., 1990), it has been considered to be a constitutively expressed housekeeping enzyme (Valerio et al., 1985). However, it is difficult to reconcile such an apparently ubiquitous role with the finding of a nearly fourteen fold difference in ADA activity between different brain regions. Therefore, either ADA has a special function in the CNS or ADA containing cells have greater housekeeping needs. Furthermore, the distribution patterns of ADA in rat are not shared by other species tested including mouse, rabbit, and guinea-pig (Yamamoto et al., 1987) and these differences must be reconciled with any proposed role(s) for this enzyme.

One approach to evaluate the degree to which ADA contributes to adenosine metabolism is through inhibition of its activity by compounds such as DCF. Clinically, DCF, has been used to limit the metabolism of some immunosuppressive, antiviral and antitumor purine analogs that are also substrates for ADA (Cummings et al., 1988; Agarwal, 1982). However even at very low doses it has side effects that are thought to originate in the CNS and that may be due to accumulations of adenosine and 2'-deoxyadenosine (Major et al., 1981). Findings in animals that DCF alters sleep function (Virus et al., 1983), desensitizes adenosine receptors (Porter et al., 1988), increases cerebral blood flow (Phillis et al., 1985), and protects against cell damage arising from cerebral ischemia or hypoxia (Phillis et al., 1988; Phillis and O'Regan, 1989), provide at least indirect support for the notion that the CNS effects of DCF might be mediated through adenosine. In order to further evaluate the potential of DCF as a therapeutic agent and as a neurochemical tool to investigate purine metabolism and neuroregulatory mechanisms we have determined the specificity, potency and disposition of DCF in brain, and the rate of recovery of ADA following DCF administration (Geiger et al., 1987; Padua et al., 1990).

Peak levels of [^{3}H]DCF in brain were achieved two-hours following a single ip. injection. Among the four brain regions examined, the greatest accumulation of [^{3}H]DCF occurred in the hypothalamus, a region with very high levels of ADA. Similarly, in peripheral tissues DCF accumulates to a greater degree in tissues with correspondingly high levels of ADA. The disposition was described kinetically by a two compartment model. Values of $t_{1/2}$ for the fast phase ranged from 0.8 hours in hypothalamus to 5.5 hours in cerebral cortex. Values for the slow phase were relatively constant among brain regions at about 50 hours (Geiger et al., 1987).

The degree of ADA inhibition by DCF in whole brain and in various brain regions at doses of DCF ranging from 0.5 to 50 mg/kg was about 90 to 95% two-hours after injection. Administration of DCF at 5 mg/kg led to a prolonged reduction in ADA activity in whole brain; activity returned to only 66% of control levels even after 50 days and to on average 59% of control levels in various brain regions after forty days; values ranged from 40% in habenula to 73% in anterior hypothalamus (Padua et al., 1990). In brain regions with lower levels of ADA, the enzyme tended to recover to a lower degree and at a slower rate. One possibility for the prolonged reduction in ADA activity was neurotoxicity leading to cell death. If DCF exerted neurotoxic effects, doses higher than 5 mg/kg might be expected to cause a greater and more prolonged loss of ADA and other enzymes particularly those colocalized in ADA-containing cells. However, we found that the degree of ADA inhibition was only slightly greater and the rate of enzyme recovery was only slightly slower in animals that received 10, 25 or 50 mg/kg. Moreover, unaltered by DCF treatment in ten brain regions examined were the activities of the neurotransmitter synthetic enzymes glutamic acid decarboxylase and choline acetyl transferase as well as histidine decarboxylase (HDC) which is expressed in virtually all ADA-containing neurons in the posterior hypothalamus. The minimal additional effect of supramaximal DCF doses on the prolonged reduction of ADA activity led us to determine the effectiveness of subsequent doses of DCF. Fourteen days after an initial injection, a second 5 mg/kg injection was administered and ADA activity was measured after 1 to 29 days. The second injection was virtually without inhibitory effect. These results suggested that the mechanism involved in the prolonged reduction of ADA activity was somehow saturated or inactivated and raised the possiblity that the returning enzyme exhibited properties different from ADA. Kinetic studies showed that the Km values of ADA for adenosine in whole brain and hypothalamus were not different between animals that received either DCF or saline vehicle. Moreover, the Ki value of DCF for ADA in animals treated with DCF were comparable to control values for hypothalamus, but were almost six-times higher in whole brain (Padua et al., unpublished observations). Further studies are presently underway to determine the biochemical bases for these differences.

In some tissues adenylate deaminase (AMPDA) has been found to be potently inhibited by DCF. In tests of the specificity of DCF for ADA versus AMPDA in brain homogenates, we found that the Ki for ADA was 23 pM while that for AMPDA was 233 μM. This very large difference in Ki indicates that in vitro ADA can be completely inhibited without any effect on AMPDA activity. In in vivo studies we found that DCF had an ED_{50} of 120 μg/kg and that at doses necessary to maximally inhibit ADA, no decrease in AMPDA activity was detected (Padua et al., 1990).

It has been suggested in recent years that adenosine exerts a tonic suppression of neural activity thereby setting a certain tone on electrical activity and neurotransmission (Newby, 1984; Dragunow and Faull, 1988). If this is indeed a fundamental role for the nucleoside, then it may be speculated that altered adenosine levels following ADA inhibition may lead to decreased production of ADA by two somewhat different mechanisms. Increased adenosine levels may reduce neural activity producing an aberrant elevation in what might be considered an 'adenosine suppression set point'. Rather than returning to some previous balance, this new set point is engrained as the status quo and requires decreased ADA activity for its continued maintenance. Alternatively, it may be speculated that in analogy with other transmitters, a DCF-induced increase in adenosine availability may result in down regulation of adenosine receptors leading to a reduction in the presumptive adenosine suppression set point. In order to explain a prolonged reduction in ADA activity, it would be necessary to postulate, in contrast to the previous scenario, that such a reduction is undesirable and compensated by decreased catabolism of adenosine. If the long-term effects of DCF on ADA involve such mechanisms then further studies of ADA may shed some light on processes governing neuromodulation by adenosine.

Acknowledgements

We wish to thank Ms. Suzanne Dambock for her excellent technical assistance. Supported by grants from the Medical Research Council of Canada (MRC). JDG is a Scholar and JIN is a Scientist of the MRC.

References

Agarwal, R.P., 1982, Inhibitors of adenosine deaminase, Pharmacol. Therap. **17**: 399-429.

Baldy, W.J., and Shank, R.P., 1988, Multiple adenosine transport systems in rat and guinea pig synaptosomes, Soc. Neurosci. Abstacts **14**: 681.

Barberis, C., Minn, A., and Gayet, J., 1981, Adenosine transport into guinea-pig synaptosomes, J. Neurochem. **36**: 347-354.

Bender, A.S., Wu, P.H. and Phillis, J.W., 1981, The rapid uptake and release of [^{3}H]adenosine by rat cerebral cortical synaptosomes, J. Neurochem. **36**: 651-660.

Caciagli, F., et al., 1988, Cultures of glial cells release purines under field electrical stimulation: The possible ionic mechanisms, Pharmacol. Res. Commun. **20**: 935-947.

Cummings, F.J. et al., 1988, Clinical, pharmacologic and immunologic effects of 2'-deoxycoformycin, Clin. Pharmacol. Therap. **44**: 501-509.

Dragunow, M., and Faull, R.L.M., 1988, Neuroprotective effects of adenosine, Trends Pharmacol. Sci. **9**: 193-194.

Geiger, J.D., 1987, Adenosine uptake and [^{3}H]nitrobenzylthioinosine binding in developing rat brain, Brain Research **436**: 265-272.

Geiger, J.D., et al., 1987, Pharmacokinetics of 2'-deoxycoformycin, an inhibitor of adenosine deaminase, in the rat, Neuropharmacol. **26**: 1383-1387.

Geiger, J.D., Johnston, M.E. and Yago, V., 1988, Pharmacological characterization of rapidly accumulated adenosine by dissociated brain cells from adult rat, J. Neurochem. **51**: 283-291.

Geiger, J.D., and Nagy, J.I., 1989, Adenosine deaminase and [^{3}H]nitrobenzylthioinosine as markers of adenosine metabolism and transport in central purinergic systems, In. Adenosine Receptors. Ed. M. Williams, Humana Press, N.J. (in press).

German, D.C., Kredich, N.M., and Bjorsson, T.D., 1989, Oral dipyridamole increases plasma adenosine levels in human beings, Clin. Pharmacol. **45**: 80-84.

Johnston, M.E., and Geiger, J.D., 1989, Sodium-dependent uptake of nucleosides by dissociated brain cells from rat, J. Neurochem. **52**: 75-81.

Lee, C.W., and Jarvis, S.M., 1988, Kinetic and inhibitor specificity of adenosine transport in guinea-pig cerebral cortical synaptosomes: Evidence for two nucleoside transporters, Neurochem. Int. **12**: 483-492.

Major, P.P., Agarwal, R.P., and Kube, D.W., 1981, Clinical pharmacology of deoxycoformycin, Blood **58**: 91-96.

Marangos, P.J., and Deckert, J., 1987, [^{3}H]Dipyridamole binding to guinea pig membranes, possible heterogeneity of central adenosine uptake sites, J. Neurochem. **48**: 1231-1237.

Morgan, P.F., and Marangos, P.J., 1987, Comparative aspects of nitrobenzylthioinosine and dipyridamole inhibition of adenosine acculmulation in rat and guinea-pig synaptosomes, Neurochem. Int. **11**: 339-346.

Nagy, J.I., Geiger, J.D., and Daddona, P.E., 1985, Adenosine uptake sites in rat brain: Identification using ^{3}H-nitrobenzylthioinosine and colocalization with adenosine deaminase, Neurosci. Lett. **55**: 47-53.

Nagy, J.I., Geiger, J.D., and Staines, W.A., 1990, Adenosine deaminase and purinergic neuroregulation, Neurochem. Int. (submitted).

Nagy, J.I., et al., 1984, Immunohistochemistry of adenosine deaminase: Implications for adenosine neurotransmission, Science **224**: 166-168.

Newby, A.C., 1984, Adenosine and the concept of "retaliatory metabolites", Trends Biochem. Sci. **9**: 42-44.

Padua, R., et al., 1990, 2'-Deoxycoformycin inhibition of adenosine deaminase in rat brain: in vivo and in vitro analysis of specificity, potency and enzyme recovery, J. Neurochem. (in press).

Phillis, J.W., DeLong, R.E., and Towner, J.R., 1985, Adenosine deaminase inhibitors enhance cerebral anoxic hyperemia in the rat, J. Cerbral Blood Flow Metab. **5**: 295-299.

Phillis, J.W., and O-Regan, 1989, Deoxycoformycin antagonizes ischemia-induced neuronal degeneration, Brain Res. Bull. **22**: 537-540.

Phillis, J.W., O'Reagan, M.H., and Walter, G.A., 1988, Effects of deoxycoformycin on adenosine, hypoxanthine, xanthine and uric acid release from the hypoxemic rat cerebral cortex, J. Cerebral Blood Flow Metab. **8**: 733-741.

Porter, N.M., Radulovacki, M., and Green, R.D., 1988, Desensitization of adenosine and dopamine receptors in rat brain after treatment with adenosine analogs, J. Pharmacol. Exp. Therap. **244**: 218-225.

Shank, R.P., and Baldy, W.J., 1989, R 70380, Soluflazine, and Mioflazine as adenosine transport inhibitors in rat and guinea-pig synaptosomes, J. Neurochem. **52**: S99.

Shimizu, H., and Daly, J., 1970, Formation of cyclic adenosine 3',5'-monophosphate from adenosine in brain slices, Biochim. Biophys. Acta **222**: 465-473.

Thampy, K.G., and Barnes Jr., E.M., 1983a, Adenosine transport by primary cultures of neurons from chick embryo brain, J. Neurochem. **40**: 874-879.

Thampy, K.G., and Barnes Jr., E.M., 1983b, Adenosine transport by cultured glial cells from chick embryo brain, Arch. Biochem. Biophys. **220**: 340-346.

Valerio, D., et al., 1985, Adenosine deaminase: Characterization and expression of a gene with a remarkable promoter, EMBO J. **4**: 437-443.

Virus, R.M., et al., 1983, The effects of adenosine and 2'-deoxycoformycin on sleep and wakefulness in rats, Neuropharmacol. **22**: 1401-1404.

Wu, P.H., and Phillis, J.W., 1984, Uptake by central nervous tissues as a mechanism for the regulation of extracellular adenosine concentrations, Neurochem. Int. **6**: 613-632.

Yamamoto, T., et al., 1987, Subcellular, regional and immunohistochemical localization of adenosine deaminase in various species, Brain Res. Bull. **19**: 473-484.

4
Transport Systems for Adenosine in Mammalian Cell Membranes

A.S. Clanachan and F.E. Parkinson

Introduction

The passage of adenosine and other nucleosides across plasma membranes may occur by simple diffusion, or may be mediated by nucleoside-specific transport systems that are intrinsic components of the plasma membrane. Both facilitated diffusion and sodium-dependent nucleoside transport (NT) systems have been described and each of these classes may possess several subtypes. It should be noted that, in some cells, adenosine fluxes are rapid. Consequently rapid sampling assays are required to measure initial rates of uptake that are representative of transport rates (see reviews by Paterson et al., 1985; Gati et al., 1989a). Interpretation of inhibition of nucleoside entry as transport inhibition requires that uptake rates be demonstrably initial rates. Thus, accurate assessment of NT and its inhibition can best be obtained using cells in culture or in fresh cells that have been dispersed enzymatically. This report will review mechanisms of adenosine transport and summarize the evidence in support of NT system heterogeneity, with particular emphasis on NT systems in cells and tissues likely to be involved in the modulation of the physiological and pharmacological actions of adenosine.

Simple Diffusion

Simple diffusion is a minor component of the membrane flux of physiological nucleosides due to their hydrophilic nature. However some synthetic, lipophilic nucleosides, such as 3'-deoxy-3'-azidothymidine (AZT) and 2',3'-dideoxythymidine readily permeate cells via a nonmediated mechanism (Zimmerman et al., 1987). N^6-phenylisopropyladenosine, an A_1 adenosine receptor agonist, also enters cells by a non-inhibitable process (Plagemann et al., 1984). Simple diffusion is recognized experimentally as its rate is directly proportional to permeant concentration gradient and depends upon permeant size, charge and lipid solubility and is not altered by other permeants or inhibitors. Corrections for nonmediated permeation, that must be made in experimental investigations of NT systems, have been facilitated by the use of radiolabelled L-adenosine. This enantiomer of the naturally occurring D-adenosine enters cells by a NT-independent route (Gati et al., 1989b).

Facilitated Diffusion NT Systems

Facilitated diffusion NT has been extensively studied in human erythrocytes (see recent review by Gati et al., 1989a). Based on studies in erythrocytes, facilitated diffusion NT systems are considered to be equilibrative, nonconcentrative processes in which the passage of nucleoside across the plasma membrane is mediated by membrane carriers that catalyse rapid fluxes. Fluxes saturate, are reversible and are inhibited competitively by related permeants and a variety of nucleoside and non-nucleoside drugs. The NT system of human erythrocytes has broad specificity and accepts as substrates physiological nucleosides and deoxynucleosides as well as many synthetic nucleosides (Gati et al., 1989a). Although studied in lesser detail, facilitated diffusion NT systems have been explored in other mammalian cells, including erythrocytes from other species and in suspensions of enzymatically dispersed cells from heart (Heaton et al, 1987; Rovetto et al., 1987; Ford et al., 1987) and brain (Johnson et al., 1989).

NT Inhibitors

NT in human erythrocytes (Cass et al., 1974), erythrocytes from some other species (Jarvis et al., 1982a) and cardiac myocytes (Heaton et al., 1987) is inhibited by low concentrations (<10nM) of some S^6-nitrobenzyl derivatives of 6-thiopurine pentofuranosides, such as nitrobenzylthioinosine (NBMPR). A significant advance in the study of NT systems arose from the observation that NBMPR binds with high affinity to specific membrane sites on human erythrocytes and that occupancy of these sites correlated with inhibition of NT (Cass et al., 1974). Radiolabelled NBMPR has been used as a binding probe to identify NT sites in many tissues, including cardiac (Williams et al., 1984) and CNS membranes (Hammond et al., 1984; 1985; Geiger et al., 1984), erythrocytes (Jarvis et al., 1982a; 1982b) and cardiac myocytes (Heaton et al., 1987). Inhibition constants obtained for drug-induced inhibition of [^{3}H]NBMPR binding and inhibition of NT are similar and the cellular abundance of NBMPR sites is predictive of transport rates (Clanachan et al., 1987). As a result of these studies [^{3}H]NBMPR has been regarded as a marker for one type of equilibrative NT system.

Heterogeneity of Facilitated Diffusion NT Systems

Not all facilitated diffusion NT systems are identical. Subtypes exist that can be identified based on functional (inhibitor sensitivity and permeant selectivity) and structural (Jarvis et al., 1986) characteristics. The detailed kinetic properties (symmetry, carrier mobilities, etc.) of most of these subtypes are unknown.

Sensitivity to NBMPR While human erythrocytes, rat and guinea pig cardiac myocytes possess NT systems that are highly sensitive to NBMPR (IC_{50}, $K_D < 10$ nM), some cultured cells (Walker 256 carcinosarcoma and Novikoff hepatoma N1S1-67) possess NT systems that are not inhibited by low concentrations of NBMPR. Although these NT systems have been termed NBMPR-insensitive (Belt, 1983), they are inhibitable by higher concentrations of NBMPR ($IC_{50} > 1\mu M$) (Paterson et al., 1987). NT systems of high and low NBMPR sensitivity co-exist in rat erythrocytes and rat cortical synaptosomes (Jarvis et al., 1986; Lee et al.,

27

1988). Thus a lack of high affinity NBMPR binding sites does not necessarily imply an absence of NT activity. Further complexity has become apparent with the identification of cells that possess high affinity NBMPR binding sites, but appear to lack NBMPR-sensitive transport (Paterson et al., 1987; Gati et al., 1989a).

Sensitivity to Dipyridamole Dipyridamole has significantly lower affinity for NT systems in rat and it does not potentiate the effects of adenosine in rat tissue. (see Hammond et al., 1985 for refs.). In addition, CNS membrane preparations from guinea pig and dog each demonstrated an apparent single class of binding sites for NBMPR, but dipyridamole produced biphasic inhibition of NBMPR binding (Hammond et al., 1985). These results suggest the existence of two subtypes of NT systems that can be distinguished by dipyridamole and possibly by other transport inhibitors, but not by NBMPR.

Dipyridamole apparently has similar affinity for NT systems that show high and low sensitivity to NBMPR. For example, in rat erythrocytes, a cell type in which NBMPR has biphasic concentration effect relationships, dipyridamole inhibition curves are monophasic (Jarvis et al., 1986). [^{3}H]Dipyridamole binding sites have been characterized in several cell and membrane preparations. In human erythrocytes (Jarvis, 1986) the numbers of sites for dipyridamole and NBMPR are similar but in other studies, the number of [^{3}H]dipyridamole binding sites exceeded that for [^{3}H]NBMPR by several fold (Woffendin et al., 1987; Marangos et al., 1987). These differences may be related to the high nonspecific binding of dipyridamole or to the labelling of additional (low affinity NBMPR) sites. Selective ligands for NT systems of low NBMPR sensitivity have yet to be identified.

Sensitivity to Other NT Inhibitors The ability of four NT inhibitors to inhibit uridine transport in erythrocytes and cultured cells from several species has been compared (Plagemann et al., 1988). Some of the differences observed were species related; eg. human cells were more sensitive to dipyridamole and dilazep than rat and mouse cells. There were also cell type differences. Human Hep-2 cells were relatively resistant to lidoflazine (IC_{50}=2 μM) compared to human erythrocytes and HeLa cells (IC_{50}=12-140 nM). These data support our findings of cell differences in affinity of lidoflazine for NBMPR sites in guinea pig heart. Using quantitative autoradiography, we have demonstrated that the distribution of NBMPR sites in guinea pig heart is not uniform; a high site density is associated with endothelial cells, as defined by co-localization with Von Willebrand Factor, and a lower number is visible over cardiac myocytes (Parkinson et al., 1989a). Comparison of inhibition constants of several NT inhibitors revealed that lidoflazine, but not close structural analogs (mioflazine, soluflazine), has an 8.2X selectivity for sites associated with coronary endothelium (Parkinson et al., 1989b).

Permeant Selectivity NT subtypes can also be distinguished by permeant selectivity. Thampy et al. (1983a; 1983b) described an NT system that has high affinity (K_m=13 μM) and selectivity for adenosine in neurons and a system with lower affinity (K_m=370 μM) for adenosine and broader substrate specificity in cultured glial cells from chick brain. It has been hypothesized that peripheral systems are characterized by broad substrate specificity, with K_m values for adenosine of 100 - 1000 μM, while neuronal systems are selective for adenosine, with K_m values of 1 - 10 μM (Geiger et al., 1985). In non-CNS tissues, adenosine is a preferred substrate among endogenous nucleosides although exogenous nucleosides (eg. 2-chloroadenosine) may have higher affinity (Heaton et al., 1987).

Sodium-Dependent Transport

Energy-dependent, concentrative NT systems have been demonstrated in several different cell types but detailed characterization of these systems has not been reported. Sodium-dependent NT has been described in rat hepatocytes and renal brush border vesicles (Angielski et al., 1983; Le Hir et al., 1984) and guinea pig (Schwenk et al., 1984), rat (Jakobs et al., 1986) and mouse intestinal epithelial cells (Vijayalakshmi et al., 1988). A small component of adenosine accumulation in dissociated rat brain cells is also sodium-dependent (Johnston et al., 1989). Such systems may constitute a vectorial transport system for nucleosides.

Heterogeneity of Sodium-Dependent NT Systems

On the basis of nucleoside-induced inhibition of [³H]formycin B or [³H]thymidine transport into mouse intestinal epithelial cells (Vijayalakshmi et al., 1988), two subtypes were identified; one with selectivity for formycin B, guanosine and inosine, and one with selectivity for thymidine and cytidine. Other nucleosides, such as adenosine and uridine, appeared to be nonselective. Although sodium-dependent NT, in most cells studied to date, is not affected by inhibitors of the facilitated diffusion processes, NBMPR has been reported to inhibit sodium-dependent nucleoside entry in a mixed population of dissociated brain cells (Johnson et al., 1989). Thus, subtypes of sodium-dependent NT systems are detectable based on both permeant and inhibitor selectivity.

Joint Presence of Facilitated Diffusion and Sodium-Dependent NT Systems

The joint presence of the two main classes of NT systems has been reported so considerable complexity may exist in the mechanisms of adenosine permeation. NT inhibitor ENHANCEMENT of nucleoside entry in mouse leukemia L1210 cells to levels five times higher than the extracellular concentration is demonstrable (Dagnino et al., 1987). The inhibitor evidently blocks efflux via the facilitated diffusion NT system without interfering with sodium-dependent influx. Similar phenomena have not been investigated in adenosine responsive cells.

NTI (Nucleoside Transport Inhibitory) Receptor Sites as Targets for New Drugs

The relationship between sites to which NBMPR, dipyridamole and other NT inhibitors bind and nucleoside permeation sites is the subject of debate in the current literature (Gati et al., 1989a, Jarvis, 1986). Evidently several subtypes of sites exist on, or are closely associated with, the NT protein. Occupancy of NTI sites inhibits NT. Study of the pharmacology and molecular biology of these sites awaits the development of NTI-selective ligands. In addition, these agents could have important therapeutic applications in several areas and may be more selective in their actions than the currently available NT inhibitors.

Nucleoside Chemotherapy

Several nucleoside drugs possess cytotoxic activity that has been exploited in the treatment of bacterial, parasitic, viral and neoplastic diseases. A "host-protection tactic" that employs a combination of cytotoxic nucleosides and NT inhibitors (to prevent entry of nucleoside into host tissues) has been shown effective in experimental therapy of several types of neoplastic (Paterson et al., 1987) and parasitic disease (El Kouni et al., 1983). A greater range of NT inhibitors is required to optimize such therapeutic strategies.

Adenosine Therapeutics

Adenosine receptor agonists elicit a wide spectrum of effects throughout the body, many of which could be used to therapeutic advantage. NT inhibitors, via potentiation of the physiological and pharmacological actions of adenosine, have many of the same properties. The CNS actions of adenosine and the potential CNS indications of NT inhibitors have been reviewed (Deckert et al., 1988). Similarly, many of the cardiovascular actions of adenosine (Belardinelli et al., 1989) may be mimicked by NT inhibitors. Selective inhibitors of the various NT systems may result in drugs that could modify adenosine actions to greater therapeutic advantage.

Acknowledgements

Work from the authors' laboratory is supported by the Alberta Heart and Stroke Foundation. F.E.P. held an Alberta Heritage Foundation for Medical Research Studentship.

References

Angielski, S., Le Hir, M., and Dubach, U.C., 1983, Transport of adenosine by renal brush border membranes, *Pflugers Arch.* **397**:75-77.
Belardinelli, L., Linden, J., and Berne, R.M., 1989, The cardiac actions of adenosine, *Prog. Cardiovasc. Dis.* **32**:73-97.
Belt, J.A., 1983, Heterogeneity of nucleoside transport in mammalian cells. Two types of transport activity in L1210 and other cultured neoplastic cells. *Mol. Pharmac.* **24**:479-484.
Cass, C.E., Gaudette, L.A., and Paterson, A.R.P., 1974, Mediated transport of nucleosides in human erythrocytes: specific binding of the inhibitor nitrobenzylthioinosine to nucleoside transport sites in the erythrocyte membrane, *Biochim. Biophys. Acta* **345**:1-10.
Clanachan, A.S., Heaton, T.P., and Parkinson, F.E., 1987, Drug interactions with nucleoside transport systems, In Gerlach, E. and Becker B.F. (eds) *Topics and Perspectives in Adenosine Research,* Springer-Verlag, Berlin pp 118-130.
Dagnino, L., Bennett, L.L., and Paterson, A.R.P., 1987, Concentrative transport of

nucleosides in L1210 mouse leukemia cells, *Proc. Am. Assoc. Cancer Res.* **28**:15.

Deckert, J., Morgan, P.F., and Marangos, P.J., 1988, Adenosine uptake site heterogeneity in the mammalian CNS? Uptake inhibitors as probes and potential neuropharmaceuticals, *Life Sci.* **42**:1331-1345.

El Kouni, M.H., Messier, N.J., and Cha, S., 1987, Treatment of schistosomiasis by purine nucleoside analogues in combination with nucleoside transport inhibitors, *Biochem. Pharmacol.* **35**:3815-3821.

Ford, D.A., and Rovetto, M.J., 1987, Rat cardiac myocyte adenosine transport and metabolism, *Am. J. Physiol.* **252**:H54-H63.

Gati, W.P., Dagino L., and Paterson, A.R.P., 1989, Enantiomeric selectivity of adenosine transport systems in mouse erythrocytes and L1210 cells, *Biochem. J.* (in press).

Gati, W.P., and Paterson, A.R.P., 1989, Nucleoside transport, In Agre, P. and Parker, J.C. (eds) *Red Blood Cell Membranes,* Marcel Decker Inc, New York pp 635-661.

Geiger, J.D., LaBella, F.S., and Nagy, J.I., 1985, Characterization of nitrobenzylthioinosine binding to nucleoside transport sites selective for adenosine in rat brain, *J. Neurosci.* **5**:735-740.

Geiger, J.D., and Nagy, J.I., 1984, Heterogeneous distribution of adenosine transport sites labelled by [³H]nitrobenzylthioinosine in rat brain: an autoradiographic and membrane binding study, *Brain Res. Bull.* **13**:657-666.

Hammond, J.R., and Clanachan, A.S., 1984, [³H]Nitrobenzylthioinosine binding to the guinea pig CNS nucleoside transport system: a pharmacological characterization, *J. Neurochem.* **43**:1582-1592.

Hammond, J.R., and Clanachan, A.S., 1985, Species differences in the binding of [³H]nitrobenzylthioinosine to the nucleoside transport system in mammalian CNS membranes: evidence for interconvertable conformations of the binding site/transporter complex, *J. Neurochem.* **45**:527-535.

Heaton, T.P., and Clanachan, A.S., 1987, Nucleoside transport in guinea pig myocytes: comparison of the affinities and transport velocities for adenosine and 2-chloroadenosine, *Biochem. Pharmacol.* **36**:1275-1280.

Jakobs, E.S., and Paterson, A.R.P., 1986, Sodium-dependent, concentrative nucleoside transport in cultured intestinal epithelial cells, *Biochem. Biophys. Res. Comm.* **140**:1028-1035.

Jarvis, S.M., 1986, Nitrobenzylthioinosine-sensitive nucleoside transport system: mechanism of inhibition by dipyridamole, *Mol. Pharmacol.* **30**:659-665.

Jarvis, S.M., Hammond, J.R., Paterson, A.R.P., and Clanachan, A.S., 1982a, Species differences in nucleoside transport: a study of uridine transport and nitrobenzylthioinosine binding by mammalian erythrocytes, *Biochem. J.* **208**:83-88.

Jarvis, S.M., McBride, D., and Young, J.D., 1982b, Erythrocyte nucleoside transport: asymmetrical binding of nitrobenzylthioinosine to nucleoside permeation sites, *J. Physiol.* **324**:31-46.

Jarvis, S.M., and Young, J.D., 1986, Nucleoside transport in rat erythrocytes: two components with differences in sensitivity to inhibition by nitrobenzyl-thioinosine and p-chloromercuriphenyl sulfonate, *J. Memb. Biol.* **93**:1-10.

Johnson, M.E., and Geiger J.D., 1989, Sodium-dependent uptake of nucleosides by dissociated brain cells from the rat, *J. Neurochem.* **52**:75-81.

Lee, C.W., and Jarvis, S.M., 1988, Nucleoside transport in rat cerebral cortical synaptosomes: evidence for two types of transporters, *Biochem. J.* **249**:557-564.

Le Hir, M., and Dubach, U.C., 1984, Sodium gradient-energized concentrative transport of adenosine in renal brush border vesicles, *Pflugers Arch* **401**:58-63.

Marangos, P.J., and Deckert, J., 1987, [^{3}H]Dipyridamole binding to guinea pig brain membranes: possible heterogeneity of central adenosine uptake sites, *J. Neurochem.* **48**:1231-1236.

Parkinson, F.E., and Clanachan, A.S., 1989a, Heterogeneity of nucleoside transport inhibitory sites in heart: a quantitative autoradiographical analysis, *Br. J. Pharmacol.* **97**:361-370.

Parkinson, F.E., and Clanachan, A.S., 1989b, Subtypes of nucleoside inhibitory sites in heart: a quantitative autoradiographical analysis, *Eur. J. Pharmacol.* **163**:69-75.

Paterson, A.R.P., Harley, E.R., and Cass, C.E., 1985, Measurement and inhibition of membrane transport of adenosine, In Paton, D.M. (ed) *Methods in Adenosine Research,* Plenum, New York pp 165-180.

Paterson, A.R.P., Jakobs, E.S., Ng, C.Y.C., Odegard, R.D., and Adjei, A.A., 1987, Nucleoside transport inhibition in vitro and in vivo, In Gerlach, E. and Becker, B.F. (eds) *Topics and Perspectives in Adenosine Research,* Springer-Verlag, Berlin pp 89-101.

Plagemann, P.G.W., and Woffendin, C., 1988, Species differences in sensitivity of nucleoside transport in erythrocytes and cultured cells to inhibition by nitrobenzylthioinosine, dipyridamole, dilazep and lidoflazine, *Biochim. Biophys. Acta* **969**:1-8.

Plagemann, P.G.W., and Wohlhuter, R.M., 1984, Inhibition of the transport of adenosine, other nucleosides and hypoxanthine in Novikoff rat hepatoma cells by methylxanthines, papaverine and N^6-phenylisopropyladenosine, *Biochem. Pharmacol.* **33**:1783-1788.

Rovetto, M.J., Ford, D.A., and Yassin, A., 1987, Cardiac myocyte and coronary endothelial cell adenosine transport, In Gerlach, E. and Becker, B.F.(eds) *Topics and Perspectives in Adenosine Research,* Springer-Verlag, Berlin pp 188-198.

Schwenk, M., Hegazy, E., and DelPino, V.L., 1984, Uridine uptake by isolated intestinal epithelial cells of guinea pig, *Biochim. Biophys. Acta* **805**:370-374.

Thampy, K.G., and Barnes, E.M. Jr., 1983, Adenosine transport by primary cultures of neurons from chick embryo brain, *J. Neurochem.* **40**:874-879.

Thampy, K.G., and Barnes, E.M. Jr., 1983, Adenosine transport by cultured glial cells from chick embryo brain, *Arch. Biochem. Biophys.* **220**:340-346.

Vijayalakshmi, D., and Belt, J.A., 1988, Sodium-dependent nucleoside transport in mouse intestinal epithelial cells. Two transport systems with differing substrate specificities, *J. Biol. Chem.* **263**:19419-19423.

Williams, E.F., Barker, P.H., and Clanachan, A.S., 1984, Nucleoside transport in heart: species differences in nitrobenzylthioinosine binding, adenosine accumulation, and drug-induced potentiation of adenosine action, *Can. J. Physiol. Pharmacol.* **62**:31-37.

Woffendin, C., and Plagemann, P.G.W., 1987, Interaction of [^{3}H]dipyridamole with the nucleoside transporters of human erythrocytes and cultured animal cells, *J. Memb. Biol.* **98**:89-100.

Zimmerman, T.P., Mahony, W.B., and Prus, K.L., 1987, 3'-azido-3'-dideoxythymidine an unusual nucleoside analogue that permeates the membrane of human erythrocytes and lymphocytes by nonfacilitated diffusion, *J. Biol. Chem.* **262**:5748-5754.

5

Formation of Adenosine in the Heart from Extracellular Adenine Nucleotides

J. Schrader, M.M. Borst, M. Kelm, B. Bading, and K.F. Bürrig

Catabolism of Adenine Nucleotides in the Coronary Circulation

Adenine nucleotides intracoronarily applied are rapidly degraded during single passage through the heart (Baer et al., 1968; Paddle et al., 1987; Schrader et al., 1982; Ronca-Testoni et al., 1982; Fleetwood et al.,1989). The pattern of catabolites formed is consistent with the major pathway of metabolism being sequential dephosphorylation of ATP -> ADP -> AMP -> adenosine (Fleetwood et al.,1989). Ecto-adenosinetriphosphatase (ATPase), -adenosinediphospha-tase (ADPase) and ecto 5'-nucleotidase have been demonstra-ted in cultured endothelial cells (Gordon et al., 1986). These enzymes are mainly responsible for the rapid catabo-lism of adenine nucleotides in the coronary circulation. To what extend cardiomyocytes can metabolize extracellular adenine nucleotides appears to be species dependent (Newby et al., 1987). Guinea pig cardiomyocytes appear not to exhibit ecto 5'-nucleotidase activity (Dendorfer et al., 1987). The further catabolism of adenosine to inosine, hypoxanthine and uric acid is exclusively by endothelial cells. Nucleoside phosphorylase (Rubio et al., 1972) and xanthine oxidase (Schoutsen et al., 1987) are cytosolic marker enzymes of coronary endothelial cells although there are pronounced species differences in the case of xanthine oxidase. In the presence of a nucleoside transport inhibitor such as dipyridamole the catabolism of extracellular ATP only proceeds to the stage of adenosine (Gordon et al., 1986), again demonstrating that the metabolism of adenosine occurs intracellularly.

In our experiments we have assessed coronary vascular ecto-phosphatase activity by perfusing isolated guinea pig hearts with 14-C-labeled ATP at 10^{-6} M. After single passage through the heart most of the radioactivity was recovered with AMP (69 %) and adenosine (24 %), while ATP and ADP comprised only 4 % and 2 %, respectively. When the experi-ments were repeated in the presence of 50 μM α,ß-methylena-denosine diphosphate (AOPCP), a potent inhibitor of ecto 5'-nucleotidase (Schütz et al., 1981) the by far major metabolic product released from the heart was AMP (89 %). Compared with data reported for the rat heart (Fleetwood et al., 1989) it appears that ecto-ATPase is more active in the guinea pig heart, whereas the rat heart converts extracellular AMP more avidly to adenosine.

Differentiation between Extra- and Intracellular Adenosine Formation

As it is schematically outlined in fig 1, adenosine can be formed intracellularly by action of 5'-nucleotidase (Schütz et al., 1981; Meghji et al., 1985) and S-adenosylhomocysteine(SAH)-hydrolase (Lloyd et al., 1988) as well as extracellularly by a cascade of ecto-phosphatases (Fleetwood et al., 1989). It is generally assumed that the major site of cardiac adenosine production is intracellular and that adenosine reaches the extracellular space by diffusion along its concentration gradient (Schütz et al., 1981; Meghji et al., 1985). The quantities of ATP liberated into the coronary effluent perfusate are considerably lower than adenosine (Paddle et al., 1987; Schrader et al., 1982) and do not appear to reach biologically active concentrations. This comparison, however, does not take into account that the majority of ATP may have been degraded during passage through the heart (see above). ATP measured in the coronary effluent perfusate may therefore constitute only the "tip of the iceberg" and extracellular ATP within the heart may be much higher.

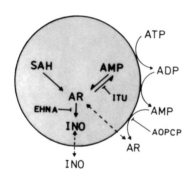

Figure 1
Schematic outline of the major routes of intra- and extracellular adenosine formation.

In order to explore this possibility we have conducted experiments in which ecto 5'-nucleotidase of isolated perfused guinea pig hearts was inhibited by AOPCP (50 μM). The sum of the adenine nucleotides (ATP + ADP + AMP) liberated into the coronary effluent perfusate was quantitated after enzymatic conversion to ATP according to the following reaction scheme: AMP + ATP $\xrightarrow{MK}$ 2ADP + 2PEP $\xrightarrow{PK}$ 2ATP + 2 pyruvate (Hampp, 1985). The rephosphorylation mixture consisted of HEPES-KOH 48 mM (pH 7.75); phospholnolepyruvate (PEP) 2.8 mM; MgSO$_4$ 20 mM; myokinase (MK) 100 U/ml and pyruvatekinase (PK) 22 U/ml. ATP was determined by a sensitive firefly-luciferine-luciferase-assay. With this technique total release of adenine nucleotides from the heart could be quantitated. Furthermore, this technique allowed to differentiate between intra- and extracellular adenosine formation.

34

Release of Adenosine and Adenine Nucleotides during Reactive
Hyperemia and Hypoxic Perfusion

In the absence of AOPCP, isolated saline perfused guinea pig
hearts release adenosine, adenine nucleotides and ATP at a
rate of 0.06 ± 0.01 (11) nmoles/min, 0.04 ± 0.01 (13) nmo-
les/min and 2 ± 1 (7) pmoles/min, respectively. In the
presence of effective inhibition of ecto 5'-nucleotidase
with 50 μM AOPCP, the release of adenine nucleotides increased about 10-fold, while adenosine release remained unaltered (fig 2). It thus appears that a metabolically stable and
well oxygenated heart releases slightly more adenine nucleotides than adenosine. Among the adenine nucleotides liberated ATP constitutes only a minor fraction and the kinetics
of ecto-phosphatases suggest that the predominant metabolite
released is AMP. This conclusion is consistent with a recent
report by Imai demonstrating AMP release into the coronary
and transmyocardial effluent which is of similar magnitude
as adenosine (Imai et al., 1989).

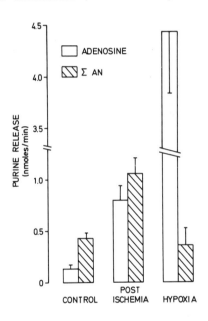

Figure 2
Release of adenosine
and adenine nucleoti-
des (AN) from the
isolated guinea pig
heart perfused with
saline medium at 10
ml/min. Ischemia was
for 30 s, and analysis
was carried out during
reactive hyperemia.
Hypoxia perfusion was
duced by perfusing
hearts with a medium
equilibrated with 10 %
O_2, n = 7 - 9, means
± SEM.

Experimental conditions of enhanced adenosine formation
include postischemic reactive hyperemia (Saito et al., 1981)
and hypoxic perfusion (Schrader et al., 1977). Occlusion of
coronary inflow for 30 s resulted not only in an augmented
adenosine release in the phase of reactive hyperemia but
also increased the release of adenine nucleotides (fig 2).
Again, the quantities of adenosine and adenine nucleotides
liberated are similar. A different picture, however, was
obtained when hearts were perfused with a medium equilibrated with 10 % oxygen. Hypoxic perfusion greatly increased
the release of adenosine (fig 2). The release of adenine
nucleotides, however, remained largely unaltered. Thus
ischemia but not hypoxia is a potent stimulus for adenine

nucleotide release. Conversively, extracellular adenine
nucleotides may contribute to the formation of adenosine
only during ischemia but not during hypoxic perfusion. This
conclusion confirms an earlier report from our laboratory
demonstrating that AOPCP did not alter the hypoxia-induced
release of adenosine (Schütz et al., 1981).

Endothelial Cells Release Adenine Nucleotides

Endothelial cells in culture have been reported to selectively release large quantities of their adenine nucleotides
when stimulated with thrombin (Pearson, 1988). In order to
investigate whether unstimulated endothelial cells normally
release adenine nucleotides, we have conducted experiments
in porcine aortic endothelial cells grown an microcarrier
beads which where superfused in a column with a HEPES-
buffered Krebs-Ringer solution (Kelm et al., 1988) Microcarrier beads were densly covered with endothelial cells (fig
3) which show the typical "cobblestone" appearance.

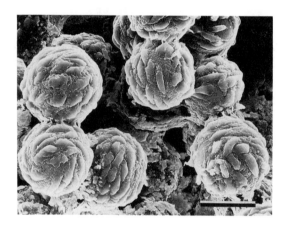

Figure 3
Scanning elec-
tron micrograph
of porcine aor-
tic endothelial
cells cultured
on microcarrier
beads (Cytodex)
bar: 100 μm

Techniques for quantification of cellular adenine nucleotides release were the same as described above for the isolated heart. As is evident from data given in the table, the
fraction of adenine nucleotides released by endothelial
cells is only small when compared with adenosine. Inhibition
of adenosine deaminase 10 μM EHNA not unexpectedly slightly
increased adenosine release but left the nucleotide release
unaltered. Inhibition of ecto 5'-nucleotidase with 50 μM
AOPCP, however, decreased adenosine by about 30 % while
adenine nucleotide release increased in turn. Since the sum
of all adenine containing purines remained constant, these
data indicate, that about 1/3 of the adenosine liberated
from endothelial cells was derived from extracellular degradation of adenine nucleotides.

Table: Release of adenosine and adenine nucleotides (Σ AN) from endothelial cells in the presence of 10 μM EHNA and after the treatment with 50 μM AOPCP.

	Adenosin	Σ AN
control	13.8 ± 1.6	2.2 ± 0.4
+ EHNA 5 x 10^{-6} M	17.2 ± 1.8*	2.8 ± 0.4
+ EHNA 5 x 10^{-6} M + AOPCP 5 x 10^{-5} M	11.9 ± 1.9	7.2 ± 1.3*

Data are normalized for 1 ml of EC-covered beads correspon-ding to 4.05 mg endothelial protein or 2.2 x 10^7 EC; n = 14 - 17, $\bar{x}$ ± SEM. Astrisks indicate significant differences vs. control evaluated by student's t-test for paired data.

Mechanisms of Release of Cellular Adenine Nucleotides and Functional Significance.

The finding that metabolically and functionally isolated hearts as well as cultured endothelial cells constantly release significant quantities of adenine nucleotides in the non-stimulated state is quite unexpected. Because of their charge adenine nucleotides are usually considered not to penetrate easily through cellular membranes unless there is cell damage. There are, however, several reasons to assume that cell damage is unlikely to explain our findings. An accelerated release of adenine nucleotides was observed in our study only during cardiac ischemia, whereas adenosine was the major purine released from the hypoxic heart (fig 2). Imai has recently found, that increased cytosolic AMP of isolated hearts induced by acetate infusion did not change the amount of AMP released into the interstitial fluid compartment (Imai et al., 1989). Again, should the adenine nucleotides released in the coronary circulation be derived from simple leakage from intracellular sources, a steepe-ning of their transmembrane gradient should have resulted in an enhanced AMP release. Another implication of this later finding is, that AMP is unlikely to be the primary adenine nucleotide released by the heart. To all likelyhood this may be ATP, which is rapidly degraded to AMP and adenosine during single passage through the heart.

In an attempt to elucidate the cellular source of the adeni-ne nucleotides liberated by the heart during ischemia, we have carried out experiments in which the adenine nucleoti-des of coronary endothelial cells were selectively prelabe-led by perfusion of isolated hearts with radioactive adeno-sine(Kroll et al., 1987). Should the ischemia-induced relea-se of nucleotides be derived from the prelabeled vascular cell compartment, this would increase the specific radioac-tivity of the released nucleotides. However, should unlabe-led cardiomyocytes preferentially contribute, this would result in a decrease in the nucleotide specific radioactivity. In the actual experiments the specific activity of adenine nucleotides released by the heart during reactive hyperemia in response to 30 s of coronary occlusion increased from 9.7 ± 1.5 to 15.2 ± 2.0 Ci/mol. This finding suggests that vascular cells most likely endothelial cells preferentially contributed to the release of adenine nucleotides under this condition. The mechanism of the augmented release, however, remains to be elucidated. It has been suggested, that in the intact heart nucleotide release from sympathetic nerves may be responsible (Imai et al., 1989), yet nerves cannot be made responsible for the nucleotide release from isolated endothelial cells.

When infused into the coronary circulation of guinea pig hearts, ATP, ADP, AMP and adenosine are equipotent in increasing coronary flow (Schrader et al., 1982). Since about equieffective concentrations of adenosine and adenine nucleotides are released during reactive hyperemia it is likely that the observed changes in coronary resistance may be due to adenosine and the adenine nucleotides. Infusion of adenosine deaminase into the coronary circulation to remove extracellular adenosine has already been shown to reduce the reactive hyperemic flow response by more than 30 %. Our finding would predict that extracellular adenine nucleotides play an equally important part in controlling vascular tone during reactive hyperemia.

References

Baer H.R., Drummond G.I.: Catabolism of adenine nucleotides by the isolated perfused rat heart. Proc.Soc.Exp.Biol.Med. 127: 33-36, 1968.

Dendorfer A., Lank S., Schaff A., Nees S. New insights into the mechanism of myocardial adenosine formation. In: E. Gerlach, B.F. Becker (Eds); Topics and perspectives in adenosine research, Springer, Berlin, p. 170, 1987.

Fleetwood G., Coade S.B., Gordon J.L., Pearson J.D.: Kinetics of adenine nucleotide catabolism in coronary circulation of rats. Am.J.Physiol 256: H 1565 - H 1572, 1989.

Gordon E.L., Pearson J.D., Slakey L.L.: The hydrolysis of extracellular adenine nucleotides by cultured endothelial cells from pig aorta: feed-forward inhibition of adenosine production at the cell surface. J.Biol.Chem. 261: 15496-15504, 1986.

Hampp R.: Adenosine 5'-triphosphat, luminometric method. In: Methods of Enxzymatic Analysis, Bermeyer H.U., Bergmeyer J., Grassl M. (Eds); Weinheim, VCH Verlagsgesellschaft, 370-379, 1985.

Imai S., Chin W.-P., Jin H., Nakazawa M.: Production of AMP and adenosine in the interstitial fluid compartment of the isolated perfused normoxic guinea pig heart. Pflugers Arch. 1989 (in press).

Kelm M., Feelisch M., Spahr R., Piper H.M., Noack E., Schrader J.: Quantitative and kinetic characterization of nitric oxide and EDRF released from cultured endothelial cells. Biochem.Biophys.Res.Com. 154: 236 - 244, 1988.

Kroll K., Schrader J., Piper H.-M., Henrich M.: Release of adenosine and cyclic AMP from coronary endothelium in isolated guinea pig hearts: relation to coronary flow. Circ.Res. 60: 659-665, 1987.

Lloyd H.G.E., Deussen A., Wuppermann H., Schrader J.: The transmethylation pathway as a source for adenosine in the isolated guinea-pig heart. Biochem.J. 252: 489 - 494, 1988.

Meghji P., Holmquist C. A., Newby A.C.: Adenosine formation and release from neonatal-rat heart cells in culture. Biochem. J. 229: 799 - 805, 1985.

Newby A.C., Worku Y., Meghji P.: Critical evaluation of the role of ecto- and cytosolic 5'-nucleotidase in adenosine formation. In: Topics and Perspectives in Adenosine Research, Gerlach E., Becher B.F. (Eds); Springer Verlag, p. 155 - 168, 1987.

Paddle B.M., Burnstock G.: Release of ATP from perfused heart during coronary vasodilatation. Blood Vessels 11: 110-119, 1987.

Pearson J.D., Gordon J.L.: Vascular endothelial and smooth muscle cells in culture selectively release adenine nucleotides. Nature 281: 384 - 386, 1979.

Pearson J.D.: Purine nucleotides as regulators of vessel tone. Biochem.Soc.Trans. 16: 480-482, 1988.

Ronca-Testoni S., Borghini F.: Degradation of perfused adenine compounds up to uric acid in isolated rat heart. J.Mol.Cell.Cardiol. 14: 177 - 180, 1982.

Rubio R., Wiedmeier T., Berne R.M.: Nucleoside phosphorylase: localization and role in the myocardial distribution of purines. Am J.Physiol. 222: 550 - 555, 1972.

Saito D., Steinhart C.R., Nixon D.G., Olsson R.A.: Intracoronary adenosine deaminase reduces reactive hyperemia. Circ.Res. 47: 875 - 882, 1981.

Schoutsen B., De Jong J.W.: Age dependent increase in xanthine oxidoreductase differs in various heart cell types. Circ.Res. 61: 604-607, 1987.

Schrader J., Haddy F.J., Gerlach E.: Release of adenosine, inosine and hypoxanthine from the isolated guinea pig heart during hypoxia, flow autoregulation and reactive hyperemia. Pflügers Arch. 369: 1 - 6, 1977.

Schrader J., Thompson C.I., Hiendlmayer G., Gerlach E.: Role of purines in acetylcholine-induced coronary vasodilation. J.Mol.Cell.Cardiol. 14: 427 - 430, 1982.

Schütz W., Schrader J., Gerlach E.: Different sites of adenosine formation in the heart. Am.J.Physiol 240: H 963 - H 970, 1981.

6
Central Nervous System Effects of Adenosine

J.W. Phillis

Introduction

Adenosine and the adenine nucleotides have been recognized to have important and potent central actions since the original studies on their intracerebroventricular administration by Feldberg and Sherwood (1954). In the brain adenosine has been shown to depress transmitter release, reduce neuronal excitability and dilate cerebral blood vessels. One of its most important roles in the brain is likely to be the matching of neuronal activity (metabolic demand) to the local blood flow (metabolic supply), to ensure that neurons and glia are able to exist in a metabolically stable state.

These actions of adenosine are mediated by extracellular receptors that have been divided into two major subclasses: A_1 receptors which mediate inhibition of adenylate cyclase, and A_2 receptors which stimulate the enzyme. At A_1 receptors N^6-[(R)-1-methyl-2-phenyl-ethyl]adenosine (R-PIA) and N^6-cyclohexyladenosine (CHA) are more potent than NECA, whereas the reverse holds for A_2 receptors. Two classes of A_2 receptor have been identified, with some A_2 receptors having EC_{50} values for adenosine in the high nanomolar range rather than the micromolar range (Daly et al., 1983). The high affinity A_2 site is present in highest amounts in the striatum and olfactory tubercle (Jarvis et al., 1989) but is also detectable at lower levels in other regions of the brain (Bruns et al., 1986).

A vigorous debate has emerged over the A_1/A_2 nature of the extracellular receptors involved in the depressant effects of adenosine on the discharge of central neurons. Electrophysiological studies on hippocampal slices have indicated that adenosine acts at A_1 receptors both presynaptically to inhibit transmitter release, and postsynaptically by interfering with membrane excitability (Jackisch et al., 1985; Proctor and Dunwiddie, 1987; Schubert and Kreutzberg, 1987; Dunwiddie and Fredholm, 1989). The evidence for an exclusively A_1 receptor mediated response in other regions of the brain is less convincing. 5'-N-ethylcarboxamidoadenosine (NECA), an adenosine analog which is equipotent at A_1 and A_2 sites (Heffner et al., 1989), was more potent as a depressant of the spontaneous firing of rat cerebral cortical neurons than R-PIA and CHA (Stone, 1982; Phillis, 1982) and in a variety of behavioral experiments involving microinjection of adenosine analogues into the brain, NECA has been more potent than the A_1 receptor agonists (Green et al., 1982; Phillis et al., 1986; Barraco and Bryant, 1987). The possibility of A_2 receptor involvement, with stimulation of adenylate cyclase, was strengthened by the observation that 2'5' dideoxyadenosine, an adenosine analog which inhibits adenylate cyclase, antagonized the responses to adenosine (Stone, 1982) and NECA (Green et al., 1982). The equipotency of NECA at the two recognized adenosine receptor sites, and the lack of other potent, selective, A_2 receptor agonists have, however, complicated any final assessment of the contribution of the A_2 receptor to the neural depressant action of adenosine.

Adenosine-evoked relaxation of cerebral blood vessels appears to be mediated by A_2 receptors, as NECA is a more potent relaxant than R-PIA and CHA (Edvinsson and Fredholm, 1983; McBean et al., 1988). This observation has been confirmed in binding studies using [^{3}H] CHA and [^{3}H] NECA, in the presence of cyclopentyladenosine (an A_1 ligand) to identify A_1 and A_2 receptors in brain microvessels (Kalaria and Harik, 1988). Whereas there was little CHA binding, indicating a lack of A_1 receptors, the blood vessels were well endowed with A_2 receptor binding sites.

CGS 21680 - a Novel A_2 Agonist

The recent availability of a novel A_2 receptor agonist, 2-[p-(2-carboxyethyl)phenethylamino-5'-N-ethylcarboxamidoadenosine (CGS 21680) offers new possibilities for the resolution of debate on the A_1/A_2 nature of the adenosine receptors associated with cerebral cortical neurons. CGS 21680 is as potent as NECA at A_2 receptors in radioligand binding studies in rat brain, and is 140 fold selective for the A_2 receptor (Williams et al., 1989). It was more potent than NECA in increasing coronary blood flow, an A_2 response, and inactive in decreasing heart rate, an A_1 response (Hutchinson et al., 1989).

In experiments on spontaneously active rat cerebral cortical neurons, iontophoretically applied CGS 21680 was observed to have a potent depressant action on spontaneous, acetylcholine-evoked, and in some instances glutamate-evoked firing. CGS 21680 was equipotent with adenosine as a depressant of spontaneous and acetylcholine-evoked firing; the inhibition was slightly slower in the onset than that observed with adenosine and recovery after termination of application was slower than following adenosine application. Confirmation that CGS 21680 exerted its depressant actions at an adenosine receptor was obtained in experiments with an adenosine antagonist. 8-p-sulphophenyltheophylline, applied iontophoretically for periods of 2-3 min, antagonized the depressant effects of CGS 21680 on all eleven of the neurons tested. Recovery of the CGS effect was apparent 3-6 min after the end of the 8-p-sulphophenyltheophylline application.

CGS 21680 depressed glutamate-evoked firing of several cortical neurons, but failed to inhibit glutamate-evoked firing of other neurons, the spontaneous and acetylcholine-evoked firing of which were clearly depressed. Inhibition of glutamate-evoked firing is indicative of a postsynaptic action of CGS 21680, whilst depression of spontaneous firing, in the absence of a reduction in glutamate-elicited excitation, is indicative of a presynaptic site of action. These findings suggest that CGS 21680, like adenosine, modulates synaptic transmission via A_2 receptors located on both the presynaptic terminals and the postsynaptic neurons.

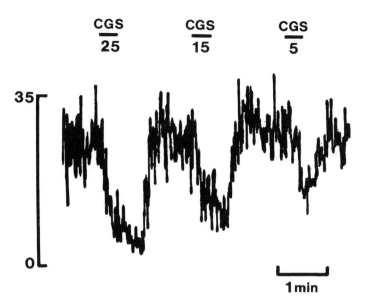

Fig. 1. Rate meter recording of a spontaneously firing rat cerebral cortical neuron. CGSS 21680 applied with currents of 25 nA, 15 nA and 5 nA for the periods shown by the horizontal bars depressed firing.

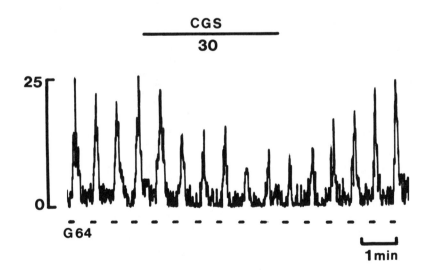

Fig. 2. The firing of this rat cortical neuron evoked by repeated brief applications of L-glutamate (G, 64 nA) was depressed by an application of CGS 21680 (30 nA for 4 min).

CGS 21680 was less effective than NECA as a depressant of cerebral cortical neuronal firing, even though it is a more potent agonist at the A_2 receptors on rat coronary arteries (Hutchinson et al., 1989). This observation suggests that both A_1 and A_2 receptor co-activation may contribute to the purinergic inhibition of cortical neuronal firing. An involvement of both A_1 and A_2 receptors would be consistent with the greater potency of NECA as a depressant of cortical neuronal firing than the selective A_1 agonists, R-PIA and CHA (Stone, 1982; Phillis, 1982).

Adenosine and Cerebral Blood Flow

The evidence for a role of adenosine in the control of the cerebral circulation has recently been reviewed (Phillis, 1989a). Adenosine is formed in the brain during hypoxia and ischemia and its release into the extracellular space has been demonstrated by several groups (Van Wylen et al., 1986; Hagberg et al., 1987; Phillis et al., 1987). Pharmacological studies have confirmed an involvement of adenosine in hypoxia-evoked cerebral vascular hyperemia. Adenosine antagonists reduce or prevent the cerebral hyperemia during hypoxia (Emerson and Raymond, 1981; Phillis et al., 1984; Morii et al., 1987). Degradation of adenosine to its inactive metabolite inosine following application of the enzyme adenosine deaminase led to almost complete inhibition of the dilatation of pial vessels during hypoxia (Kontos and Wei, 1981).

Conversely, a number of agents that inhibit adenosine transport including dipyridamole, papaverine, lidoflazine were able to potentiate the increase in cerebral blood flow during hypoxic episodes in the rat (Phillis et al., 1984; Phillis et al., 1985a). Administration of the potent inhibitors of adenosine deaminase, deoxycoformycin and erythro-hydroxynonyladenine, potentiated hypoxia-evoked increases in cerebral blood flow (Phillis et al., 1985b). Adenosine also appears to be involved in the increases in cerebral blood flow observed when animals inhale carbon dioxide (Phillis and DeLong, 1987). The CO_2-induced hyperemia is reduced by the adenosine antagonist, caffeine, and potentiated by transport and adenosine deaminase inhibitors. An increase in adenosine release from the isolated rabbit heart perfused with fluids made acidic with carbon dioxide has been reported (Mustafa and Mansour, 1984) and the release of adenosine from vascular endothelial cells and cultured aortic smooth muscle cells is substantially increased upon elevations of pCO_2 in the perfusion mixture (Nees and Gerlach, 1983; Belloni et al., 1989). An localized increase in adenosine release into the vicinity of cerebral blood vessels could account for carbon dioxide-induced increases in cerebral blood flow.

Adenosine as a Cerebroprotective Agent

Adenosine and its more stable analogs have recently received attention in the treatment of cerebral ischemic damage. To circumvent the blood-brain barrier, these compounds have been applied locally into the brain. CHA injected intracerebroventricularly into gerbils prior to a 30 min period of bilateral carotid occlusion had a protective action on CA1 hippocampal pyramidal neurons (Von Lubitz et al., 1988). Similar protection against ischemic damage was observed when 2-chloroadenosine was injected into the rat hippocampus prior to the onset of forebrain ischemia (Evans et al., 1987). Conversely, pretreatment with the adenosine antagonist theophylline enhanced both the neurological symptoms and neuronal damage observed in gerbils following forebrain ischemia (Rudolphi et al., 1987).

These beneficial effects of adenosine analogs on ischemic brains may result from the several actions of this purine, including increased blood flow to the ischemic areas; decreased release of the excitotoxic amino acids glutamate and aspartate; lowering membrane calcium permeability and reduced adhesion of platelets and neutrophils. Adenosine release may thus constitute an integral part of the brain's defense mechanism against ischemic insults.

The inability of the currently available adenosine analogs to readily penetrate the blood-brain barrier mitigates against the use of these compounds for the prevention and treatment of the neuronal damage associated with a stroke or cardiac arrest. An alternate strategy is the utilization of agents which affect its metabolism to potentiate the protective actions of endogenously released adenosine. The adenosine deaminase inhibitor deoxycoformycin elevates hypoxia-evoked levels of adenosine in the extracellular space and depresses the formation of the oxypurines (Phillis et al., 1988). Deoxycoformycin pretreatment reduced ischemic damage to the CA1 region of the gerbil hippocampus and at the same time prevented the associated increase in locomotor activity associated with forebrain ischemia in this species (Phillis and O'Regan, 1989). Oxypurinol, an inhibitor of xanthine oxidase which also increases hypoxia-elicited levels of interstitial adenosine (O'Regan et al., 1989), significantly protected gerbil hippocampal CA1 pyramidal cells against ischemia damage (Phillis, 1989b).

Conclusions

Adenosine appears to play an important role in the regulation of cerebral blood flow and by virtue of its ability to increase local blood flow whilst depressing neuronal discharges, it regulates the critical balance between metabolic demand and neuronal activity. In the ischemic brain adenosine exerts a protective function by virtue of its ability to increase cerebral blood flow, depress excitotoxic amino acid release, reduce membrane calcium permeability and prevent platelets and neutrophil aggregation. Potentiation of endogenously released adenosine levels may provide new therapeutic approaches to the prevention and treatment of cerebral ischemic damage.

References

Barraco, R.A., and Bryant, S.D., 1987, Depression of locomotor activity following bilateral injections of adenosine analogs into the striatum of mice, Med. Sci. Res. **15**:421-422.

Belloni, F.L., et al., 1986, Uptake and release of adenosine by cultured rat aortic smooth muscle, Microvasc. Res. **32**:200-210.

Bruns, R.R., Lu, G.H., and Pugsley, T.A., 1986, Characterization of the A_2 adenosine receptor labeled by [^{3}H] NECA in rat striatal membranes, Molec. Pharmacol. **29**:331-346.

Daly, J.W., Butts-Lamb, P., and Padgett, W., 1983, Subclasses of adenosine receptors in the central nervous system: interaction with caffeine and related methylxanthines, Cell. Mol. Neurobiol. 3:69-80.

Dunwiddie, T.V., and Fredholm, B.B., 1989, Adenosine A_1 receptors inhibit adenylate cyclase activity and neurotransmitter release and hyperpolarize pyramidal neurons in rat hippocampus, J. Pharmac. Exp. Ther. 249:31-37.

Edvinsson, L., and Fredholm, B.B., 1983, Characterization of adenosine receptors in isolated cerebral arteries of cat, Br. J. Pharmac. 80:631-637.

Emerson, T.E., and Raymond, R.M., 1981, Involvement of adenosine in cerebral hypoxic hyperemia in the dog, Am. J. Physiol. 241:H134-H138.

Evans, M.C., Swan, J.H., and Meldrum, B.S., 1987, An adenosine analogue, 2-chloroadenosine, protects against long term development of ischaemic cell loss in the rat hippocampus, Neurosci. Letts. 83:287-292.

Feldberg, W., and Sherwood, S.L., 1954, Injections of drugs into the lateral ventricle of the cat, J. Physiol. (Lond.) 123:148-167.

Gerlach, E., Becker, B.F., and Nees, S., 1987, Formation of adenosine by vascular endothelium: a homeostatic and antithrombogenic mechanism?, In: Topics and Perspectives in Adenosine Research, Gerlach, E., and Becker, B.F. (Eds), Springer-Verlag, Berlin, pp 309-319.

Green, R.D., Proudfit, H.K., and Yeung, S.-M.H., 1982, Modulation of striatal dopaminergic function by local injection of 5'-N-ethylcarboxamide adenosine, Science 218:58-61.

Hagberg, H., et al., 1987, Extracellular adenosine, inosine, hypoxanthine, and xanthine in relation to tissue nucleotides and purines in rat striatum during transient ischemia, J. Neurochem. 49:227-231.

Heffner, T.G., et al., 1989, Comparison of the behavioral effects of adenosine agonists and dopamine antagonists in mice, Psychopharmacol. 98:31-37.

Hutchinson, A.J., et al., 1989, CGS 21680, an A_2 selective adenosine (ADO) receptor agonist with preferential hypotensive activity, FASEB J. 3:A281.

Jackisch, R., et al., 1984, Endogenous adenosine as a modulator of hippocampal acetylcholine release, Naunyn-Schmiedebergs Arch. Pharmacol. 327:319-325.

Jarvis, M.F., Jackson, R.H., and Williams, M., 1989, Autoradiographic characterization of high-affinity adenosine A_2 receptors in the rat brain, Brain Res. 484:111-118.

Kalaria, R.N., and Harik, S.I., 1988, Adenosine receptors and the nucleoside transporter in human brain vasculature, J. Cereb. Blood Flow Metab. 8:32-39.

Kontos, H.A., and Wei, E.P., 1981, Role of adenosine in cerebral arteriolar dilation from arterial hypoxia, Fed. Proc. 40:454.

McBean, D.E., Harper, A.M., and Rudolphi, K.A., 1988, Effects of adenosine and its analogues on porcine basilar arteries: are only A_2 receptors involved?, J. Cereb. Blood Flow. Metab. 8:40-45.

Morii, S., et al., 1987, Role of adenosine in regulation of cerebral blood flow: effects of theophylline during normoxia and hypoxia, Am. J. Physiol. 253:H165-H175.

Mustafa, S.J., and Mansour, M.M., 1984, Effect of perfusate pH on coronary flow and adenosine release in isolated rabbit heart, Proc. Soc. Exp. Biol. Med. 176:22-26.

Nees, S., and Gerlach, E., 1983, Adenine nucleotide and adenosine metabolism in cultured coronary endothelial cells: formation and release of adenine compounds and possible functional implications, In: Regulatory Function of Adenosine, Berne, R.M., Rall, T.W., and Rubio, R. (Eds), Martinus Nijhoff, Boston, pp 347-360.

O'Regan, M.H., Phillis, J.W., and Walter, G.A., 1989, The effects of the xanthine oxidase inhibitors, allopurinol and oxypurinol on the pattern of purine release from hypoxic rat cerebral cortex, Neurochem. Int. **14**:91-99.

Phillis, J.W., 1982, Evidence for an A_2-like adenosine receptor on cerebral cortical neurons, J. Pharm. Pharmacol **34**:453-454.

Phillis, J.W., 1989a, Adenosine in the control of the cerebral circulation, Cerebrovasc. Brain Metabol. Rev. **1**:26-54.

Phillis, J.W., 1989b, Oxypurinol attenuates ischemia-induced hippocampal damage in the gerbil, Brain Res. Bull. (in press).

Phillis, J.W., et al., 1986, Behavioral characteristics of centrally administered adenosine analogs, Pharmacol. Biochem. Behav. **24**:261-270.

Phillis, J.W., and DeLong, R.E., 1987, An involvement of adenosine in cerebral blood flow regulation during hypercapnia, Gen. Pharmacol. **18**:133-139.

Phillis, J.W., DeLong, R.E., and Towner, J.K., 1985a, The effects of lidoflazine and flunarizine on cerebral reactive hyperemia, Europ. J. Pharmac. **112**:323-329.

Phillis, J.W., DeLong, R.E., and Towner, J.K., 1985b, Adenosine deaminase inhibitors enhance cerebral anoxic hyperemia in the rat, J. Cereb. Blood Flow Metab. **5**:295-299.

Phillis, J.W., and O'Regan, M.H., 1989, Deoxyformycin antagonizes ischemia-induced neuronal degeneration, Brain Res. Bull. **22**:537-540.

Phillis, J.W., O'Regan, M.H., and Walter, G.A., 1988, Effects of deoxycoformycin on adenosine, inosine, hypoxanthine, xanthine and uric acid release from the hypoxemic rat cerebral cortex, J. Cereb. Blood Flow Metab. **8**:733-741.

Phillis, J.W., Preston, G., and DeLong, R.E., 1984, Effects of anoxia on cerebral blood flow in the rat brain: evidence for a role of adenosine in autoregulation, J. Cereb. Blood Flow Metab. **4**:586-592.

Phillis, J.W., et al., 1987, Increases in cerebral cortical perfusate adenosine and inosine concentrations during hypoxia and ischemia, J. Cereb. Blood Flow Metab. **7**:679-686.

Proctor, W.R., and Dunwiddie, T.V., 1987, Pre- and postsynaptic actions of adenosine in the in vitro rat hippocampus, Brain Res. **426**:187-190.

Rudolphi, K.A., Keil, M., and Hinze, H.-J., 1987, Effect of theophylline on ischemically induced hippocampal damage in Mongolian gerbils: a behavioral and histopathological study, J. Cereb. Blood Flow Metab. **7**:74-81.

Schubert, P., and Kreutzberg, G.W., 1987, Pre- versus postsynaptic effects of adenosine on neuronal calcium fluxes, In: Topics and Perspectives in Adenosine Research, Gerlach, E., and Becker, B.F. (Eds), Springer-Verlag, Berlin, pp 521-531.

Stone, T.W., 1982, Purine receptors involved in the depression of neuronal firing in cerebral cortex, Brain Res. **248**:367-370.

Van Wylen, D.G.L., et al., 1986, Increases in cerebral interstitial fluid adenosine concentration during hypoxia, local potassium infusion, and ischemia, J. Cereb. Blood Flow Metab. **6**:522-528.

Von Lubitz, D.K.J.E., et al., 1988, Cyclohexyladenosine protects against neuronal death following ischemia in the CA1 region of the gerbil hippocampus, Stroke **19**:1133-1139.

Williams, M., et al., 1989, [^{3}H] CGS 21680, an A_2 selective adenosine (ADO) receptor agonist directly labels A_2 receptors in rat brain tissue, FASEB J. **3**:A1047.

7

Adenosine Receptor Sub-types in Isolated Tissues: Antagonist Studies

M.G. Collis

Introduction

Adenosine mediates many of its effects via cell surface receptors. These receptors have been divided into A_1 and A_2 sub—types, based on the original observation by Van Calker et al (1979) that adenosine could either inhibit or stimulate adenylate cyclase activity in a brain cell culture. Subsequent studies utilising the measurement of cyclic nucleotide levels or ligand binding techniques in membrane preparations have established that there are different structure activity relationships for a series of agonists at A_1 and A_2 receptors (see Collis, 1985).

The sub—classification of adenosine receptors in isolated tissues has relied upon the order of potency of agonists. This has lead to the proposal that cardiac receptors, such as those which mediate a decrease in rate or force of contraction of the guinea—pig atrium, are similar to the A_1 receptor identified in isolated or broken cell preparations. In contrast, smooth muscle receptors such as those which evoke relaxation of the pre—contracted guinea—pig aorta or trachea, are similar to the A_2 receptor (Brown and Collis, 1982; Collis, 1983; Collis and Brown, 1983; Collis and Saville, 1984). The use of agonist potency in receptor classification is not ideal as it does not provide an estimate of affinity. Thus, the A_1/A_2 sub—classification of adenosine receptors in isolated tissues cannot be accepted until selective antagonism of the receptor sub—types has been demonstrated (Collis, 1985). Some 8—aryl substituted xanthine antagonists have been claimed to exhibit selective affinity for sub—types of the adenosine receptor in ligand binding studies. Their effects on adenosine evoked responses in isolated tissues are described in this paper.

Methods

The apparent affinity of adenosine receptor antagonists was evaluated using the negative chronotropic response of guinea—pig or rat atria (putative A_1 receptor), and the relaxant response of pre—contracted aortic and tracheal preparations from guinea—pigs (putative A_2 receptor). 2—chloroadenosine or adenosine were used as the agonists, with the addition of dipyridamole or dipyridamole plus EHNA, respectively, to prevent uptake and metabolism (Brown and Collis, 1982; Collis and Brown

1983). Dose—response curves to these agonists were generated both before and after equilibration of the isolated tissues with an antagonist .

Results and Discussion

8—phenyltheophylline (8PT) has been reported to exhibit moderate (at most 10 fold) A_1 selectivity in ligand binding or adenylate cyclase based assay systems using rat tissues (Bruns et al., 1988; Oei et al., 1988; Ukena et al., 1986a). However, 8PT caused equivalent degrees of antagonism of adenosine evoked responses in atria, and in aortic and tracheal preparations from guinea—pigs (Fig 1). Analysis of the data produced a Schild plot (Arunlakshana and Schild, 1959) with a slope close to unity, which is indicative of competitive antagonism. The pA_2 values for this antagonist did not differ between the three tissues (6.3—6.7, Collis et al., 1985; Collis et al., 1989). Thus, the results do not support adenosine receptor sub—type selectivity for this xanthine in guinea—pig isolated tissues. 8PT has also been reported to be non—selective in studies using ligand binding techniques or adenylate cyclase levels in guinea—pig cells (Daly et al., 1985; Ukena et al., 1986b; Ferkany et al., 1986). 8PT exhibits a 5 fold greater affinity for adenosine receptors in rat atria than in atria from guinea—pigs (Table 1) which suggests that further evaluation of the compound in isolated tissues from the rat is warranted and might reveal moderate A_1 selectivity.

Table 1

Antagonist affinity of 8—substituted xanthines in atria from rat and guinea—pig, using 2—chloroadenosine as the agonist.(Collis et al., 1988)

Antagonist	Species	pA_2	Schild Slope	n	pA_2 Rat/G.pig.
8PT	Rat	7.16	—0.95	13	
	G.Pig	6.48	—0.86	14	4.7
XAC	Rat	7.82	—0.88	20	
	G.Pig	7.22	—1.03	11	4.0
CPX	Rat	8.24	—0.93	13	
	G.Pig	8.14	—0.90	15	1.3

Propyl group substitution in the 1 and 3 position of the xanthine ring enhances affinity for adenosine receptors. 8—(4—amino—2—chlorophenyl)—1,3—dipropylxanthine (PACPX) has been reported to be an A_1 selective antagonist with high affinity in ligand binding and cyclase based studies (Bruns et al., 1988; Oei et al., 1988; Martinson et al., 1987; Daly et al., 1985) using guinea—pig or rat cells. Attempts to accurately assess the affinity of this compound for A_1 and A_2 receptors in isolated tissues have been prevented by the poor aqueous solubility of this xanthine. The limited data obtained using low concentrations of PACPX does not, however, support marked A_1 receptor selectivity for this compound since similar degrees of antagonism were observed in guinea—pig atria and aorta (Collis et al., 1987).

The limited aqueous solubility of the 8—aryl—1,3—dipropylxanthines can be overcome by substitution of a hydrophilic group at the para position of

FIGURE 1

Effect of 8PT (■ 10⁻⁶M, ● 3x10⁻⁶M, ▼10⁻⁵M) on Responses Evoked by Adenosine Receptor Stimulation in Guinea-Pig Atria (A), Aorta (B), Trachea (C).

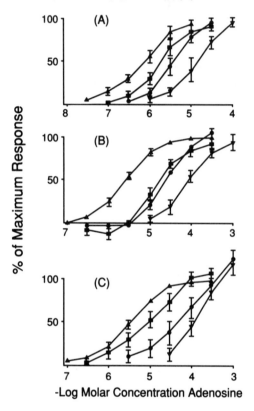

the phenyl ring as in 8-(4-((2-aminoethylamino)carbonylmethoxy)phenyl)-1,3-dipropylxanthine (XAC, Jacobson et al, 1985). Although XAC is a potent antagonist in isolated tissues from guinea-pigs it does not exhibit receptor sub-type selectivity (Collis et al., 1987). XAC is also non-selective in studies using ligand binding or adenylate cyclase based techniques in guinea-pig cells (Ukena et al.,1986b; Jacobson et al.,1985). As with a number of 8-substituted xanthines, the affinity of XAC in rat atria is somewhat higher than in the guinea-pig (Table 1). In ligand binding or cyclase based studies using rat cells, XAC exhibits moderate A_1 receptor selectivity (Schwabe, 1988; Jacobson et al., 1986; Ukena et al.,1986a,b; Oei,et al.,1988; Lohse et al., 1987).

Recently, an 8-cyclopentyl substituted 1,3-dipropylxanthine (CPX) has been reported to exhibit selective A_1 receptor affinity in a rat isolated heart preparation (Haleen et al., 1987). This selectivity has also been supported by studies using guinea-pig isolated tissues. CPX is a more potent antagonist of the bradycardia induced by adenosine in guinea-pig atria than of the relaxations evoked by the purine in aortic and tracheal preparations (Fig 2 from Collis et al., 1989). The antagonism appears to be competitive and the compound has a pA_2 of 8.1–8.4 (depending on experimental conditions) in atria compared to values of 6.6 in aorta and trachea (Collis et al., 1989). Thus CPX is the first compound to show unequivocal A_1 receptor selectivity in isolated tissues from both the rat and the guinea-pig.

FIGURE 2 (from Collis et al., 1989)

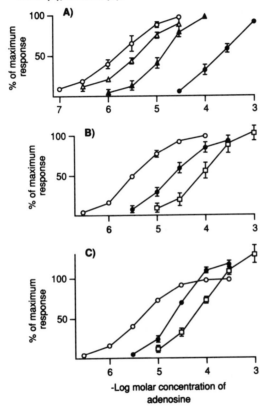

Effect of CPX ($\triangle$ 10^{-8}M, $\blacktriangle$ 10^{-7}M, $\bullet$ 10^{-6}M, $\square$ 5×10^{-6}M) on Adenosine Evoked Responses in Guinea-pig Atria (A), Aorta (B), Trachea (C).

The marked A_1 receptor selectivity of CPX has been noted in all studies of the compound published to date, irrespective of the species or technique used to assess affinity (Bruns et al., 1988; Martinson et al., 1987; Lohse et al., 1987; Haleen et al., 1987; Ukena et al., 1986b; Collis et al.,1989). In addition, the species differences in affinity noted with other xanthines seem to be less pronounced with CPX (Table 1 and Ukena et al., 1986b).

51

The majority of the adenosine antagonists published to date have been xanthines. CGS 15943A, a triazoloquinazoline, is chemically related to a series of benzodiazepine inverse agonists and is a potent adenosine receptor antagonist (Williams et al., 1987). pA_2 values of 10.8 and 10.1 have been reported for CGS 15943A in isolated coronary arteries and guinea-pig trachea respectively, compared to a value of 7.4 on guinea-pig atria (Ghai et al., 1987). These results suggest a marked A_2 selectivity for this antagonist. In our studies of this compound we have obtained evidence of competitive antagonism in the guinea-pig atrium (pA_2=7.6, Schild plot slope = 0.92) however in the aorta, Schild plots were not linear and similar tendancy was noted in the guinea-pig trachea. The compound did not appear to exhibit marked A_2 selectivity although a slightly greater antagonism of A_2 than of A_1 mediated responses was seen at some concentrations (Fig 3). The results of these experiments are in agreement with ligand binding studies in which CGS 15943A has been reported to be a competitive inhibitor at A_1 binding sites but a non-competitive inhibitor at A_2 binding sites in rat brain. In these binding studies CGS 15943A also exhibited only moderate (7 fold) A_2 selectivity.

FIGURE 3

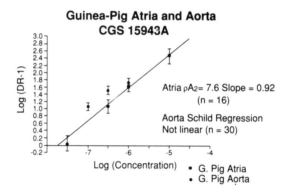

Conclusions

A number of adenosine receptor antagonists have been proposed to exhibit selectivity between the proposed sub-types of adenosine receptors in broken cell preparations. Many of these compounds have also been evaluated in isolated tissues. In a number of cases the selectivity observed in broken cell preparations from the rat has not been confirmed in isolated tissues from guinea-pigs. This may be a consequence of species differences since A_1 affinity is often higher in cells or tissues from the former species. The limited solubility of some xanthines has also complicated the evaluation of their affinity. CPX, however, is an antagonist with adequate solubility that exhibits marked A_1 receptor selectivity in isolated tissues from the rat and guinea-pig. This observation greatly strengthens the hypothesis that two sub-types of adenosine receptors exist in isolated intact tissues, a proposal which was previously based solely on the relative potency of agonists. The discovery of an antagonist capable of demonstrating A_2 selective effects in isolated tissues would give further support to the sub-classification of adenosine receptors, however, compounds exhibiting this selectivity in a consistent fashion have not been reported to date.

References

Arunlakshana,O. and Schild,H.O. (1959). Brit.J.Pharmacol. 14:48–58.

Brown,C.M. and Collis, M.G. (1982). Brit.J.Pharmacol. 76:381–387.

Bruns,R.F., Davis,R.E., Ninteman,F.W., Poschel,B.P.H., Wiley,J.N. and Heffner,T.G. (1988). In " Adenosine and Adenine Nucleotides: Physiology and Pharmacology " ed. D.M.Paton, Taylor and Francis, pp39–49.

Bruns,R.F., Fergus,J.H., Badger,E.W., Briston,J.A, Santay,L.A, Hartman,J.D., Hays,S.J. and Huang,C.C. (1987). Naunyn–Schmeidebergs Arch. Pharmacol. 335:59–63.

Collis,M.G. (1983). Brit.J.Pharmacol. 78:207–212.

Collis,M.G. (1985). In "Purines: Pharmacology and Physiological Roles" ed T.Stone. Macmillan, pp75–84.

Collis,M.G. and Brown, C.M. (1983). Europ.J.Pharmacol. 96:61–69.

Collis,M.G., Culver,J.C. and Holmes,S. (1988). Brit. J. Pharmacol. 94:423P.

Collis,M.G., Jacobson,K.A. and Tomkins,D.M. (1987). Brit.J.Pharmacol. 92:69–75.

Collis,M.G., Palmer,D.B. and Saville, V.L. (1985). J.Pharm.Pharmacol. 37: 278–280.

Collis,M.G. and Saville,V.L. (1984). Brit.J.Pharmacol. 83:413P.

Collis,M.G., Stoggall,S.M. and Martin,F.M. (1989). Brit.J.Pharmacol. 97:1274–1278.

Daly,J.W., Padgett,W., Shamim, M.T., Butts–Lamb,P. and Waters , J.(1985). J.Med.Chem. 28:1334–1340.

Ferkany,J.W., Valentine,H.L., Stone,G.A. and Williams, M. (1986). Drug.Dev.Res. 9:85–93.

Ghai,G., Francis,J.E., Williams,M., Dotson,R.A, Hopkins,M.F., Cote,D.T., Goodman,F.R. and Zimmerman,M.B. (1987). J.Pharmacol.Exp.Ther. 242:784–790.

Haleen,S.J., Steffen,R.P. and Hamilton,H.W. (1987). Life Sci. 40:555–561.

Jacobson,K.A., Kirk,K.L., Padgett,W.L. and Daly, J.W. (1985). J.Med.Chem. 28:1334–1340.

Jacobson,K.A., Ukena,D., Kirk,K.L. and Daly,J.W. (1986). Proc.Natl. Acad. Sci. 83:4089–4093.

Lohse,M.J., Klotz,K–N., Lindenborn–Fotinos,J., Reddington,M., Schwabe,U. and Olsson,R.A.(1987). Naunyn–Schmeidebergs Arch.Pharmacol.336:204–210.

Martinson,E.A., Johnson,R.A. and Wells,J.N. (1987). Molec.Pharmacol.31: 247–252.

Oei,H.H., Ghai,G.R., Zoganas,H.C., Stone,G.A., Zimmerman,M.B., Field,F.P. and Williams, M. (1988). J.Pharmacol.Exp.Ther. 247:882–888.

Ukena,D., Daly,J.W., Kirk,K.L. and Jacobson,K.A. (1986a). Life Sci. 38:797–807.

Ukena,D., Jacobson.K.A., Padgett,W.L., Ayala,C., Shamim,M.T., Kirk,K.L., Olsson,R.O. and Daly,J.W. (1986b) . Febs.Lett. 209:122–128.

Van Calker,D., Muller,M. and Hamprecht,B. (1979). J.Neurochem. 33:999–1005.

Williams,M., Francis,J., Ghai,G., Braunwalder,A., Psychoyos,S., Stone,G.A and Cash,W.D. (1987). J.Pharmacol.Exp.Ther. 241: 415–420.

Acknowledgements

The author thanks Drs R James and R Hargreaves, Chemistry Department 2, ICI, for synthesis of PACPX, CPX and CGS 15943A.

8
Molecular Probes for Adenosine Receptors

K.A. Jacobson

Introduction

High affinity adenosine agonists and antagonists have been developed as radioligands for characterizing adenosine receptors in binding studies. Previous studies have utilized reversibly binding tritiated ligands such as the A_1-selective agonists (Table 1) N^6-cyclohexyladenosine, *1* (Bruns et al, 1983), and R-N^6-(phenylisopropyl)adenosine, *2* (Schwabe and Trost, 1980), or the A_1-selective antagonists (Table 2), XAC, *7* (xanthine amine congener, 8-((2-aminoethyl)-aminocarbonylmethyloxyphenyl)-1,3-dipropylxanthine) (Jacobson et al, 1986) and CPX, *6* (1,3-dipropyl-8-cyclopentylxanthine) (Bruns et al, 1987). [^{3}H]CPX has a K_d at rat brain A_1 receptors of 0.42 nM and a K_i at A_2 receptors of 410 nM. At present, [^{3}H]CPX is the most A_1-selective among the commonly used antagonist radioligands. [^{3}H]XAC is 60-fold A_1-selective in the rat brain and is of nearly the same affinity (K_d = 1.2 nM) as CPX.

Table 1. Structures of adenosine agonist radioligands.

Figure 1. Structures of tritiated xanthines as adenosine antagonist radioligands.

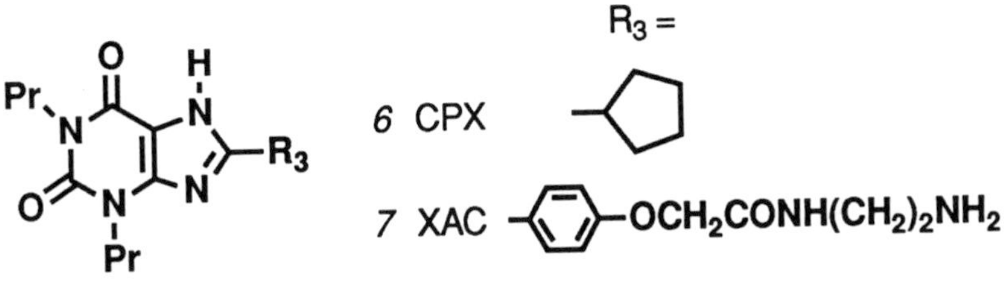

$R_3 =$

6 CPX

7 XAC —⟨benzene⟩—$OCH_2CONH(CH_2)_2NH_2$

Radioiodinated adenosine derivatives have been introduced as high affinity probes at A_1-receptors (Schwabe et al, 1985, Stiles et al, 1985; Linden et al, 1984, Munshi and Baer, 1982). ^{125}I-R-N^6-(p-Hydroxyphenylisopropyl)-adenosine, 3, has a K_d of 1 nM at rat A_1 adenosine receptors. Iodinated adenosine receptor ligands have been particularly useful for peripheral adenosine receptors of relatively low density (Friessmuth et al, 1987).

Probes for A_2-Adenosine Receptors

A satisfactory ligand for A_2 receptors has been more elusive. Binding assays utilizing [^3H]NECA (5'-N-ethylcarboxamidoadenosine), 5, were developed by Yeung and Green (1984) and by Bruns et al. (1986), for the characterization of central A_2 receptors. Since NECA is nonselective, it was necessary to add an unlabeled A_1-selective agonist, N^6-cyclopentyladenosine, in order to block the A_1 component of binding. Numerous structure activity studies have been carried out using the [^3H]NECA binding assay at A_2 receptors.

Although moderately A_1-selective in the rat brain, [^3H]XAC binds in a saturable manner to A_2-adenosine receptors in human platelets (Ukena et al, 1986). Substituted xanthines displaced [^3H]XAC binding in platelets with inhibition constants which closely approximated functional potencies of the xanthines as inhibitors of NECA-induced stimulation of adenylate cyclase activity.

Recently, Hutchison, Williams, Jarvis, and co-workers at CIBA-Geigy synthesized and characterized an A_2-selective agonist, CGS21680, 8 (Figure 2), 2-(carboxyethylphenylethylamino)adenosine-5'-carboxamide (Hutchison et al, 1990). In competitive binding studies against [^3H]CHA and [^3H]NECA, CGS21680 was 170-fold A_2 selective. This analog combines the uronamide feature responsible for the high A_2 affinity of NECA with a 2-substitution of the purine ring, reminiscent of the slightly (5-fold) A_2 selective agonist CV-1808 (2-phenylaminoadenosine). [^3H]CGS21680 bound specifically to A_2 receptors in rat brain with the expected pharmacology (Jarvis et al, 1989).

The carboxylic acid group of CGS21680, 8, may be extended as a functionalized chain for affinity labeling of the A_2 receptor protein (Barrington et al, 1989b, Jacobson et al, 1989b). An amine derivative, 2-[(2-aminoethyl-amino)carbonylethylphenylethylamin, 9 (APEC), was condensed with p-aminophenylacetic acid. The conjugate, PAPA-APEC, *10*, served as a high affinity, A_2-selective and iodinatable ligand. The aryl amine of PAPA-APEC was derivatized in the manner shown in Fig. 3 (see Barrington et al, this volume), to achieve photoaffinity cross-linking to the A_2 adenosine receptor protein. The A_1 and A_2 receptors were found to be distinct glycoproteins having different molecular weights.

Figure 2. Structures of 2-substituted adenosine 5'-carboxamide derivatives as A_2-adenosine receptor probes.

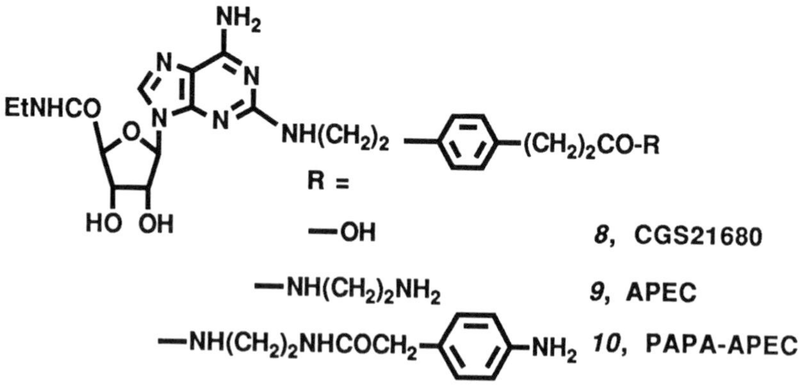

Functionalized Congeners

XAC, 7, and the agonists ADAC, 4, (adenosine amine congener) and APEC, 9, were designed as functionalized congeners, i.e. they contain a chemical functional group that can be coupled covalently to other molecules to produce biologically active conjugates (Jacobson et al, 1985; 1986). The receptor affinity in a given series has been enhanced by coupling of functionalized congeners to certain chemical groups, for example, long chain fatty acids (Jacobson et al, 1987b). Conjugates have been prepared that contain prosthetic groups for the incorporation of radioisotopes (Stiles and Jacobson, 1987), or fluorescent groups for spectrophotometric detection (Jacobson et al, 1987a). An example of a prosthetic group for radioiodination is p-hydroxyphenylpropionic acid (Bolton and Hunter reagent), which is used routinely for labeling proteins via reaction at available amino sites. The p-hydroxyphenylpropionic conjugate of ADAC bound to A_1 receptors with K_i of 4 nM. We have introduced the iodinatable p-aminophenylacetic acid (PAPA) group as an alternative to p-hydroxyphenylpropionic acid, in which the amino group serves as a site for additional reaction steps, resulting in covalent labeling of the receptor (see below).

The ability to couple xanthine functionalized congeners to relatively bulky attached groups ("carriers") with the retention of receptor affinity has led to the synthesis of spectrophotometric probes for adenosine receptors. The functionalized chains of XAC and ADAC have served as points of attachment for fluorescent dye moieties. For example the conjugate of ADAC and fluorescein isothiocyanate displays a K_i value of 2.85 nM in the displacement of [^{3}H]PIA at bovine brain A_1 receptors (Jacobson et al, 1987). Such fluorescent ligands of high affinity hold promise for the development of novel methods of receptor detection and characterization which will obviate the need for radioisotopes.

Affinity Labels - Photoactivated

A general scheme for use of chemically functionalized adenosine and xanthine derivatives for photoaffinity labeling of adenosine receptors (Figure 3) has been explored in collaboration with Dr. Gary Stiles of Duke University Medical Center. A purine functionalized congener, *11* (for example XAC), is condensed with an aryl-amine containing prosthetic group to yield conjugate *12*, for the dual purpose of radioiodination and receptor cross-linking. After radioiodination, the ^{125}I-labeled aryl amine, *13*, is either converted to the photoreactive aryl azide, *14*, for direct photoaffinity labeling (Barrington et al, 1989a) or first bound to the receptor and then exposed to a reagent for photoaffinity cross-linking, such as SANPAH, N-succinimidyl-6-(4';-azido-2'-nitrophenylamino)-

Figure 3. Photoaffinity labeling - and - photoaffinity crosslinking using functionalized congeners.

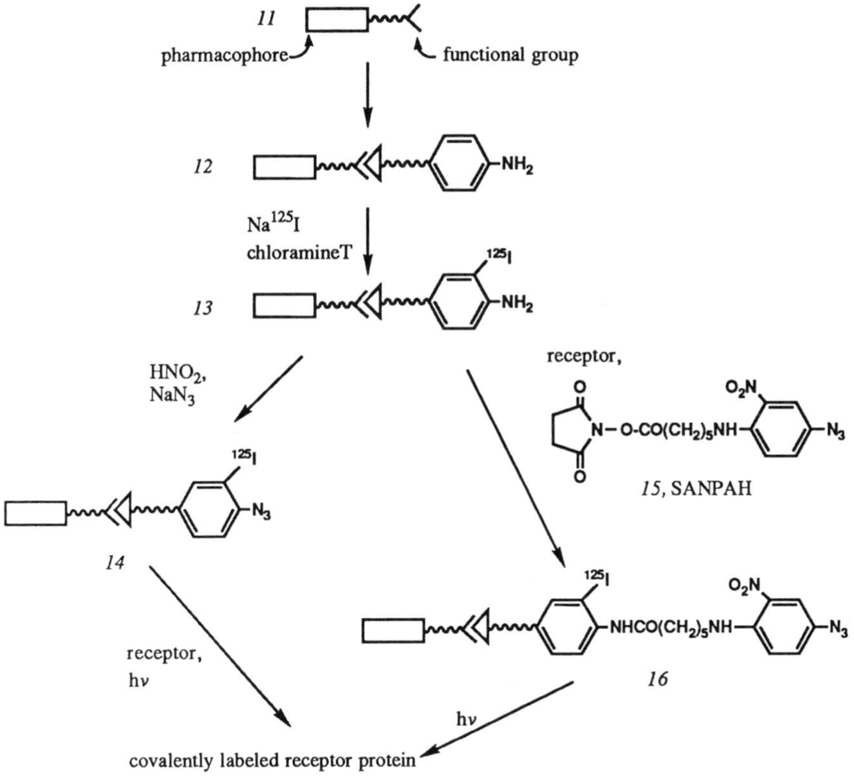

hexanoate, *15* (Stiles and Jacobson, 1987). The SANPAH conjugate may be either formed *in situ* at the receptor site or synthesized and purified separately (Barrington et al, 1989a).

The amide conjugate of XAC and p-aminophenylacetic acid (PAPAXAC) was iodinated with [125]I. The resulting radioligand bound to bovine brain A_1-adenosine receptors with a K_d of 0.1 nM. The aryl amine was crosslinked to the receptor through photoaffinity crosslinking, by SANPAH in the presence of the receptor (Stiles and Jacobson, 1987). The ^{125}I-labeled A_1 receptor displayed a molecular weight of 38,000-40,000 by SDS gel electrophoresis, i.e. nearly identical to the molecular weight of the A_1-receptor as determined through agonist labeling.

Other photoreactive purine derivatives which have been used to label the A_1 receptor include agonist (Stiles et al, 1986, Choca et al, 1985, Lohse et al, 1985) or antagonist ligands (Linden et al, 1989).

Chemically-Reactive Affinity Labels

A general scheme is presented for the chemical affinity labeling of receptors using functionalized congeners (Figure 4). An amino group (for example, the primary amine of XAC, *7*, or ADAC, *4*) is coupled to a crosslinking reagent,

Figure 4. Chemical affinity labeling using amine functionalized congeners.

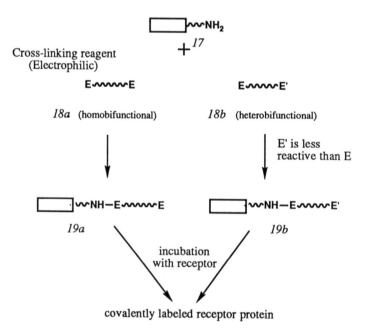

covalently labeled receptor protein

Scheme. Synthesis of a purine isothiocyanate, *21*, an A_1-receptor affinity label.

18, bearing two identical (homobifunctional) or dissimilar (heterobifunctional) electrophilic groups of the type known to react effectively with protein nucleophiles at neutral pH in aqueous solution. A potential affinity label, *19*, designed to react with nucleophilic groups of the receptor protein which are in proximity during binding, is then isolated.

Table 2. Electrophilic purines tested as irreversible affinity labels at A_1-adenosine receptors.

▭⸺NH⸺E~~E	% Inhibition[a]	Reactivity
21, XAC ⸳⫶⸺CSNH⸺◯⸺NCS	29	=NH or -SH
22, ADAC ⸳⫶⸺ "	>80[b]	"
23, XAC ⸳⫶⸺CSNH⸺◯⸺NCS	67	"
24, ADAC ⸳⫶⸺ "	>90[b]	"
25, XAC ⸳⫶⸺COCH₂NHCSNH⸺◯⸺NCS	50	"
26, XAC ⸳⫶⸺COCH₂Br	16	"
27, ADAC ⸳⫶⸺ "	38	"
28, XAC ⸺⫶⸺CO(CH₂)₂SS⸺◯(N)	3	-SH
29, XAC ⸳⫶⸺CO(CH₂)₃⸺N(maleimide)	15	-SH
30, XAC ⸳⫶⸺CO(CH₂)₆COO-N(succinimide)	32	-NH₂
31, XAC ⸳⫶⸺SO₂⸺◯⸺SO₂Cl	23	=NH or -SH

a percent inhibition following a 45 min incubation with the compound at 50 nM, unless noted, and washout, in rat brain membranes, against ^{125}I-APNEA, data from Jacobson et al., 1989.
b compound present at 10 nM.

59

For chemical crosslinking to A_1 receptors (Stiles and Jacobson, 1988; Jacobson et al, 1989a), XAC was coupled to meta- and para-phenylene diisothiocyanates, e.g. 20. The resulting conjugates, m- and p-DITC-XAC, respectively, in submicromolar concentrations incubated with brain membranes selectively inactivated adenosine receptor sites. When the DITC-XAC isomers were synthesized using [^{3}H]XAC (a simple, single step reaction), the xanthine specifically labeled the receptor protein for detection on SDS gel electrophoresis.

The agonist-derived isothiocyanate derivatives, 22 and 24 (Table 2) were the most effective irreversible inhibitors of A_1-adenosine receptors. N-Bromoacetyl derivatives, 26 and 27, of XAC and ADAC, respectively, were synthesized, but only 27 resulted in appreciable irreversible inhibition. Amine-reactive affinity labels were more effective at inhibiting A_1 receptors than those (e.g. maleimides, disulfides) known to react preferentially with thiol nucleophiles (Jacobson et al, 1989a).

The A_1 selectivity of the purine isothiocyanates (Table 3) suggests that these compounds are potentially useful in pharmacological or physiological studies to delineate adenosine-mediated effects from other pathways or to separate A_1- and A2-mediated biological effects. The receptor inactivation occurs efficiently and in the absence of light. m-DITC-ADAC at 10 nM, during a 45 minute incubation with brain membranes, caused a greater than 90% specific loss of A_1 binding sites.

Table 3. Selectivity of Purine Aryl Isothiocyanate Derivatives
for Central Adenosine A_1 and A_2 Receptors
(K_i-values in nM)*

Compound	A_1	A_2	A_2/A_1
p-DITC-XAC, 21	6.6	321	49
p-DITC-ADAC, 22	0.47	191	407
m-DITC-XAC, 23	2.4	343	144
m-DITC-ADAC, 24	0.87	176	203

* binding in rat cortex (A_1) against [^{3}H]PIA
  or in rat striatum (A_2) against [^{3}H]NECA.

Immobilized Ligands and Biotin-Avidin Probes

Functionalized congeners may be coupled to a solid support, commonly an agarose matrix, for receptor isolation by affinity chromatography. XAC has been linked via an amide bond to an agarose resin (Affigel 10), activated in the form of an N-hydroxysuccinimide ester, for the partial purification of A_1-adenosine receptors (Nakata, 1989 and this volume; Olah, 1989). The drug-spacer arm of this resin consists of XAC-CO(CH$_2$)$_2$CONH(CH$_2$)$_2$NHCOCH$_2$O-agarose. 150-fold or greater purification of the A_1 receptor protein has been achieved using this resin.

Table 4. Biotin-avidin probes for A1-adenosine receptors

| | spacer group | | K_i at A_1 [a] | |
			no avidin	+ avidin
32, ADAC		"	11	36
33, ADAC	CO(CH$_2$)$_5$NH	"	18	35
34, ADAC	COCHNH$_2$(CH$_2$)$_4$NH	"	8.9	-
35, XAC		"	54	>500
36, XAC	CO(CH$_2$)$_5$NH	"	50	>500
37, XAC	(COCH$_2$NH)$_3$CO(CH$_2$)$_5$NH	"	50	260

[a] in nM, inhibition of binding of [^{3}H]PIA to rat brain membranes

The biotin-avidin complex is also useful in receptor purification, as was demonstrated for insulin receptors (Hofmann and Finn, 1985). A biotin moiety has been incorporated in adenosine and xanthine analogs (Jacobson et al, 1985), to allow complexation to avidin for purposes of receptor detection and isolation (Table 4). Biotin probes are also useful for assessing indirectly the dimensions of the binding pockets for agonists and antagonists at adenosine receptors.

In vivo Activity of XAC and Its Derivatives

In addition to its use a receptor probe, XAC has been used widely for in vivo studies. XAC was the first adenosine antagonist shown to be bioavailable and A_1-selective in vivo, in a cardiovascular model in the rat (Fredholm et al, 1987) due to its favorable aqueous solubility. In rats, intraperitoneally administered XAC antagonized the cardiac depressant (A_1) effect of NECA, at an EC50 dose for XAC (60 μg/kg) which was 20-fold lower than the dose required for reversal of the hypotensive effect (A_2). Various peptide conjugates of XAC, which were shown to be approximately as potent as XAC in competitive binding at A_1 receptors, were also highly potent in vivo (Fredholm et al, unpublished). D-Lys-XAC and D-tyr-D-lys-XAC were each roughly an order of magnitude less potent than XAC in reversing the bradycardiac effects of NECA. D-Lys-XAC was 9-fold cardioselective, and D-tyr-D-lys-XAC was equipotent in heart rate and blood pressure effects.

XAC is active in vivo in the kidney, where it acts as a diuretic. It reversed the reduction in urine flow and sodium excretion produced by CHA (Barone et al, 1989).

Seale et al (in press) found that XAC (administered intraperitoneally in a vehicle of 15% dimethylsulfoxide, 15% emulphor, and 70% saline) acted as a peripherally-selective adenosine antagonist in DBA/2 inbred mice. The adenosine agonists NECA and CHA caused profound hypothermia with a reduction of core temperature up to 10°C, with ED50 values of 8 and 200 μg/kg, respectively. XAC reversed this hypothermic effect in a dose dependent manner. XAC, at a dose of 0.1 mg/kg, administered 15 min before the agonist, reduced the hypothermic effect by 30% produced by an ED50 dose of NECA. The ED50 dose for XAC inhibition of NECA-induced core temperature depression was 0.16 mg/kg. At a high dose of XAC (1 mg/kg), there was a complete hypothermia-blocking effect of either agonist. The effect of XAC on agonist-induced hypo-thermia was overcome, in part, by increased doses of CHA, more readily than by increased doses of NECA.

In contrast to the potent peripheral effects of XAC was a lack of effect on locomotor depression (measured over a 60 min period in 5 min increments in an Omnitech Digiscan apparatus) elicited by adenosine agonists. The ED50 values for locomotor depression by NECA and CHA were 0.7 and 140 μg/kg, respectively. Pretreatment of the mice with XAC, at the same doses as above, failed to reverse the locomotor depression elicited by either NECA or CHA. Thus, XAC behaved as a peripheral adenosine antagonist, similar to p-sulfophenyltheophylline, but in contrast to the centrally-active caffeine and theophylline. XAC alone at a dose of 1 mg/kg had no effect on either locomotor activity or core temperature.

References

Barone, S., Churchill, P.C., Jacobson, K.A. (1989) Adenosine receptor prodrugs: towards kidney-selective dialkylxanthines. J. Pharm. Exp. Therap. 250, 79-85.

Barrington, W.W., Jacobson, K.A., and Stiles, G.L. (1989a) Demonstration of distinct agonist and antagonist conformations of the A_1 adenosine receptor, J. Biol. Chem., 264, 13157-13164.

Barrington,W.W., Jacobson, K.A., Williams, M., Hutchison, A.J. and Stiles,G.L. (1989b) Identification of the A_2 adenosine receptor binding subunit by photoaffinity crosslinking. Proc. Natl. Acad. Sci USA 86, 6572-6576.

Bruns,R.F.,Daly,J.W. and Snyder,S.H. (1980) Adenosine receptors in brain membranes: binding of N^6-cyclohexyl[^{3}H]adenosine and 1,3-diethyl-8-[^{3}H]phenylxanthine. Proc. Natl. Acad. Sci. U.S.A. 77, 5547-5551.

Bruns, R.F., Daly, J.W. and Snyder, S.H. (1983) Adenosine receptor binding: structure-activity analysis generates extremely potent xanthine antagonists. Proc. Natl. Acad. Sci. U.S.A. 80, 2077-2080.

Bruns, R.F. Lu, G.H. and Pugsley, T.A. (1986) Characterization of the A_2 adenosine receptor labeled by [^{3}H]NECA in rat striatal membranes. Mol. Pharmacol. 29, 331-346.

Bruns, R.F.,Fergus, J.H.,Badger, E.W.,Bristol,J.A.,Santay,L.A., Hartman,J.D., Hays, S.J. and Huang, C.C. (1987).Binding of the A_1-selective adenosine antagonist 8-cyclopentyl-1,3-dipropylxanthine to rat brain membranes. Naunyn-Schmiedeberg's Arch. Pharmacol. 335,59-63.

Choca, J.I., Kwatra, M.M., Hosey, M.M. and Green R.D.(1985) Specific photoaffinity labeling of inhibitory adenosine receptors. Biochem. Biophys. Res. Comm. 131, 115-121.

Earl, C.Q., Patel, A.,Craig, R.H., Daluge, S.M. and Linden, J.L. (1988) Photoaffinity labeling adenosine A_1 receptors with an antagonist ^{125}I-labeled aryl azide derivative of 8-phenylxanthine. J.Med.Chem. 31, 752-756.

Fredholm, B.B., Jacobson, K.A., Jonzon, B., Kirk, K.L., Li, Y.O., Daly, J.W. (1987) Evidence that a novel 8-phenyl-substituted xanthine derivative is a cardioselective adenosine receptor antagonist in vivo. J. Cardiovasc. Pharmacol. 9, 396-400.

Friessmuth, M., Hausleithner, V., Tuisl, E., Nanoff, C., and Schutz, W. (1987) Naunyn-Schmiedeberg's Arch. Pharmacol. 335, 438.

Hofmann, K. and Finn, F.M. (1985) Affinity chromatography of ther insulin receptor based on the avidin-biotin interaction. Ann. N.Y. Acad. Sci. 447, 359-372.

Hutchison, A.J.,Williams, M.deJesus, R., Oei, H.H., Ghai, G.R., Webb, R.L., Zoganas, H.C.,Stone,G.A. and Jarvis,M.F.(1990) 2-Arylalkylamino-adenosine 5'-uronamides: a new class of highly selective adenosine A_2 receptor agonists. J.Med. Chem., in press.

Jacobson, K.A., Kirk, K.L., Padgett, W., Daly, J.W. (1985) Probing the adenosine receptor with adenosine and xanthine biotin conjugates. FEBS Lett. 184, 30-35.

Jacobson, K.A., Ukena, D., Kirk, K.L., Daly, J.W. (1986) [3H]xanthine amine congener of 1,3-dipropyl-8-phenylxanthine: an antagonist radioligand for adenosine receptors. Proc. Natl. Acad. Sci. USA 83, 4089-4093.

Jacobson, K.A., Ukena, D., Padgett, W., Kirk, K.L., Daly, J.W. (1987) Molecular probes for extracellular adenosine receptors. Biochem. Pharmacol. 36, 1697-1707.

Jacobson, K.A., Barone, S., Kammula, U., Stiles, G.L. (1989a) Electrophilic derivatives of purines as irreversible inhibitors of A_1-adenosine receptors. J. Med. Chem. 32, 1043-1051.

Jacobson, K.A., Barrington, W.W., Pannell, L.K., Jarvis, M.F., Ji, X.-D., Williams, M., Hutchison, A.J., Stiles, G.L. (1989b) Agonsist-derived molecular probes for A_2-adenosine receptors. J. Mol. Recognition, in press.

Jarvis, M.F, Schutz, R., Hutchison, A.J., Do, E., Sills, M.A. and Williams, M. (1989) [^{3}H]CGS 21680, an A_2 selective adenosine receptor agonist directly labels A_2 receptors in rat brain tissue. J.Pharmacol. Exp. Ther. in press.

Klotz, K.-N., Cristalli, G., Grifantini, M., Vittori, S. and Lohse, M.J. (1985) Photoaffinity labeling of A_1-adenosine receptors. J. Biol. Chem. 260, 14659-14664.

Linden, J., Patel, A. and Sadek, S. (1985) [^{125}I]Aminobenzyl adenosine, a new radioligand with improved specific binding to adenosine receptors in heart. Circ. Res. 56, 279-284.

Munshi,R. and Baer,H.P. (1982) Radioiodination of p-hydroxylphenylisopropyl adenosine: development of a new ligand for adenosine receptors. Can.J. Physiol. Pharmacol. 60, 1320-1322.

Nakata, H. (1989) Affinity chromatography of A_1 adenosine receptors of rat brain membranes. Mol. Pharmacol. 35, 780-786.

Olah, M.E., Jacobson, K.A. , and Stiles, G.L.(1989) FEBS Letters, in press.

Schwabe, U.and Trost, T. (1980) Characterization of adenosine receptors in rat brain by (-)[^{3}H]N^6-phenylisopropyl adenosine Naunyn-Schmiedeberg's Arch. Pharmacol. 313, 179-187.

Schwabe, U.,Lenschow, V., Ukena, D.,Ferry, D.R. and Glossman, H. (1982) [^{125}I]N^6-p-hydroxyphenylisopropyl adenosine , a new ligand for Ri receptors. Naunyn-Schmiedeberg's Arch. Pharmacol. 321,84-87.

Seale, T.W., Abla, K.A., Jacobson, K.A., Carney, J.M. Xanthine amine congener of 1,3-dipropyl-8-phenylxanthine (XAC): a peripherally-selective adenosine receptor blocking agent. Pharmacol. Biochem. Behav., in press.

Stiles, G.A.and Jacobson, K.A. (1987). A new high affinity iodinated adenosine receptor antagonist as a radioligand/photoaffinity crosslinking probe. Mol. Pharmacol. 32, 184-188.

Stiles,G.A.and Jacobson,K.A. (1988). High affinity acylating antagonists for the A_1 adenosine receptor: identification of binding subunit. Mol. Pharmacol. 34, 724-728.

Stiles, G.A., Daly, D.T. and Olsson, R.A. (1985) The A_1 adenosine receptor . Identification of the binding subunit by photoaffinity crosslinking.J.Biol. Chem. 260, 10806-10811.

Ukena, D.,Jacobson, K.A., Kirk, K.L. and Daly, J.W. (1986) A [3H]amine congener of 1,3-dipropyl-8-phenyl xanthine - a new radioligand for A2 adenosine receptors of human platelets. FEBS Lett. 199, 269-274.

Yeung, S.M.H. and Green, R.D. (1984) [^{3}H]5'-N-ethylcarboxamide adenosine binds to both R_a and R_i receptors in rat striatum. Naunyn-Schmiedeberg's Arch. Pharmacol. 325,218-225.

9

Purification of A₁ Adenosine Receptor of Rat Brain Membranes by Affinity Chromatography

H. Nakata

Introduction

Adenosine receptors are one of the receptors which are coupled with adenylate cyclase system via G proteins. A_1 adenosine receptor (ADOR) which is coupled to adenylate cyclase in the inhibitory manner has been well characterized pharmacologically and biochemically in various tissues. However, more detailed molecular characterization of ADOR has been difficult mainly due to the lack of a purified receptor preparation. Recently, an affinity chromatography system was developed for the purification of ADOR from rat brain membranes using xanthine amine congener (XAC) as an immobilized ligand (Nakata, 1989a), although only 150-fold purification of ADOR was achieved in that study. In this report, an improvement of the affinity chromatography and the purification of ADOR from rat brain membranes to apparent homogeneity are presented (Nakata, 1989b).

Materials and Methods

Materials - XAC-agarose was prepared by coupling XAC to Affi-Gel 10 (Bio-Rad) in dimethylsulfoxide as described previously (Nakata, 1989a). Rat brain membranes were prepared from whole rat brains and solubilized with digitonin/sodium cholate as described previously (Nakata, 1989a). Adenosine ligands were purchased from RBI. 1,4-phenylenediisothiocyanate (p̲-DITC) was obtained from Fluka.

Purification of ADOR - All the purification procedures were performed at 4-8°C. The solubilized receptor (1000 ml) was applied to an XAC-agarose column (70 ml). The column was then washed with 50 mM Tris-acetate buffer, pH 7.2, containing 100 mM NaCl, 1 mM EDTA and 0.1% digitonin (buffer A). The bound receptor was eluted with 0.1 mM 8-cyclopentyltheophylline(CPT)/buffer A. The active eluate from the XAC-agarose column was applied to a hydroxylapatite column (0.5 ml). After the column was washed with 100 mM phosphate buffer, the receptor was eluted with 500 mM phosphate buffer, pH 7.0, containing 100 mM NaCl and 0.1% digitonin. The eluate was then applied to an XAC-agarose column (4 ml). After washing with buffer A, the column was eluted with 0.1 mM CPT. The purified receptor can be stored at -85°C for 1-2 weeks.

Binding assays - ADOR binding assays were performed in 0.25-ml total volume using 8-cyclopentyl-1,3-[^{3}H]dipropylxanthine (DPCPX) as a radiolabeled ligand as described previously (Nakata, 1989a). For the assay of the highly purified preparations, pass-through fractions (0.005 mg of protein) of the first XAC-agarose chromatography, which had been heated at 80°C for 3 min, centrifuged, and desalted into buffer A, were added to the assay mixture.

Affinity labeling of ADOR - [^{3}H]p-DITC-XAC was synthesized as reported (Stiles, et al., 1988) and was reacted with the purified ADOR followed by SDS-PAGE analyses

Results

Purification of ADOR - ADOR assessed with [^{3}H]DPCPX binding was purified about 50,000-fold from rat brain membranes using three chromatographic procedures. The most effective step is the first XAC-agarose chromatography which resulted in a 2,500-fold purification over the solubilized preparation. A typical ADOR purification is summarized in table 1.

Table 1. Purification of ADOR from rat brain membranes

Step	Total activity	Total protein[a]	Specific activity	Yield	Purification
	pmol	mg	pmol/mg	%	-fold
Membranes	3830	8700	0.44	100	1
Solubilized	1150	2800	0.41	30	0.9
XAC-agarose	460	0.418	1100	12	2500
Hydroxylapatite	300	0.030	10000	7.8	22700
Re-XAC-agarose	160	0.0073	21900	4.2	49800

[a]Protein content of the membrane or solubilized preparations was determined by the Bradford method. Protein content of the receptor preparations after the XAC-agarose chromatography was determined by the Amido-Schwarz method.

Purity and identity of ADOR - Figure 1A shows the silver-stained pattern of SDS-PAGE of the receptor preparations from several stages of purification. After the reaffinity chromatography, a single protein band of Mr=34,000 was revealed. This peptide band was also specifically labeled by an affinity labeling reagent, [^{3}H]p-DITC-XAC, showing that this peptide has a ligand binding activity (figure 1B, lane C-D). The purity of the purified ADOR was confirmed by autoradiogram after radioiodination (figure 1B, lane E-F).

Binding properties of the purified ADOR - The purified ADOR showed a typical ADOR pharmacological specificity similar to that of unpurified receptor preparations as summarized in table 2, although the Ki values for the agonists binding to the purified ADOR were significantly higher than the corresponding values for the unpurified ADOR. [^{3}H]DPCPX binding to the purified ADOR was specific and saturable. Parameters obtained from the linear Scatchard plot were Kd=1.4 nM and Bmax=24 nmol/mg of protein. It should be noted that the addition of the "pass-through fraction" to the assay mixture was essential to obtain a measurable and reproducible binding activity of the highly purified ADOR under the assay conditions employed.

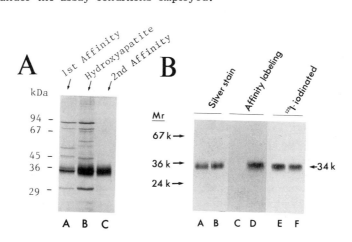

Figure 1. SDS-PAGE analyses of ADOR. A, SDS-PAGE of receptor preparations from various stages of purification stained by silver. B, SDS-PAGE of purified ADOR after several treatments. Lane A-D, purified ADOR labeled with [^{3}H]p-DITC-XAC in the presence of (lane A and C) or in the absence (lane B and D) of 0.001 mM XAC. Lane A and B were silver stained and lane C and D were fluorographs. Lane E and F are autoradiograms of ^{125}I-labeled purified ADOR by the chloramine-T method. About 5,000 (lane F) and 10,000 cpm (lane E) of the radioiodinated receptors were loaded in the gel. Figures are taken from Nakata (1989b).

Discussion

In this work, the purification of the ADOR from rat brain membranes to apparent homogeneity has been described. The purified ADOR gave a Bmax value of about 24 nmol[^{3}H]DPCPX/mg of protein and displayed a single band of Mr=34,000 in SDS-PAGE. The Bmax value obtained is consistent with the theoretical specific binding activity (29.4 nmol/mg) for a protein with a molecular mass of 34,000 Da if it is assumed that there is

Table 2. Affinities of solubilized and purified ADOR for various adenosine ligands.

Ligand	Ki (nM)	
	Solubilized	Purified
Agonists		
(R)-PIA	1.3	23
Adenosine	-	100
NECA	7.0	124
(S)-PIA	59	950
Antagonists		
DPCPX	0.46	1.5
XAC	1.4	4.6
CPT	4.0	6.6
8PT	-	40
IBMX	2400	5600

PIA, N^6-phenylisopropyladenosine; NECA, 5'-N-ethylcarboxamido-adenosine; 8PT, 8-phenyltheophylline, IBMX, 3-isobutyl-1-methylxanthine.

one ligand binding site/receptor molecule. The Mr=34,000 band was also specifically labeled with an affinity labeling reagent. These results indicate that the Mr=34,000 peptide is the minimum unit of the ADOR itself and no larger subunits of the receptor exist, although it is still to be determined whether the ADOR consists of a monomer or polymer of Mr=34,000 peptide. From the competition experiments with various adenosine agonists and antagonists, the purified ADOR appear to exist as a low agonist-high antagonist affinity state. This low affinity state for agonists may be induced by the dissociation of G proteins from the receptor protein during the purification.

In summary, these studies present the first purification of ADOR, which has appropriate properties as an intact ADOR. The molecular weight of the receptor is estimated to be about 34,000 from SDS-PAGE. These findings will facilitate the detailed molecular characterization of the ADOR including amino acid sequencing of the receptor protein, cloning of the gene and reconstitution studies with individual G proteins.

Acknowledgement

The author would like to thank Dr. D. M. Jacobowitz for his support during this work.

References

Nakata,H.,1989a, Affinity chromatography of A_1 adenosine receptors of rat brain membranes, Mol. Pharmacol. 35:780-786.
Nakata, H.,1989b, Purification of A_1 adenosine receptor from rat brain membranes, J. Biol. Chem. in press.
Stiles,G.L., and Jacobson,K.A.,1988, High affinity acylating antagonists for the A_1 adenosine receptor: identification of binding subunit, Mol. Pharmacol. 34:724-728.

10

Characteristics of A_1 Adenosine Receptors and Guanine Nucleotide Binding Proteins Co-Purified from Bovine Brain

J. Linden, R. Munshi, M.L. Arroyo, and A. Patel

Introduction

A_1 adenosine receptors can be solubilized in complexes with guanine nucleotide binding proteins (G proteins), as first shown by Gavish et al., (1982). The existence of complexes of receptors and G proteins (R-G complexes) is inferred from the presence of high affinity agonist binding sites which display reduced affinity if GTP is added. R-G complexes comprise 40-45% of bovine brain receptors solubilized by CHAPS and can be purified >2,000-fold by agonist affinity chromatography (Munshi et al., 1989a). The degree of receptor purification may be underestimated since most receptors (75-80%) appear to lose the ability to bind radioligands during the purification process. The addition of phospholipid to the elution buffer is critical for the preservation of some radioligand binding. We have attempted to calculate the purification of R-G complexes by quantitating α-subunits of G proteins which co-purify with receptors. Based on the binding of $[^{35}S]GTP\gamma S$, putative R-G complexes appear to be purified 10,000-fold to a specific activity of 4.6 nmol/mg protein which is > 50% homogeneous if it is assumed that R-G complexes have molecular masses of about 120 kDa and that receptors and G proteins are eluted in a 1:1 stoichiometry.

Forty % of CHAPS solubilized receptors adhere to agonist affinity columns. These correspond to R-G complexes since very few high affinity agonist radioligand binding sites are found in the column passthrough (Fig. 1). Elution of receptors and G proteins can be achieved by adding GTP or N-ethylmaleimide (NEM) to R-G complexes on an agonist affinity column. These agents uncouple receptors from G proteins, greatly reduce the affinity of the agonist for receptors, and thereby cause the elution of receptors and G proteins. A drawback of this procedure is that those solubilized receptors which are uncoupled from G proteins do not bind to the affinity column and hence are not purified. However, two major advantages are: (1) high specificity of elution can be achieved by the addition of GTP or NEM; and (2) the nature of G proteins which are co-purifed with receptors can be ascertained.

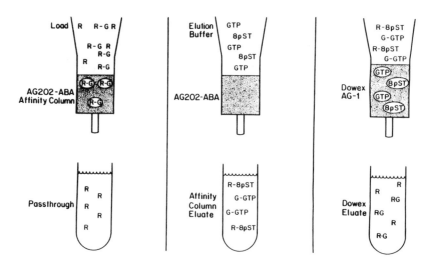

Figure 1. Schematic diagram of the method used to purify complexes of A_1 adenosine receptors and G proteins. Coupled (R-G) and uncoupled (R) solubilized receptors are loaded onto an affinity column composed of an agonist, N^6-aminobenzyladenosine (ABA), coupled to agarose beads. After extensive washing, receptors are eluted by the addition of GTP and the antagonist ligand 8-p-sulfophenyltheophylline (8pST). GTP and 8pST can be removed from purified receptors by adsorption to an anion exchange column (Dowex AG 1x8).

Properties of Purified A_1 Receptors

Purified A_1 receptors can be conveniently monitored with the A_1 selective ^{125}I-labeled antagonist ^{125}I-BW-A844U (Patel et al., 1988) which binds coupled and uncoupled forms of receptors with equivalent affinities. In competition assays, the binding affinities of R-phenylisopropyladenosine, cyclopentyladenosine, 5'-N-ethyl-carboxamidoadenosine, 8-p-sulfophenyltheophylline and theophylline for crude soluble and purified receptors have been found to be similar. If receptors are eluted from the affinity column with GTP (Fig. 2), subsequent removal of the guanine nucleotide restores some high affinity agonist binding which implies that the purified receptors and G proteins are capable of forming functional complexes following their purification. R-G complexes are conveniently detected by the binding of subnanomolar concentrations of the ^{125}I-labeled agonist, ^{125}I-aminobenzyladenosine (Linden et al., 1988). The molecular masses of crude (Earl et al., 1988; Linden et al., 1987) and purified receptors on SDS polyacrylamide gels under reducing or nonreducing conditions is 35 kDa as calculated from photoaffinity labeling by both agonist and antagonist photoaffinity labels. The molecular masses of crude and purified receptors is reduced to 32 kDa by deglycosylation (Klotz et al., 1987; Munshi et al., 1989a).

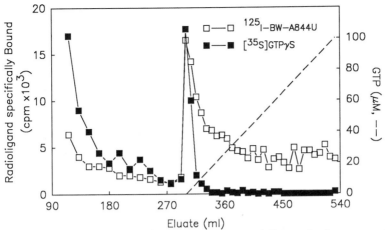

Figure 2. Elution profile of A_1 adenosine receptors and G proteins from an amino-benzyladenosine-agarose column. Solubilized receptors from bovine brain were applied to an affinity column as shown schematically in Fig. 1. Following washing with 190 ml of buffer for 19 h, receptors and G proteins were eluted by the addition of a GTP gradient (0-100 μM and 5 mM $MgCl_2$) and detected by the binding of ^{125}I-BW-A844U and [^{35}S]GTPγS, respectively. Eight ml fractions were eluted at 21° C at a rate of 10 ml/h.

Properties of Purified G proteins and a 34 kD Nonreceptor Polypeptide

GTP does not elute receptors from affinity columns previously eluted with NEM. This suggests that all of the R-G complexes which adhere to affinity columns contain G proteins which are uncoupled from receptors by NEM. Purified receptors eluted with NEM have been reconstituted into membranes and phospholipid vesicles along with added purified G proteins. Such reconstituted receptors can form functional complexes with G proteins inasmuch as high affinity GTP-sensitive agonist radioligand binding can be restored (Munshi et al., 1989b).

G proteins which are co-eluted by GTP with A_1 receptors from affinity columns have been [^{32}P]ADP-ribosylated by pertussis toxin and have α-subunits with molecular masses of 39 and 41 kDa (Munshi et al. 1989a). Specific antisera (provided by S. Mumby and A. Gilman) were used to detect $G_{o\alpha}$, $G_{i\alpha1}$, $\beta 35$, and $\beta 36$. $G_{i\alpha3}$ reacted weakly and $G_{i\alpha2}$ was absent. Small molecular mass (21-29 kDa) G protein α subunits assayed with [α-^{32}P]GTP following Western blotting also were not detected. Purified A_1 receptors have been reconstituted with individual species of G proteins in order to determine which of them can functionally couple to receptors as indicated by the restoration of GTP-sensitive high affinity agonist binding sites. Such reconstitution experiments confirm that A_1 receptors can interact with $G_{o\alpha}$ or a mixture of $G_{i\alpha1}$ and $G_{i\alpha3}$, but not $G_{i\alpha2}$ (Munshi, R., Pang, I-H., Sternweis, P.C and Linden, J., manuscript in preparation).

Affinity purified R-G complexes contain unidentified polypeptides. We have prepared monoclonal antibodies to the most prominent unidentified polypeptide which has a molecular mass of 34 kDa (p34). Unlike the receptor it is not a glycoprotein, it adheres weakly to DEAE columns, and it does not bind A_1 receptor radioligands. Antisera to p34 do not cross react with adenosine receptors. Also, p34 is found in some tissues which do not contain A_1 receptors. Binding of p34 to affinity columns is reduced by the addition of R-phenylisopropyladenosine. This polypeptide may correspond to an adenosine binding protein described by Ravid, Cann and Lowenstein in this volume.

References

Earl, C.Q., Patel, A., Craig, R.H., Daluge, S.M. and Linden, J., 1988. Photoaffinity labeling adenosine A_1 receptors with an antagonist ^{125}I-labeled aryl azide derivative of 8-phenylxanthine, *J. Med. Chem.* **31**:752-756.

Gavish, M., Goodman, R.R. and Snyder, S.H., 1982. Solubilized adenosine receptors in the brain. Regulation by guanine nucleotides, *Science* **215**:1633-1635.

Klotz, K-N., Lohse, M.J. 1986. The glycoprotein nature of A_1 adenosine receptors. *Biochem. Biophys. Res. Commun.* **140**:406-413.

Linden, J., Earl, C.Q., Patel, A., Craig, R.H. and Daluge, S.M., 1987. Agonist and antagonist radioligands and photoaffinity labels for the adenosine A_1 receptor. In E. Gerlach and B.F. Becker (Eds.), Topics and Perspectives in Adenosine Research, Springer-Verlag, Berlin Heidelberg, pp. 3-14.

Linden, J., Patel, A., Earl, C.Q., Craig, R.H. and Daluge, S.M., 1988. ^{125}I-Labeled 8-phenylxanthine derivatives: antagonist radioligands for adenosine A_1 receptors, *J. Med. Chem.* **31**:745-751.

Munshi, R. and Linden, J., 1989a. Co-purification of A_1 adenosine receptors and guanine nucleotide binding proteins from bovine brain, *J. Biol. Chem.* **264**:14853-14859.

Munshi, R. and Linden, J., 1989b. Functional reconstitution of purified bovine brain adenosine A_1 receptors into human platelet membranes. *FASEB J.* **3**:A280.

Patel, A., Craig, R.H., Daluge, S.M. and Linden, J., 1988. ^{125}I-BW-A844U, an antagonist radioligand with high affinity and selectivity for adenosine A_1 receptors, and ^{125}I-azido-BW-A844U, a photoaffinity label, *Mol. Pharmacol.* **33**:585-591.

11

Antagonist Radioligand Binding to Solubilized Porcine Atrial A_1 Adenosine Receptors

M. Leid, P.H. Franklin, and T.F. Murray

Introduction

The negative chronotropic properties of adenosine were initially described in the seminal investigations of Drury and Szent-Gyorgyi (1929). Adenosine and adenine nucleotides are now generally recognized to have physiologically significant and potent cardiovascular actions. In the heart, adenosine exerts a multiplicity of actions including vasodilation of coronary vasculature, depression of sinoatrial and atrioventricular nodal activity, inhibition of atrial contractility, attenuation of the stimulatory effects of catecholamines in the ventricular myocardium, and a depression of ventricular automaticity (Belardinelli et al., 1989).

The cardioinhibitory responses of adenosine are believed to be mediated through an activation of the A_1 subtype of adenosine receptors (Lohse et al., 1985). The pharmacological profile of adenosine receptors labeled with either an antagonist (Lohse et al., 1987; Martens et al., 1987; Leid et al., 1988) or agonist (Linden et al., 1985; Leid et al., 1988) radioligand suggest that the receptor present in cardiac membranes is of the A_1 subtype. We have previously used the agonist radioligand $(-)N^6$-$[^{125}I]$-p-hydroxyphenylisopropyladenosine ($[^{125}I]$HPIA) to characterize adenosine A_1 receptors in porcine atrial membranes (Leid et al., 1988). $[^{125}I]$HPIA bound saturably to an apparently homogeneous population of sites with a maximum binding capacity of 35 fmol/mg protein and a K_D of 2.5 nM. Guanine nucleotides negatively modulated $[^{125}I]$HPIA binding by enhancing its rate of dissociation. More recently, the porcine atrial A_1 adenosine receptor has been solubilized and labeled with $[^{125}I]$HPIA (Leid et al., 1989). This agonist radioligand labeled a single homogeneous population of solubilized receptors with a density of 88 fmol/mg protein and a K_D of 1.4 nM. Thus, solubilization afforded a 2.5 fold enrichment of A_1 adenosine receptor specific activity.

The solubilized receptor retained its functional integrity with respect to coupling to guanine nucleotide binding proteins inasmuch as addition of guanine nucleotides to solubilized preparations resulted in a rapid and complete dissociation of $[^{125}I]$HPIA (Lied et al., 1989). Thus, the A_1 adenosine receptor-G protein complex solubilized from porcine atria appears to provide an excellent system in which A_1 receptor-G protein interactions can be further studied. This report summarizes our recent characterization of the interactions of the antagonist radioligand $[^3H]$8-cyclopentyl-1,3-dipropylxanthine $[^3H]$DPCPX with the solubilized porcine atrial A_1 receptor.

Methods

The porcine atrial P_3 membrane preparation and solubilization in a digitonin and sodium cholate detergent system were performed as previously described (Leid et al., 1988; 1989). Briefly, digitonin and sodium cholate were added to a porcine atrial membrane preparation to give a final concentration of 0.4% w/v, digitonin, 0.08% w/v, sodium cholate and the mixture was centrifuged at 100,000 x g for 60 min. The supernatant was discarded and the pellet was resuspended in 25 mM imidazole buffer with 5 mM $MgCl_2$ and preincubated on ice for 15 min. Digitonin and sodium cholate were then added to give final concentrations of 0.8% w/v digitonin, and 0.16% w/v sodium cholate. Following a 10 min incubation at room temperature, the mixture was diluted 2-fold and centrifuged as above. The supernatant from this second extraction was used directly in radioligand binding experiments. The binding of [^{3}H]DPCPX to solubilized atrial A_1 receptors was carried out at 20^{0}C for varying times and terminated by filtration over Schleicher and Schuell #32 glass fiber filters which had been presoaked in 0.5% w/v polyethyleneimine (Leid et al., 1989). Specific binding was defined as total binding minus that occurring in the presence of 100μM cyclopentyladenosine or 3 mM theophylline.

Results

As shown in figure 1, [^{3}H]DPCPX exhibited saturable binding to a homogeneous population of solubilized atrial receptors with a B_{max} of 60 fmol/mg protein and K_D of 4.7 nM as determined by nonlinear regression analysis. This specific activity of [^{3}H]DPCPX binding sites in the solubilized preparation represents approximately a 2-fold enrichment over that previously reported for the membrane-bound adenosine receptor of porcine atria (Leid et al., 1988).

The pharmacological profile of the solubilized [^{3}H]DPCPX binding sites was ascertained by comparing the potency of 3 reference adenosine analogs in competition experiments. The results of these initial studies depicted in figure 2 revealed that the rank order potency for the adenosine receptor agonists tested was R-PIA > NECA > S-PIA. This rank order potency is consistent with the pharmacological signature of the A_1 subtype of adenosine receptor. R-PIA was 15-fold more potent as an inhibitor of [^{3}H]DPCPX binding to the solubilized porcine atrial adenosine receptor than its less active diastereomer, S-PIA.

Discussion

The A_1 selective antagonist radioligand [^{3}H]DPCPX has previously been shown to be a useful probe in the labeling of membrane-bound cardiac adenosine receptors. The present results demonstrate that the utility of this radioligand extends to the solubilized porcine atrial A_1 adenosine receptor. The specific activity of [^{3}H]DPCPX binding sites in this preparation confirms our previously reported enrichment of A_1 adenosine receptors from porcine atria using this mixed detergent system. This enrichment and stability of the solubilized receptor render this preparation useful for further biochemical characterization.

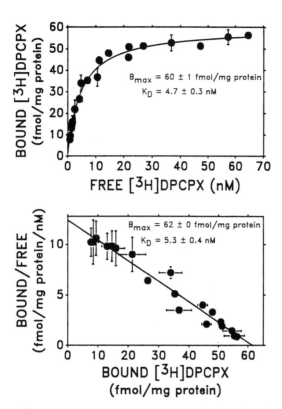

Figure 1. [³H]DPCPX saturation isotherm and Scatchard replot in solubilized porcine atrial preparations. Binding parameter estimates derived from nonlinear and linear regression analysis are provided.

Table 1. Comparison of the Binding Properties of [³H]DPCPX to membrane-bound and solubilized porcine atrial A_1 Adenosine Receptors

	MEMBRANE-BOUND	SOLUBILIZED RECEPTOR
1. Mechanism of binding	Simple Bimolecular	Unknown
2. $k_{+1}(M^{-1}min^{-1})$	$(4.1 \pm 0.7) \times 10^7$	n.d.
3. k_{-1}(extrapolated, min^{-1})	$(3.8 \pm 2.6) \times 10^{-2}$	n.d.
4. k_{-1}(measured, min^{-1})	$(4.6 \pm 0.2) \times 10^{-2}$	$(2.07 \pm 0.05) \times 10^{-2}$
5. K_D(equilibrium, $\underline{M}$)	$(4.0 \pm 0.5) \times 10^{-10}$	$(4.7 \pm 0.3) \times 10^{-9}$
6. K_D(kinetic, $\underline{M}$)	9.3×10^{-10}	n.d.
7. B_{max}(fmol/mg of protein)	32 ± 1	60 ± 1
8. Specific binding at K_D	85%	80%
9. Pharmacological specificity	A_1 subtype	A_1 subtype
10. Diastereomer potency ratio	18-fold	15-fold

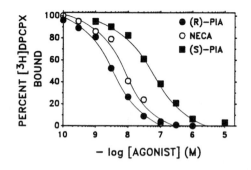

COMPOUND	$K_I \pm$ S.E. (nM)	$n_H \pm$ S.E.
(R)–PIA	0.89 ± 0.14	0.97 ± 0.11
(S)–PIA	13.7 ± 1.1	0.82 ± 0.04
NECA	2.1 ± 0.3	1.09 ± 0.14

Figure 2. Agonist titration of [^{3}H]DPCPX binding to solubilized porcine atrial adenosine receptor. Parameter estimates derived from nonlinear regression analysis are provided.

Although we have previously reported that agonist radioligand affinity is unaltered in digitonin/cholate extracts, the affinity of [^{3}H]DPCPX for A_1 receptors was reduced by approximately 10-fold (table 1). This reduction in affinity did not, however, compromise the signal to noise ratio of this ligand with specific binding representing 80% of the total amount bound at the [^{3}H]DPCPX K_D. Thus, the mixed-detergent solubilization procedure appears to be suitable for biochemical characterization of cardiac A_1 adenosine receptors and the coupling between the A_1 receptor and guanine nucleotide binding proteins may now be assessed in this preparation using agonist competition of [^{3}H]DPCPX binding.

References

Belardinelli, L., Linden, J., and Berne, R.M., 1989, The cardiac effects of adenosine. Prog. Cardiovascular Dis. 32:73-97.

Drury, A.N. and Szent-Gyorgyi, A., 1929, The physiological activity of adenine compounds with special reference to their action upon the mammalian heart. J. Physiol. (London) 68:213-226.

Leid, M., Franklin, P.H., Murray, T.F., 1988, Labeling of A_1 adenosine receptors in porcine atria with the antagonist radioligand 8-cyclopentyl-1,3-[^{3}H]dipropylxanthine. Europ. J. Pharmacol. 147:141-144.

Leid, M., Schimerlik, M.I. and Murray, T.F., 1988, Characterization of agonist radioligand interactions with porcine atrial A_1 adenosine receptors. Mol. Pharmacol. 34:334-339.

Leid, M., Schimerlik, M.I. and Murray, T.F., 1989, Agonist radioligand interactions with the solubilized porcine atrial A_1 adenosine receptor. Mol. Pharmacol. 35:450-457.

Martens, J., Lohse, M.J., and Schwabe, U., 1987, Pharmacological characterization of A_1 adenosine receptors in isolated rat ventricular myocytes. Nauyn-Schmiedeberg's Arch. Pharmacol. 336:342-348.

Linden, J., Patel, A., and Sadek, S., 1985, [^{125}I]aminobenzyladenosine, a new radioligand with improved specific binding to adenosine receptors in heart. Circ. Res. 56:279-284.

Lohse, M., Ukena, D., and Schwabe, U., 1985, Demonstration of R_i-type adenosine receptors in bovine myocardium by radioligand binding. Naunyn-Schmiedeberg's Arch. Pharmacol. 328:310-316.

Lohse, J.J., Klotz, K.N., Fotinos, J.L., Reddington, M., Schwabe, U., and Olsson, R.A., 1987, 8-Cyclopentyl-1,3-dipropylxanthine (DPCPX): a selective high affinity antagonist radioligand for A_1 adenosine receptors. Naunyn-Schmiedeberg's Arch. Pharmacol. 336:204-210.

Martens, D., Lohse, M.J., Rauch, B., and Schwabe, U., 1987, Pharmacological characterization of A_1 adenosine receptors in isolated rat ventricular myocytes. Naunyn-Schmiedeberg's Arch. Pharmacol. 336:342-348.

12

Isolation of an Adenosine Binding Protein Which Has Properties Expected of the A_1 Adenosine Receptor

K. Ravid and J.M. Lowenstein

SUMMARY

Affinity chromatography has been used to isolate an adenosine-binding protein in pure form with a molecular weight of 35K. The protein has properties consistent with its being the A_1 adenosine receptor or part of this receptor.

INTRODUCTION

Three types of adenosine receptors have been characterized on the basis of their effects on adenylate cyclase: inhibitory (A_1) and stimulatory receptors (A_2) which occur on the cell surface, and an intracellular adenosine receptor (P site). The adenosine analog affinity for the A_1 receptor is PIA > adenosine > NECA, while that for the A_2 receptor is NECA > adenosine > PIA (Londos et al., 1983; Wolff et al., 1981). The P-site can be distinguished from A_1 and A_2, because it binds analogs modified in the 2'- and 3'-positions, such as 2',3'-dideoxyadenosine, and requires an unaltered adenine moiety (Londos et al., 1983). All adenosine receptors bind analogs modified in the 5'-position. A_1 receptors are abundant in brain and adipocytes (Londos et al., 1983; Garcia-Sainz and Torner, 1985). A_2 receptors are abundant in liver (Londos and Wolff 1977), fibroblasts (Rozengurt, 1982; Ravid and Lowenstein, 1988), and vascular smooth muscle (Anand-Srivastava et al., 1982). The P-site is believed to exist in all cell types (Londos et al., 1983). Affinity labeling with ^{125}I-labeled, A_1-specific receptor analogs, yielded a radioactive protein with M_r of 38K (Stiles et al., 1985), 35K (Klotz et al., 1985), and 36K (Choca et al., 1985). Labeling of this protein was blocked by L-PIA and theophylline.

We have used 5A5D-Ado agarose affinity columns, prepared by reacting Affi-Gel 10 with 5A5D-Ado, to isolate adenosine binding proteins. Londos and Preston (1977) found that 5A5D-Ado inhibits adenylate cyclase in the micromolar range.

---

[1]Supported by grant GM 07261 from the National Institutes of Health. Abbreviations used: CHA, N^6-cyclohexyladenosine; NECA, 5'-N-ethyl-carboxamidoadenosine; PIA, N^6-(R-phenylisopropyl)adenosine; 5A5D-Ado, 5'-amino-5'-deoxyadenosine; K, 1,000.

RESULTS

Calf brain membranes were extracted with 4.3% Na cholate; 50 mM Na Hepes buffer, pH 7.4; 150 mM NaCl (Buffer). Affinity chromatography on 5A5D-Ado agarose yielded a pure adenosine-binding protein with a molecular weight of 28K (Fig. 1) (Ravid et al., 1989). Under these conditions, no 35K protein is detected among the proteins eluted by the adenosine gradient.

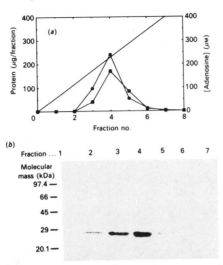

Fig. 1. Elution of calf brain membrane protein solubilized with cholate from affinity column by adenosine gradient. a, Cholate-solubilized proteins from calf brain membrane (572 mg; 5.6 mg/ml) were applied to the column. In a separate run the detergent was dialyzed from the solubilized proteins before applying them to the column. The column was washed with 450 ml of Buffer and eluted with a linear gradient of 0 to 400 μM adenosine in Buffer over 50 ml. The Buffer wash and adenosine gradient were run either in the presence (o) or in the absence ( ) of detergent. Fractions of 7 ml were collected and concentrated to 0.4 ml. b, Gel electrophoresis of 12 μl of fractions from (o). The gradient started with lane 1. The gel was silver stained. (Reproduced from Ravid et al., 1989, with permission of the publishers.)

The 28K membrane protein was not eluted from the column by cAMP, inosine, PIA, NECA, ADP, ATP, NAD, or GTP. It had no adenosine kinase, adenosine deaminase, AMPase, and IMPase activity. Lack of elution by inosine indicates that it is not part of the nucleoside transporter, which has the same affinity for adenosine and inosine (Plagemann and Wohlhueter, 1983). Two peptide fragments of 22 and 25 amino acids were sequenced. Comparison with a sequence library of known proteins including adenosine deaminase, HGPRT, and adenylate kinase showed no significant homology.

A different extraction and elution strategy yielded different results. When calf brain membranes were extracted with 1% Triton X-100 and applied to the 5A5D-Ado affinity column, elution with an adenosine gradient in 0.1% Triton X-100 yielded two major proteins with molecular weights of 28 and 35K (Fig. 2). The 35K protein emerged at 250 μM adenosine. When the Triton X-100 concentration was raised to 0.25%, the 35K protein was eluted from the column in the absence of adenosine, indicating that 0.25% Triton X-100 decreases the binding of the 35K protein to the affinity column sufficiently to cause its elution.

Linear gradients of different adenosine analogs were tested for their effectiveness in eluting the 35K protein from the column. PIA yielded a protein peak in which some fractions contained pure 35K protein, which emerged at 90-100 μM PIA (Fig. 3). CHA eluted pure 35K protein between 20 and 40 μM CHA (not shown). NECA eluted a broad protein peak between 200 and 400 μM. This peak contained two proteins with molecular masses of 35K and 33K (not shown). When elution of the column with PIA or CHA or NECA was followed with a linear gradient of adenosine from 0 to 400 μM, the 28 kda protein was eluted from the affinity column (not shown).

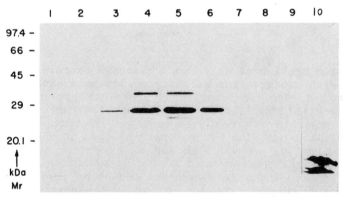

Fig. 2. Elution of calf brain membrane proteins solubilized with Triton X-100 from 5A5D-Ado affinity column by adenosine gradient. Calf brain membrane proteins were solubilized with 1% Triton X-100 (218 mg; 2.9 mg/ml) and applied to the column. The column was washed with 150 ml of a mixture containing 0.1% Triton X-100, 50 mM Na Hepes buffer, pH 7.4, 150 mM NaCl, and 1 mM MgCl$_2$ (Buffer) followed by a linear gradient from 0 to 500 μM adenosine in Buffer over 40 ml. Fractions of 4.5 ml were collected. The column was then washed with 45 ml of 7 M urea/1 M NaCl. Each fractions was concentrated to 0.5 ml and 10 μl was subjected to gel electrophoresis. The adenosine gradient started with fraction 1. Lane 10 contained 12 μl from the urea wash of the column which had been concentrated to 0.5 ml by vacuum dialysis. (Reproduced with permission of the publishers.)

80

Comparison of the concentrations of adenosine analogs which eluted the 35K protein showed that its affinity follows the order: CHA > PIA > adenosine > NECA. The elution order and the molecular weight of 35K are properties expected of the A_1-adenosine receptor. The 35K protein had a K_D for adenosine of 0.53 μM in the presence of 0.08% Triton X-100.

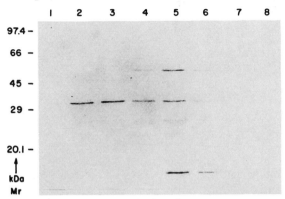

Fig. 3. Elution of calf brain membrane proteins solubilized with Triton X-100 from 5A5D-Ado affinity column by a gradient of PIA. Details were as for Fig. 2, except that the linear gradient went from 0 to 300 μM PIA in buffer C over 40 ml. Fractions of 5 ml were collected, concentrated, and subjected to gel electrophoresis. The gradient started with lane 1. The gel was silver stained.

REFERENCES

Anand-Srivastava, M. B., Douglas, J. F., Cantin, M., and Genest,J. (1982) Biochem. Biophys. Res. Commun. **108**, 213-219

Choca, J. I., Kwatra, M. M., Hosey, M. M., and Green, R. D. (1985) Biochem. Biophys. Res. Comm. **131**, 115-121

Garcia-Sainz, J. A., and Torner, M. L. (1985) Biochem. J. **232**, 439-443

Klotz, K.-N., Cristalli, G., Grifantini, M., Vittori, S., and Lohse, M. J. (1985) J. Biol. Chem. **260**, 14659-14664

Londos, C., Wolff, J., and Cooper, M. F. (1983) Regulatory Function of Adenosine, Berne, R. M., Rall, T. W., and Rubio, R., eds., pp. 17-32

Londos, C., and Wolff, J. (1977) Proc. Natl. Acad. Sci. USA **74**, 5482-5486

Londos, C., and Preston, M. S. (1977) J. Biol. Chem. **252**, 5951-5956

Plagemann, P. G. W., and Wohlhueter, R. M. (1983) Regulatory Function of Adenosine, Berne, R. M., Rall, T. W., and Rubio, R., eds., pp. 179-201

Ravid, K., Rosenthal, R. A., Doctrow, S. R., and Lowenstein, J. M. (1989) Biochem. J. **258**, 653-661

Ravid, K., and Lowenstein, J. M. (1988) Biochem. J. **249**, 377-381

Rozengurt, E. (1982) Experimental Cell Research **139**, 71-78

Stiles, G. L., Daly, D. T., and Olsson, R. A. (1985) J. Biol. Chem. **260**, 10806-10811

Wolff, J., Londos, C., and Cooper, M. F. (1981) Adv. Cyclic Nucleotide Res. **14**, 199-214

13

Identification of the A_2 Adenosine Receptor Binding Subunit by Photoaffinity Crosslinking

W.W. Barrington, K.A. Jacobson, A.J. Hutchison, M. Williams, and G.L. Stiles

Introduction

Adenylate cyclase coupled adenosine receptors are divided into two main subtypes (Stiles, 1986). The A_1 subtype (A_1AR) is inhibitory to adenylate cyclase and exhibits an agonist potency order where $\underline{R}$-PIA > NECA > $\underline{S}$-PIA. The rapid accumulation of information concerning the structure, function and regulation of the A_1AR can be directly attributed to the development of a large number of selective, high affinity radioligands, photoaffinity and affinity probes for this receptor subtype. The A_2 adenosine receptor (A_2AR), in contrast, is stimulatory to adenylate cyclase and exhibits a distinctly different agonist potency order with NECA > $\underline{R}$-PIA > $\underline{S}$-PIA. In this case, the dearth of useful A_2AR probes has lead to an almost complete lack of information on the structure, function and regulation of this physiologically important receptor subtype.

We now report on the development of ^{125}I-PAPA-APEC (^{125}I- 2- [4- [2- [2- [(4-aminophenyl) methylcarbonyl- amino] ethylaminocarbonyl] ethyl] phenyl] ethylamino-5′ -N- ethylcarboxamido adenosine), the first high affinity, highly selective, iodinated radioligand for the A_2AR.

Saturation binding experiments (figure 1A) were performed in bovine striatal membranes (Barrington et al., 1989). Binding parameters, determined both in the presence ($K_d = 1.8 \pm 0.8$ nM, $B_{max} = 222 \pm 95$ fmol/mg) and absence ($K_d = 1.6 \pm 0.7$ nM, $B_{max} = 203 \pm 77$ fmol/mg) of 50 nM CPA were statistically identical (P > 0.8, n=3) and indicate that ^{125}I-PAPA-APEC is very selective for the A_2AR (Barrington et al., 1989; Bruns et al., 1986). This selectivity was confirmed by competitive binding studies where PAPA-APEC failed to decrease A_1AR binding at concentrations less than 1000 nM, suggesting that PAPA-APEC is at least 500 fold selective for the A_2 AR.

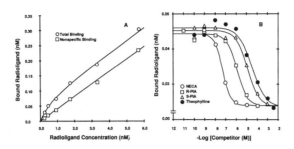

Figure 1: ^{125}I-PAPA-APEC Binding Characteristics: **A**: ^{125}I-PAPA-APEC saturation binding to bovine striatal membranes in the presence of 50 nM CPA. This curve is typical of three experiments and yielded a K_d = 1.8 ± 0.8 nM and a B_{max} = 222 ± 95 fmol/mg protein. **B**: Competitive binding curves of NECA, $\underline{R}$-PIA, $\underline{S}$-PIA and theophylline with ^{125}I-PAPA-APEC. Refer to the text for details.

The agonist potency order observed in competitive binding experiments (figure 1B) is appropriate for the A_2AR with NECA (IC_{50} = 86 ± 41 nM) being more potent than $\underline{R}$-PIA (IC_{50} = 1,350 ± 423 nM) which is more potent that $\underline{S}$-PIA (IC_{50} = 16,000 ± 4,480 nM) which is more potent still than the antagonist theophylline (IC_{50} = 31,600 ± 4,240 nM).

Adenylate cyclase assays (in platelet membranes) demonstrated the expected A_2AR mediated agonist stimulation of cAMP production with a dose dependent increase of 10% above basal activity at a 10^{-8} M concentration of PAPA-APEC, 43% at 10^{-6} M and 52.5% at 10^{-5} M. NECA (10^{-6} M) in the same preparation resulted in a 35% increase over basal cAMP activity.

The presence of both A_1AR and A_2AR subtypes in bovine striatal membranes allowed a direct comparison between the A_1AR and A_2AR binding subunits (figure 2). In these membranes, NECA was more potent at inhibiting the photoaffinity crosslinking of ^{125}I-PAPA-APEC into the A_2AR than it is at inhibiting the labelling of the A_1AR with [^{125}I]PAPAXAC (figure 2). The A_1AR binding subunit (38 kDa) is shown in figure 2A lane 1 and the distinct A_2AR binding subunit is shown in lanes 2-4 as the 45 kDa labelled protein. This 45 kDa band exhibits the appropriate A_2AR pharmacology with 10^{-6} M NECA (lane 3) being a more potent inhibitor of labelling than 10^{-6} M $\underline{R}$-PIA (lane 4). The full pharmacology of the ^{125}I-PAPA-APEC labelled A_2AR is shown in figure 2B. The 45 kDa band exhibits all the appropriate A_2AR pharmacology. Specific labelling was decreased 47% by 10^{-6} M NECA (lane 2), 36% by 10^{-6} M R-PIA (lane 3), 8% by 10^{-6} M S-PIA (lane 4) and completely blocked by 5 mM theophylline (lane 5). Binding was also decreased ~10% in the presence of 10^{-4} M Gpp(NH)p (lane 6).

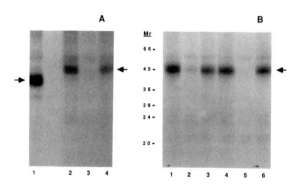

Figure 2: Autoradiograph of Photoaffinity Crosslinked Bovine Striatal Membranes: **A**: The A_1AR binding subunit (indicated by left hand arrow) is photoaffinity crosslinked with SANPAH and the A_1AR selective antagonist [^{125}I]PAPAXAC in lane 1. Crosslinking ^{125}I-PAPA-APEC with SANPAH (lanes 2-4) reveals the A_2AR binding subunit (righthand arrow) with an apparent molecular weight of 45,000 daltons. This specific band exhibits the appropriate A_2AR pharmacology when binding is performed in the presence of 10^{-6} M NECA (lane 3) and 10^{-6} M R-PIA (lane 4). This is typical of the results seen in four experiments. **B**: The full pharmacology of photoaffinity crosslinking the A_2AR binding subunit with ^{125}I-PAPA-APEC. Lane 1 (control) exhibits the specific 45 kDa protein of the A_2AR binding subunit. The presence of 10^{-6} M NECA (lane 2), 10^{-6} M R-PIA (lane 3), 10^{-6} M S-PIA (lane 4) and 10^{-3} theophylline (lane 5) demonstrate the appropriate A_2AR pharmacology (see text for details). Lane 6 shows an ~10% decrease in labelling with the presence of 10^{-4} M Gpp(NH)p.

We believe that these photoaffinity crosslinking experiments demonstrate the first successful labelling of the A_2AR binding subunit and demonstrate that ^{125}I-PAPA-APEC is a powerful new tool for further evaluation of the A_2AR.

References

Barrington, W.W., Jacobson, K.A., Hutchison, A.J., Williams, M., and Stiles, G.L., 1989, Identification of the A_2 adenosine receptor binding subunit by photoaffinity crosslinking, in *Proc. Natl. Acad. Sci. USA* **86**:6572-6576.

Bruns, R.F., Lu, G.H., and Pugsley, T.A., 1986, Characterization of the A_2 adenosine receptor labeled by [^{3}H]NECA in rat striatal membranes, in *Mol. Pharmacol.* **29**:331-346.

Stiles, G.L., 1986, Adenosine receptors: structure, function and regulation in *TIPS* **7**:486-490.

Stiles, G.L., and Jacobson, K.A., 1987, A new high affinity, iodinated adenosine receptor antagonist as a radioligand/photoaffinity crosslinking probe, in *Mol. Pharmacol.* **32**:184-188.

14
Human Placental Adenosine A_2-Like Binding Sites Properties and Homology with Mammalian and Avian Stress Proteins

K.A. Hutchison, B. Nevins, F. Perini, and I.H. Fox

Introduction

Adenosine is a biologically active compound at two cell surface receptors, the adenosine A_1 and the adenosine A_2 receptors. The adenosine A_1 receptor has been labeled and characterized with selective adenosine analogs. The adenosine A_2 receptor has been labeled using N-ethylcarboxamidoadenosine (NECA). However, in most tissues the majority of definable specific NECA binding is to the ubiquitous adenosine A_2-like binding sites which has properties that distinguish it from the adenosine A_2 receptor (Lohse et al., 1988). One of the major differences in binding relates to the high affinity binding of N^6 substituted adenosine analogs and C^8 substituted xanthines to the adenosine A_2 receptor and the lack of binding of these compounds to the adenosine A_2-like binding site.

Recently, this ubiquitous adenosine A_2-like binding site was separated chromatographically and distinguished pharmacologically from the adenosine A_2 receptor (Lohse et al., 1988). The function of this adenosine A_2-like binding site has remained unclear. To clarify the properties of the adenosine A_2-like binding site, we have purified and characterized it from the membranes and the soluble fractions of the human placenta.

Materials and Methods

Membranes and supernatants were prepared from donor human placentas. Supernatant was separated from the particulate components by centrifugation at 48,000 x g for 15 min (Hutchison and Fox, 1989).

Purification was carried out by first solubilizing the placental membranes with 0.1% CHAPS. Purification was performed for both the solubilized membrane fraction and the soluble fraction by ammonium sulfate precipitation, lectin affinity chromatography on concanavalin A, ion exchange chromatography on

DEAE sephadex, and gel filtration chromatography on sepharose 6B (Hutchison and Fox, 1989). Amino acid analysis, carbohydrate analysis, and amino terminal sequence determination, sodium dodecyl sulfate polychromate gel electrophoresis were performed (Hutchison and Fox, 1989; Hutchison et al., 1989). Binding assays were performed using 10 to 20 nM (^3H)NECA. Stokes radius determination, sedimentation constants, partial specific volume, frictional ratio, and the native molecular weight were estimated (Hutchison and Fox, 1989).

Results

The membrane bound adenosine A_2-like binding site was purified 127 fold to final specific activity of 1.5 to 1.9 nmol/mg protein (Bmax) (Table 1). Sodium dodecyl sulfate polyacrylamide gel electrophoresis (SDS-PAGE) revealed a single band with a molecular mass of 98 kDa. Soluble placental adenosine A_2-like binding sites were purified 134 fold to a final specific activity of 2.5 nmol/mg protein (Bmax) with a yield of 47% sodium SDS-PAGE revealed a 98 kDa band comprising 89% of the stained protein (Table 2).

Table 1. Properties of Homogeneously Purified Adenosine A_2-Like Membrane Binding Site

- Purified 127-fold to final specific activity of 1.5 to 1.9 nmol/mg (Bmax)
- Subunit is 98 kDa
- 5.5% carbohydrate by weight
- Detergent-protein complex has molecular mass 230 kDa, Stokes radius 70 A, S_{20w} 6.9S, partial specific volume 0.698 ml/g
- f/fo is 1.5
- Almost 1% of membrane protein

The membrane bound purified material was 5.5% carbohydrate by weight while the soluble purified material was 43% carbohydrate by weight. Both purified proteins had an identical Stokes radius. Careful studies of the molecular conformation of the detergent to membrane protein complex was performed. The complex has a Stokes radius of 70 A, an S_{20w} of 6.9S, and a partial specific

Table 2. Properties of Homogeneously Purified Soluble Adenosine A_2-Like Binding Site

- Purified 134-fold to final specific activity of 2.5 nmol/mg (Bmax)
- Subunit 98 kDa; minor form subunit 74 kDa
- 4.3% carbohydrate by weight
- Stokes radius identical to membrane
- More acidic than membrane form
- Almost 1% of soluble protein
- Kinetic properties identical to membrane form

volume of 0.698 ml/g. The native molecular mass of the detergent to protein complex is 230 kDa and the frictional ratio is 1.5 (Table 1). This suggests that both proteins are highly asymmetrical. The soluble protein was more acidic than the membrane derived protein as determined by their relative elution from an ion exchange column.

Both proteins are kinetically identical. The kD for NECA was 210 nmol for the membrane protein and the displacement by other analogs was typical for the adenosine A_2-like binding site (Table 3).

Table 3. Properties of Homogeneously Purified Membrane Adenosine A_2-Like Binding Site

- Kd NECA 210 nM
- Ki (µM)

R-PIA	$\geq$ 1000
2-Chloroadenosine	1.8
Adenosine	20
IBMX	22
Theophylline	220

The amino acid composition of the two proteins was virtually identical. Forty-six amino acids were sequenced from the unblocked amino terminus of the soluble adenosine A_2-like binding site and 29 amino acids were sequenced from the amino terminus of the membrane adenosine A_2-like binding site. Except for two amino acids from the membrane, which were indeterminate, the amino acid sequences were identical. There were unexpected homologies found with four other proteins which can be classified as stress proteins. With the exception of two indeterminate amino acids in the soluble human adenosine A_2-like binding site, the murine endoplasmic reticulum protein (ERP99) and the murine tumor rejection antigen (GP96) were absolutely homologous. There was 83% homology to hamster glucose regulator protein (GRP 94) and 87% homology to chicken heat shock protein (HSP 108).

Table 4. Aminoterminal Sequence of Soluble Adenosine A_2-Like Binding Site

- Identical to membrane Adenosine A_2-like binding site
- 94% homology to murine ERP 99 and GP 96
- 83% homology to hamster GRP 94
- 87% homology to chicken HSP 108

Discussion

Our observations indicate that the adenosine A_2-like binding site exists in large quantities in both membrane and soluble fractions of human placenta. Both proteins have identical properties with respect to kinetics, hydrodynamic properties, and subunit molecular weight. They do vary by virtue of charge in that the soluble protein is more acidic than the membrane protein. In addition, there is a difference in the carbohydrate content of these two proteins. The basis for this difference in properties remains unclear. We favor the hypothesis that these proteins are the products of two highly homologous genes or alternative splicing of the same gene. However, other hypothesis involving posttranslational modification of sialic acid content or different states of phosphorylation remain plausible explanations.

The striking amino terminal homology of these proteins with stress protein sequences in mammalian and avian tissue indicates that these binding sites are highly conserved and may be important in cellular metabolism. Although the function of stress proteins remain unclear, they have been implicated in a number of activities ranging from protection during thermal shock, glucose starvation, viral infection, metabolic insult from amino acid analogs to prevention of transformation to the DNA binding state of steroid receptors and to the regulation of protein synthesis initiation. The possibility is raised by this homology that the adenosine A_2-like binding site is a human stress protein, but the relevance of these properties to the adenosine A_2-like binding site remains to be determined.

Acknowledgment

The authors wish to thank Jeanne Schmaltz ane Emberly Cross for excellent typing of this manuscript and the patients and nurses in the Women's Hospital delivery room for providing placenta. This work was supported by grants 5 M01 RR00042 and 2P60AR-20557 from the United States Public Health Service.

References

Hutchison, K.A. et al., 1989, Soluble human adenosine A_2-like binding site: Properties and homology with human membrane adenosine A_2-like binding site, hamster GRP 94, murine ERP99 and GP96, and chicken HSP 108. *J. Biol. Chem.* (Submitted for publication).

Hutchison, K.A. and I.H. Fox, 1989, Purification and characterization of the adenosine A_2-like binding site from human placental membranes. *J. Biol. Chem.* (In press).

Lohse, M.J. et al., 1988, Separation of solubilized A_2 adenosine receptors of human platelets from non-receptor [3H]NECA binding sites by gel filtration. *Naunyn Schmiedeberg's Arch. Pharmacol.* 337:64-68.

15

Cardiac Interstitial Fluid Adenosine in Normoxia and Hypoxia

R.M. Berne

Adenosine is involved in many physiological processes, particularly in the cardiovascular and nervous systems. In the heart adenosine is an important regulator of coronary vascular resistance, and plays a significant role in matching myocardial oxygen supply to oxygen needs under different physiological and pathophysiological conditions. Adenosine also has a depressant effect on conduction through the atrioventricular (AV) node and on sinoatrial node excitation, and has a negative inotropic effect on atrial myocardium.

The sensitivity of these effects of adenosine are dissimilar. For instance, the coronary vessels become dilated to exogenously administered adenosine before any effect is detectable on AV conduction; in fact, with increasing doses of adenosine, maximal coronary vasodilation is reached before significant prolongation of atrial to His bundle conduction occurs. These responses can be thought of as a protective action of the nucleoside. As oxygen demands of the heart are increased the coronary resistance vessels progressively dilate to supply additional oxygen to the myocardium. When the coronary reserve is exhausted (as might occur during exercise in patients with coronary artery disease) AV conduction can become depressed to the point where second degree heart block occurs, which results in a decrease in myocardial oxygen needs.

The differences in sensitivity of the AV node and the coronary vascular smooth muscle to adenosine have been mainly gleaned from studies in which different concentrations of adenosine are added to the perfusion fluid of the isolated perfused guinea pig heart. However, the concentrations of endogenous adenosine that elicit responses in these tissues is not known. The effective interstitial fluid concentrations of adenosine cannot be estimated from exogenous adenosine experiments because of the tremendous uptake of administered adenosine by the endothelial cells (Nees, et al.,1985). This latter observation has led to the suggestion that adenosine acts on the endothelial cells which then produce vascular smooth muscle relaxation either by releasing a vasodilator substance or by an electrical coupling of the endothelial with the surrounding vascular smooth muscle cells (Daut, et al., 1988). Whether endogenous adenosine may also act via the endothelium is doubtful because l) adenosine can relax vascular rings from which the endothelium is removed (Rubanyi, et al., 1985) and 2) endogenous adenosine arises from myocardial cells (Deussen, et al., 1986) and would reach the vascular smooth muscle before it reaches the endothelium.

These questions point up the importance of knowing the concentration of adenosine to which the nodel tissue and vascular smooth muscle are exposed-namely the interstitial fluid concentration. Early attempts to achieve this goal consisted of plaing 25-40 ml of Krebs-Henseleit solution in the pericardial sac of the dog heart for several minutes during control periods and during various interventions, and then withdrawing the solution for determination of its adenosine concentration (Miller, et al., 1979 and Knabb, et al.,1983). With this method only directional changes in the interstitial fluid adenosine concentration could be measured since equilibrium between the adenosine concentration in the interstitial fluid and the pericardial infusate could not be attained because of the large pericardial volume to epicardial surface area ratio. In an attempt to overcome this problem and obtain a more accurate index of interstitial fluid adenosine levels, a 1 cm^2 chamber was placed on the intact left ventricular surface of the dog heart, secured firmly with remote sutures and a grease seal and filled with 100 µl of Krebs-Henseleit solution (Gidday, et al., 1988). The adenosine concentration in the chamber increased from 0 to 0.15 µM over a period of 2-4 minutes and then, reached a plateau value which remained unchanged for at least one hour (Berne, et al., 1986). Although this method had great advantages over the pericardial infusate in that 1) the volume/area ratio was reduced, 2) the parietal pericardium was eliminated as a source or sink for adenosine and 3) the area examined was only left ventricle without possible contributions (or subtractions) from the other cardiac chambers, it still required a relatively large volume of fluid over a small area of the exposed left ventricle.

Consequently we undertook another refinement of the method to estimate the interstitial fluid adenosine concentration. This consisted of 6 or 9 mm porous nylon discs, applied to the epicardial surface of the left ventricle of the open chest dog and the isolated perfused guinea pig heart (Phillips, et al., 1987 and Soracco, et al., 1989). The discs were presoaked in Krebs-Henseleit solution and after two minutes of contact with the epicardium were removed and washed free of adenosine three times with distilled water (total volume of 200 µl). The washes were pooled and analyzed for adenosine by HPLC. Contact time controls in which the discs were placed on µl Millipore filters which lay on the surface of a solution of known adenosine content indicated that the adenosine concentration of the disc plateaued at the concentration of the solution within one minute.

In situ dog hearts

In the open chest dog experiments, three 9 mm discs were applied simultaneously to areas of the left ventricular epicardium that did not have visible coronary blood vessels. After two minutes of contact, the discs were removed, washed and analyzed individually for adenosine. Three new discs were then applied to the same spots on the epicardium and this process was repeated several times with each dog heart. In other experiments, the volume of fluid in contact with the epicardium was increased without increasing the area of contact. This was accomplished by stacking two discs or by attaching one or two 9 mm meshes of non-absorbent interwoven polyester with large interstices to a single nylon disc which was in contact with the epicardium. After the two minute contact period, the stacked or "modified" discs were removed, the components separated and analyzed individually for their adenosine content.

When compared to the results of similar experiments with sucrose (an inert diffusable substance of about the same molecular weight as adenosine), the adenosine experiments showed variations in the nucleoside concentration of discs applied simultaneously to adjacent areas of the left ventricular epicardium and also among discs applied sequentially to the same area of the epicardium. Thus there exists a temporal heterogeneity of interstitial fluid adenosine concentration in the heart and in all probability a spatial heterogeneity as well. It is unlikely that this heterogeneity of cardiac interstitial adenosine concentration is a function of the reported heterogeneity of blood flow (Steenbergen, et al., 1977)   because the disc sucrose concentration did not show either temporal or spatial heterogeneity. Were differences in blood flow to different areas of the epicardium responsible for their wide variation in adenosine concentrations, then one would expect the same to be true for sucrose distribution. A more likely explanation is that there are local differences in the production of adenosine by the myocytes and its removal by the endothelial cells, myocytes and the flowing blood.

With two stacked discs the adenosine concentration was equal in the two discs and equal to that of a single disc. However, when the fluid volume/area ratio was increased by adding one or two polyester meshes to single discs, the concentration of adenosine in the intact modified disc was significantly less then that of single disc but the concentration of adenosine was the same in the component parts of that modified disc. Hence, diffusion of adenosine through the interstices of the modified disc was not responsible for its lower adenosine concentration. With very large volume/area ratios as with chambers placed on the heart and filled with 100 µl of Krebs-Henseleit solution (volume/area ratio 13 times greater then that of single disc), the adenosine concentration was only 8% of that measured with single discs. These results suggest that the epicardial methods measure the adenosine concentration in only the most superficial layer of the epicardium rather then the average concentration of the interstitial fluid of the entire wall of the left ventricle. If the ventricular muscle served as an infinite source of adenosine, then the equilibrium concentration measured would not be affected by the 3.5 to 5 fold increase in the volume/area ratio of the modified discs over the single simple discs.

Isolated perfused guinea pig hearts

The porous nylon disc technique was used in the isolated perfused guinea pig heart to determine the basal level of interstitial fluid adenosine and the effect of hypoxia on this adenosine concentration (Matherne, G.P., et al., unpublished observations).

To validate the use of the epicardial discs for measurement of interstitial fluid adenosine, a series of experiments was carried out with infusion of two concentrations of adenosine and one of 9-ß-D arabinofuranosyl hypoxanthine (Ara H) in the presence and absence of 0.5 µM dipyridamole and 5.0 µM EHNA. Hearts were perfused by the Langendorff technique at 8 ml/min. The pulmonary artery was cannulated for measurement of coronary flow and for sampling of coronary venous effluent for its adenosine concentration. Pairs of discs were placed on the left ventricular epicardium for two minutes and then removed and analyzed for their adenosine content. The concentration of adenosine in the discs was the same for epicardial contact times of 0.5 to 4 minutes. The heart was supported by a plastic frame so that the left ventricle was uppermost and contamination of the left ventricular epicardium with leakage from the base of the heart was ruled out by the absence of fluorescein, isothiocyanato-Dextran (W.W. 150,000) on the disc after its injection into the aortic cannula.

During basal conditions, the mean concentrations of adenosine in the discs was 0.28 ± 0.03 µM whereas that of the venous effluent was 0.004 ± 0.001 µM. In the presence of dipyridamole to   block cellular uptake of adenosine and EHNA to inhibit adenosine deaminase, the disc (interstitial fluid) concentrations increased to 1.19±.09 µM and that of the venous effluent to 0.027±0.004 µM. These results indicate that venous effluent concentration greatly underestimates the interstitial fluid concentration even when adenosine uptake is blocked and adenosine deaminase is inhibited. The reasons for this large difference in interstitial fluid and venous effluent adenosine concentrations are probably a combination of endothelial cell uptake and dilution in the vascular compartment by the perfusion fluid. In the presence of dipyridamole and EHNA, the difference must be attributable to dilution alone.

With arterial infusions of 6 and 12 µM adenosine, venous and interstitial fluid concentrations were less then 0.3 µM and were not significantly different from each other. However in the presence of dipyridamole and EHNA, the adenosine concentrations in the interstitial fluid and venous effluent during infusion of 6 or 12 µM adenosine were equal and not significantly different from that in the arterial infusate. Under these conditions (dipyridamole and EHNA) adenosine behaves as an inert tracer and, both the discs and the venous effluent accurately estimate the interstitial fluid concentration. When the adenosine analogue AraH (which is restricted to the extracellular space) was infused at a 6 µM concentration, the AraH concentration in the discs and venous effluent was the same as the arterial concentration in the presence or absence of dipyridamole.   From these observations it is concluded that the discs provide an accurate estimate of interstitial fluid adenosine and that in the absence of an adenosine uptake blocker the nucleoside is so rapidly taken up by endothelial cells that little endogenous adenosine reaches the vascular compartment and that little exogenous adenosine reaches the interstitial fluid compartment.

Table 1: Effect of Hypoxia on Interstitial Fluid and Effluent Adenosine, Inosine and Hypoxanthine Concentrations -(nM)

Adult Guinea Pig

	Interstitial Fluid (n=8)			Effluent (n=4)		
	Ado	In	Hx	Ado	In	Hx
Normoxia	261±19	915±40	380±25	48±3	209±21	92±9
Hypoxia	928±36	4855±110	940±141	253±23	834±40	380±40

Adult Rabbit (n=11)

Normoxia	228±35	1154±126	287±30	61±13	225±23	41±5
Hypoxia	1255±300	5220±1217	876±147	836±102	705±78	185±16

Neonatal Rabbit (n=10)

Normoxia	130±16	699±88	392±80	30±9	134±26	55±8
Hypoxia	1180±231	4049±500	1099±98	1398±207	1589±188	476±58

All adenosine, inosine and hypoxanthine levels were significantly increased by hypoxia (p< 0.05).

Ado, adenosine; In, inosine; Hx, hypoxanthine.

In hypoxia, the adenosine, inosine and hypoxanthine concentrations in the interstitial fluid and venous effluents of the isolated perfused guinea pig heart increased substantially (Table 1). A large gradient was observed between interstitial fluid and venous effluent concentrations during normoxia and hypoxia but the gradient was less during hypoxia than during normoxia. Similar results were obtained in the isolated perfused rabbit heart (Table 1) except that the gradient of the adenosine concentrations between interstitial fluid and venous effluent was abolished during hypoxia and was associated with a lesser increase in inosine and hypoxanthine than was observed in the hypoxic guinea pig heart. This finding was even more pronounced in the heart of neonatal rabbits (Table 1). The reason for this difference between rabbit and guinea pig hearts is not known.

In summary, the porous nylon disc technique appears to be a reliable method for the determination of interstitial fluid concentrations of adenosine and hence provides a feasible approach to a better understanding of the role of adenosine in the regulation of coronary blood flow and in conduction and excitation in the heart.

References

Berne, R.M., et al., 1986, Measurement of femtomolar concentrations of adenosine, J. Liq. Chromotog. 9:113-119.
Daut, J., et al., 1988, Passive electrical properties and electrogenic sodium transport of cultured guinea-pig coronary endothelial cells, J. of Physiol. 402:237-254.
Deussen, A., Moser, G., and Schrader, J., 1986, Contribution of coronary endothelial cells to cardiac adenosine production. Pfluegers Arch. 406:608-614.
Gidday, J.M., et al., 1988, Estimates of left ventricular interstitial fluid adenosine during catecholamine stimulation, Am. J. Physiol. 254:H207-H216.
Knabb, R.M., et al., 1983, Consistent parallel relationships among myocardial oxygen consumption, coronary blood flow, and pericardial infusate adenosine concentration with various interventions and ß-blockade in the dog, Circ. Res. 53:33-41.
Matherne, G.P., et al., 1990, Effect of hypoxia on cardiac interstitial fluid adenosine (unpublished data).
Miller, W.L., et al., 1979, Canine myocardial adenosine and lactate production, oxygen consumption, and coronary blood flow during stellate ganglia stimulation, Circ. Res. 45: 708-718.
Nees, S., et al., 1985, The coronary endothelium: a highly active metabolic barrier for adenosine, Basic Res. Cardiol. 80:515-529.
Phillips, C.L., et al., 1987, Estimates of interstitial fluid adenosine concentration with epicardial porous discs, Physiologist 30:216.
Rubanyi, G., and Vanhoutte, P.M., 1985, Endothelium removal decreases relaxations of canine coronary arteries caused by ß-adrenergic agonists and adenosine, , J. Cardiovasc. Pharmacol. 7:139-144.
Soracco, C.A., et al., 1989, Validation of porous discs for estimation of cardiac interstitial fluid adenosine, FASEB J. 3:A406.
Steenbergen, L., et al., 1977, Heterogeneity of the hypoxic state in perfused rat heart, Circ. Res. 41:606-615.

16

Rationale for the Use of Adenosine in the Diagnosis and Treatment of Cardiac Arrhythmias

L. Belardinelli and A. Pelleg

<u>Introduction</u>

The endogenous purine nucleoside, adenosine, exerts pronounced effects on the cardiovascular system including vasodilation, negative chronotropic action on pacemakers, negative dromotropic action on atrio-ventricular (AV) conduction, negative inotropic effect on atrial myocardium and attenuation of the effects of catecholamines. These effects are mediated by specific cell surface receptors coupled to various cellular transduction systems. Stimulation or inhibition of the adenylyl cyclase system leading to altered cellular levels of cyclic adenosine monophosphate (cAMP) have been shown to be mechanistically involved in, at least, some of adenosine's effects. Whereas, many other actions of this nucleoside are independent of cellular cAMP.

The following is a brief outline of the rationale for the use of exogenous adenosine as a cardioactive drug for the treatment and diagnosis of cardiac arrhythmias.

<u>Electrophysiologic Actions of Adenosine</u>

The negative chronotropic action of adenosine in the sinus node (bradycardia) and dromotropic action (AV block) are independent of each other. However, the ultimate action of adenosine appears to be regulating ventricular rate either indirectly by affecting atrial rate in the absence of AV block and by causing AV block or, alternately, directly via its negative chronotropic action on ventricular pacemaker activity (Heller et al, 1985; Pelleg et al, 1986).

In atrial myocardium, adenosine exerts both electrophysiological and inotropic effects.   Adenosine shortens the atrial action potential and causes hyperpolarization effects which are accompanied by a decrease in contractility. These actions of adenosine can be demonstrated under basal conditions (e.g., in the absence of catecholamine stimulation) and are independent of changes in cellular cAMP.   In contrast, in ventricular myocardium, neither the electrophysiological nor inotropic effects of adenosine are observed unless the adenylyl cyclase-cAMP system is stimulated.   In ventricular myocytes, adenosine attenuates the electrophysiological and positive inotropic effects of catecholamine stimulation (for reviews see Pelleg, 1985; Belardinelli et al, 1989).

The direct electrophysiological actions of adenosine in supraventricular tissues (i.e., atrial myocardium, SA and AV nodes) are mediated by the activation of a specific subset of K^+ channels resulting in increased K^+ outward current. These channels are also regulated by acetylcholine (Belardinelli et al, 1983; Kurachi et al, 1986; Belardinelli et al, 1988). The activation of K^+ channels is independent of cAMP, but involves a pertussis toxin sensitive guanine regulatory protein (Kurachi et al, 1986; Belardinelli et al, 1989).

Adenosine antagonizes the electrophysiologic effects of catecholamines in supraventricular tissues.   Specifically, in SA node cells, adenosine also attenuates catecholamine-induced stimulation of the hyperpolarization activated pacemaker current (I_F) and calcium inward current (I_{Ca}) (Belardinelli et al, 1988).   Similarly, in atrial myocytes, adenosine, attenuates catecholamine-stimulated I_{Ca}. The latter is associated with decrease in cellular cAMP levels (Belardinelli et al, 1989; Linden et al, 1985). Likewise, in SA node cells, the attenuation by adenosine of catecholamine-induced stimulation of I_F and I_{Ca} is also probably due to inhibition of adenylyl cyclase activity. Thus, it appears that in supraventricular tissue, activation of adenosine receptors results in two distinct electrophysiological signals, i.e., a direct action resulting in the activation of I_k and indirect cAMP-dependent action mediated by the inhibition of catecholamine-stimulated I_{Ca} and I_F.   In contrast, the electrophysiological actions of adenosine in ventricular myocardium are due to antagonism of the stimulatory effects of catecholamines, and are cAMP-dependent.   That is, in ventricular myocytes, adenosine attenuates catecholamine-stimulated I_{Ca} and the transient inward current (I_{Ti}), which has been implicated in the genesis of delayed afterpotentials that, when reach threshold, lead to triggered activity (Belardinelli et al, 1989). Less is known about the ionic currents which mediate the action of adenosine in the ventricular specialized tissue (His-Purkinje system). The fact that adenosine suppressed spontaneous activity of "normal" Purkinje fibers even prior to catecholamine challenge, suggests that direct action on I_k (activation) and/or I_f (attenuation) could be mechanistically involved in this action.

Antiarrhythmic Actions of Adenosine

Although most, if not all, cardiac actions of adenosine were known by the mid-1930's, it was not until the late 70's and throughout the 80's that their mechanisms were elucidated and their significance fully appreciated (Belardinelli et al, 1989).

The above mentioned electrophysiological actions of adenosine form the basis for its presently known anti-arrhythmic properties (Belhassen and Pelleg, 1984). The most notable example has been the potent depressant effects of adenosine on the AV nodal conduction, which selectively impairs and blocks impulse propagation through the AV node. Hence, adenosine has the ability to effectively slow and/or terminate abnormal cardiac rhythms such as AV nodal reciprocating tachycardias and other types of reentrant tachycardias which depend on the AV node for electrical impulse propagation. Based on this action, adenosine has now been proven highly efficacious (>90%) in terminating supraventricular tachycardias in which the AV node is part of the reentrant circuit (DiMarco et al, 1985). Because of its short half-life, adenosine, when given in intravenous boluses, transiently (<10 sec) interrupts conduction of electrical impulses through the AV node and terminates these types of supraventricular tachycardias. The short half life of adenosine and its transient mild side effects together with its comparable efficacy to verapamil make it a drug of choice in the acute management of supraventricular tachycardia involving the AV node (Sellers et al., 1987). In addition, the transient AV block caused by adenosine has been shown useful to reveal atrial rhythms such as atrial flutter and fibrillation which are otherwise masked by the much larger electrical signals generated by the depolarization of the ventricles (DiMarco et al, 1985; Griffith et al, 1988).

In atrial myocardium, the shortening of the action potential, which is due to an increase in I_k, is expected to decrease the refractory period and in turn facilitate intra-atrial reentry, i.e., atrial flutter. In fact, adenosine as well as ATP have been documented to induce transient episodes of atrial flutter (Belhassen et al.1984; DiMarco et al, 1985; Griffith et al, 1988). On the other hand, the hyperpolarization caused by adenosine in atrial myocytes has the effect of stabilizing the membrane potential (i.e., decrease excitability) and hence, has the potential to stop automatic rhythms (Belardinelli et al, 1983). Alternatively, the hyperpolarization could lead to an increase in the rate of rise of phase 0 of the action potential and, thereby, to an increase in conduction velocity (West et al, 1985). This latter effect would lessen the likelihood of reentry due to slow conduction. Thus, in atrial myocardium, adenosine has the potential to either terminate or facilitate intra-atrial reentrant rhythms.

In keeping with the findings that in isolated ventricular myocytes adenosine abolishes isoproterenol-and forskolin-facilitated triggered activity (Belardinelli et al, 1989), intravenous boluses of adenosine have been shown to rapidly terminate isoproterenol-induced ventricular tachycardia in patients without structural heart disease who experienced exercise-induced sustained ventricular tachycardia (Lerman et al, 1986). This effect seems to be specific to this subset of patients since ventricular tachycardias of which reentry or enhanced automaticity are the probable mechanisms (e.g., during myocardial infarction) were unresponsive to adenosine (Lerman et al, 1986; Griffith et al, 1988). The antagonism by adenosine of the electrophysiological effects of catecholamines and its lack of effect in the absence of ß-adrenergic stimulation, combined with adenosine's somewhat selective effect on AV nodal conduction, makes this nucleoside a very useful and unique diagnostic tool for atrial and ventricular arrhythmias. In fact, adenosine has been shown to provide useful diagnostic information in patients presenting with wide complex tachycardias (Griffith et al, 1988).

In conclusion, the rationale for the use of adenosine as anti-arrhythmic agent is based on its cellular electrophysiologic of actions. Additional investigation of the antiarrhythmic properties of this nucleoside and determination of which of the electrophysiological mechanisms of action are clinically relevant will certainly result in further therapeutic and diagnostic applications. The development of specific and selective adenosine related compounds (agonists and antagonists) should therefore lead not only to better tools to determine the role of adenosine in the genesis of cardiac rhythm disturbances, but also to new therapeutic approaches for management of cardiac arrhythmias.

References

Belhassen, B., and Pelleg, A., 1984, Electrophysiologic effects of adenosine triphosphate and adenosine in the mammalian heart: Clinical and experimental aspects, J. Am. Coll. Cardiol. 4:414-424.

Belhassen B., Pelleg, A., Shoshani, O., Laniado, S., 1984, Atrial fibrillation induced by adenosine triphosphate, Am. J. Cardiol. 53:1405-1406.

Belardinelli, L., Giles, W., and West A., 1988, Ionic mechanism of adenosine actions in pacemaker cells from rabbit heart, J. Physiol.(Lond) 405:615-633.

Belardinelli, L., and Isenberg G., 1983, Isolated atrial myocytes: adenosine and acetylcholine increase potassium conductance, Am. J. Physiol. 244:H734-H737.

Belardinelli, L., Linden, J., and Berne, R.M., 1989, The cardiac effects of adenosine, Prog. Cardiovasc. Dis. 32:73-97.

DiMarco, J., et al., 1985, Diagnostic and therapeutic use of adenosine in patients with supraventricular tachyarrhythmias, J. Am. Coll. Cardiol.6:417-425.

Griffith, M.J., et al., 1988, Adenosine in the diagnosis of broad complex tachycardia, Lancet 1:672-676.

Heller, L.J., and Olsson, R.A., 1985, Inhibition of rat ventricular automaticity by adenosine, Am. J. Physiol. 248:H907-H913.

Kurachi, Y., Nakajima, T., and Sugimoto, T., 1986, On the mechanism of activation of muscarinic K^+ channels by adenosine in isolated atrial cells:involvement of GTP-binding proteins, Pflügers Arch 407:264-274.

Lerman, B.B., et al., 1986, Adenosine-sensitive ventricular tachycardia: evidence suggesting cyclic AMP-mediated triggered activity, Circulation 74:270-280.

Pelleg, A., 1985, Cardiac cellular electrophysiologic effects of adenosine and adenosine triphosphate, Am. Heart J. 110:688-693.

Pelleg, A., Mitamura, H., Mitsuaka, T., et al., 1986, Effects of adenosine and adenosine 5'-triphosphate on ventricular escape rhythm in the canine heart, J. Am. Col. Cardiol. 8:1145-1151.

Sellers, T.D., Kirchhoffer, J.B., Modesto, T.A., 1987, Adenosine: A clinical experience and comparison with verapamil for the termination of supraventricular tachycardias, In: Cardiac Electrophysiology of Adenosine and ATP: Basic and Clinical Aspects, A. Pelleg, E.L. Michelson and L.S. Dreifus, eds., Alan R. Liss, Inc. 1987, pp. 285-299.

West, G.A., and Belardinelli, L., 1985, Correlation of sinus slowing and hyperpolarization caused by adenosine in sinus node, Pflügers Arch 403:75-81.

17
The Central Adenosine System as a Therapeutic Target in Stroke

P.J. Marangos, D. Von Lubitz, L.P. Miller, J.L. Daval, and J. Deckert

Adenosine is a major non-peptide neuromodulator in the central nervous system (CNS). Numerous reviews have been written on this subject with detailed accounts of the receptor heterogeneity, localization and functionality (Fredholm et.al 1980, Marangos et.al 1985). Existing information paints a clear picture regarding the importance of the adenosine system in CNS functions. There are few neurotransmitter or neuromodulator systems that have been characterized as well as that for adenosine. The prevailing view is that not only do adenosine receptors modulate c-AMP levels in both directions but that perhaps more importantly they regulate both K^+ and Ca^{++} ion fluxes across neural membranes (Michaelis et.al 1988). It appears that these latter adenosine mediated processes (i.e. stimulation of K^+ efflux or inhibition of Ca^{++} influx) serve to shorten the action potential and thereby inhibit the release of a number of neurotransmitters. The inhibition of Ca^+-dependent neurotransmitter release by adenosine has been consistently observed in a number of laboratories (for review see Marangos et.al 1985, Fredholm et.al, 1980) and is thought to be the most functionally relevant action of the purine riboside.

Behaviorally adenosine and metabolically stable adenosine agonists have marked sedating and anti-convulsant properties (Dunwiddie 1985) which make them of potential clinical interest. Adenosine has in fact been referred to as an endogenous naturally occurring sedative and anticonvulsant agent. These CNS depressant actions of adenosine probably represent a more global function of the purine which involves the regulation or adjustment of cellular activity to a level consistent with energy supply. Newby has termed adenosine a retaliatory metabolite (Newby 1984) with recuperative effects on the functions of metabolically compromised cells. In this analysis metabolic trauma would constitute any situation where adenosine triphosphate (ATP) utilization exceeds the ability of the cellular synthesis rate which would result in elevated adenosine levels. Such a situation certainly occurs at seizure foci as well as during brain ischemia and brain trauma (Dragunow et al 1988). Adenosine, via interactions with specific ectocellular receptors therefore appears to serve the homeostatic role of sparing cells during periods of generalized trauma. The important question from a drug development perspective is whether adenosine receptors might be amenable to targeting for various CNS disease indications which are characterized by a decrease in energy reserves such as seizures, ischemia and generalized trauma.

Our thinking in this regard was prompted by the relatively recent findings showing that the increased release of various excitatory amino acids (EAA) such as glutamate and aspartate post ischemia is a major cause of neural degeneration (Rothman et al 1986). This phenomenon is referred to as delayed excitotoxicity and is now being implicated as a major determinant in the neural pathology of not only cerebral ischemia (stroke) but also in head trauma (Faden et al 1989) and neurodegenerative disease (Sonsalla et al 1989). The importance of delayed excitotoxicity in these brain pathologies and its mediation by the EAA neurotransmitters has resulted in the development of drugs which block EAA receptors. Agents such as these have now been clearly documented to have efficacy in a number of animal model systems of ischemic stroke, the best known of which is the Merck compound MK-801 (Ozyurt et al 1988). The EAA blocker approach to the treatment of stroke has, however, raised important issues regarding the consequences of such a complete inhibition of a neurotransmitter system that subserves so many important roles in brain function. In fact a number of rather troubling untoward affects of EAA receptor blockers have been observed in animals such as psychosis and cytological abnormalities (Olney et al 1989). In this context it is important to realize that the psychoactive drug PCP (phencyclidine) is in fact active as an EAA receptor blocker and that many of its undesirable effects have been ascribed to this.

Since adenosine has a well documented effect on the release of a number of neurotransmitters (Fredholm et al 1980) including glutamate and aspartate (Burke et al 1988) it follows that adenosine receptor agonists would reduce the degree of post-ischemic EAA release. Such a hypothesis was supported by a recent study (Rudolphi 1987) showing that the adenosine receptor antagonist theophylline was capable of worsening the post-ischemic hippocampal neuronal damage in gerbils. These findings prompted a number of studies which now constitute a rather persuasive body of work supporting the role of adenosine as a cerebral neuroprotective agent post-ischemic (von Lubitz et al 1988, Evans et al 1987, Daval et al 1989).

TABLE 1

EFFECT OF CYCLOHEXYLADENOSINE (CHA) ON
GLOBAL CEREBRAL ISCHEMIA IN (GERBIL)

PARAMETER EVALUATED	DOSE	RESULT
I. Hippocampal CA-1 Cell counts, (30 min. Bilateral)	30 ng ICV (15 min. Post)	Protection P<.001
II. Receptors, Hippocampus (20 min.) Adenosine A$_1$ Forskolin GppNHp	2 mg/kg IP (5 min. Post)	Protection P<.01 P<.05 P<.05
III. Survival 30 min.	2 mg/kg IP (5 min. Post)	Protection 53% vs 10% P<.01
20 min.		66% vs 14% P<.01

Table 1. The table summarizes the results of several different studies performed over a period of several years. In each case the doses of CHA given and the route of administrations are given. The results obtained and statistical significance are indicated. In the first column the time refers to the duration of ischemia.

The studies done to date in our laboratory are summarized in Table I. The key features of these studies are that the adenosine A$_1$ receptor agonist cyclohexyladenosine (CHA) can afford marked neuroprotective effects in gerbils when administered post-reperfusion. In gerbils we have established this by both intracerebroventricular (ICV) as well as systemic (IP) administration. The mechanistic rationale for the protective effect of CHA in post ischemic injury in our view and that of others probably involves an inhibition of the release of EAA transmitters. The mediation of the effect by adenosine receptors is probable although few studies have addressed this issue directly. Recent data from our laboratory does indicate that the protective effects of CHA (2mg/kg IP) in gerbils subjected to 5 minutes of global ischemic can be reversed by theophylline. (Table 2).

TABLE 2

EFFECT OF ADENOSINE ANTAGONISTS ON POST-ISCHEMIC PROTECTION BY CHA

TREATMENT	PROTECTION
I. 5 min. ischemia	1/27
II. 5 min. + CHA	19/20
III.5 min. + CHA + THEO	0/6
IV.5 min. + CHA + 8SPT	5/5

Table 2. In each case protection is defined by greater than 90% preservation of CA-1 hippocampal cells 7 days post-ischemia.   CHA (2 mg/kg) was given 5 min. post-ischemia. The dose of Theophylline (theo) was 50 mg/kg and of 8 sulphophenyltheophylline was 100 mg/kg.

It is especially interesting that the peripheral (non CNS penetrating) adenosine antagonist 8-sulphophenyltheophylline at rather high doses (100 mg/kg) does not block the protective effect of CHA suggesting that the effects are in fact centrally mediated.

Other laboratories have also shown a protective effect of adenosine receptor agonists in gerbils and it now appears that his type of treatment strategy is effective when administered after the re-establishment of blood flow. In an effort to generalize the beneficial effects of adenosinergic agents we have also explored the rat model of ischemia. In our laboratory we used a three vessel (Both carotids and the right middle cerebral artery) occlusion model which results in focal cerebral ischemia (Chen et al 1986). Table 3 summarizes the results we have obtained with CHA in this model.

In the first experiment we gave one dose of CHA I.P. (0.5 mg/kg) ten minutes after re-establishing blood flow and employed a 60 minute occlusion. The infarct size (determined by tetrazolium staining) 24 hours post surgery was markedly reduced in the CHA treated animals. The second experiment employed a shorter ischemia time (45 minutes) and a different dosing schedule. The dose was lowered to 50 micrograms per kg with the first one given 10 minutes post reperfusion I.V. (jugular) and the second I.P. 2 hours after. Again CHA treatment resulted in a reduction of infarct size.

TABLE 3

EFFECT OF CYCLOHEXYLADENOSINE (CHA) ON
BRAIN INFARCT SIZE IN RAT FOCAL ISCHEMIA

CONDITIONS	INFARCT VOL.(MM³)	PERCENT REDUCTION
Bilateral CA,RMCA (60 min.)		
I. CHA dose = 0.5 mg/kg i.p 10 min. postreperfusion	Control -163 $\pm$ 18 (9) CHA - 79 $\pm$ 21 (7)**	52%
II. Bilateral CA, RMCA (45 min.) CHA dose = 0.05 mg/kg X 2 Postreperfusion 10 min. i.v. 2 hrs. i.p.	Control - 88 $\pm$ 4 (10) CHA - 62 $\pm$ 7 (9)*	30%

* $P<.02$
**$P<.01$

Table 3. Two studies are summarized. In both cerebral ischemia was accomplished by reversible occlusion of both carotid arteries (CA) and the right middle cerebral artery (RMCA).

Adenosine agonists appear to be beneficial when given acutely after experimentally induced brain ischemia. The potential clinical relevance of these results is apparent but many questions remain to be answered. The proper dosing regimen and dose response for the neuroprotective effect remain to be worked out. Also the rather substantial peripheral effects of adenosine agonists (hypotension, bradycardia, renal inhibition) must be assessed in relation to the cerebroprotective actions. It is also important to recognize that CHA has very limited blood brain barrier penetration and agonists with improved brain accessibility may prove even more beneficial since lower doses would be attainable. The fact that the adenosine agonists have efficacy when administered post-ischemia is of potential therapeutic interest since it suggests that these agents may be of value in post stroke syndrome interventions. Recent results from our laboratory show efficacy of CHA when given 60 minutes post-ischemia in gerbils (Miller et al unpublished).

Direct comparisons between adenosine agonists, EAA receptor blockers, calcium channel blockers and free radical scavengers as regard their efficacy in various animal model systems will be useful in assessing the therapeutic potential of adenosine in stroke. Also the evaluation of other adenosinergic agents such as uptake blockers and inhibitors of adenosine metabolizing enzymes are of potential value in stroke therapeutics. Deoxycoformycin has in fact recently been shown to reduce post-ischemic damage in gerbil hippocampus (Phillis et al 1989) suggesting that the enhanced preservation afforded by adenosine deaminase inhibition can be of therapeutic benefit in stroke. The adenosine uptake blocker approach has yet to be tried in stroke but one might predict efficacy with such agents if their accessibility to the brain was assured. Drugs targeted at enzymes such as adenosine deaminase or at the uptake site might however, only be effective when present during ischemia, i.e. when ATP levels are rapidly decreasing. In these situations the ischemic foci serves to generate a local increase in adenosine levels which is further potentiated by the uptake blocker or enzyme inhibitor, by increasing the halflife of the purine. In this regard it is of interest that the one report showing efficacy for the adenosine deaminase inhibitor, deoxycoformycin employed a pre-ischemic (15 minutes) dosing schedule (Phillis et al 1989).

Until recently the major potential therapeutic indications for adenosine agonists in the CNS were as sedatives and anticonvulsants. The necessity for chronic treatment in both these indications coupled with the rather substantial side effects of these agents raises questions regarding the feasibility of this approach. The neuroprotective post-stroke indication is different from the sedative or seizure situation in that an acute short term intervention is envisioned. This may prove to be a more advantageous situation for an adenosingeric therapy i.e one or at most several doses of an agonist during or shortly after the occurrence of a stroke. Adenosine uptake blockers and inhibitors of adenosine metabolism are two additional adenosinergic approaches that might be taken with their advantage being much less pronounced side effects since they are limited by endogenous adenosine levels. The efficacy of such agents post-reperfusion has however, not yet been established.

REFERENCES

Burke, S.P., and Nadler, J.V., 1988, Regulation of glutamate and aspartate release from slices of the hippocampal CA-1 area: effects of adenosine and baclofen., J. of Neurochem. 1541-1551.

Chen, S.T., et al., 1986, A model of focal ischemic stroke in the rat: reproducible extensive cortical infarctions, Stroke 17:738-743.

Daval, J.L., et al., 1989, Protective effect of cyclohexyladenosine on adenosine A_1 receptors, guanine nucleotide and forskolin binding sites following transient brain ischemia: a quantitative autoradiography study, Brain Res. 491 212-226.

Dragunow, M., and Faull, R.L.M., 1988, The neuroprotective effects of adenosine. Trends in Pharmacol. Sci. 7:194-1965.

Dunwiddie, T.V. (1985), The physiological role of adenosine in the central nervous system. Int. Rev. Neurobiol. 27:63-139.

Evans, M.C., Swan, J.H. and Meldrum, B.S., 1987, An adenosine analogue, 2-chloradenosine, protects against long term development of ischemic cell loss in the rat hippocampus, Neurosci. Lett. 83:287-292.

Faden, A.I., et al., 1989, The role of Excitatory amino acids and NMDA receptors in traumatic brain injury, Science 244:798-800.

Fredholm, B.B. and Hedqvist, P., 1980, Modulation of neurotransmission by purine nucleotides and nucleosides, Biochem. Pharmacol. 29:1635-1643.

Marangos, P.J. and Boulenger, J.P., 1985, Basic and clinical aspects of denosinergic neuromodulation, Neurosci. & Biobehav. Reviews 9:421-430.

Michaelis, N.L., et al., 1988, Studies on the ionic mechanism for the neuromodulatory action of adenosine in the brain, Brain Res. 473 249-260.

Newby, A.C., 1984, Adenosine as a retaliatory metabolite, Trends in Biochem. Sci. 9:42-44.

Olney, J.W., Labruyere, J. and Price, M.T., 1989, Pathologic changes induced in cerebrocortical nervous by phencyclidine and related drugs, Science 244 1360-1362.

Ozyurt, E., et al., 1988, Protective effect of the glutamate antagonist MK 801 in focal cerebral ischaemia in the cat, J. Cereb Blood Flow Metab. 8:138-143.

Phillis, J.W. and O'Regan, M.H., 1989, Deoxycoformycin antagonizes ischemia-induced neuronal degeneration, Brain Res.Bull. 22 537-540.

Rothman, S.M and 1986, Glutamate and the pathophysiology of hypoxic-ischemic brain damage, Ann Neurol. 19:105-111.

Rudolphi, K.A, Keil., and Hinze, H.J., 1987, Effect of theophylline on ischemically induced hippocampal damage in Mongolian gerbils: a behavioral and histopathological study, J. Cereb Blood Flow and Metab. 7:74-81.

106

Sonsalla, P.K., Nicklas, W.J. and Heikkila, R.E., 1989, Role for excitatory amino acids in methamptamine-induced nigrostriatal dopamine toxicity, Science 243:398-401.

von-Lubitz, D., et al., 1988, Post-ischemically applied cyclohexyladenosine protects against neuronal death in the CA-1 region of the hippocampus in gerbil, Stroke 19:1133-1139.

18
Adenosine-Mediated Vasoconstriction in the Skin Microcirculation

K.G. Proctor and I. Stojanov

Introduction

Although adenosine (ADO) is a physiologically important vasodilator in virtually every vascular bed, it causes vasoconstriction in the femoral and tail arteries of rats (Brown & Collis, 1981; Sakai & Akima, 1978) and in the kidneys of most species. ADO-induced vasoconstriction is usually attributed to indirect effects, such as the release of serotonin (Brown & Collis, 1981; Sakai & Akima, 1978) or angiotensin (Thurau, 1964; Spielman & Thompson, 1982). In the kidney, the response can be distinguished from the renin-angiotensin system (Churchill & Churchill, 1988); it causes renal vasoconstriction by a receptor-mediated mechanism that is concentration dependent and coupled to calcium movements through potential-operated channels (Murray & Churchill, 1985; Churchill & Bidani, 1987; Rossi et al, 1987; Rossi et al, 1988). Nevertheless, ADO-induced vasoconstriction is independent of angiotensin generation in the isolated perfused kidney, but not the kidney *in situ* (Rossi et al, 1987; Rossi et al, 1988). This present report will summarize recent evidence for a receptor-mediated (and apparently *direct*) vasoconstrictor action of ADO in the skin microcirculation *in situ*.

Methods

All these studies were performed in the sub-cutaneous microcirculation of pentobarbital-anesthetized male hamsters (Stojanov & Proctor, 1989a). Briefly, the femoral artery and vein were cannulated for measurement of systemic blood pressure and the administration of supplemental fluids. A tracheostomy assured a patent airway for spontaneous respiration on room air + oxygen. The animal was positioned laterally on a plexiglass viewing platform that attached to the stage of an intra-vital microscope. The dorsal skin along the mid-line was shaved, elevated using several sutures for traction, and a 1 cm circular section of epidermis was excised. After removing the underlying connective tissue, a circular cover-slip was placed over the wound. Throughout the surgical and experimental procedure, the tissue was continuously suffused with a temperature-controlled Ringers-bicarbonate solution. In most experiments, skin temperature was maintained at 32°C with the heated suffusate, but in one series, it was varied ± 5°C to examine whether changes in skin temperature influenced ADO-mediated responses. Various vasoactive substances (ADO or its synthetic analogs, norepinephrine, angiotensin or serotonin) or receptor antagonists (methylxanthines, phentolamine, saralasin, methysergide) were added to the suffusate.

The preparation was transilluminated with filtered white light and visualized with a closed-circuit color video system. Red blood cell velocity was measured with an optical detector that attached to the camera. Arteriolar diameter was measured with a electronic video-micrometer. Blood flow was calculated from the product of mean blood velocity and vascular cross-sectional area. All dynamic variables were continuously recorded on a polygraph.

Results & Discussion

Pharmacologic evidence for ADO receptors: Because ADO is a non-selective receptor agonist, as well as a metabolic substrate, it is difficult to attribute a particular response to a receptor-mediated mechanism if ADO is used as the only agonist (Daly, 1982). On the other hand, using the potency order of a series of synthetic, non-metabolized agonists, and a nM vs µM receptor affinity, ADO receptors can be classified according to A_1 (or Ri) and A_2 (or Ra) subtypes (Daly, 1982; Londos & Cooper, 1980). In addition, methylxanthines are antagonists at both types of receptors (Daly, 1982; Londos & Cooper, 1980). This basic logic was used to pharmacologically characterize ADO receptors in the skin microcirculation. It was assumed that ADO receptor stimulation would elicit changes in diameter of small arterioles because small arterioles control blood flow to most tissues by changing diameter and because ADO receptor stimulation elicits parallel changes in adenylate cyclase and blood flow in the heart, kidney, and brain. In these experiments, synthetic A_1 and A_2 agonists were continuously applied by a topical route and responses were observed in arterioles whose resting diameter ranged from 20-60 µm (Stojanov & Proctor, 1989a).

At > 10nM, the potency order for causing sustained vasodilation (max. = 170-190% of control) was N-ethyl carboxamido adenosine (NECA) > 2-chloro adenosine (2CA) > ADO. The ED_{50}'s were 0.22 ± 0.04 µM, 1.3 ± 0.3 µM, and 12.9 ± 3.0 µM, respectively. These responses were attenuated by 10 µM 8-phenyl theophylline (8pTHEO). In contrast, 10 µM dipyridamole (DIPYRID), a cellular uptake inhibitor, potentiated ADO-induced vasodilation ($ED_{50} = 2.7 \pm 0.8$ µM) to a response that was similar to that evoked by 2CA alone or 2CA + DIPYRID. These data suggested the presence of A_2 receptors in this tissue and that cellular uptake reduced vasodilation caused by exogenous ADO (Stojanov & Proctor, 1989a).

From 0.1-10 nM, cyclohexyl adenosine (CHA) and ADO were equipotent for causing vasoconstriction (min. = 80-90% of control). The ED_{50}'s were 7 ± 5 nM and 0.3 ± 0.2

nM. These responses were not altered by DIPYRID, but were completely antagonized by 8pTHEO. At μM or higher concentrations, CHA caused vasodilation, which is not surprising because CHA loses its A_1 selectivity in the μM range (Daly, 1982). Norepinephrine (NOREPI) was a more potent vasoconstrictor (the vessel essentially closed at >10 μM) and the response was not altered by 8pTHEO. These data suggested the presence of A_1 receptors in this tissue and that cellular uptake does not alter the vasoconstriction caused by exogenous ADO (Stojanov & Proctor, 1989a).

A major fraction of the pre-capillary resistance is partitioned in small arterioles (< 100 μm diameter) in most vascular beds. However, in some tissues, larger blood vessels upstream from the microcirculation can limit blood flow (Segal & Duling, 1986); in this condition, arteriolar diameter changes alter blood flow *distribution*, rather than blood flow, *per se* (Proctor & Busija, 1985). To establish whether the major resistive elements were proximal to the skin microcirculation, red blood cell velocity was measured along with arteriolar diameter so that volume blood flow could be calculated (Stojanov & Proctor, 1989a). The blood flow changes evoked by NECA, 2CA, ADO, and CHA were equal to, or greater than, the diameter changes, which verified that the 20-60 μm arterioles observed in these experiments were important resistive elements in the skin microcirculation.

Interactions between ADO and other vasoconstrictors To rule out the possibility that ADO-mediated vasoconstriction was caused by the release of other substances, CHA responses were observed with various receptor antagonists in the suffusate. In pilot experiments, a dose-response curve to each of the antagonists was determined; the highest concentration that had no effect on baseline diameter was chosen for subsequent tests. CHA, instead of ADO, was used to minimize the effect of non-receptor-mediated actions of the parent compound on arteriolar diameter changes.

Angiotensin II (ANG) caused a reduction in diameter to $59 \pm 2\%$ of control at 0.1 nM. This response was antagonized by 10 μM saralasin (diameter = $95 \pm 2\%$) but CHA + 10 μM saralasin evoked dose-related constriction ($ED_{50} = 16 \pm 6$ nM; half-max diameter = $85 \pm 2\%$ of control). Thus, CHA-induced vasoconstriction cannot be attributed to the actions of ANG.

Serotonin (5HT) caused a reduction in diameter to $83 \pm 2\%$ of control at 10 μM. This response was antagonized by 10 μM methysergide (diameter = $94 \pm 3\%$). However, CHA + 10 μM methysergide evoked dose-related vasoconstriction ($ED_{50} = 1 \pm 1$ nM; half-max diameter = $89 \pm 2\%$ of control), which argues against a role for 5HT.

NOREPI (1 μM) reduced diameter to $48 \pm 12\%$ of control. This response was antagonized by 10 μM phentolamine (diameter = 85 $\pm$ 2%). However, CHA + 10 μM phentolamine evoked dose-related vasoconstriction (ED_{50} = 6 $\pm$ 6 nM; half-max diameter = 89 $\pm$ 2% of control) showing that CHA-induced vasoconstriction cannot be attributed to NOREPI.

In summary, these results are consistent with a direct vasoconstrictor action of CHA, but the possible actions of vasoactive subtances other than ANG, NOREPI, or 5HT cannot be excluded.

Temperature sensitivity of ADO responses Recent evidence has shown that temperature alters responses mediated by A_1, but not A_2, receptors in vitro (Broadley et al, 1985). To examine whether A_1 vascular responses were temperature sensitive in the skin microcirculation, ADO or its synthetic analogs were applied at three different solution temperatures (Stojanov & Proctor, 1989b).

The magnitude of the responses and the potency order for vasodilation evoked by 0.01-1 μM NECA, 2CA, and ADO at 27°C, 32°C, or 37°C were identical to that described earlier, so A_2-mediated vasodilation was not changed by alterations in local skin temperature (Stojanov & Proctor, 1989b).

In contrast, vasoconstrictions evoked by ADO and CHA were enhanced at high temperature and unchanged or slightly reduced at low temperature. At 0.2-0.4 nM CHA, diameter was 79 $\pm$ 1% of control at 37°C, 93 $\pm$ 1% at 32°C, and 97 $\pm$ 2% at 27°C. At 0.1-1 nM ADO, diameter was 84 $\pm$ 2% of control at 37°C, 93 $\pm$ 1% at 32°C, and 95 $\pm$ 1% at 27°C (Stojanov & Proctor, 1989b).

These temperature-sensitive ADO responses probably cannot be attributed to a non-specific alteration in vascular reactivity for two reasons. *First*, 10 μM 8pTHEO completely antagonized the CHA and ADO responses, which is consistent with the effect being receptor-mediated. *Second*, sub-maximal vasoconstrictions evoked by NOREPI or ANG did not follow a similar temperature pattern. NOREPI evoked responses that were reduced at low temperature, but not enhanced at high temperature; the application of 100 nM NOREPI caused diameter decreases to 72 $\pm$ 4% at 37°C, 68 $\pm$ 4% at 32°C, and 84 $\pm$ 2% at 27°C. ANG evoked responses that were temperature-insensitive; the application of 10 nM ANG caused diameter decreases to 83 $\pm$ 2% at 37°C, 82 $\pm$ 1% at 32°C, and 83 $\pm$ 2% at 27°C.

In summary, increases in local skin temperature potentiated the vasoconstrictions caused by nM concentrations of ADO or CHA, but there was no alteration in the vasodilations evoked by μM concentrations of ADO, NECA, or 2CA. Even though the temperature sensitivity appeared selective for A_1-mediated vasoconstriction and not responses to ANG or NOREPI, it is possible that this phenomena simply reflects a physical effect, such as the solubility of the various agonists or diffusion through the interstitial space. Alternatively, changes in skin blood flow and temperature have an important role for regulating body core temperature in man, so it is conceivable that these may serve a similar homeostatic role in hamsters even though these small rodents would not normally control body temperature by varying skin blood flow.

Effects of chronic caffeine consumption Methylxanthines are present in therapeutic concentrations in coffee, tea, soft drinks, and over-the-counter medications and are among the most widely consumed drugs in the world (Fredholm, 1985). For many years the action of methylxanthines was attributed to phosphodiesterase inhibition, but it is now clearly established that the principal mechanism of action is to antagonize the action of endogenous ADO, particularly the receptors in the central nervous system (Fredholm, 1985; Daly, 1982). To examine possible effects on ADO receptors in the skin, hamsters were exposed to drinking water that contained caffeine (1 g/l) for 30 days. The skin microcirculation was prepared for observation during the treatment period and for 9 days after the animals were returned to plain drinking water.

For the first three days of caffeine consumption, vasodilation evoked by NECA and vasoconstriction evoked by CHA were unaltered. On days 5-9, both responses were virtually eliminated. For the remaining 21 days of caffeine treatment, the vasodilator response returned; in fact, NECA evoked vasodilator responses whose magnitude was equal to, or greater than that, evoked during the pre-treatment period. In contrast, after day 17, the the vasoconstrictor response was absent.

Immediately upon restoration of normal drinking water, CHA-induced vasoconstriction returned, and the magnitude of the response increased with each post-treatment day. A peak response was recorded 10 days after caffeine withdrawal, which was the last day of the study. In contrast, the vasodilation evoked by NECA declined to the original pre-treatment response for the first 8 days after caffeine withdrawal. However by the 10th post-treatment day, NECA-evoked vasodilation also reached peak values. Thus, both vasodilation and vasoconstriction were enhanced when caffeine exposure stopped.

These responses are complex and difficult to interpret, but might lead to interesting new ideas. It is speculated that NECA and CHA responses were not altered for the first few days because caffeine levels had not reached steady state plasma concentrations. For days 5-9, both responses were blocked because caffeine reached a therapeutic concentration. For the remaining 21 days of treatment, the vascular effect evoked by NECA or CHA reflected a balance in the net synthesis of A_1 and/or A_2 receptor protein, an alteration in coupling between the receptors and adenylate cyclase, or an alteration in the affinity of the receptors for the agonists. Upon washout, the gradual potentiation in the responses evoked by both CHA and NECA reflected a net up-regulation in ADO receptors.

Conclusion   Overall, these results show that exogenous ADO can cause vasoconstriction or vasodilation in the skin microcirculation depending upon concentration, that the responses are probably mediated by A_1 and A_2 receptors, and that the responses can be altered by chronic caffeine consumption. It is unknown if endogenous ADO exerts a similar action.   The location of the receptors is unknown, but A_1-mediated vasoconstriction was not altered by antagonists of ANG, 5HT, or NOREPI receptors, which is consistent with direct activation of vascular smooth muscle. The function of the receptors is unknown, but A_1-mediated vasconstriction was enhanced at high skin temperature, which may suggest a role in the control of body temperature. Additional study is required to explore the physiologic significance of A_1 and A_2 receptors in the skin microcirculation.

Supported by NIH-grant HL #30663 and Amer. Heart Ass'n grant #86-1071.

References

Broadley KJ, Broome S, Paton DM. *Br J Pharmacol* 1985; 84:407-415

Brown CM, Collis MG. *Eur J Pharmacol* 1981; 76: 275-277

Churchill PC, Bidani AK. *Am J Physiol* 1987; 252:F299-303

Churchill PC, Churchill MC *ISI Atlas of Sci: Pharmacol* 1988: 367-373

Daly JW. *J Med Chem* 1982; 25:197-207

Fredholm BB *Acta Med Scand* 1985; 217:149-153

Londos C, Cooper DMF, Wolff J. *Proc Nat'l Acad Sci USA* 1980: 77:2551-2554

Murray RD, Churchill PC. *J Pharmacol Exp Therap* 1985; 232:189-193

Proctor KG, Busija DW. *Am J Physiol* 1985; 249:H34-H41

Rossi N, Churchill P, Ellis, Amore B. *Am J Physiol* 1988; 255:H885-H890

Rossi NF et al. *J Pharmacol Exp Therap* 1987; 240:911-915

Sakai K, Akima M. *Naunyn-Schmiedeberg's Arch Pharmacol* 1978; 302:55-59

Segal SS, Duling BR. *Circ Res* 1986; 59:283-290

Spielman WS, Thompson CI. *Am J Physiol* 1982; F423-F435

Stojanov I, Proctor KG. *Circ Res* 1989a; 65:176-184

Stojanov I, Proctor KG. *J Pharmacol Exp Therap* 1989b; (submitted for publication)

Thurau K. *Am J Med* 1964; 36:698-719

19

Adenosine is an Antiinflammatory Autocoid: Adenosine Receptor Occupancy Promotes Neutrophil Chemotaxis and Inhibits Superoxide Anion Generation

B.N. Cronstein, L. Angaw-Duguma, D. Nicholls, A. Hutchison, and M. Williams

INTRODUCTION

While migrating to sites of infection or tissue injury neutrophils and other inflammatory cells are in close contact with tissue cells. Migrating neutrophils are under the influence of chemoattractants which may also stimulate the neutrophil to generate potentially toxic oxygen metabolites.

We and, subsequently, others have shown that adenosine, a purine released by many tissues, inhibits superoxide anion generation and adherence to endothelium by activated neutrophils yet, paradoxically, promotes neutrophil chemotaxis (Roberts et al.1985; Skubitz et al.1988; Grinstein et al.1986; Schmeichel et al.1987; Cronstein et al.1983; Rose et al.1988; Cronstein et al.1987; Cronstein et al.1986; Cronstein et al.1985; Schrier et al.1986; Iannone et al.1985). These effects on neutrophil function, mediated through occupancy of specific neutrophil receptors for adenosine, are unlike those of any other physiologic modulator of neutrophil function. Even more surprisingly, we have found that adenosine and its analogues modulate chemotaxis at concentrations markedly lower than those required to inhibit superoxide anion generation, an observation which could be accounted for by the presence of two distinct neutrophil adenosine receptors.

Since there are at least two types of adenosine receptor, A_1 and A_2, which can be distinguished by different agonists and the lower concentrations of adenosine required for saturation of A_1 receptors (van Calker et al.1979; Londos et al.1980) we have studied the effect of specific adenosine receptor agonists on superoxide anion generation and chemotaxis in response to the surrogate bacterial chemoattractant, N-formyl-methionyl-leucyl-phenylalanine or FMLP. We have previously reported that 5'N-ethylcarboxamidoadenosine or NECA, an agonist equally potent at A_2 and A_1 receptors, is the most potent inhibitor of superoxide anion generation and promoter of chemotaxis (Rose et al.1988; Cronstein et al.1985). CV-1808 and CGS-21680 are more highly selective A_2 agonists which are 7 and 180 fold A_2 specific (Hutchison et al.1989). N^6-cyclopentyladenosine (or CPA), in contrast, is a 600-fold more specific agonist at A_1 receptors. Using these agonists we now report that adenosine inhibits superoxide anion generation by occupying an A_2 receptor but promotes chemotaxis by occupying a previously unrecognized A_1 receptor on the neutrophil. We have also found that occupancy of adenosine A_1 receptors promotes chemotaxis by a mechanism which is dependent upon intact G proteins and microtubules. In contrast occupancy of adenosine A_2 receptors inhibits O_2^- generation by promoting association of chemoattractant receptors with a cytoskeletal fraction in a process associated with desensitization of the neutrophil to FMLP.

METHODS
We studied superoxide anion generation by determining the superoxide dismutase inhibited reduction of cytochrome C by neutrophils after a 5 minute incubation with FMLP in the presence of cytochalasin B (Goldstein et al.1975). Chemotaxis to FMLP was studied by the technique of chemotaxis under agarose (Nelson et al.1975). A chemotactic index was calculated following computerized analysis of directed migration and subtraction of random movement. Association of chemoattractant, presumably bound to receptors, with the cytoskeletal preparations was determined by incubating neutrophils with [^{3}H]-FMLP followed by washing, cell lysis with Triton-X and quantitation of radioactivity in the triton insoluble material (Jesaitis et al.1984). After subtraction of non-specific binding, data are presented as counts per minute. Non-specific binding accounted for 15% or less of total binding and was unaffected by any of the compounds used here. All of the results shown here represent the means and standard errors of 3-8 different experiments.

RESULTS AND DISCUSSION
We first studied the effect on superoxide anion generation of two relatively specific A_2 receptor agonists, CV-1808 and CGS-21680, the non-selective agonist NECA and the A_1 selective agonist CPA. Not surprisingly CV-1808, CGS-21680 and NECA were similar with respect to their inhibition of O_2^- generation in response to the chemoattractant FMLP (Table I). In contrast, the A_1 agonist CPA did not inhibit O_2^- generation (Table I). When we studied the effects of these same agonists on chemotaxis we found, again, that CV-1808, CGS-21680 and NECA were all similar with respect to promotion of chemotaxis (Table I). Surprisingly, though, CPA was significantly more potent as a promoter of chemotaxis than any of the other compounds tested (Table I). These observations are consistent with our previous findings and those of others that adenosine inhibits O_2^- generation by occupying A_2 receptors. These results also suggest that adenosine promotes chemotaxis by occupying specific A_1 receptors on the neutrophil.

Table I

AGONIST	IC$_{50}$ FOR O$_2^-$ (pM)	EC$_{50}$ FOR CHEMOTAXIS
NECA	23,000	9
CV-1808	17,000	7
CGS-21680	55,000	11
CPA	>1,000,000	2

The observation that an adenosine receptor agonist strongly promotes chemotaxis yet does not inhibit superoxide anion generation suggests adenosine promotes chemotaxis by occupying a different receptor from that which inhibits superoxide anion generation. Indeed, the markedly lower concentrations of adenosine receptor agonists required to promote chemotaxis and the greater specificity and potency of CPA for promotion of chemotaxis are most consistent with the hypothesis that adenosine occupies an A_1 receptor to promote chemotaxis and an A_2 receptor to inhibit superoxide anion generation.

The mechanism of stimulus-response coupling differs for adenosine A_1 and A_2 receptors. We therefore sought to confirm the presence of two different adenosine receptors on the neutrophil by examining the mechanisms by which adenosine modulates chemotaxis and superoxide anion generation. Previous studies have shown that stimulus-response coupling at adenosine A_1 receptors, in other cell types, is transduced through a GTP-binding or G protein. The alpha subunit of G proteins is irreversibly ADP-ribosylated by pertussis toxin and, thereby, rendered incapable of coupling ligand-receptor binding to its appropriate cellular response. We therefore determined whether pertussis toxin alters the effect of adenosine receptor occupancy on chemotaxis. We found that pertussis toxin at both 400 and 1000ng/ml (concentrations which did not affect chemotaxis alone) completely abrogated the effect of CPA on chemotaxis at all concentrations studied (10^{-12}-10^{-9}M, p<0.0001). In contrast pertussis toxin (700ng/ml) partially inhibited O_2^- generation in response to FMLP (0.1μM, $36+7\%$, p<0.02) but did not prevent NECA (10^{-9}-10^{-5}M) from inhibiting O_2^- generation. These results indicate that adenosine receptor occupancy promotes chemotaxis via a G_i protein and are consistent with the hypothesis that adenosine promotes chemotaxis by occupying an A_1 receptor. Moreover, these results demonstrate that occupancy of A_2 receptors does not inhibit O_2^- generation by a G protein dependent mechanism.

To date the intracellular mechanisms by which adenosine either promotes chemotaxis or inhibits O_2^- generation remain unknown. We sought to determine whether microtubules play a role in stimulus-response coupling at the neutrophil adenosine receptor by studying the effect of NECA on chemotaxis in the presence of vinblastine and colchicine at concentrations previously shown by this laboratory to disrupt neutrophil microtubules (10μM). NECA increased chemotaxis to a maximum of $156+14\%$ of control and both vinblastine and colchicine completely abrogated the effect of NECA on chemotaxis (p<0.001). While colchicine alone, similar to many previous reports, increased chemotaxis to approximately 122% of control in these experiments vinblastine alone did not affect neutrophil chemotaxis. Similar to pertussis toxin, neither vinblastine nor colchicine interfered with the capacity of NECA to inhibit O_2^- generation in response to FMLP.

The mechanism by which adenosine inhibits superoxide anion generation remains unclear. We designed experiments to explore the hypothesis that occupancy of adenosine receptors inhibits superoxide anion generation by promoting more rapid desensitization of the neutrophil to stimulation by chemoattractants. This process, desensitization of one receptor as a result of occupancy of a different receptor is more formally known as heterologous desensitization. Jesaitis and colleagues have recently reported that bound FMLP receptors associate with the cytoskeleton in domains of the plasma

116

membrane relatively depleted of GTP binding proteins (Jesaitis et al.1984; Jesaitis et al.1988; Jesaitis et al.1986). Thus, desensitization of the neutrophil to FMLP results from segregation of bound receptors to microdomains of the plasma membrane incapable of stimulating an appropriate response.

To determine whether adenosine receptor occupancy inhibits or promotes desensitization of the neutrophil we have studied the effect of adenosine receptor occupancy on association of [³H]-FMLP with a cytoskeletal preparation. We found that association of [³H]-FMLP with cytoskeletal preparations increases over time and is significantly greater in the presence of NECA at all time points tested from 30 seconds to 10 minutes after addition of stimulus (p<0.0001). Conclusion: adenosine receptor occupancy promotes association of bound FMLP receptors with the cytoskeleton, a process previously associated with desensitization of the neutrophil to FMLP.

Is increased binding of [³H]-fmlp to a cytoskeletal preparation associated with inhibition of superoxide anion generation or promotion of chemotaxis? To resolve this question we studied the effect of vinblastine and colchicine on binding of [³H]-FMLP to the cytoskeleton. We found that preincubation with neither colchicine nor vinblastine affects association of [³H]-FMLP with the cytoskeleton (Figure 2). Neither colchicine nor vinblastine alone significantly affected the association of FMLP-receptors with the cytoskeleton. Thus, adenosine receptor occupancy increases association of FMLP receptors with the cytoskeleton by a microtubule independent mechanism in parallel with inhibition of superoxide anion generation.

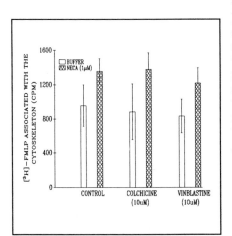

Figure 1

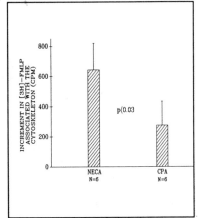

Figure 2

We were interested to confirm the link between binding of [³H]-FMLP to cytoskeletal preparations and inhibition of superoxide anion generation. Therefore, we compared the association of [³H]-FMLP with the cytoskeleton in the presence of NECA which both inhibits superoxide anion generation and promotes chemotaxis, or N^6-cyclopentyladenosine (CPA), a selective adenosine A_1 receptor agonist which is a potent promoter of chemotaxis but which only minimally inhibits superoxide anion generation. We found that NECA promotes association of [^{3}H]-FMLP with the cytoskeleton significantly better than CPA. These results confirm that adenosine A_2 receptor occupancy increases association of [^{3}H]-FMLP with the cytoskeleton in parallel with inhibition of superoxide anion generation. Moreover, these observations are consistent with the hypothesis that adenosine A_2 receptor occupancy inhibits superoxide anion generation by uncoupling bound FMLP receptors from the proteins required for stimulus transduction.

In summary, we found that 1. the specific adenosine A_2 agonists CV-1808 and CGS-21680 inhibit superoxide anion generation and promote chemotaxis as well as NECA, an agent which is equally potent at A_1 and A_2 receptors. 2. The highly selective A_1 receptor agonist CPA is a more potent promoter of chemotaxis than NECA, CGS-21680 or CV-1808 but is only a very weak inhibitor of superoxide anion generation. 3. Occupancy of adenosine A_1 receptors promotes neutrophil chemotaxis by a pertussis toxin sensitive mechanism, most likely via a G protein while occupancy of adenosine A_2 receptors inhibits superoxide anion generation by a pertussis toxin insensitive mechanism. 4. Intact microtubules are required for adenosine receptor occupancy to promote chemotaxis since both vinblastine and colchicine completely reverse the effect of 5'N-ethylcarboxamidoadenosine on chemotaxis. In contrast, intact microtubules are **not** required for adenosine receptor occupancy to inhibit superoxide anion generation. 5. Adenosine receptor occupancy promotes association of the chemoattractant FMLP with the cytoskeleton by a microtubule independent mechanism linked to inhibition of superoxide anion generation.

In conclusion, occupancy of adenosine A_1 receptors promotes chemotaxis by a mechanism which requires intact microtubules and G proteins. In contrast, our results suggest that occupancy of A_2 receptors inhibits superoxide anion generation by uncoupling chemoattractant receptors from the proteins required for stimulus transduction in a novel form of heterologous desensitization.

REFERENCES

Cronstein, B.N., Kramer, S.B., Weissmann, G.and Hirschhorn, R., 1983, Adenosine:a physiological modulator of superoxide anion generation by human neutrophils. *J.Exp.Med.* **158**: 1160-1177.

Cronstein, B.N., Rosenstein, E.D., Kramer, S.B., Weissmann, G.and Hirschhorn,R., 1985, Adenosine; a physiologic modulator of superoxide anion generation by human neutrophils. Adenosine acts via an A2 receptor on human neutrophils. *J.Immunol.* **135**: 1366-1371.

Cronstein, B.N., Levin, R.I., Belanoff, J., Weissmann, G.and Hirschhorn, R., 1986, Adenosine: an endogenous inhibitor of neutrophil-mediated injury to endothelial cells. *J.Clin.Invest.* **78**: 760-770.

Cronstein, B.N., Kubersky, S.M., Weissmann, G.and Hirschhorn, R., 1987, Engagement of adenosine receptors inhibits hydrogen peroxide (H2O2-) release by activated human neutrophils. *Clin. Immunol.Immunopathol.* **42**: 76-85.

Goldstein, I.M., Roos, D., Kaplan, H.B.and Weissmann, G., 1975, Complement and immunoglobulins stimulate superoxide production by human leukocytes independently of phagocytosis. *J. Clin. Invest.* **56**:

1155-1163.

Grinstein, S.and Furuya, W., 1986, Cytoplasmic pH regulation in activated human neutrophils: effects of adenosine and pertussis toxin on Na+/H+ exchange and metabolic acidification. *Biochim Biophys Acta* **889**: 301-309.

Hutchison, A.J., Oei, H.H., Ghai, G.R.and Williams, M., 1989, CGS21680, an A2 selective adenosine (ADO) receptor agonist with preferential hypotensive activity. *Faseb J.* **3**: A281.

Iannone, M.A., Reynolds-Vaughn, R., Wolberg, G.and Zimmerman, T.P., 1985, Human neutrophils possess adenosine A2 receptors. *Fed. Proc.* **44**:580.

Jesaitis, A.J., Naemura, J.R., Sklar, L.A., Cochrane, C.G.and Painter, R.G., 1984, Rapid modulation of N-formyl chemotactic peptide receptors on the surface of human granulocytes: formation of high affinity ligand-receptor complexes in transient association with cytoskeleton. *J. Cell Biol.* **98**: 1378-1387.

Jesaitis, A.J., Tolley, J.O.and Allen, R.A., 1986, Receptor-cytoskeleton interaction and membrane traffic may regulate chemoattractant-induced superoxide production in human granulocytes. *J.Biol.Chem.* **261**: 13662-13669.

Jesaitis, A.J., Bokoch, G.M., Tolley, J.O.and Allen, R.A., 1988, Lateral segregation of neutrophil chemotactic receptors into actin- and fodrin-rich plasma membrane microdomains depleted in guanyl nucleotide regulatory proteins. *J. Cell Biol.* **107**: 921-928.

Londos, C., Cooper, D.M.F.and Wolff, J., 1980, . *Proc.Natl.Acad.Sci.USA* **77**: 2551-2554.

Nelson, R.D., Quie, P.G.and Simmons, R.L., 1975, Chemotaxis under agarose: a new and simple method for measuring chemotaxis and spontaneous migration of human polymorphonuclear leukocytes and monocytes. *J.Immunol.* **115**: 1650-1656.

Roberts, P.A., Morgan, B.P.and Campbell, A.K., 1985, 2-Chloroadenosine inhibits complement-induced reactive oxygen metabolite production and recovery of human polymorphonuclear leukocytes attacked by complement. *Biochem. Biophys. Res. Comm.* **126**: 692-697.

Rose, F.R., Hirschhorn, R., Weissmann, G.and Cronstein, B.N., 1988, Adenosine promotes neutrophil chemotaxis. *J.Exp.Med.* **167**: 1186-1194.

Schmeichel, C.J.and Thomas, L.L., 1987, Methylxanthine bronchodilators potentiate multiple human neutrophil functions. *J.Immunol.* **138**: 1896-1903.

Schrier, D.J.and Imre, K.M., 1986, The effects of adenosine agonists on human neutrophil function. *J.Immunol.* **137**: 3284-3289.

Skubitz, K.M., Wickham, N.W.and Hammerschmidt, D.E., 1988, Endogenous and exogenous adenosine inhibit granulocyte aggregation without altering the associated rise in intracellular calcium concentration. *Blood* **72**: 29-33.

van Calker, D., Muller, M.and Hamprecht, B., 1979, Adenosine regulates, via two different types of receptors, the accumulation of cyclic AMP in cultured brain cells. *J. Neurochem.* **33**: 999-1005.

20
Beta Adrenergic Receptor Mediated Stimulation of Adenine Nucleotide Catabolism and Purine Release in Human Adipocytes

H. Kather

Introduction

Many hormones exert their biological effects via increases in cyclic AMP. Hydrolysis of cyclic AMP yields AMP which is the first substrate of irreversible catabolic reactions. Any increase in its rate of synthesis can, therefore, lead to its further metabolism by dephosphorylation or deamination (Arch et al. 1978, Bidger et al. 1983). Among the products of AMP-catabolism, adenosine has a variety of biological effects, whereas inosine and hypoxanthine are relatively inert in biological terms (Arch et al. 1978). In adipose tissue adenosine influences glucose transport and acts as an inhibitor of cyclic AMP accumulation and lipolysis, both basal- and hormone-activated (Schwabe et al. 1973, Londos et al. 1980, Kather et al. 1985).It has therefore been proposed that adenosine is produced from cyclic AMP via the sequence cyclic AMP $\rightarrow$ AMP $\rightarrow$ adenosine and is utilized as an endogenous regulator of fat cell function.

This report summarizes recent findings demonstrating that human adipocytes release inosine and hypoxanthine rather than adenosine in response to ß-adrenergic agonists. The hormone-induced increase in purine release results in an irreversible loss of adenine nucleotides which is largely determined by the velocity of cyclic AMP hydrolysis.

Materials and Methods

Fat cells were isolated from adipose tissue of surgical subjects and incubated in Krebs-Henseleit bicarbonate buffer, pH 7.4, containing 20 g/l human serum albumin and 5 mmol/l glucose, as described (Kather et al. 1985, 1987; Kather 1988). Cell number was determined by counting all cells in appropriately diluted aliquots of suspensions. Adenosine, inosine, and hypoxanthine/xanthine were measured by a chemiluminescent method based on the determination of hydrogen peroxide formed by sequential catabolism of purines to uric acid (Kather et al. 1987a). The determination of glycerol, free fatty acids, adenine nucleotides, and cyclic AMP have been described (Kather et al. 1984,1985a, Kather 1988, Spielmann et al. 1981).

Results

Human adipocytes contained 20-40 nmol/10^6cells adenine nucleotides (figure 1). In controls, the adenine nucleotides
were 74 + 6% ATP, 16 + 7% ADP, and 10 + 5% AMP. Cyclic AMP
levels were below 1%. At densities of 30,000-40,000 cells/ml,
the suspensions accumulated adenosine, inosine, and hypoxanthine, and their concentrations were 38 + 8, 120 + 10, and
31 + 7 nmol/l after 3 h of incubation.

In the presence of isoproterenol (1 µmmol/l), ATP became depleted to less than one half of control levels (figure 1a).
Concomitantly, cyclic AMP concentrations rose to 8 + 1.5
nmol/10^6cells (figure 1b). The initial rise in cyclic AMP was
followed by a gradual decline lasting more than two hours.
The concentrations of ADP and AMP were not substantially influenced by the β-adrenergic agonist. Therefore, the isoproterenol-induced drop in ATP was reflected by a corresponding
decrease in total adenine nucleotide contents of suspensions
(figure 2b). The isoproterenol-induced increase in ATP-turnover via adenylate cyclase was associated with a marked increase in inosine and hypoxanthine production, whereas the
accumulation of adenosine in the media was not influenced at
all by the β-adrenergic agonist.

Human adipocytes are fragile. In standard incubations, 8+2%
of total cellular lactate dehydrogenase activity was recovered in the media at the start of incubations. A further
8+2% was released per hour of incubation. The cells possess
ectophosphatases that are capable of sequentially degrading
ATP to adenosine (Kather 1988). Considering the short-lived

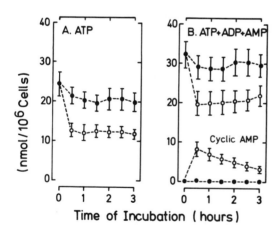

Time of Incubation (hours)

Figure 1: Influence of a maximal concentration of isoproterenol on cyclic AMP accumulation and adenine nucleotide contents of suspensions.

Values are means + SE of 4 separate experiments carried out
in the absence (closed symbols) or presence (open symbols) of
1 µmol/l isoproterenol. Incubations were carried out in the
absence of inhibitors of ectophosphatase activities.

integrity of the cells, it appeared important to discriminate between adenine nucleotide catabolism from broken cells, occurring extracellularly, and intracellular adenine nucleotide breakdown. Therefore, ectophosphatase activities were blocked by β-glycerophosphate, an inhibitor of nonspecific phosphatases, and antibodies against ecto-5'-nucleotidase or the highly selective inhibitor α,β-methylene adenosine 5'-diphosphate, respectively (Kather, 1988, 1989). Nonspecific losses of ATP resulted in a corresponding accumulation of ADP and AMP in the media under these conditions that was closely related to lactate dehydrogenase release. Concomitantly, adenosine concentrations dropped beyond detectable levels regardless of whether ATP-turnover via adenylate cyclase was increased by β-adrenergic catecholamines or not, indicating that the adenosine that inevitably accumulates in fat cell suspensions is derived from adenine nucleotides that are extruded by broken or damaged cells. Experiments with inhibitors of adenosine deaminase and adenosine kinase indicated that small amounts of adenosine ($\sim$1 nmol/10^6cells per hour) were also produced inside the cells (Kather, unpublished). However, adenosine production was unrelated to ATP-turnover via adenylate cyclase, and the nucleoside could not escape because any adenosine formed was reconverted to adenine nucleotides by adenosine kinase in the absence and presence of β-adrenergic agonists.

In contrast to adenosine, inosine and hypoxanthine were released by intact cells (figure 2). Their production was related to ATP-turnover via adenylate cyclase (Kather 1988), and proceeded exclusively by way of AMP-deamination (Kather, unpublished). The stimulatory effects of isoproterenol and

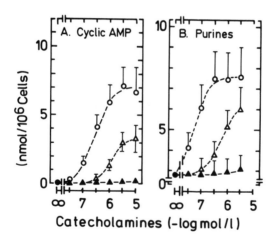

Figure 2: Effects of increasing concentrations of different catecholamines on cyclic AMP accumulation and purine release.

Symbols refer to the effects of isoproterenol (O), or epinephrine either alone (▲), or in combination with 1 μmol/l yohimbine (△), and are means + SE of 7 separate experiments for each condition. The media contained 10 mM β-glycerophosphate and 10 μM α,β-methylene adenosine 5'-diphosphate.

epinephrine (either alone, or in combination with the alpha$_2$-adrenergic blocking agent, yohimbine) on purine release resembled their cyclic AMP-elevating properties with respect to relative effacies and rank order of potency. Inosine and hypoxanthine cannot be re-utilized for adenine nucleotide synthesis (Kather 1989). Therefore, one third of total cellular adenine nucleotides was irreversibly lost in the presence of 1 µmol/l isoproterenol.

The catecholamine-induced activation of purine release could be blocked by phosphodiesterase inhbitors, suggesting that cyclic AMP is the main precursor of purines in the presence of β-adrenergic agonists (Kather 1989). However, the catecholamines were 10-times more potent in stimulating purine release than in elevating cyclic AMP. In addition, purine release ceased - within 20-30 min - when cyclic AMP was still markedly increased.

Discussion

The findings summarized in this report demonstrate that human white adipocytes have an enormous potential for dissipating energy via the adenylate cyclase-phosphodiesterase couple. In addition, the current results are the first to document that the pathway involving the formation and hydrolysis of cyclic AMP not only assumes a major role in ATP-utilization, but also constitutes the principle route of irreversible adenine nucleotide catabolism in the presence of β-adrenergic agonists. Unexpectedly, β-adrenergic catecholamines caused a selective increase in AMP-deamination to yield inosine and hypoxanthine, whereas the accumulation of adenosine reflected adenine nucleotide catabolism from broken cells. Inosine and hypoxanthine cannot be re-utilized for adenine nucleotide synthesis by human adipocytes. A prolonged β-adrenergic stimulation, therefore, leads to an irreversible loss of cellular adenine nucleotides that can be prevented by phosphodiesterase inhibitors, indicating that the rate of AMP-deamination is largely determined by the velocity of cyclic AMP hydrolysis in the presence of β-adrenergic agonists

Phosphodiesterase exists in multiple forms which vary widely in substrate affinities and mode of regulation (Beavo 1988, Manganiello et al. 1988). Among the at least three isoforms that are present in fat cells, a low k_m cyclic AMP phosphodiesterase has been identified which is activated by catecholamines. The issue whether or not the increase in enzyme activity observed in broken-cell preparations relates to the intact cell remains to be clarified. It is, therefore, difficult to decide whether the catecholamine-induced acceleration of AMP catabolism is solely due to increased availability of its precursor cyclic AMP, or reflects regulatory influences at the level of hormone-sensitive cyclic AMP phosphodiesterase. The striking parallelism between the cyclic AMP-elevating properties of epinephrine and isoproterenol and their stimulatory effects on purine release suggests that increased substrate availabilty is one of the factors that are causally involved. However, both catecholamines were more potent in stimulating purine release than in elevating cyclic AMP, suggesting a compartmentation of the cyclic nucleotide and/or involvement of the hormone-sensitive phosphodiesterase. The

observation that purine release ceased as soon as cyclic AMP
had attained relatively levels can be interpreted as reflecting a homologous type of desensitization. Intriguingly, partially purified preparations of hormone-sensitive fat cell
phosphodiesterase display the same type of response during
sustained β-adrenergic stimulation (Beavo 1988, Manganiello
et al. 1988). The question as to whether and to what extent
the stimulatory effects of catecholamines on purine release
reflect their effects on hormone-sensitive phosphodiesterase,
therefore, deserves further investigation.

Acknowledgements

The studies summarized in this report were supported by
grants of the Deutsche Forschungsgemeinschaft and of the Bun-
desministerium für Forschung und Technologie (0704721 AZ),
Bonn, W.-Germany. The author is indepted to B. Löser, E. Mess-
mer, B. Nessel, and B. Sattel for expert technical assistance.
Cyclic AMP determinations were kindly performed by V. Herr-
mann.

References

Arch, J.R.S., and Newsholme, E.A. 1987, The control of the
    metabolism and the hormonal role of adenosine. Essays Bio-
    chem. 14: 82-123.
Beavo, J.A., 1988, Multiple isozymes of cyclic nucleotide
    phosphodiesterase. Adv. Cycl. Nucleotide. Protein Phosphor-
    ylation Res. 16: 1-11
Bidger, W.A., and Henderson, 1983, Cell ATP. Wiley, New York,
    170 pp.
Kather, H., and Wieland, E., 1984, Glycerol, luminometric
    method. In: Methods of Enzymatic Analysis. Vol VI, Berg-
    meyer, H.U., ed., Verlag Chemie, Weinheim. 510-518.
Kather, H., Bieger, W., Michel, G., Aktories, K., and Jakobs,
    K.H., 1985, Human fat cell lipolysis is primarily regula-
    ted by inhibitory regulators acting through distinct mecha-
    nisms. J. Clin. Invest. 76: 1559-1565.
Kather, H., and Wieland, E., 1985a, Free fatty acids, lumino-
    metric method. In: Methods of Enzymatic Analysis. Vol VIII,
    Bergmeyer, H.U., ed., Verlag Chemie, Weinheim, 25-34.
Kather, H., Wieland, E., Scheurer, A., Vogel, G., Wildenberg,
    U., and Joost, C., 1987, Influences of variation in total
    energy intake and dietary composition on regulation of fat
    cell lipolysis in ideal weight subjects. J. Clin. Invest.
    80: 566-572.
Kather, H., Wieland, E., and Waas, W., 1987a, Chemilumines-
    cent determination of adenosine, inosine and hypoxanthine/
    xanthine. Anal. Biochem. 163: 45-51.
Kather, H., 1988, Purine accumulation in human fat cell sus-
    pensions: evidence that human adipocytes release inosine
    and hypoxanthine rather than adenosine. J. Biol. Chem. 263:
    8803-8809.
Kather, H., 1989, Beta-adrenergic receptor-mediated stimu-
    lation of adenine nucleotide catabolism and purine release
    in human adipocytes. J. Clin. Invest. in press.

Londos, C., Cooper, D.M.F., and Wolff, J., 1980, Subclasses of external adenosine receptors. Proc. Natl. Acad. Sci. USA. 77: 2551-2554.

Manganiello, V., Degerman, E., and Elks, M., 1988, Selective inhibitors of specific phosphodiesterases in intact adipocytes. Methods in Enzymology 188: 504-521.

Schwabe, U., Ebert, R., and Erbler, H.G., 1973, Adenosine release from isolated fat cells and its significance for the effects of hormones on cyclic 3,5'-AMP levels and lipolysis. Naunyn Schmiedeberg's Arch. Pharmacol. 276: 133-148.

Spielmann, H., Jacob-Müller, U., and Schultz, P., 1981, Simple assay of o.1 - 1 pmol ATP, ADP, and AMP in single somatic cells using purified luciferin luciferase. Anal. Biochem. 193: 172-178.

21
Structure Activity Relationships for Adenosine Antagonists

R.F. Bruns

Introduction

An important milestone in adenosine research was the discovery by Sattin and Rall in 1970 that theophylline and other methylxanthines could block responses to adenosine. With the availability of competitive adenosine antagonists, it was possible for the first time to investigate the role of endogenous adenosine by blocking adenosine receptors under various physiological and pathological conditions. However, although theophylline was an effective adenosine antagonist *in vivo*, its use as a tool was hampered by its relatively low affinity for adenosine receptors and by effects that were unrelated to adenosine, such as phosphodiesterase inhibition and release of intracellular calcium (Rall, 1982).

These problems stimulated the exploration of structure-activity relationships for adenosine antagonists. Important objectives of these studies have been greater receptor affinity, improved ability to differentiate between A_1 and A_2 adenosine receptor subtypes, and greater activity *in vivo*. Both xanthine and non-xanthine antagonists have been reported, and several nucleosides with adenosine antagonist activity have also been described. The present article reviews the structure-activity relationships for adenosine antagonists, with an emphasis on the issues involved in designing compounds that are useful as pharmacological tools. Other topics discussed include the general structural requirements for adenosine antagonism and the similarities between the receptor domains involved in the binding of agonists and antagonists.

Xanthines

Theophylline (figure 1) is a relatively weak adenosine antagonist with an A_1 affinity of 8.5 μM (table 1). The great advantage of theophylline is its good activity *in vivo*, which in turn is probably related to its excellent water solubility of 46 mM. The ratio of solubility to A_1 affinity is 5,400. This ratio, which seems to be an important prognosticator of *in vivo* activity (Bruns and Fergus, 1989a), determines the maximum rightward shift that a blocker can achieve (*i.e.*, as a saturated solution).

The first adenosine antagonist with substantially improved affinity compared to theophylline was 8-phenyltheophylline (8-PT) (Bruns, 1981). Unfortunately, although the A_1 affinity of this compound was 100 times better than that of theophylline, its water solubility was 5,000 times worse (table 1). Because of its poor solubility/affinity ratio of only 63, 8-PT shows only marginal activity *in vivo* (Lautt and Legare, 1985). 8-PT has been found to precipitate on intraperitoneal injection, resulting in undetectable blood levels (Wormald et al., 1989). 8-(2-Amino-4-chlorophenyl)-1,3-dipropylxanthine (PACPX) carries this trend even further: a 34-fold improvement in A_1 affinity compared to 8-PT is more than offset by a 38-fold decrease in solubility (table 1). However, there is a small silver lining to this otherwise negative outcome: PACPX also shows 37-fold selectivity for the A_1 receptor over the A_2, providing the first indication that A_1 and A_2 blocking activity can be separated.

Subsequent efforts have focused on breaking the futile trend towards ever greater receptor affinity coupled with ever poorer solubility. Two main strategies have emerged. In the "additive" strategy, the unsatisfactory solubility of the 1,3-dipropyl-8-phenylxanthine core is ameliorated by appending a hydrophilic side-chain, generally one with a positive charge. Typifying this approach is xanthine amine congener (XAC) (figure 1) (Jacobson, et al., 1985). While retaining the high A_1 affinity of PACPX, XAC possesses a much more satisfactory solubility of 90 μM (table 1). XAC's solubility/affinity ratio of 105,000 is actually higher than that of theophylline. Similar compounds with even greater solubility/affinity ratios (see Bruns and Fergus, 1989a) include the D-Lys derivative of XAC (Jacobson et al., 1986) and PD 113,297 (Hamilton et al., 1985). These compounds usually show good activity by the intravenous route, but may have poor oral absorption and poor brain penetration because of their positive charges.

The "subtractive" or "minimalist" approach towards improving the solubility/affinity ratio essentially involves making maximum use of the binding potential of hydrophobic groups, thereby minimizing the amount of solubility that must be sacrificed. Groups that reduce solubility more than they increase receptor affinity (*e.g.*, 8-phenyl) are omitted or replaced with groups that trade off solubility for affinity more efficiently. This approach is illustrated by 8-cyclopentyltheophylline (CPT) (not shown) and 8-cyclopentyl-1,3-dipropylxanthine (CPX) (figure 1). [It should be noted that CPT was originally synthesized almost twenty years ago as a potential phosphodiesterase inhibitor (Goodsell et al., 1971). Since CPT was not specifically made as an adenosine antagonist, the approach discussed here is really a *post hoc* explanation of the favorable activity of CPT and its homolog CPX; however, this strategy should also be applicable to the design of new antagonists.] Because the cyclopentyl group of CPX is not nearly as damaging to solubility as the 2-amino-4-chlorophenyl group of PACPX, CPX retains moderate water solubility (17 μM, table 1), while the close fit of the cyclopentyl group to the A_1 receptor results in an extremely good A_1 affinity of 0.46 nM and a solubility/affinity ratio of 37,000. CPT and CPX both block behavioral effects of adenosine A_1 agonists, and CPT is orally active (CPX has not yet been tested orally) (Bruns et al., 1988). The ability of these two compounds to cross the intestinal and blood-brain barriers is probably due to their neutral charge and moderate degree of hydrophobicity.

CPX also shows very high (740-fold) selectivity for the A_1 receptor over the A_2, and is by far the most A_1-selective antagonist reported to date. Highly A_2-selective xanthines have not yet been reported, although 1,3-dimethyl-7-propylxanthine and 3,7-dimethyl-1-propargylxanthine have been reported to possess a moderate degree of A_2-selectivity (Ukena et al., 1986). Surprisingly, 8-cyclohexylcaffeine (CHC) has been reported to show up to 140-fold selectivity for the A_2 receptor, although the degree of selectivity is highly dependent on the particular A_1 and A_2 assays utilized (Shamim et al., 1989).

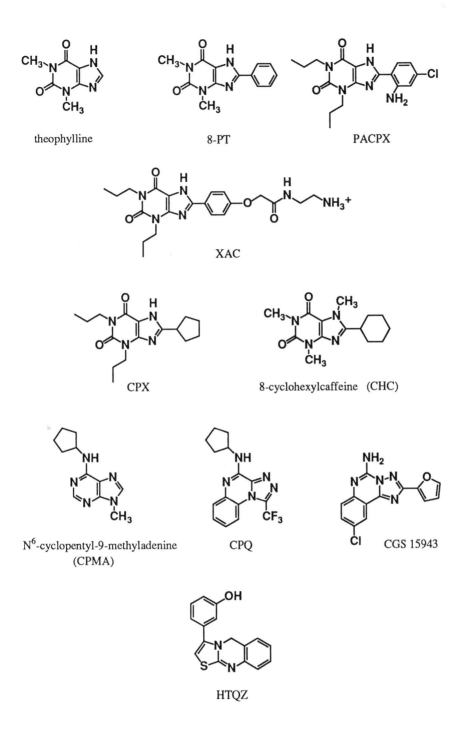

theophylline

8-PT

PACPX

XAC

CPX

8-cyclohexylcaffeine (CHC)

N⁶-cyclopentyl-9-methyladenine
(CPMA)

CPQ

CGS 15943

HTQZ

Figure 1. Structures of adenosine antagonists.

Table 1. Affinities and solubilities of adenosine antagonists. Structures of antagonists are given in figure 1. A_1 affinities are K_i values for inhibition of [³H]CHA binding in rat brain membranes, and A_2 affinities are K_i values for inhibition of [³H]NECA binding to rat striatal membranes in the presence of 50 nM unlabeled N^6-cyclopentyladenosine (Bruns et al., 1986). For CHC and CPMA, A_1 affinities are from [³H]R-PIA binding and A_2 affinities are from NECA-stimulated human platelet adenylate cyclase (Ukena et al., 1987; Shamim et al., 1989). Solubilities of adenosine antagonists were determined by dissolving the compound in DMSO, diluting 100-fold into buffer to a final concentration above the solubility limit, incubating overnight, centrifuging, and measuring the amount of antagonist remaining in the supernatant by radioreceptor assay using [³H]CHA (Bruns and Fergus, 1989a).

compound	K_i or solubility (µM)			ratio		
	A_1	A_2	solub	A_2/A_1	sol/A_1	sol/A_2
theophylline	8.5	25	46,000	2.9	5,400	1,850

compound	K_i or solubility (nM)			ratio		
	A_1	A_2	solub	A_2/A_1	sol/A_1	sol/A_2
8-PT	86	850	5,380	10	63	6.3
PACPX	2.5	92	141	37	56	1.53
XAC	0.86	27	90,000	31	105,000	3,300
CPX	0.46	340	17,100	740	37,000	50
CHC	28,000	190		0.0068		
CPMA	540	4,900		9.1		
CPQ	7.3	1,000	4,000	137	550	4.0
CGS 15943	3.9	1.51	1,740	0.39	450	1,150
HTQZ	4,000	124	1,960	0.031	0.65	15.8

Comparison of the Binding Modes of Xanthines and Adenosine Analogs. Although xanthines and adenosine are both purines, structure-activity relationships at corresponding positions of these two series are markedly different. For instance, 8-position substitution increases xanthine affinity but destroys the affinity of adenosine. These results indicate that the two types of molecules do not line up in the same orientation in relation to the receptor. Some intriguing hints about these different binding modes arise from a comparison of structure-activity relationships for cycloalkyl substitution at the 8-position of xanthine and the N^6-position of adenosine. In both cases, cyclopentyl substitution provides the most potent compound and cyclopropyl the weakest, with cyclobutyl and cyclohexyl showing intermediate affinity (Moos et al., 1985; Martinson et al., 1987; unpublished observations). Another point of similarity is the improvement of A_1 affinity by distal amino groups (for instance, in XAC and adenosine amine congener) in both series. These parallels suggest that the 8-position substituent of xanthines may extend into the N^6-domain of the A_1 receptor.

Nonxanthine Adenosine Antagonists

9-Methyladenine Derivatives. 9-Methyladenine was originally shown to be a weak antagonist at A_2 receptors (Bruns, 1981). More recently, Ukena et al. (1987) revisited this series with the aim of determining whether structure-activity relationships at the N^6 position of 9-methyladenine agreed with those at the N^6 position of adenosine. Interesting parallels were seen: N^6-cyclopentyl-9-methyladenine (CPMA) was the most potent and A_1-selective N^6-cycloalkyl derivative (K_i 540 nM, table 1), in agreement with the pattern seen for adenosine. The N^6-[(R)-1-methyl-2-phenylethyl] derivative was 4-fold more potent than the (S) derivative, indicating that the N^6-position stereoselectivity of adenosine also carries over to the 9-methyladenine series, although to a lesser degree. In general, the structure-activity relationships for the 9-methyladenine series mirrored the pattern for adenosine, albeit in a somewhat distorted manner.

[1,2,4]Triazolo[4,3-a]quinoxalines. This series was identified from a Pfizer antidepressant patent (Sarges, 1985) based on its similarity to adenine. The N^4-cyclopentyl derivative CPQ (table 1) was synthesized in the expectation that the structure-activity relationships for adenosine would also apply to this series (Trivedi and Bruns, 1988). CPQ showed an A_1 affinity of 7.3 nM and 137-fold A_1-selectivity.

[1,2,4]Triazolo[1,5-c]quinazolines. The ring structure for this series is the same as for CPQ except that the atoms of the 5-membered ring are shifted one unit counterclockwise (as drawn in figure 1). Originally synthesized as potential benzodiazepine antagonists, these compounds showed unexpected potency as adenosine antagonists (Williams et al., 1987). CGS 15943 is the most potent A_2 antagonist yet reported, with a K_i of 1.51 nM (table 1). The 2-fold A_2-selectivity of CGS 15943 is in striking contrast to the 137-fold A_1-selectivity of CPQ. The high A_2 affinity of this compound combined with a rather poor solubility of 1.74 μM results in a solubility/affinity ratio of 1,150. CGS 15943 blocks A_2 responses *in vivo* and shows some activity by the oral route (Ghai et al., 1987).

Thiazolo[2,3-b]quinazolines. HTQZ (figure 1) was found in a broad screening program (Bruns and Coughenour, 1987). The major point of interest about this compound is its high A_2-selectivity (25-fold). However, its poor solubility results in a solubility/affinity ratio of only 15 (table 1). Little is know about structure-activity relationships of this series other than the fact that the *meta* hydroxyl group seems essential for high A_2 affinity (unpublished results).

Other nonxanthine adenosine antagonists. Several other structural classes of nonxanthine adenosine antagonists have been reported, including pyrazolo[3,4-b]pyridines, pyrazolo[3,4-d]pyrimidines, pyrazolo[4,3-d]pyrimidines, quinazolines, and benzo[g]pteridines (see Trivedi et al., 1989 for a review).

Recently, several 2-amino-3-benzoylthiophenes were reported to block adenosine A_1 receptors (Bruns and Fergus, 1989b). However, the main interest in these compounds resides in their ability to increase the binding and functional activity of adenosine A_1 agonists by an apparent allosteric mechanism, a property that was fortuitously discovered during testing of this series for antagonist affinity. The allosteric enhancing activity of this series (including a summary of structure-activity relationships) is described in Bruns and Fergus (1989b).

Nucleoside Adenosine Antagonists

Attempts to design adenosine antagonists by removing the intrinsic activity of adenosine have been surprisingly few. However, several 5'-modified adenosine analogs, including 5'-methylthioadenosine (K_i 8.2 μM), were shown to block an A_2 response in human fibroblasts (Bruns, 1980). 5'-Methylthioadenosine was later found to block A_2-mediated stimulation of adenylate cyclase in neuroblastoma cells, but to act as a full agonist for inhibition of adenylate cyclase in rat cerebellar membranes (an A_1 response) (Munshi et al., 1988).

N^6-cyclohexyl-2',3'-dideoxyadenosine was recently shown to possess weak A_1 antagonist activity, with a K_i of 4.8 μM (Lohse et al., 1988).

General Structural Requirements for Adenosine Antagonist Activity

What are the common structural features of the diverse xanthine and nonxanthine adenosine antagonists described above? All of the above molecules possess flat, nitrogen-containing, fused heterocyclic rings. In all cases the ring and the groups immediately adjacent to the ring are uncharged. Once these basic structural requirements are met, there seems to be rather wide latitude for structural variations, lending hope that, with sufficient ingenuity, new adenosine antagonists with greater specificity and better *in vivo* activity can be discovered.

References

Bruns, R.F., 1980, Adenosine receptor activation in human fibroblasts: Nucleoside agonists and antagonists, *Canad. J. Physiol. Pharmacol.* **58**:673-691.

Bruns, R.F., 1981, Adenosine antagonism by purines, pteridines, and benzopteridines in human fibroblasts, *Biochem. Pharmacol.* **30**:325-333.

Bruns, R.F., and Coughenour, L.L., 1987, New non-xanthine adenosine antagonists, *Pharmacologist* **29**:146.

Bruns, R.F., and Fergus, J.H., 1989a, Solubilities of adenosine antagonists determined by radioreceptor assay, *J. Pharm. Pharmacol.* **41**:590-594.

Bruns, R.F., and Fergus, J.H., 1989b, Allosteric enhancers of adenosine A_1 receptor binding and function, In: *Adenosine Receptors in the Nervous System*, ed. Ribeiro, J.A., Taylor and Francis, London, pp 53-60.

Bruns, R.F., and Pugsley, T.A., 1986, Characterization of the A_2 adenosine receptor labeled by [^{3}H]NECA in rat striatal membranes, *Mol. Pharmacol.* **29**:331-346.

Bruns, R.F., et al., 1988, Adenosine antagonists as pharmacologic tools, In: *Adenosine and Adenine Nucleotides: Physiology and Pharmacology*, ed. Paton, D.M., Taylor and Francis, London, pp 39-49.

Ghai, G., et al., 1987, Pharmacological characterization of CGS 15943A: A novel nonxanthine adenosine antagonist, *J. Pharmacol. Exp. Ther.* **242**:784-790.

Goodsell, E.B., Stein, H.H., and Wenzke, K.J., 1971, 8-Substituted theophyllines. *In vivo* inhibition of 3',5'-cyclic adenosine monophosphate phosphodiesterase and pharmacological spectrum in mice, *J. Med. Chem.* **14**: 1202-1205.

Hamilton, H.W., et al., 1985, Synthesis of xanthines as adenosine antagonists, a practical quantitative structure-activity relationship application, *J. Med. Chem.* **28**:1071-1079.

Jacobson, K.A., Kirk, K.L., and Daly, J.W., 1985, Functionalized congeners of adenosine: Preparation of analogues with high affinity for A_1-adenosine receptors, *J. Med. Chem.* **28**:1341-1346.

Jacobson, K.A., Kirk, K.L., and Daly, J.W., 1986, A functionalized congener approach to adenosine receptor antagonists: Amino acid conjugates of 1,3-dipropylxanthine, *Mol. Pharmacol.* **29**:126-133.

Lautt, W.W., and Legare, D.J., 1985, The use of 8-phenyltheophylline as a competitive antagonist of adenosine and an inhibitor of the intrinsic regulatory mechanism of the hepatic artery, *Canad. J. Physiol. Pharmacol.* **63**:717-722.

Lohse, M.J., et al., 1988, 2',3'-Dideoxy-N^6-cyclohexyladenosine: an adenosine derivative with antagonistic properties at adenosine receptors, *Eur. J. Pharmacol.* **156**:157-160.

Martinson, E.A., Johnson, R.A., and Wells, J.N., 1987, Potent adenosine receptor antagonists that are selective for the A_1 receptor subtype, *Mol. Pharmacol.* **31**:247-252.

Moos, W.H., Szotek, D.S., and Bruns, R.F., 1985, N^6-Cycloalkyladenosines: Potent, A_1-selective adenosine agonists, *J. Med. Chem.* **28**:1383-1384.

Munshi, R., Clanachan, A.S., and Baer, H.P., 1988, 5'-Deoxy-5'-methylthioadenosine: a nucleoside which differentiates between adenosine receptor types, *Biochem. Pharmacol.* **37**:2085-2089.

Rall, T.W., 1982, Evolution of the mechanism of action of methylxanthines: From calcium mobilizers to antagonists of adenosine receptors, *Pharmacologist* **24**:277-287.

Sarges, R., 1985, Method of using [1,2,4]triazolo[4,3-*a*]quinoxaline-4-amine derivatives as antidepressant and antifatigue agents, US Patent 4,547,501.

Sattin, A., and Rall, T.W., 1970, The effect of adenosine and adenine nucleotides on the cyclic adenosine 3',5'-phosphate content of guinea pig cerebral cortex slices, *Mol. Pharmacol.* **6**:13-23.

Shamim, M.T., Ukena, D., and Daly, J.W., 1989, Effects of 8-phenyl and 8-cycloalkyl substituents on the activity of mono-, di-, and trisubstituted alkylxanthines with substituents at the 1-, 3-, and 7-positions, *J. Med. Chem.* **32**:1231-1237.

Trivedi, B.K., and Bruns, R.F., 1988, [1,2,4]Triazolo[4,3-*a*]quinoxaline-4-amines: A new class of A_1 receptor selective adenosine antagonists, *J. Med. Chem.* **31**:1011-1014.

Trivedi, B.K., Bridges, A.J., and Bruns, R.F., 1989, Structure-activity relationships of adenosine A_1 and A_2 receptors. In: *The Adenosine Receptor*, ed. Williams, M., Humana Press, Clifton, New Jersey, USA, (in press).

Ukena, D., Shamim, M.T., and Daly, J.W., 1986, Analogs of caffeine: Antagonists with selectivity for A_2 adenosine receptors, *Life Sci.* **39**:743-750.

Ukena, D., et al., 1987, N^6-substituted 9-methyladenines: A new class of adenosine receptor antagonists, *FEBS Lett.* **215**:203-208.

Williams, M., et al., 1987, Biochemical characterization of the triazoloquinazoline, CGS 15943, a novel, non-xanthine adenosine antagonist, *J. Pharmacol. Exp. Ther.* **241**:415-420.

Wormald, A., Bowmer, C.J., and Collis, M.G., 1989, Pharmacokinetics of 8-phenyltheophylline in the rat, *J. Pharm. Pharmacol.* **41**:418-420.

22
Structure Activity Relationships for Adenosine Agonists

B.K. Trivedi

Introduction

Adenosine elicits a wide variety of physiological responses (Williams, 1987) via interactions with two major subtypes of extracellular adenosine receptors, designated as A_1 and A_2. These two receptor subtypes can be distinguished on the basis of structure-activity relationships (Hamprecht et al., 1985) and have opposite effects on adenylate cyclase (Londos et al., 1980). With the advent of an A_2 binding assay (Bruns et al., 1986, Yeung et al., 1984), it has become possible to quantitatively evaluate the subtype selectivity of adenosine agonists. Recently, considerable effort has been devoted to the search for adenosine agonists with high selectivity for either A_1 or A_2 receptors. Our interest in adenosine receptors began with earlier disclosures that adenosine agonists such as R-PIA (Snyder et al., 1981), NECA (Prasad et al., 1980) or CV-1808 (Kawazoe et al., 1980) had potential utility as therapeutic agents (Figure 1). Thus, we set out to develop structure-activity relationships (SAR) for both A_1 and A_2 receptors and especially to identify receptor selective agonists.

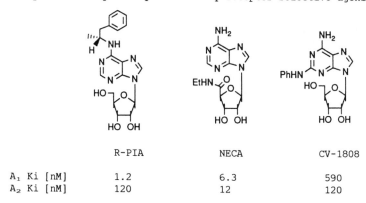

	R-PIA	NECA	CV-1808
A_1 Ki [nM]	1.2	6.3	590
A_2 Ki [nM]	120	12	120

Figure 1. Receptor Binding Affinities for Reference Adenosine Agonists

Structure-Activity Relationships

A series of N^6-alkyladenosines (Daly et al., 1986) and N^6-cycloalkyladenosines (Moos et al., 1985) have been reported to be A_1 receptor selective agonists. Among these, N^6-cyclopentyladenosine (CPA) showed high affinity (0.59 nM) and 780 fold selectivity for the A_1

136

receptor. We further explored the SAR by preparing a series of
N^6-bicycloalkyladenosines (Table 1). Interestingly, the epimeric endo-
and exo-norbornyladenosines showed differences in potency and
selectivity for the A_1 receptor. Furthermore, one of the diastereomers
of endo-2-norbornyladenosine was more potent and selective for the A_1
receptor than the other suggesting a high degree of stereospecific
interaction at the receptors (Trivedi et al., 1989). This selectivity
was significantly enhanced by converting the 5'-hydroxyl group to a
5'-chloro function. Thus, the 5'-chloro derivative of
$(2S)-N^6-(2\text{-endo-norbornyl})$adenosine is the most potent and selective
agonist for the A_1 receptor reported to date.

Table 1. Affinities of N^6-Bicycloalkyladenosines in A_1 and A_2 Receptor
Binding

Example	R	R_1	A_1	A_2	A_2/A_1
			Ki(nM)		
1	cyclopentyl	OH	0.59	462	780
2	2-endo-norbornyl	OH	0.42	750	1790
3	2-exo-norbornyl	OH	0.91	970	1070
4	1R,2S,4S isomer of 2	OH	0.30	1390	4700
5	1S,2R,4R isomer of 2	OH	1.65	610	370
6	cyclopentyl	Cl	0.72	1870	2600
7	2-endo-norbornyl	Cl	0.42	2100	4900
8	1R,2S,4S isomer of 7	Cl	0.24	3900	16000
9	1S,2R,4R isomer of 7	Cl	1.86	3000	1600

Based on this and other SAR studies (Daly et al., 1986), it was apparent
that N^6-alkyl and cycloalkyladenosines were, in general more potent and
selective for the A_1 receptor. Thus, in order to identify a series of
agonists with high affinity for the A_2 receptor, we turned our attention
towards the N^6-arylalkyladenosines. Initial study (Bridges et al.,
1987) of simple arylalkyladenosines showed that these derivatives,
although exhibiting moderate affinity for the A_2 receptor, were still
selective for the A_1 receptor. Furthermore, substitution in the alpha
position of the N^6 side chain as in R-PIA, enhanced A_1 receptor
selectivity. Effects of substitution in the beta position had not been
thoroughly examined. Thus, we prepared a series of
N^6-diarylalkyladenosines (Table 2) and evaluated them in the receptor
binding assays. Among this series of compounds,
$N^6-(2,2\text{-diphenylethyl})$adenosine was found to have significant affinity
for the A_2 receptor (Ki=25 nM). The lower and the higher homologs had
only moderate affinity for the A_2 receptor. From this short series of
compounds, it was evident that certain hydrophobic interactions at an
optimal distance from the N^6 nitrogen were required for greater affinity

Table 2. Affinities of N^6-Diarylalkyladenosines in A_1 and A_2 Receptor Binding

R	A_1	A_2	A_2/A_1
		Ki(nM)	
$CH(Ph)_2$	480	3800	7.9
$CH_2CH(Ph)_2$	6.8	25	3.7
$(CH_2)_2CH(Ph)_2$	79	290	3.7
$(CH_2)_3CH(Ph)_2$	290	4700	16.2
$(CH_2)_4CH(Ph)_2$	146	2200	15.0
$CH_2C(Ph)_3$	14300	26000	1.8

at the A_2 receptor. In order to better define the structural requirements for high A_2 affinity, we prepared several rigid analogs of N^6-(2,2-diphenylethyl)adenosine (Table 3) (Trivedi et al., 1988). Interestingly, N^6-(9-fluorenylmethyl)adenosine exhibited high potencies for both A_1 and A_2 receptors, with Ki values of 5.1 and 4.9 nM respectively. Once again, the lower and higher homologs had significantly less affinity for the A_2 receptor, further confirming the specific spatial arrangement required of the phenethyl moiety for high affinity at the A_2 receptor. A comparison between the highly active fluorenyl analog and the inactive anthracenyl analog provides some information about the geometry of the interaction between the adenosine receptor and the N^6 side chain. Both compounds possess a planar tricyclic aryl moiety. However, in the former this is at an angle of about 50° with a spacer carbon, whereas it is coplanar with the spacer carbon in the latter. Thus, the disposition of the tricyclic ring relative to N^6 must be important.

These SAR results along with earlier observation that N^6-(2,2-diphenylethyl)adenosine had an excellent profile in behavioral tests predictive of antipsychotic-like activity (Bridges et al., 1988), convinced us that a detailed examination of this series might be fruitful in a search for potent A_2 selective adenosine agonists with potential therapeutic utility. Thus, we focused our attention on the more synthetically amenable N^6-(2,2-diarylethyl)adenosines. A general synthetic route leading to the synthesis of this series of analogs has been described (Bridges, 1989). Monosubstitution in the ortho and meta positions improved the A_2 affinity while in the para position, substitution was detrimental. Substitutions in both rings, especially in the meta positions, afforded analogs which retained the binding affinity of the corresponding monosubstituted analogs (Bridges et al., 1988). However, substitutions in both the para positions significantly lowered the binding affinity not only for the A_2 receptor but also for the A_1 receptor. This is primarily due to a steric interaction with the receptor.

Table 3. Affinities of Rigid Analogs of N^6-Diarylalkyladenosines in A_1 and A_2 Receptor Binding

R	A_1	A_2	A_2/A_1
		Ki(nM)	
2,2-diphenylethyl	6.8	25	3.7
9-fluorenylmethyl	5.1	4.9	0.96
9-fluorenyl	6200	42000	6.7
9-fluorenylethyl	320	4300	13.0
1-naphthylmethyl	24	9.1	0.38
9-anthracenylmethyl	9000	29000	3.2
9-xanthenylmethyl	62	48	0.77

During the course of the SAR studies for the N^6-diarylalkyladenosines, we began developing a pharmacophore model for the A_2 receptor specifically for the N^6 region, using the SYBYL molecular modeling system. This model was further refined using the existing SAR data which suggested that, for high affinity at the A_2 receptor, there exists a primary binding site for an aryl function along with an accessory binding site located at a certain distance from the N^6 nitrogen into which these diaryl substituted analogs project (D. Ortwine et. al., 1990). In order to better understand and also validate this pharmacophore model, we prepared a series of analogs in which one of the phenyl rings was polysubstituted (Table 4). Interestingly, polysubstituted analogs, such as 2,6- or 3,5-disubstituted compounds began to show, for the first time, A_2 selectivity. Analogs substituted in the 3,5-positions had significant affinity for the A_2 receptor regardless of the functional groups examined. For example, the 3,5-bis(trifluoromethyl) analog had 10.9 nM affinity at the A_2 receptor and was about 13 fold selective. However, substitutions in the 2,6-positions had a deleterious effect on both A_1 and A_2 receptor affinities which was prominent with a bulkier substituent such as methoxy group in these positions.

Based on this SAR, and the modeling analysis, factors responsible for the deleterious effect on the A_1 receptor affinity were defined and several targets were proposed for synthesis. Evaluation of these analogs gave, for the first time, analogs with low nanomolar affinity and 10 to 30 fold selectivity for the A_2 receptor. As predicted by the modeling analysis, analogs containing 2- or 2,6-disubstitution in the phenyl ring had a deleterious effect on the A_1 affinity. The phenyl ring containing these substituents is postulated to bind at the primary

Table 4. Affinities of N^6-(2,2-diphenylethyl)adenosines in A_1 and A_2 Receptor Binding (Polysubstitution in One Ring).

R	A_1	A_2	A_2/A_1
		Ki(nM)	
2,5-di-Cl	7.8	26.3	3.4
2,6-di-Cl	61	19	0.31
3,5-di-Cl	18.4	6.4	0.34
2,5-di-CH$_3$	29	68	2.3
2,6-di-CH$_3$	229	62	0.27
3,5-di-CH$_3$	23	6.1	0.26
2,5-di-OCH$_3$	162	32	0.20
2,6-di-OCH$_3$	2170	1060	0.49
3,5-di-OCH$_3$	30	6.1	0.20
3,5-di-CF$_3$	147	10.9	0.074

binding site at the A_2 receptor. Furthermore, incorporation of substituents, such as methoxy groups, which enhance the binding affinity at the A_2 receptor, in the same molecule afforded analogs with both high affinity and selectivity for the A_2 receptor. The optimal substitution seems to be the one in which one of the phenyl rings is substituted by a single ortho substituent and the other ring is polysubstituted in the 3,5- or 3,4,5-positions by an electron donating substituent such as methoxy (Table 5). Thus, N^6-[2-(3,5-dimethoxyphenyl)-2-(2-methylphenyl)ethyl]adenosine is one of the most potent (Ki = 4.4nM) and A_2 receptor selective (30 fold) agonists reported to date.

Simultaneously, we were also interested in studying the effects of multiple substitutions on the adenosine molecule towards the binding affinity at the A_2 receptor. Adenosine derivatives such as NECA, a balanced agonist at both the receptors, and CV-1808, an A_2 receptor selective agonist had been reported. Thus, we became interested in the possibility that the functional groups responsible for conferring the selectivity on these and other N^6 substituted analogs might interact with independent sites on the adenosine receptor, thereby allowing additive enhancement of selectivity by combining the structural modifications at these positions. Initially we prepared and tested a series of 5'-modified analogs of N^6-[2-(3,5-dimethoxyphenyl)-2-(2-methylphenyl)ethyl]adenosine (Table 6) (Bridges et. al., 1988). The corresponding NECA derivative had similar affinity at both the receptors compared to the parent compound and thus did not show enhanced A_2 selectivity. The other 5'-amide and alkylamide analogs showed significantly lower A_1 and A_2 affinity and thus, were all as selective as the parent compound.

Table 5. Affinities of N^6-(2,2-diphenylethyl)adenosines in A_1 and A_2 Receptor Binding (Polysubstitutions in Both the Phenyl Rings).

R	R'	A_1	A_2	A_2/A_1
		Ki(nM)		
2-OCH$_3$	3'-OCH$_3$	28.8	5.2	0.18
2-OCH$_3$	3'-CF$_3$	91	10.9	0.12
2,6-di-CH$_3$	3',5'-di-OCH$_3$	140	19.5	0.14
2-OCH$_3$	3',5'-di-OCH$_3$	42	3.6	0.086
2-OCH$_3$	3',5'-di-CF$_3$	282	12.6	0.045
2-CH$_3$	3',4',5'-tri-OCH$_3$	159	5.5	0.035
2-CH$_3$	3',5'-di-OCH$_3$	142	4.4	0.031

Table 6. Affinities of 5'-Modified N^6-[2-(3,5-Dimethoxyphenyl)-2-(2-methylphenyl)ethyl]adenosines in A_1 and A_2 Receptor Binding.

R	A_1	A_2	A_2/A_1
	Ki(nM)		
CH$_2$OH	142	4.4	0.031
CONH$_2$	873	27	0.032
CONHCH$_3$	1010	43	0.042
CONHC$_2$H$_5$	207	5.6	0.027
CONH(c-C$_3$H$_5$)	232	8.7	0.038

141

We then focused our attention on the modifications in the C2 and
N^6-positions. We envisioned synthesizing a variety of doubly modified
analogs including the ones in which the anilino moiety would be
incorporated at the C2-position of various N^6 substituted agonists.
Initial evaluation of the literature revealed that one of the syntheses
described for 2-(phenylamino)adenosine (CV-1808) (Marumoto et. al.,
1981) was not amenable to the synthesis of a wide variety of
C2,N^6-disubstituted analogs. The alternate synthesis (Marumoto et. al.,
1975) involved starting with a fermentation product, AICA-riboside which
was not easily accessible. Thus we began investigating the chemistry of
guanosines with the aim of developing not only an economical synthesis
of CV-1808 but also a synthesis which would allow us to prepare a wide
range of targeted analogs. Details of our efforts in this regard have
recently been reported (Trivedi, 1988; Trivedi and Bruns, 1989). We
first evaluated a series of analogs in which various substituents were
incorporated into the C2-position of A_1 receptor selective agonists such
as CPA and R-PIA (Table 7). These substituents had deleterious effect
on both A_1 and A_2 receptor affinities. However, this effect was more
prominent at the A_1 receptor than at the A_2 receptor and thus, these
analogs were significantly less A_1 selective. This suggested that there
is a more tolerance at the 2-position domain of the A_2 receptor than at
the A_1 receptor. Furthermore, the sulfone derivatives had greater

Table 7. Affinities of C2, N^6-Disubstituted Adenosines in A_1 and A_2
Receptor Binding.

R	X	A_1	A_2	A_2/A_1
		Ki(nM)		
cyclopentyl	H	0.59	460	430
	NH_2	8.3	6100	730
	SPh	37	4000	107
	SO_2Ph	96	2300	24
(R)-1-methyl-2-	H	1.17	124	106
phenylethyl	NH_2	19.2	1530	80
	SPh	210	1000	4.7
2,2-diphenylethyl	H	6.8	25	3.6
	NH_2	61	135	2.2
	SPh	840	800	0.95
1-naphthylmethyl	H	24	9.4	0.38
	SPh	1470	610	0.42
	SO_2Ph	2000	270	0.133
	$NH(c-C_6H_{11})$	45000	10300	0.23
	$NHCH_2Ph$	530	350	0.66

affinity at the A_2 receptor and were more selective than the corresponding phenylthio analogs. The enhancement in the A_2 affinity was certainly greater with the N^6-(1-naphthylmethyl)adenosine analog than with the corresponding R-PIA analog. The improvement in selectivity for these sulfone derivatives is primarily due to the greater loss of affinity at the A_1 receptor. Interestingly, the 2-(cyclohexylamino) derivative of N^6-(1-naphthylmethyl)adenosine, although A_2 selective, had a very weak affinity at both the receptors, especially compared to the corresponding 2-phenylthio and 2-phenylamino analogs. The major difference between these analogs is that the saturated six-membered ring in the former occupies a larger space than the planar aromatic ring present in the latter two compounds. This suggests that a limited pocket favorable for aromatic hydrophobic interactions may exist near the C2-position of adenosine at the A_2 receptor.

Finally, our particular interest was to evaluate a series of CV-1808 analogs having a wide variety of N^6 substituents in the A_1 and A_2 receptor binding assays. We first studied the analogs in which, the N^6 side chains of highly A_2 selective agonists were incorporated (Table 8) (Trivedi and Bruns, 1989). Disappointingly, the anticipated additivity was not observed in these molecules. Most of these anilino derivatives lost significant affinity at both the receptors. The anilino derivative

Table 8. Effects of 2-(Phenylamino) Substitution on A_1 and A_2 Affinities of N^6-Modified Adenosines.

R	X	A_1 Ki(nM)	A_2	A_2/A_1
H	NHPh	600	116	0.19
1-naphthylmethyl	H	24	9.4	0.38
	NHPh	560	230	0.41
2,2-diphenylethyl	H	6.8	25	3.6
	NHPh	2700	650	0.24
9-fluorenylmethyl	H	5.2	4.9	0.94
	NHPh	8900	2100	0.24
2-(3,5-dimethoxyphenyl)-2-(phenyl)ethyl	H	30	6.1	0.20
	NHPh	9000	470	0.052
2-(3,5-dimethoxyphenyl)-2-(2-methylphenyl)ethyl	H	142	4.4	0.031
	NHPh	10300	340	0.034

of N^6-(9-fluorenylmethyl)adenosine showed the highest loss of affinity (400 fold) at the A_2 receptor. Following these results, we synthesized and evaluated a series of 2-(phenylamino) derivatives of A_1-selective N^6-modified adenosines in order to determine whether the lack of additivity seen with A_2-selective N^6 derivatives would also pertain to A_1-selective compounds. Most of these analogs showed increased A_2 affinity and, although A_1 selective, were much less so than the parent N^6 substituted compounds (Table 9). The corresponding R-PIA analog, however, did not show the increase in affinity suggesting that the lack of additivity seen with A_2 selective agonists is not due to the A_2 selectivity per se, but rather to structural features that the A_2 selective compounds share with R-PIA (i.e. phenethyl side chain) but not with other A_1 selective agonists. One of the possible explanations for the lack of additivity would involve direct steric interference

Table 9. Effects of 2-(Phenylamino) Substitution on A_1 and A_2 Affinities of A_1 Selective N^6-Modified Adenosines.

R	X	A_1	A_2	A_2/A_1
			Ki(nM)	
H	NHPh	600	116	0.19
cyclopropyl	H	3.2	1240	390
	NHPh	68	960	14.1
cyclopentyl	H	0.59	460	780
	NHPh	12.4	144	11.6
cyclohexyl	H	1.4	609	430
	NHPh	54	450	8.4
2-endo-norbornyl	H	0.42	750	1790
	NHPh	7.7	472	61
(S)-2-hydroxypropyl	H	5.0	9300	1870
	NHPh	50	1700	34
(R)-1-methyl-2-phenylethyl	H	1.17	124	106
	NHPh	152	240	1.56

between the C2 and N^6 aryl-binding pockets. Thus, distal aryl groups at C2 and N^6 would occupy their respective pockets when present alone, but could not both occupy the overlapping portion of the two pockets when present in the same molecule. Another possibility may be that, the N^6 side chain could induce an allosteric change in the receptor, resulting in closure of the C2 pocket. Alternatively, the presence of a substituent at C2 or N^6 might cause a shift in the position of adenosine on the receptor, thereby displacing the other side chain to an unfavorable location. Nonetheless, this study suggests that, the affinity of these analogs at both the receptors is quite sensitive to the modifications in the adenosine molecule.

Conclusion

In summary, like the N^6-cycloalkyladenosines, the N^6-bicycloalkyladenosines have high affinity and selectivity for the A_1 receptor. The N^6-diarylalkyladenosines have high affinity for the A_2 receptor which can be modulated by substitutions at specific positions in both the phenyl rings. These modifications, lead to very potent and highly A_2 receptor selective adenosine agonists. Finally, we have also demonstrated that, there is a hydrophobic binding site near the C2-position of adenosine that is specific for the aromatic function. Occupation of this site increases the A_2 affinity and selectivity when N^6 is occupied by an alkyl or cycloalkyl group, but not when N^6 is occupied by an arylalkyl side chain.

References

Bridges, A.J., **1989**, Nucleosides and Nucleotides 8(3):357.
Bridges, A. J., Moos, W. H., Szotek, D. L., Trivedi, B. K., Bristol, J. A., Heffner, T. G., Bruns, R. F., Downs, D. A., **1987**, J. Med. Chem. 30:1709.
Bridges, A. J., Bruns, R. F., Ortwine, D. F., Priebe, S. R., Szotek, D. L., Trivedi, B. K., **1988**, J. Med. Chem. 31:1282.
Bruns, R. F., Lu, G. H., Pugsley, T. A., **1986**, Mol. Pharmacol. 29:331.
Daly, J. W., Padgett, W., Thompson, R. D., Kusachi, S., Bugni, W. J., Olsson, R. A., **1986**, Biochem. Pharmacol. 35:2467.
Hamprecht, B., van Calker, D., **1985**, Trends Pharmacol. Sci. 6:153.
Kawazoe, K., Matsumoto, N., Tanabe, N., Fujiwara, S., Yanajimoto, M., Hirata, M., Kikuchi, K., **1980**, Arzneim.-Forsch. 30:1083.
Londos, C., Cooper, D. M. F., Wolff, J., **1980**, Proc. Natl. Acad. Sci. U.S.A. 77:2551.
Marumoto, R., Yoshikoto, Y., Miyashita, O., Shima, S., Imai, K., Kawazoe, K., Honjo, M., **1975**, Chem. Pharm. Bull. 23:759.
Marumoto, R., Omura, K., Furukawa, Y., **1981**, Chem. Pharm. Bull. 29:1870.
Moos, W. H., Szotek, D. L., Bruns, R. F., **1985**, J. Med. Chem. 28:1383.
Ortwine, D. F., Bridges, A. J., Humblet, C., Trivedi, B. K., **1990**, see chapter in this book.
Prasad, R. N., Bariana, D. S., Fung, A., Savic, M., Tietje, K., Stein, H. H., Brondyk, H., Egan, R. S., **1980**, J. Med. Chem. 23:313.
Snyder, S. H., Katims, J. J., Annau, Z., Bruns, R. F., Daly, J. W., **1981**, Proceedings of the National Academy of Sciences USA 78:3260.
Trivedi, B. K., Bristol, J. A., Bruns, R. F., Haleen, S. J., Steffen, R. P., **1988**, J. Med. Chem. 31:271.
Trivedi, B. K., **1988**, Nucleosides and Nucleotides 7:393.
Trivedi, B. K., Bridges, A. J., Patt, W. C., Priebe, S. R., Bruns, R. F., **1989**, J. Med. Chem. 32:8.
Trivedi, B. K., Bruns, R. F., **1989**, J. Med. Chem. 32:1667.
Williams, M., **1987**, Annu. Rev. Pharmacol. 27:315.
Yeung, S. M. H., Green, R. D., **1984**, Naunyn-Schmiedeberg's Arch. Pharmacol. 325:218.

23
The Design of a Series of Highly A$_2$ Selective Adenosine Agonists

A.J. Hutchison, H. Oei, M. Jarvis, M. Williams, and R.L. Webb

Introduction

The purine nucleoside, adenosine has been extensively studied as a modulator of cardiovascular function since it was shown to have potent hypotensive and bradycardic activity some 60 years ago (Drury et al., 1929). The hypotensive actions of adenosine occur via several mechanisms among which are direct regulation of blood flow via vasodilation of the peripheral vasculature, including the coronary arteries (Berne et al., 1983). Adenosine also produces sinus bradycardia and prolongation of impluse conduction in the atrioventricular node (Bellardinelli et al., 1983). In addition, adenosine has the ability to inhibit neurotransmitter release (Fredholm et al., 1988) and possesses potent central nervous system depressant and anticonvulsant activity (Dunwiddie, 1985).

The vasodilator and conduction effects of adenosine are mediated through different receptor subtypes. In the heart, A$_1$ receptors present on nodal cells and cardiac myocytes are responsible for the negative dromo-, chrono- and inotropic actions of adenosine (Bellardinelli et al., 1983). Activation of A$_2$ receptors located on coronary smooth muscle results in vasodilation (Berne et al., 1983). The potential use of an adenosine agonist as an antihypertensive agent has been limited by this spectrum of actions, non-selective agonists producing vasodilation that can be associated with cardiac depression as well as marked angina (Sylven et al., 1982). Selective A$_2$ receptor agonists may provide more viable agents as potential therapeutic candidates possessing effective vasodilatory hypotensive actions without the detrimental effects on cardiac conduction and renal function observed with currently available agonists. Whereas many highly selective agonists for the A$_1$ receptor have been described (Daly, 1982), the prototypical A$_2$ agonist NECA **(1)** (Bruns et al., 1986) show little or no A$_2$ selectivity. Until recently the most selective A$_2$ agonist described was CV1808 **(2)** being approximately 5-10 fold selective for the A$_2$ receptor (Bruns et al., 1986). More recently several N-6 substituted purine ribosides and NECA analogs with about 40 fold selectivity for the A$_2$ receptor have been described (Bridges et al., 1988). However most of these analogs still possess reasonably potent A$_1$ receptor affinity (K$_i$~200nM). In the present manuscript, the synthesis and receptor binding profiles at adenosine receptor subtypes for a series of 2-arylalkylamino-

146

affinity ($K_i \sim 200nM$). In the present manuscript, the synthesis and receptor binding profiles at adenosine receptor subtypes for a series of 2-arylalkylamino-adenosine-5'-uronamides are described. Some of the analogs described possess as much as 200 fold selectivity for the A2 receptor based on rat brain receptor binding and in addition have negligible affinity for A1 receptors (>1uM) in an absolute sense. In addition functional data in a perfused working rat heart model (Neely et al., 1967; Hutchison et al., 1989) on several of these selective ligands is presented which shows them to possess full agonist properties at A2 receptors as well as greater than 2000 fold separation between A2 (coronary vasodilatory) and A1 (negative chronotropic) receptor mediated events.

The geneology of the 2-arylalkylamino-adenosine-5'-uronamides is shown in **Figure 1**. Since at the initiation of this work the two most A2 selective agonists described were NECA (**1**) and CV1808 (**2**), we sought to combine the structural features of both series to generate the general structure **3a-k**. Optimzation of R1, R2 and R3 in this series led to a number of highly potent and selective A2 receptor ligands among which CGS 21680A (**3h**) was selected for extensive biolgical evaluation (Hutchison et al., 1989) as well as tritiation (Williams et al., 1989) for use as a A2 selective ligand for receptor binding.

Figure 1

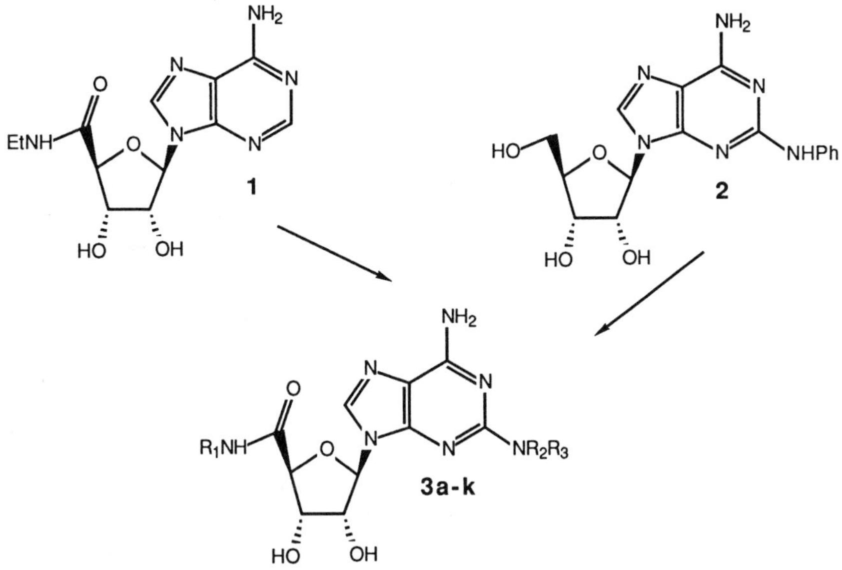

Chemistry

The general synthetic route employed for the preparation of the 2-arylalkylamino-adenosine-5'-uronamides **3a-k** is outlined in **Scheme I**. 2-CADO (**4**) was converted to the amides **6a-c** in good overall yields.by modification of methods found in the patent literature (Stein et al., 1979). The acetonides **6a-c** were then reacted with substituted arylalkylamines in excess at 130°C to afford the corresponding substituted derivatives **8a-j**.

Unfortunately this methodology failed for preparation of anilino analogs (R$_2$=Aryl) such as the NECA analog of CV1808 (**2**) but works well for unhindered primary and secondary amines. Treating the acetonides **8a-j** with 1N HCl afforded the desired analogs **3a-j** in good overall yields from the acetonides **6a-c**. The t-butyl ester protecting group of the derivatives **8h-j** was simultaneously cleaved during acetonide removal in the conversion of **8h-j** to **3h-j**. Finally the analog **3k** was prepared by esterification of the free base derived from **3h** with diazomethane in THF in 66% yield.

Scheme 1

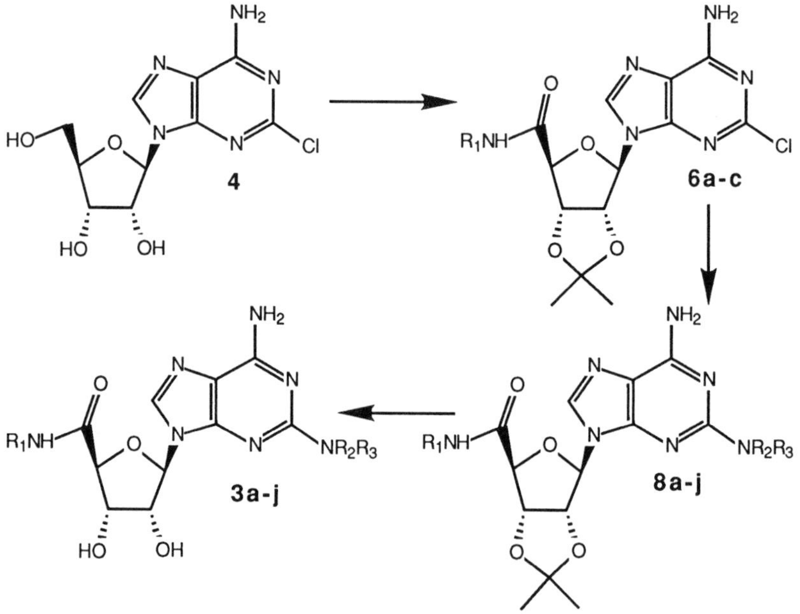

Results and discussion

The A$_1$ and A$_2$ receptor binding affinities for the analogs **7a-b** and **3a-k** along with relevant standards are collected in **Table II**. A$_1$ binding was measured in adenosine deaminase (ADA) pretreated rat cortical membranes using [3H]cyclohexyladenosine (CHA; specific activity 25 Ci/mmole) in the presence of 10 μM 2-chloroadenosine (2-CADO) to define specific binding as previously reported (Williams et al., 1986). Binding to A$_2$ receptors was measured in ADA-pretreated rat striatal membranes using [3H]CGS 21680(specific activity 30-80 Ci/mmol) (Williams et al., 1989).

As shown in **Table 1,** NECA (**1**) and MECA are virtually equipotent on A$_1$ and A$_2$ receptors in these assay systems whereas CV1808 (**2**) is about 10 fold selective for A_2 receptors. However NECA is about 7 fold more potent at A_2 receptors than CV1808. The corresponding 2-phenethylamino analogs **3a** (R_1=Me) and **3b** (R_2=Et) showed good affinity at A_2 receptors (K_i's=77 and 9.7nM respectively) while being only weakly active at A_1 receptors (K_i's=5350 and 473nM respectively). The 2-phenethylamino cyclopropyl amide **3c** (R_1=c-C_3H_5) showed no enhancement of either potency or selectivity over **3b** and therefore the R_1=Et series was chosen for a more extensive optimization of R_2 and R_3 since the R_1=Et substitution showed the highest affinity for A_2 receptors of the analogs **3a-c**.

Table 1

Number	R_2-R_3[a] K_i nM$\pm$sem	R_1 K_i nM$\pm$sem	A_1 Binding	A_2 Binding	A_1/A_2 Ratio
NECA (1)			5.8$\pm$.3	11.7$\pm$.6	.50
CV1808 (2)			689$\pm$31	71$\pm$.2	9.7
3a	R_2=-$(CH_2)_2$Ph	Me	5350$\pm$300	77$\pm$4	69
3b	R_2=-$(CH_2)_2$Ph	Et	473$\pm$18	9.7$\pm$1	49
3c	R_2=-$(CH_2)_2$Ph	c-C_3H_5	502$\pm$14	14$\pm$1.1	36
3d	R_2=-CH_2Ph	Et	8604$\pm$169	1130$\pm$53	7.6
3e	R_2=p-Cl-Ph$(CH_2)_2$-	Et	1505$\pm$81	7.5$\pm$1	201
3f	R_2=p-F-Ph$(CH_2)_2$-	Et	1184$\pm$32	7.7$\pm$.2	154
3g	R_2=-$(CH_2)_2$Ph, R_3=Me	Et	12100$\pm$700	71$\pm$2	170
3h	R_2=p-[HOOC$(CH_2)_2$]- Ph$(CH_2)_2$-	Et	1408$\pm$63	19$\pm$2	74
3i	R_2=p-[HOOCCH$_2$O]- Ph$(CH_2)_2$-	Et	4250$\pm$81	42$\pm$2	101
3j	R_2=p-[HOOCCH$_2$]- Ph$(CH_2)_2$-	Et	4880$\pm$240	45$\pm$2	109
3k	R_2=p-[MeO$_2$C$(CH_2)_2$] Ph$(CH_2)_2$-	Et	1257$\pm$97	9$\pm$.4	140

The corresponding benzyl analog of **3b**, analog **3d**, was virtually inactive at A_2 receptors and addition of an N-2 methyl group to analog **3b**, analog **3g**, results in an approximately 10 fold loss in A_2 receptor affinity. However, introduction of substituents on the aromatic ring of **3b** could be utilized to enhance the potency and selectivity in this series. In particular the p-chlorophenyl analog **3e** and the corresponding p-fluoro analog **3f** show 150-200 fold selectivity for the A_2 receptor subtype as well as greater affinity than NECA (K_i~7.5nM). Compound

3e currently represents the most selective ligand for the A2 receptor yet described on the basis of receptor binding.

Having solved the problem of A2 receptor affinity and selectivity we sought to introduce substiuents onto **3b** which would increase hydrophilicity and hence limit blood brain barrier penetration as well as limiting potential intracellular actions of these compounds. One such successful stategy was the introduction of a carboxylic acid moiety to the template **3b** to afford the analogs **3h-j** among which **3h** (CGS 21680A) shows the highest affinity for A2 receptors (K_i=15nM) as well as possessing excellent (74-fold) selectivity. The methyl ester of this compound, **3k**, shows even higher affinity and selectivity for the A2 receptor subtype but is considerably more lipophilic.

Intrinsic activity data for analogs **3b** and **3h** in a perfused working heart model are shown in **Table 2** among comparable data on CV1808 (**2**) and NECA (**1**) (Hutchison et al., 1989). Heart rate and coronary flow were measured as described by Neely (Neely et al., 1967) in hearts obtained from male Sprague-Dawley rats (Tac:N(SD)fBR; weight 275-350 g; Taconic Inc.,Germantown, NY). The relative negative chronotropic or vasodilatory potency of each agonist was expressed as the EC_{25} value, the mean concentration of agonist that produced a 25% decrease in heart rate or a 25% increase in coronary flow.

Table 2

| Compound | Perfused Working Heart Model[a] | | |
	EC_{25} CF[b] A2	EC_{25} HR[c] A1	A1/A2 Ratio[d]
NECA (1)	5nM	60nM	12
CV1808	110nM	>3000nM	>28
3b	3nM	>3000nM	>1000
3h	2nM	>3000nM	>1500

In this model, analog **3h** effectively increased coronary flow with an ED_{25} value of 2 nM. The corresponding value for **3b** was 3 nM while that for CV 1808 (**2**) was110 nM. The EC_{25} for eliciting bradycardia for all three compounds was >3000 nM. The effects of all three compounds could be reversed by treatment with the xanthine adenosine antagonist, XAC. Both compounds were full agonists at the rat coronary A2 receptor relative to NECA (Hutchison et al., 1989). Thus the analog **3h** showed greater than 1500 fold separation between A2 and A1 receptor mediated effects. On the basis of this data coupled with its high hydrophilicity and lack of effects on the adenosine transport system (Krulan et al., 1989) the analog **3h** was selected as an ideal compound for extensive exploration of A2 receptor pharmacology. The structure activity relationships derived for the 2-substituted ribose uronamides would indicate that bulky substituents in the 2-position are uniformly effective in reducing A_1 receptor activity and functional efficacy while certain of these substituents still retain very potent A_2 receptor affinity and efficacy as agonists .

In conclusion it has been shown that 2-arylakylamino-adenosine uronamides such as **3b** and **3h** are potent and extremely selective agonists for the adenosine A_2 receptor subtype and analog **3e** represents the most selective ligand for the A_2 receptor reported to date.

References

Bellardinelli, L., West, A., Crampton, R., and Berne, R.M., 1983, In *Regulatory Function*
  of *Adenosine*. (Eds. Berne, R.M.; Rall, T.W.; Rubio, R.) Nihoff, Boston, p.378.
Berne, R.M., Winn, H.R., Knabb, R.M., Ely, S.W., and Rubio, R., 1983, In *Regulatory*
  Function of Adenosine. (Eds. Berne, R.M.; Rall, T.W.; Rubio, R.) Nihoff, Boston, p.293.
Bridges, A.J., Bruns, R.F., Ortwine, D.F., Priebe, S.R., Szotek, D.L., and Trivedi, B.K.,
  1988, *J. Med. Chem.* **31**:1282.
Bruns, R.F., Lu, G.H., and Pugsley, T.A., 1986, *Mol. Pharmacol.* **29**:331.
Daly, J.W., 1982, *J. Med. Chem.* **25**:197.
Drury, A.N., and Szent-Gyorgyi, A., 1929, *J. Physiol. (Lond.))* **68**:214.
Dunwiddie, T.V., 1985, *Inter. Rev. Neurobiol.* **25**:63.
Fredholm, B., and Dunwiddie, T.V., 1988, *Trends Pharmacol. Sci.* **9**:130.
Hutchison, A.J., Oei, H.H., Ghai, G.R., and Williams M., 1989, *FASEB J.* **3**:A281.
Krulan, C., Wang, Z.C., Jeng, A.Y., Chen, J., and Balwierczak, J.L., 1989, *FASEB J.*
  3:A1048.
Neely, J.R., Liebermeister, H., Battersby, E.J., and Morgan, H.E., 1967, *Am. J. Physiol.*
  212:804.
Sylven, C., Beermann, B., Jonzon, B., and Brandt, R., 1986, *Br. Med. J.* **293**:227.
Stein, H.H., Prasad, R.N., Tietje, K.R., and Fung, A.K.L., 1979, *U.S. Patent 4,167,565* .
Williams, M.,Braunwalder, A., and Erickson, T.E., 1986, *Naunyn-Schmiedberg's Arch.*
  Pharmacol. **332**:179.
Williams, M., Schulz, R., Hutchison, A.J., Do, E., Sills, M.A., and Jarvis, M.F., 1989,
  FASEB J. **3**:A1047.

24

Adenosine Agonists. Characterization of the N$_6$-Subregion of the Adenosine A$_2$ Receptor via Molecular Modeling Techniques

D.F. Ortwine, A.J. Bridges, C. Humblet, and B.K. Trivedi

Introduction

Biological effects of adenosine are mediated by extracellular adenosine receptors that are divided into two main subtypes, A$_1$ and A$_2$, which inhibit and stimulate, respectively, adenylate cyclase (Van Calker et al., 1979; Londos et al., 1980). A$_2$ receptors have been further subdivided into A$_{2a}$ and A$_{2b}$ forms, based on their high (10^{-7} M) and low (10^{-5} M) affinities for adenosine. Highly selective ligands such as R-PIA (Vapaatalo et al., 1975) and CHA (Moos et al., 1985) are known for the A$_1$ receptor; however, only weak, 5-fold selective (CV-1808) or nonselective (NECA) ligands are known for the A$_{2a}$ receptor (Kawazoe et al., 1980; Prasad et al., 1980). Previous research (Heffner et al., 1985) demonstrated that A$_{2a}$-selective compounds such as CV-1808 had a desirable antipsychotic profile in secondary pharmacological tests such as sidman avoidance. It was hoped that a potent, highly selective A$_{2a}$ agonist would enhance these antipsychotic effects and lead to a clinically useful drug.

	(R)-PIA	CHA	CV-1808	NECA
A$_1$ K$_i$ (nM)	1.2	1.4	600	6.3
A$_{2a}$ K$_i$ (nM)	120	610	120	12

Because the N^6-arylalkyl derivative R-PIA possessed higher A$_{2a}$ affinity than N^6-alkyl and N^6-cycloalkyl analogs (Bruns et al., 1986; Kusachi et al., 1985), a number of N^6-arylalkyl substituted adenosines were synthesized in an attempt to identify potent, more selective A$_{2a}$ agonists. The requirement of an aryl ring within the N^6 substituent for high (K$_i$ <100 nM) A$_{2a}$ affinity emerged early in the work. However, seemingly small changes in the aryl ring environment led to large changes in binding affinity. For example, fusing a phenyl ring onto the potent N^6-naphthylmethyl analog (2, K$_i$ = 9.4 nM) resulted in an anthracenylmethyl derivative with markedly reduced affinity (K$_i$ = 29 µM) for the A$_{2a}$ receptor (Trivedi et al., 1988a). In an attempt to rationalize the observed SAR for A$_{2a}$ binding, a number of N^6-arylalkyl analogs were examined via a molecular modeling approach.

152

Materials and Methods

The synthesis and A_1 and A_{2a} receptor binding affinities of N^6-substituted adenosine analogs have been reported elsewhere (Bridges et al., 1987 and 1988; Bristol et al., 1988; Trivedi et al., 1988a and b). Modeling analyses were carried out using version 3.5 of the SYBYL program package (Tripos Associates, St Louis, MO) operating on a VAX 11/785. Structures were built from available fragments within the program, and minimized where necessary using the MAXIMIN procedure (Labanowski et al., 1986). The adenosine portion of the molecules was held fixed throughout the modeling study; only the N^6 substituent was examined. After applying systematic conformational searching (Motoc et al., 1986) to three of the most potent, rigid analogs (**1-3**, table 1) and superimposing their allowed aryl ring orientations, a specific region was located where all three could project aryl rings, with the constraint that these aryl rings be coplanar (figure 1). The center of this region was then chosen as a putative aryl pocket, and it was demonstrated that other potent, N^6-arylalkyl derivatives (**4-15**, table 1) could project aryl rings into this site without significant increases in energy. These 15 potent analogs were used to define a tolerated volume in the N^6-subregion of the A_{2a} receptor (figure 1). Less potent analogs were then examined in the model and hypotheses were generated regarding conformational, steric, and electronic tolerances surrounding this aryl pocket. The model was further extended and refined by the addition of a number of substituted analogs from a series of 2,2-(diphenyl)ethyl derivatives, examples of which appear in table 2.

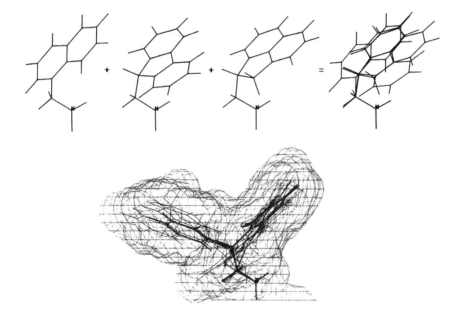

Figure 1. (top) N^6-sidechains of three initial structures (**1-3**, table 1) used to locate a common aryl binding site. Their superposition is shown at the right. (bottom) Overlay of the N^6-sidechains all 15 potent analogs (**1-15**, table 1) and the composite volume occupation.

Table 1. Adenosine Agonists Used In the Construction of the A_{2a} Receptor Pharmacophore Model

	1	2	3
A_1 K_i (nM)	5.2	24	52
A_{2a} K_i (nM)	4.9	9.4	43

no.	R	K_i (nM) A_1	A_{2a}	no.	R	K_i (nM) A_1	A_{2a}
4		61	19	10		5.2	67
5		4.0	19	11		8.5	20
6		16	52	12		14	33
7		37	30	13		6.8	25
8		56	41	14		62	45
9		39	12	15		1.2	120

ADENOSINE RECEPTOR PHARMACOPHORE MODEL

N6 - SUBREGION

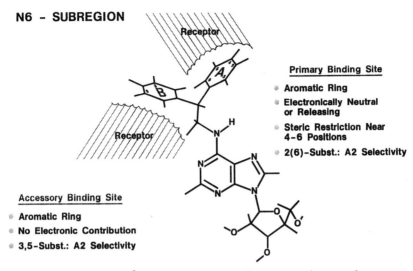

Primary Binding Site

- Aromatic Ring
- Electronically Neutral or Releasing
- Steric Restriction Near 4-6 Positions
- 2(6)-Subst.: A2 Selectivity

Accessory Binding Site

- Aromatic Ring
- No Electronic Contribution
- 3,5-Subst.: A2 Selectivity

Figure 2. Summary of the N⁶-subregion of the adenosine receptor pharmacophore model.

Table 2. Effects of Ring Substitution On Receptor Affinity and Selectivity of 2,2-(Diphenyl)ethyl Adenosines

no.	Substituents		K_i (nM)			Comments
	A Ring	B Ring	A_1	A_{2a}	A_{2a}/A_1	
16	H	H	6.8	25	3.7	Reference compound.
17	H	4-Cl	15	61	4.0	One Cl is tolerated.
18	4-Cl	4-Cl	290	480	1.6	Projection of A ring Cl into a sterically disallowed region.
19	3-Cl	H	1.7	5.6	3.3	Meta substitution
20	3-Cl	3-Cl	2.6	6.0	2.3	by Cl increases potency.
21	3,5-Cl₂	H	11	6.7	0.61	Two Cl's are tolerated.
22	H	3,5-(OCH₃)₂	30	6.1	0.10	Two OCH₃ groups are tolerated in B ring.
23	3,5-(OCH₃)₂	4-OCH₃	320	160	0.50	Steric intolerance of two OCH₃ groups in A ring.
24	2,6-Cl₂	H	61	19	0.31	Stereoisomer mixture.
25	2,6-Cl₂	H	31	9.4	0.30	R isomer. (Isomers were assigned by the
26	2,6-Cl₂	H	210	700	3.3	S isomer. modeling analysis.)

155

Results

The pharmacophore model is summarized in figure 2. The primary site aryl pocket (A ring) is centered 4 angstroms from the N^6 nitrogen and is flanked by receptor walls next to the para, and to a lesser extent, one of the ortho and meta positions on the aryl ring. Electron releasing rings (thiophene, furan) or electron releasing substituents in this pocket increase A_{2a} affinity, while electron withdrawing rings (pyridine) or substituents (CF_3) markedly reduce affinity. In addition, an accessory, A_{2a} potency-enhancing, aryl binding site was located into which diaryl substituted analogs project. This pharmacophore model was completely consistent with the SAR of all N^6-substituted analogs studied, including potency differences among resolved stereoisomers.

Design of New Structures

Among analogs containing N^6-arylethyl fragments, A_1 and A_{2a} binding affinities were generally parallel. However, by focusing on newly synthesized analogs within the diaryl series that were moderately (5-fold) A_{2a} selective (table 2), factors responsible for decreased A_1 affinity relative to A_{2a} were defined. It was found that 2- or 2,6-disubstitution of the ring in the aryl pocket and 3,5-disubstitution of the ring in the accessory binding site selectively diminished A_1 affinity.

This information was used in the design of new analogs within the diaryl series (table 3). Compounds 27-29 are examples from an initial set of compounds that were designed to test and extend the pharmacophore model. From the modeling analysis, any 2,6-disubstituted phenyl ring with substituents larger than fluorine was forced to lie in the aryl pocket. A_{2a} selectivity was maintained with **27**, while decreased potency resulted from too large a group at one of the ortho positions in the case of **28**. The 4-substituent of **29** projected into a sterically disallowed region, and thus this analog was less potent, as predicted. Compounds **30-33** illustrate the results of combining features thought to impart A_{2a} selectivity.

Table 3. 2,2-(Diphenyl)ethyl Adenosines Prepared Subsequent to the Modeling Analysis

no.	Substituents A Ring	Substituents B Ring	K_i (nM) A_1	A_{2a}	A_{2a}/A_1
16	H	H	6.8	25	3.7
27	2,6-$(CH_3)_2$	H	230	62	0.27
28	2,6-$(OCH_3)_2$	H	2200	1100	0.50
29	2,4,6-$(CH_3)_3$	H	520	510	0.98
30	2,6-$(CH_3)_2$	3,5-$(OCH_3)_2$	140	20	0.14
31	2-CH_3	3,5-$(OCH_3)_2$	140	4.4	0.031
32	2-CH_3	3,4,5-$(OCH_3)_3$	160	5.5	0.034
33	2-OCH_3	3,5-$(OCH_3)_2$	42	3.6	0.086

As predicted, the combination of 2 and 2,6-disubstitution of the A ring with 3,5-disubstitution of the B ring provided compounds with high potency and 30 fold selectivity for the A_{2a} receptor. Efforts to further enhance this selectivity through additional substitution at the 2 position have been reported elsewhere (Trivedi et al., 1989).

Conclusion

Development of a detailed pharmacophore model of the N^6-subregion has allowed the refinement of a series of diaryladenosines, producing several agonists with 4-5 nM affinity and 20-40 fold selectivity for the A_{2a} receptor. This potency and selectivity should make these compounds useful probes for adenosine pharmacology.

Acknowledgements

We would like to thank R.F. Bruns and coworkers for the biological data, J. Bristol for his steadfast support of the program, and H. Hamilton, W. Moos, W. Patt, S. Priebe, and D. Szotek for their synthesis efforts.

References

Bridges, A.J., et al., 1987, *J. Med. Chem* **30**:1709.

Bridges, A.J., et al., 1988, *J. Med. Chem.* **31**:1282.

Bristol, J.A., et al., 1988, In *Adenosine and Adenine Nucleotides: Physiology and Pharmacology*; Paton, D.M., Ed.; Taylor & Francis: London, pp. 17-26.

Bruns, R.F., Lu, G.H., and Pugsley, T.A., 1986, *Mol. Pharmacol.* **29**:331.

Heffner, T.G., et al., 1985, *Pharmacologist* **27**:293.

Kawazoe, K., et al., 1980, *Arzneim.-Forsch.* **30**:1083.

Kusachi, S., et al., 1985, *J. Med. Chem.* **28**:1636.

Labanowski, J., et al., 1986, *Quant. Struct.-Act. Relat.* **5**:138.

Londos, C., Cooper, D.M.F., and Wolff, J., 1980, *Proc. Natl. Acad. Sci. U.S.A.* **77**:2551.

Moos, W.H., Szotek, D.L., and Bruns, R.F., 1985, *J. Med. Chem.* **28**:1383.

Motoc, I., et al., 1986, *Quant. Struct.-Act. Relat.* **5**:99.

Prasad, R.N., et al., 1980, *J. Med. Chem.* **23**:313.

Trivedi, B.K., et al., 1988a, *J. Med. Chem.* **31**:271.

Trivedi, B.K., et al., 1988b, *Abstr. of Papers, 195th ACS National Meeting, Toronto, Canada*, June, MEDI 32.

Trivedi, B.K., and Bruns, R.F., 1989, *J. Med. Chem.* **32**:1667.

Van Calker, D., Muller, M., and Hamprecht, B., 1979, *J. Neurochem.* **33**:999.

Vapaatalo, H., et al., 1975, *Arzneim.-Forsch.* **25**:407.

25

Physiological and Biochemical Aspects of "P"-Site-Mediated Inhibition of Adenylyl Cyclase

R.A. Johnson, S.-M.H. Yeung, M. Bushfield, D. Stübner, and I. Shoshani

Introduction

Adenosine modulates adenylyl cyclase (ATP:pyrophosphate lyase, cyclizing; EC 4.6.1.1.) activity through two G-protein-linked cell-surface receptors (A_1 and A_2) and through an inhibitory site ("P"-site) on the cytoplasmic face of the enzyme's catalytic moiety (Londos et al. (1977). Inhibition of adenylyl cyclase via the "P"-site is characteristically dependent on divalent cations and is non-competitive with respect to metal-ATP (Johnson et al., 1979). With the exception of forskolin, reversible and irreversible activators of adenylyl cyclase enhance the enzyme's sensitivity to "P"-site-mediated inhibition (Stübner et al., 1989). The "P"-site thus provides an interesting and unique target for probing structural and functional aspects of the enzyme's catalytic moiety. Uncertain, though, is the physiological relevance of the "P"-site for regulation of adenylyl cyclase. In recent years our laboratory has approached the "P"-site with biochemical and physiological questions in view. Biochemical approaches have been to identify potent "P"-site agonists, to synthesize agonists for use in affinity chromatography matrices, in the development of radio-ligand binding assays, or as covalent affinity ligands. Radioactively labeled covalent probes could be used to identify the residues with which they bind and the structural/functional relationships of the enzyme could be better understood. Physiological approaches have been to develop a method for quantitating cellular levels of the more potent "P"-site agonists and to determine whether and to what extent their levels might change in response to conditions known to alter sensitivity of adenylyl cyclase to hormonal regulation. Results from both biochemical and physiological studies are presented here.

Methods and Materials

<u>Preparation and Assay of Adenylyl Cyclases</u> Particulate and detergent- dispersed adenylyl cyclase from rat or bovine brain were prepared as described by Stübner et al., 1989. The catalytic moiety was purified from bovine brain to near homogeneity on forskolin- and wheat-germ lectin-agarose columns as described by Smigel, 1986. Proteolytic activation and assay of adenylyl cyclase were as previously described (Johnson et al., 1989b).

<u>Inactivation of adenylyl cyclase by 2'5'dd3'FSBA (2'5'dideoxy-,3'-*p*-fluoro-sulfonyl-benzoyl-adenosine)</u> Detergent-dispersed adenylyl cyclase from rat brain was dialyzed overnight against 10 mM MOPS, pH 7.5, 1 mM $MgCl_2$, 0.1% Lubrol-PX, 5% glycerol,

principally to remove dithiothreitol. For purified enzyme 0.1% Lubrol-PX was replaced by 0.1% Tween 60. Crude or purified enzyme was incubated with 100 μM 2'5'dd3'FSBA (45 min/30°C) in buffer containing 10 mM MOPS, pH 7.5, 1 mM $MnCl_2$, 10 mM $MgCl_2$, 0.1% Tween 60, 5% glycerol, 0.1 mM ATP, 0.3 mg bovine serum albumin/ml, in a final volume of 220 μl. Bovine serum albumin was then added (20 μl $\rightarrow$ 1 mg/ml final) and 130 μl of this was dialyzed overnight at 4°C as above but with 10 mM $MgCl_2$. For crude adenylyl cyclase Tween 60 was 0.2%. The rest of the samples were stored at 4°C overnight. Enzyme activity was determined on aliquots of enzyme before and after dialysis. For controls 2'5'dd3'FSBA was replaced with either solvent (2% dimethylsulfoxide) or 100 μM 2'5'dideoxy-adenosine.

Determination of 3'AMP Levels Tissues were removed from decapitated rats, were immediately frozen by clamping at liquid N_2 temperature, and were stored at -70°C until extraction. Tissues were extracted by boiling for 15 min in a buffer containing 1% Lubrol-PX, 10 mM TEA•COOH, 5 mM EDTA, pH 7.5, and [^{3}H]3'AMP, to monitor recovery of 3'AMP. Samples were clarified by centrifugation and extracts were incubated with 5'nucleotidase. 3'AMP was purified by sequential chromatography on DEAE-Sephadex, HPLC anion exchange, and C18 reverse-phase columns. Extracted 3'AMP was determined from both the HPLC peak area and inhibition of Mn^{2+}-activated rat brain adenylyl cyclase, compared to a standard inhibition curve for authentic 3'AMP. For estimating formation of 3'AMP from poly(A), homogenates were incubated (20 min/37°) with 0.2 - 5 μM poly(A), 0.1 μCi [^{3}H]poly(A), 1 mM 3'AMP, 167 mM NH_4acetate, pH 5.7, and 0.7 mM EDTA, without and with 5'nucleotidase. Formation of 3'AMP, with [^{3}H]3'AMP to monitor recovery, from 5 μM unlabeled poly(A) was determined in separate experiments with similar results.

Streptozotocin-Induced Diabetes Diabetes was induced in male Sprague-Dawley rats by intraperitoneal injection of streptozotocin (80 mg/kg) in 0.1 M sodium citrate, pH 4.5. Control animals received only buffer. Blood glucose was determined four days later and animals whose level exceeded 450 mg/dl were considered diabetic. Insulin-treated rats were given a 10 U injection of insulin followed by a 6 U dose every 6 hours thereafter.

Materials Reagents used in the purification and assay of adenylyl cyclase and adenosine analogs used as "P"-site agonists were obtained or synthesized as reported (Johnson et al., 1989b; Stübner et al., 1989). Tritiated nucleosides, nucleotides, and poly(A) were from Amersham or ICN. Nucleic acids and nucleotides were either from Sigma or Boehringer-Mannheim. 5'Nucleotidase (Crotalus atrox) and streptozotocin were from Sigma. Other reagents were from commercial sources and were of the highest quality available.

Results

Structure/activity relationship of the "P"-site Sensitivity to "P"-site-mediated inhibition of brain adenylyl cyclase was enhanced by increasing concentration of divalent cation and by a variety of activators, whether reversible or irreversible (table 1). Although the IC_{50} for 2'5'ddAdo decreased upon activation with cholera toxin, GTPγS, Mn^{2+}, and Ca^{2+}/calmodulin, forskolin decreased "P"-site sensitivity, in contrast with its effects on enzyme from liver and S49 cyc⁻ membranes (Florio et al.,1983; Stübner et al.,1989). The effect of forskolin on the brain enzyme was seen with purified enzyme and was found to be independent of G-proteins or calmodulin. These observations suggest that the different responses of the several adenylyl cyclases to forskolin reside in structural differences in the catalytic units of these enzymes. To define the agonist structural requirements for the "P"-site a detergent-dispersed adenylyl cyclase from rat brain was used that was activated by Mn^{2+} or by proteolysis in the presence of GTPγS (Johnson et al.,1989b). The structures and IC_{50}s of the principal "P"-site agonists are shown in table 2. The data suggest a strict requirement for an intact adenine moiety and a β-glycosidic linkage for the ribosyl moiety. 2'Deoxy- and especially 2'5'dideoxy-ribosyl moieties enhanced sensitivity and a strong preference for phosphate at the 3'-position was exhibited. Substitutions at the 5'ribose position impaired sensitivity, but large substitutions at the 3'ribose position were tolerated.

Table 1. Effects of Various Stimuli on the Sensitivity of Adenylyl Cyclase to Inhibition by 2'5'ddAdo. Particulate adenylyl cyclase from rat brain was pretreated with GTPγS (15 min/30°C) or cholera toxin (50 μg/ml + 500 μM GTP; 30 min/20°C), or was assayed directly. The reaction mixture contained 100 μM GTP and varying concentrations of 2'5'ddAdo. Values are averages ± range from two experiments, each assayed in duplicate. Modified from Stübner et al., 1989.

IC_{50}'s for 2'5'ddAdo

$MgCl_2$	Control	Forskolin (100 μM)	GTPγS (10 μM)	CTx	Mn^{2+} (1 mM)	Forskolin + Mn^{2+}
(mM)			(μM)			
10	19± 1	44 ± 12	20± 9	10± 0	4.9 ± 1.4	4.3 ±0.3

2'5'dd3'FSBA, a Specific "P"-Site Affinity Ligand 2'5'dd3'FSBA was synthesized as a potential covalent probe for the "P"-site (Yeung et al., 1989). The sulfonyl fluoride moiety of 2'5'dd3'FSBA is a reactive functional group that can act as an electrophilic agent in covalent reactions with several classes of amino acids, including tyrosine, lysine, histidine, serine, and cysteine (Coleman, 1983). A covalent reaction with the catalytic moiety of adenylyl cyclase might be indicated by irreversible inactivation of the enzyme. This was tested with both crude and purified preparations of the enzyme from bovine brain, with which 2'5'dd3'FSBA exhibited IC_{50}s ~10-30 μM. Inhibition of the purified enzyme was time dependent, suggestive of both rapid and slow phases (not shown). The time-dependent inhibition was not reversed by dialysis (table 3). In the experiments shown here the purified enzyme was incubated with 100 μM 2'5'dd3'FSBA in the presence of 100 μM ATP. Irreversible inactivation by 2'5'dd3'FSBA also was observed in the presence of ATP from 10 μM to 3 mM (Yeung et al., 1989), but not in the presence of 100 μM 2'5'ddAdo (table 3). The effectiveness of 2'5'ddAdo to prevent inactivation was concentration-dependent. These observations strongly suggest that inactivation by 2'5'dd3'FSBA occurs at the enzyme's "P"-site and not at its catalytic active site, a conclusion also supported by inhibition kinetics of the enzyme (Johnson et al., 1989a).

Table 2. Structures of Principal "P"-Site Agonists. Adapted from Johnson et al. (1989b)

Nucleoside	R_1	R_2	R_3	IC_{50} (μM)
Adenosine	-OH	-OH	-OH	82
2'dAdo	-H	-OH	-OH	15
3'dAdo	-OH	-H	-OH	13
5'AMP	-OH	-OH	$-OPO_3$	150
3'AMP	-OH	$-OPO_3$	-OH	8.9
2'3'ddAdo	-H	-H	-OH	9
2'5'ddAdo	-H	-OH	-H	2.7
2'd3'AMP	-H	$-OPO_3$	-OH	1.2
2'd5'AMP	-H	-OH	$-OPO_3$	>300
2'5'dd3'AMP	-H	$-OPO_3$	-H	<0.1
2'5'dd3'FSBA	-H	$-FSO_2Bz$	-H	30

Table 3. Protection by 2'5'ddAdo against Inactivation of Adenylyl Cyclase by 2'5'dd3'FSBA. Partially purified adenylyl cyclase was incubated with 100 μM 2'5'dd3'FSBA and/or 2'5'ddAdo, as indicated. Samples were dialyzed and then assayed with 50 μM forskolin. Values are percentages ($\pm$ SEM) or ($\pm$ range for n=2) of activities with inhibitor relative to activities without inhibitor, for the number of experiments shown in parentheses. Control activity before dialysis was 41 pmol cAMP(15 min$\bullet$10 μl enzyme)$^{-1}$ and 10.6 after dialysis. Adapted from Yeung et al., 1989.

Preincubation	Activity before[a] Dialysis	Activity after Dialysis
	% Control	% Control
2'5'dd3'FSBA	46.5 ± 3.4 (6)	46.7 ± 4.7 (4)
2'5'ddAdo	49.9 ± 6.1 (3)	108 ± 8 (2)
2'5'ddAdo + 2'5'dd3'FSBA	40.5 ± 5.0 (2)	114 ± 1 (2)

[a] Due to carryover into adenylyl cyclase assays, concentrations of agents in the assay were 3/10 of those during preincubation.

Other investigators have demonstrated that dithiothreitol partially reactivated enzymes inactivated by various FSO_2Bz-substituted ligands (Coleman, 1983). Similarly, dithiothreitol protected adenylyl cyclase from irreversible inactivation by 2'5'dd3'FSBA and substantially (~80%) reactivated enzyme that had been inactivated by 2'5'dd3'FSBA (Yeung et al., 1989). These data suggest that inactivation of adenylyl cyclase by

2'5'dd3'FSBA may be due to two actions of the ligand, one sensitive and one insensitive to dithiothreitol, perhaps indicative of interactions with more than one residue, one of which is likely cysteine.

Potential sources for cellular "P"-site agonists Since 3'AMP and 2'd3'AMP are the most potent naturally occuring "P"-site inhibitors of adenylyl cyclase, we attempted to determine their source. We found no evidence that either nucleotide was a product of a unique cyclic nucleotide phosphodiesterase or of a unique adenosine kinase (e.g. → 2'd3'AMP or 3'AMP instead of 2'd3'AMP or 5'AMP) (Bushfield et al., 1989). However, 2'd3'AMP and 3'AMP were formed from added DNA and RNA, respectively, suggesting that nucleic acid degradation may be their likely *in vivo* source. In particular, the polyadenylate tracts at the 3'-end of mammalian mRNA could provide a rich cytosolic source of 3'AMP. With assumptions that 5% of total RNA (20 mg/g wet weight; Chirgwin et al., 1979) is mRNA averaging 2000 nucleotides with steady state polyadenylate tracts of ~100 nucleotides (Brawerman et al., 1975) the intracellular mRNA would be ~2.4 μM. Thus, cellular concentrations of poly(A)$^+$mRNA would be sufficient as substrate for the formation of ample 3'AMP for inhibition of adenylyl cyclase. Incubation of rat liver or spleen homogenates with 0.2 to 5 μM poly(A) resulted in the formation of substantial quantities of 3'AMP (table 4). Formation of 3'AMP was linear with time to 30 min and with tissue to 1 mg wet weight. The obvious conclusions are that the enzymic machinery exists in these tissues to make 3'AMP from poly(A) and that the amount formed would be sufficient to inhibit adenylyl cyclase (Bushfield et al., 1989).

Tissue content of "P"-site agonists In preliminary experiments we have attempted to determine levels of 3'AMP in several rat tissues (table 5). Spleen had the highest content of 3'AMP while 3'AMP was not measureable (i.e. <0.1 nmol/g) in skeletal muscle. In tissues with measureable amounts of 3'AMP levels would have been adequate to inhibit the respective adenylyl cyclases *in vitro*, suggesting a possible role for 3'AMP as a "P"-site inhibitor *in vivo* (Bushfield et al., 1989).

Table 4. Formation of $[^3H]3'AMP$ from $[^3H]poly(A)$ by spleen and liver homogenates. Values are from a representative experiment.

Tissue	[Poly(A)] (μM)	Formation of 3'AMP nmol(min•g)$^{-1}$
Spleen	0.2	6.0
	1.2	32.0
	5.0	39.0
Liver	0.2	4.7
	1.2	13.3
	5.0	37.0

Table 5. Content of 3'AMP in several rat tissues. Values are means ± SEM (± range for n=2) for the number of tissues extracted (in parentheses).

Tissue	[3'AMP] nmol/gram	(n)
Spleen	280 ± 47	(5)
Kidney	130 ± 37	(4)
Liver	46.7 ± 7.6	(8)
Heart	3.7 ± 1.0	(3)
Brain	1.8 ± 0.4	(2)
Skeletal muscle	<0.1	(4)

Diabetes significantly distorts regulatory processes in cells and is known to decrease sensitivity of hepatic adenylyl cyclase to stimulation by glucagon (Pilkis et al., 1974, Dighe et al., 1984). Consequently, Streptozotocin-induced diabetes was used as a model to determine whether changes in hepatic 3'AMP content could be induced and whether they could be opposed by maintenance with insulin. 3'AMP content of livers from diabetic rats increased 75%, from 47 ± 8 in sham-injected animals to 84 ± 18 nmol/g. This increase was partially reversed by maintenance of diabetic animals with insulin; 3'AMP content was 57 ± 15 nmol/g. These findings suggest that levels of 3'AMP may vary due to changes in circulating hormones, conceiveably sufficiently to change sensitivity of adenylyl cyclase to stimulatory agents.

Discussion

Most adenylyl cyclases exhibit a characteristic inhibition by adenosine and certain ribose-modified analogs. An apparent requirement for an intact purine moiety led to the designation of this inhibitory site as the "P"-site (Londos et al., 1977). From structure/activity relationships determined with the brain enzyme rank order of potency was 2',5'dd3'AMP > 2'd3'AMP > 2'5'ddAdo > 3'AMP > 2'dAdo > adenosine (Johnson et al., 1989b). Large substitutions were tolerated at the 3'-position, whereas substitutions at the 5'position typically impaired potency, and modifications in the adenine moiety were not tolerated. These relationships suggest that this inhibitory site might be more correctly designated the "dAP"-site (deoxy-adenosine-phosphate) (Johnson et al. 1989b).

Interesting aspects of the "P"(read "dAP")-site are that only stimulated forms of the enzyme are sensitive to inhibition, that it is localized on the catalytic moiety of adenylyl cyclase at a site distinct from the catalytic active site, and that regulation of the enzyme via the "P"-site may be physiologically relevant. Although activation of adenylyl cyclase sensitizes the enzyme to inhibition by "P"-site agonists, the efficacy of the various stimuli differed. The greatest sensitivity to inhibition of the brain enzyme was observed with activation by: ninhibin/GTPγS ~ Mn^{2+} ~ Mn^{2+}/forskolin > cholera toxin ~ GTPγS > basal > forskolin. Common aspects of activation of adenylyl cyclase by hormone/GTP, by stable GTP analogs, and by proteolysis in the presence of GTPγS are that the apparent affinity of the enzyme (and/or α_s) for Mg^{2+} is substantially increased and that association of α_s with the catalytic moiety (C) is enhanced. Since the "P"-site is localized on the catalytic moiety of adenylyl cyclase (Premont et al.,1979; Florio et al.,1983; Henry et al., 1986; Yeager et al., 1986; Minocherhomjee et al., 1987; and Johnson et al., 1989b), activation of the enzyme by GTPγS•α_s or Mn^{2+} may be viewed as conformationally altering C, i.e. → GTPγS•α_s•C^{Mg2+} or C^{Mn2+}, such that these are the most active forms of the enzyme and are most sensitive to "P"-site-mediated inhibition. Forskolin evidently does not induce this conformational change even though it substantially activates the enzyme (table 1 and Stübner et al., 1989).

The locus of the "P"-site within the catalytic moiety is uncertain. Kinetic studies from our laboratory suggest that it is distinct from the catalytic active site (Johnson et al., 1989a). This conclusion is supported by experiments reported here with the "P"-site affinity ligand, 2'5'dd3'FSBA. Inhibition by 2'5'dd3'FSBA was not reversed upon dialysis. Irreversible inactivation occurred in the presence of ATP, but was prevented in a concentration dependent manner by 2'5'ddAdo, a potent "P"-site agonist. Targets for the reactive sulfonyl fluoride moiety of FSO_2Bz-containing compounds are generally thought to be nucleotphiles, either amino or sulfhydryl groups located at or near the ligand-binding domain (Coleman, 1983). The fact that dithiothreitol reactivated the 2'5'dd3'FSBA-modified enzyme (Yeung et al., 1989) suggests that the ligand may form a covalent link to a cysteine in proximity to the binding domain. However, since inactivation by 2'5'dd3'FSBA was only partially reversed by dithiothreitol, attachment through other residues also may have occurred. Our data would support the suggestion that of the two cytosolic domains of adenylyl cyclase exhibiting interdomain homology as well as homology with guanylyl cyclase (Krupinski et al., 1989), one may be the catalytic active site and the other the "P"-site, possibly containing the cysteine through which 2'5'dd3'FSBA inactivates the enzyme. Once radioactively labeled 2'5'dd3'FSBA has been synthesized it will be possible for the relationship between the "P"-site and the catalytic unit of adenylyl cyclase to be probed directly, to determine the stoichiometry of interaction of 2'5'dd3'FSBA with adenylyl cyclase, to ascertain whether it inactivates the enzyme through reactive groups other than the sulfhydryl moeity of cysteine, and to identify the enzyme's residues and domains which constitute the "P"-site.

Whether the "P"-site is involved in physiological regulation of adenylyl cyclases remains to be established. In a physiological context adenylyl cyclase activity could be regulated via the "P"-site if: a) the enzyme is activated and hence sensitized to inhibition by an on-board concentration of "P"-site agonist, b) the level of "P"-site agonist changes and thereby alters activity of the stimulated enzyme, or c) both occur. The potencies of 2'd3'AMP and 3'AMP (IC_{50}s ~1.2 and 9 µM) for inhibition of the brain enzyme are substantially lower than of adenosine (IC_{50} ~80 µM). This suggests to us that the 3'phosphorylated analogs are the more likely naturally occurring inhibitors of the enzyme. Although tissue levels of 3'AMP varied considerably among tissues (table 5), levels were well within those necessary to inhibit activated forms of adenylyl cyclase. The tissue differences suggested, though, that synthetic and degradative pathways for 3'AMP varied significantly. Since 3'AMP can be formed from poly(A), for which poly(A)$^+$mRNA would be a rich *in vivo* source, it is conceiveable that nuclease(s) catalyzing this process differ among tissues and that their activity might respond to hormones or changes in cell function or metabolism. That 3'AMP may change in response to such stimuli is suggested from the studies on streptozotocin-induced diabetes.

In summary, the data suggest that the "P"("dAP")-site is a distinct binding domain on the catalytic unit of adenylyl cyclase and mediates inhibition of stimulated forms of the enzyme. The domain can be probed biochemically with specific affinity ligands and and it may be a physiological target for 2'd3'AMP or 3'AMP, the cellular levels of which may be regulated. (Supported by NIH grant DK 38828)

References

Brawerman, G., and Diez, J., 1975, Metabolism of the polyadenylate sequence of nuclear RNA and messenger RNA in mammalian cells. *Cell* 5: 271-280.
Bushfield, M., Shoshani, I., and Johnson, R.A., 1989, Tissue levels, source, and regulation of 3'AMP, an intracellular regulator of adenylyl cyclase. *In review*.

Chirgwin, J.M., Przybla, A.E., MacDonald, R.J., and Rutter, W.J., 1979, Isolation of biologically active ribonucleic acid from sources enriched in ribonuclease. *Biochemistry* **18**: 5294-5299.

Coleman, R.F., 1983, Affinity labeling of purine nucleotide sites in proteins. *Ann. Rev. Biochem.* **53**: 67-91.

Dighe, R.R., Rojas, F.J., Birnbaumer, L., and Garber, A.J., 1984, Glucagon-stimulable adenylyl cyclase in rat liver: The impact of streptozotocin-induced diabetes mellitus. *J. Clin. Invest.* **73**: 1013-1023.

Florio, V.A., and Ross, E.M., 1983, Regulation of the catalytic component of adenylate cyclase: Potentiative interaction of stimulatory ligands and 2',5'-dideoxyadenosine. *Molec. Pharmacol.* **24**: 195-202.

Henry, D., Ferino, F., Tomora, S., Ferry, N., Stengel, D., and Hanoune, J., 1986, Inhibition of the catalytic subunit of ram sperm adenylate cyclase by adenosine. *Biochem. Biophys. Res. Commun.* **137**: 970-977.

Johnson, R.A., Saur, W., and Jakobs, K.H., 1979, Effects of prostaglandin E_1 and adenosine on metal and metal-ATP kinetics of platelet adenylate cyclase. *J. Biol. Chem.* **254**: 1094-1101.

Johnson, R.A., and Shoshani, I., 1989a, Kinetics of "P"-site-mediated inhibition of adenylyl cyclase and the requirement for substrate. *In review*.

Johnson, R.A., Yeung, S.-M.H., Stübner, D., Bushfield, M., and Shoshani, I., 1989b, Cation and structural requirements for "P"-site-mediated inhibition of adenylate cyclase. *Molec. Pharmacol.* **35**: 681-688.

Krupinski, J., Coussen, R., Bakalyar, H.A., Tang, W.J., Feinstein, P.G., Orth, K., Slaughter, C., Reed, R.R., and Gilman, A.G., 1989, Adenylyl cyclase amino acid sequence: Possible channel- or transporter-like structure. *Science* **244**: 1558-1564.

Londos, C., and Wolff, J. (1977) Two distinc adenosine-sensitive sites on adenylate cyclase. *Proc. Natl. Acad. Sci. USA* **74**: 5482-5486.

Minocherhomjee, A.M., Selfe, S., Flowers, N.J., and Storm, D.R., 1987, Direct interaction between the catalytic subunit of the calmodulin-sensitive adenylate cyclase from bovine brain with ^{125}I-labeled calmodulin. *Biochemistry* **26**: 4444-4448.

Pilkis, S.J., Exton, J.H., Johnson, R.A., and Park, C.R., 1974, Effects of glucagon on cyclic AMP and carbohydrate metabolism in livers from diabetic rats. *Biochim. Biophys. Acta* **343**: 250-267.

Premont, J., Guillon, G., and Bockaert, J., 1979, Specific Mg^{2+} and adenosine sites involved in a bireactant mechanism for adenylate cyclase inhibition and their probable localization on this enzyme's catalytic component. *Biochem. Biophys. Res. Commun.* **90**: 513-519.Smigel, M.D., 1986, Purification of the catalyst of adenylate cyclase. *J. Biol. Chem.* **261**: 1976-1982.

Stübner, D., and Johnson, R.A., 1989, Forskolin decreases sensitivity of brain adenylate cyclase to inhibition by 2',5'-dideoxyadenosine. *FEBS Lett.* **248**: 155-161.

Yeager, R.E., Nelson, R., and Storm, D.R., 1986, Adenosine inhibition of calmodullin-sensitive adenylate cyclase from bovine cerebral cortex. *J. Neurochem.* **47**: 139-144.

Yeung, S.M.H., and Johnson, R.A., 1989, Irreversible inactivation of adenylyl cyclase by the "P"-site agonist 2'5'dideoxy-, 3'-p-fluorosulfonylbenzoyl-adenosine. *In review*.

26
Adenosine Receptors Mediating Inhibition of Peripheral and Central Neurons

J.A. Ribeiro and A.M. Sebastiao

Introduction

The inhibition caused by adenosine on neurotransmitter release from rat phrenic nerve terminals, described by Ginsborg and Hirst (1972), constitutes the first of a series of observations demonstrating that adenosine decreases the release of excitatory neurotransmitters (acetylcholine, serotonin, noradrenaline, glutamate, aspartate) (see e.g. Ribeiro and Sebastião, 1989). No consistent descriptions appeared on reduction by adenosine of the release of inhibitory neurotransmitters such as GABA or glycine.

The adenosine receptors that mediate inhibition of transmitter release in peripheral and central neurons will be discussed in relation to: **1)** adenylate cyclase/cyclic AMP coupling, **2)** agonist profile, **3)** antagonist profile.

Adenylate Cyclase/Cyclic AMP Coupling

Neither activators (e.g. forskolin) nor inhibitors (e.g. MDL 12,330A) of adenylate cyclase modify the presynaptic inhibitory effect of adenosine on transmission at the innervated frog sartorius muscle (Sebastião, 1989). This suggests that the inhibitory effect of adenosine on neurotransmitter release is not mediated through the adenylate cyclase transducing system. Also, at the central nervous system, Fredholm et al. (1989) could not find a clear correlation between activation of the presynaptic adenosine receptor and modifications in adenylate cyclase activity.

Agonist Profile

The criteria defined to distinguish adenosine receptors into A_1 and A_2 subtypes on the basis of different agonist profiles (see e.g. Daly, 1983) became difficult to reconcile with the agonist profile obtained for the adenosine receptor mediating inhibition of transmitter release.

The A_1 adenosine receptor mediating inhibition of lipolysis in rat adipocytes has an agonist profile with L-PIA > CADO $\geq$ NECA. It seems clear that activation of this receptor in this preparation causes inhibition of adenylate cyclase (Londos et al., 1980). The A_2 adenosine receptor present in human platelets and in other tissues (e.g. blood vessels) has a clear cut definition either in terms of agonist profile with NECA > CADO > L-PIA or in terms of positive coupling to the adenylate cyclase/cyclic AMP system (see references in Ribeiro and Sebastião, 1986).

One of the difficulties with the characterization of the adenosine receptor mediating inhibition of neurotransmitter release is that several authors found an agonist profile with L-PIA being about equipotent or only a little more potent than NECA, and both more potent than CADO. This different agonist profile (see table 1) prompted the

Table 1. Agonist profiles of the presynaptic adenosine receptor.

Preparation	Effect	Order of Potencies (EC_{50};uM)
Peripheral nervous system:		
Frog neuromuscular junction	↓Nerve-evoked contractions and EPPs	L-PIA(0.08), NECA(0.11) > CADO(0.35)[a]
Rat neuromuscular junction	↓Nerve-evoked contractions and EPPs	L-PIA(0.11), NECA(0.11) > CADO(3.8)[b]
Rat neuromuscular junction	↓Nerve-evoked ^{3}H-ACh release	L-PIA(0.1), NECA(0.1) > CADO[c]
Mouse vas deferens	↓Electrically-evoked EJPs	NECA(0.20), L-PIA(0.38) > CADO(2.1)[d]
Central nervous system:		
Rat hippocampus	↓EPSPs	L-PIA(0.02-0.09), NECA(0.15-0.21) > CADO(0.68-0.80)[e]
Rabbit hippocampus	↓Electrically-evoked ^{3}H-ACh release	L-PIA(0.1) $\geq$ NECA(0.45) > CADO(1.5)[f]
Rabbit hippocampus	↓Electrically-evoked ^{3}H-NA release	L-PIA(0.06) $\geq$ NECA(0.30) > CADO(0.90)[g]
Guinea-pig olfactory cortex	↓EPSPs	NECA(0.17) > L-PIA(0.78) $\geq$ CADO(0.96)[h]

↓Decrease or inhibition; $\geq$ at least 2 times more potent; [a]Ribeiro and Sebastião (1985); [b]Sebastião and Ribeiro (1988); [c]Correia-de-Sá et al. (1989); [d]Blakeley et al. (1988); [e]Dunwiddie and Fredholm (1984), Reddington et al. (1985); [f]Jackisch et al. (1984); [g]Jackisch et al. (1985); [h]McCabe and Scholfield (1985). Citations are by no means comprehensive.

proposal that an A_3 adenosine receptor mediates inhibition of neuro-transmitter release (Ribeiro and Sebastião, 1986). Different lipophilicity of the agonists, and non-equilibrium conditions, have been advanced to explain why the presynaptic adenosine receptor has an agonist profile distinct either from that of A_1 or A_2 adenosine receptors. The questions raised by these possibilities are: 1) Why in adipocytes the agonist profile is that expected for an A_1 adenosine receptor? 2) Why the relative potency for the A_2 adenosine receptors is not affected by the differences in lipophilicity of the agonists? 3) How to explain that different laboratories, using different preparations and eventually working in non-equilibrium conditions obtained a similar agonist profile for the adenosine receptor mediating inhibition of neurotransmitter release?

Fredholm and Dunwiddie (1988) postulated the existence of a receptor entity with great plasticity, which might operate the adenylate cyclase/cyclic AMP system, the potassium channel in the case of the heart, and the calcium channel in the case of the nerve endings. These operations would involve G-proteins, which would be different for the different mechanisms triggered by the activated receptor. This would mean that a different agonist profile does not reflect a different receptor but different G proteins, an idea theoretically supported after the development by Kenakin et al. (1989) of a mathematical model that simulates the steady state kinetics of agonist interactions with a promiscuous receptor, i.e. a receptor that can interact with different G-proteins. This model predicts that different relative potencies of agonists for the same receptor would be obtained in different organs, if 1) they are coupled to different G-proteins in these organs, 2) the different agonists are G-protein selective, i.e. if the receptor when bound to one agonist binds preferentially to one G-protein and when bound to other agonist binds preferentially to the other G-protein, and 3) these two G-proteins could activate separately the same cellular response. Applying this concept to the adenosine receptor one has to assume 1) that this receptor can couple to at least two G-proteins (G_i, G_o or G_p?), 2) that L-PIA is selective for one G-protein and NECA is selective for the other G-protein, and 3) that these two G-proteins can both induce inhibition of neurotransmitter release. Coupling of one receptor to different G-proteins does not preclude that the A_1 adenosine receptor is coupled to G_i and inhibits adenylate cyclase, the A_2 adenosine receptor is coupled to G_s and stimulates adenylate cyclase, and the A_3 adenosine receptor is coupled to G_o and/or G_p and inhibits transmitter release independently of modifications in adenylate cyclase activity.

Antagonist Profile

Receptor antagonists are the most important tools to identify pharmacologically different receptors because they avoid problems with efficacy and signal transduction.

Xanthine-derivatives with some antagonist selectivity for A_1 and A_2 adenosine receptors have been developed (e.g. Daly et al.[1], 1987)[2]. Besides being selective for the different receptors, a useful antagonist should not be species selective (i.e. should have a similar potency for similar receptors in different animal species), and within the same animal species, should have similar potency for similar receptors in different preparations. So, it was decided to study the relative potency

Table 2   Potencies of 1,3,8-subtituted xanthines as antagonists of the adenosine receptor at peripheral (neuromuscular junction) and central (hippocampus) synapses, as well as of the A_1 (brain membranes and fat cells) and A_2 (platelets) adenosine receptors.

Preparation	Ki (nM)					
	XAC	DPCPX	8-PT	DPX	XCC	PACPX
Frog neuromuscular junction[a]	23	35	200	295	1905	2291
Rat neuromuscular junction	11	0.54	25	22	10	13
Rat hippocampus	8–11[b]	0.42			5.4	
Rat brain membranes[c]	1.2–3.5	0.3–1.2	76	70	58	2.5
Rat fat cells[d]	4.4–15	0.4–0.5	17	310	83	3.0
Human platelets[e]	24–25	390	1900–4100	210	2400	470

Data from: [a]Sebastião and Ribeiro (1989); [b]Sebastião et al. (1989); [c]Ukena et al. (1986a), Daly et al. (1987), Lohse et al. (1987); [d]Ukena et al. (1986b), Lohse et al. (1987), Martinson et al. (1987); [e]Schwabe et al. (1985), Ukena et al. (1986b), Lohse et al. (1987).

of some of the substituted xanthines as antagonists of the adenosine receptor that mediates inhibition of neurotransmission at the innervated sartorius muscle of the frog, at the innervated diaphragm of the rat and at hippocampal slices of the rat. Using this strategy one can compare the potency of the same antagonist in similar preparations from different species (rat and frog neuromuscular junctions) and in different preparations from the same species (neuromuscular junction and hippocampus of the rat). These preparations seem to posses similar adenosine receptors, operating the same physiological effect — inhibition of neurotransmitter release, and with identical agonist profiles — L-PIA, NECA > CADO.

In table 2 are shown the Ki values determined for the 1,3,8-substituted xanthines, XAC, XCC, DPCPX, PACPX, DPX and 8-PT in both frog and rat neuromuscular junctions. XAC and DPCPX were the most potent to antagonize the inhibitory effect of CADO on nerve-evoked twitch responses in the frog. XCC and PACPX were the least potent, and DPX and 8-PT had intermediate potency at the frog neuromuscular junction. At the rat neuromuscular junction DPCPX was the most potent to antagonize the inhibitory effect of CADO on nerve-evoked twitch responses. XAC, XCC and PACPX were equipotent, and DPX and 8-PT were only slightly less potent than XAC, XCC or PACPX.

Comparing the Ki values obtained for each xanthine in the rat and in the frog neuromuscular junctions (table 3) it can be seen that XAC has about the same potency in both preparations, the greatest discrepancy in the Ki values being obtained with PACPX and XCC. Marked species differences also seem to occur in relation to DPCPX.

Table 3 Ratios between the Ki values obtained for the 1,3,8-substituted xanthines in the frog and in the rat neuromuscular junctions (frog NMJ/rat NMJ) and between the Ki values obtained in the rat neuromuscular junction and in the rat hippocampal slices (rat NMJ/rat hippocampus).

	Ki ratios					
	XAC	8–PT	DPX	DPCPX	PACPX	XCC
Frog NMJ/rat NMJ	2.1	8	13	65	176	190
Rat NMJ/rat hippocampus	1–1.4[a]			1.3		1.8

[a]Data from Sebastião et al. (1989)

To compare the potency of adenosine receptor antagonists at peripheral and central synapses in the same animal species, it was investigated the potency of some of the xanthine derivatives as antagonists of the inhibitory effect of CADO on the amplitude of orthodromically-evoked population spikes recorded from CA_1 pyramidal neurons of rat hippocampal slices. It was used DPCPX, the most potent antagonist at the rat neuromuscular junction, XAC, which showed small species differences, and XCC, which exhibited great species differences. The affinities of these xanthines as antagonists of the adenosine receptor at the rat hippocampus are shown in table 2. These affinities are similar to the affinities observed at the rat neuromuscular junction (see table 3) suggesting that when the comparisons are made within the same animal species the antagonist profile of the presynaptic adenosine receptor in peripheral and central synapses is similar.

In table 2 are also shown the potencies of the xanthines as antagonists of the A_1 adenosine receptor, determined either by their ability to inhibit binding of [^{3}H]CHA or [^{3}H]L-PIA to rat brain membranes (Ukena et al., 1986a; Daly et al., 1987; Lohse et al., 1987) or by their ability to inhibit L-PIA induced inhibition of adenylate cyclase activity in rat adipocytes (Ukena et al., 1986b; Lohse et al., 1987; Martinson et al., 1987). The Ki values obtained for XAC and DPCPX in rat brain membranes, rat adipocytes, rat neuromuscular junction or rat hippocampus are similar, but XCC is 5 to 15 times less potent in rat brain membranes and rat adipocytes than in the rat neuromuscular junction or rat hippocampus.

Comparing the Ki values of the xanthines for the A_2 adenosine receptor in human platelets, determined as their ability to antagonize NECA-induced stimulation of adenylate cyclase (Schwabe et al., 1985; Ukena et al., 1986b; Lohse et al., 1987) with the Ki values obtained in the rat neuromuscular junction or rat hippocampus (table 2), marked differences exist in relation to most of the xanthines, the smaller differences beeing observed with XAC, which is only 2 to 3 times less potent in human platelets than in the rat neuromuscular junction or rat hippocampus. It is of interest to note that it was also XAC which had less species differences, as assessed by the ratio between its Ki values in the frog and in the rat neuromuscular junctions (table 3). This raises the question of whether the A_1/A_2 selectivity of some of the xanthines is not due, at least in part, to 'species selectivity'. Also, one might ask whether the similarity between the potencies of some of the xanthines as antagonists of the A_1 adenosine receptor (in rat brain membranes and rat adipocytes) and as antagonists of the adenosine receptor mediating inhibition of transmission at the rat neuromuscular junction and rat hippocampus, is only a consequence of the comparisons being made in the same animal species. In fact, no correlation could be found between the potencies of these xanthines as antagonists of the pre-synaptic adenosine receptor at the frog neuromuscular junction and as antagonists of the A_1 adenosine receptor in rat preparations or the A_2 adenosine receptor in human platelets (Sebastião and Ribeiro, 1989).

Conclusions

From the three criteria proposed to classify adenosine receptors, 1) the coupling to the adenylate cyclase, 2) the agonist profile, and 3) antagonist profile, one can say that the first two indicate that the adenosine receptor mediating inhibition of transmitter release in peripheral and central synapses is an entity distinct from the A_1 or A_2 adenosine receptors, and seem to be species independent. In relation to the antagonists, the similarities or differences between their affinities for the pre-synaptic adenosine receptor and for the A_1 or A_2 adenosine receptors seem to depend more upon the similarities of differences between species than upon similarities or differences between receptors.

Acknowledgements

We thank Drs J.W. Daly and K. Jacobson for gifts of some of the antagonists.

Abbreviations

ACh: acetylcholine; CADO: 2-chloroadenosine; DPCPX: 1,3-dipropyl-8-cyclopentylxanthine; DPX: 1,3-diethyl-8-phenylxanthine; EJPs: excitatory junction potentials; EPPs: end-plate potentials; EPSP: excitatory posd-synaptic potentials; GABA: gama-aminobutyric acid; G-protein: guanine nucleotide binding protein; L-PIA: L-N^6-phenylisopropyladenosine; NA: noradrenaline; NECA: 5'-N-ethylcarboxamide adenosine; NMJ: neuromuscular junction; PACPX: 1,3-dipropyl-8-(2-amino-4-chlorophenyl)-xanthine; 8-PT: 8-phenyltheophylline; XAC: 1,3-dipropyl-8-(4-((2-aminoethyl)-amino)-carboxymethyloxyphenyl)-xanthine; XCC: 1,3-dipropyl-8-(4-carboxymethyl-oxyphenyl)-xanthine.

References

Blakeley, A.G.H., Dunn, P.M. and Petersen, S.A., 1988, A study of the actions of P$_1$-purinoceptor agonists and antagonists in the mouse vas deferens in vitro, Br. J. Pharmac. **94**:37–46.
Correia-de-Sá, P., et al., 1989, Are there inhibitory and excitatory adenosine receptors modulating acetylcholine release from the phrenic nerve endings? In: Adenosine Receptors in the Nervous System, ed. J.A.

Ribeiro, pp. 196, Taylor & Francis, London.

Daly, J.W., 1983, Role of ATP and adenosine receptors in physiologic processes: summary and prospectus, In: Physiology and Pharmacology of Adenosine Derivatives, ed. J.W. Daly, et al., pp. 275–290, Raven Press, New York.

Daly, J.W., Ukena, D. and Jacobson, K.A., 1987, Analogues of adenosine, theophylline, and caffeine: selective interactions with A_1 and A_2 adenosine receptors, In: Topics and Perspectives in Adenosine Research, ed. E. Gerlach and B.F. Becker, pp. 23–36, Springer-Verlag, Berlin.

Dunwiddie, T.V. and Fredholm, B.B., 1984, Adenosine receptors mediating inhibitory electrophysiological responses in rat hippocampus are different from receptors mediating cyclic AMP accumulation, Naunyn-Schmiedeberg's Arch. Pharmac. 326:294–301.

Fredholm, B.B. and Dunwiddie, T.V., 1988, How does adenosine inhibit transmitter release? TIPS 9:130–134.

Fredholm, B.B., et al., 1989, Mechanism(s) of inhibition of transmitter release by adenosine receptor activation, In: Adenosine Receptors in the Nervous System, ed. J.A. Ribeiro, pp. 123–130, Taylor & Francis, London.

Ginsborg, B.L. and Hirst, G.D.S, 1972, The effect of adenosine on the release of the transmitter from the phrenic nerve of the rat, J. Physiol. Lond. 224:629–645.

Jackisch, R., et al., 1984, Endogenous adenosine as a modulator of hippocampal acethylcholine release, Naunyn-Schmiedeberg's Arch. Pharmac. 327:319–325.

Jackisch, R., Fehr, R. and Herting, G., 1985, Adenosine: an endogenous modulator of hippocampal noradrenaline release, Neuropharmacology 6:499–507.

Kenakin, T.P. and Morgan, P.H., 1989, Theoretical effects of single and multiple transducer receptor coupling proteins on estimates of the relative potency of agonists, Mol. Pharmac. 35:214–222.

Lohse, M.J., et al., 1987, 8-Cyclopentyl-1,3-dipropylxanthine (DPCPX) - a selective high affinity antagonist radioligand for A_1 adenosine receptors, Naunyn-Schmiedeberg's Arch. Pharmac. 336:204–210.

Londos, C., Cooper, D.M.F. and Wolff, J., 1980, Subclasses of external adenosine receptors, Proc. Natl. Acad. Sci. U.S.A. 77:2551–2554.

Martinson, E.A., Johnson, R.A. and Wells, J.N., 1987, Potent adenosine receptor antagonists that are selective for the A_1 receptor subtype, Mol. Pharmac. 31:247–252.

McCabe, J. and Scholfield, C.N., 1985, Adenosine-induced depression of synaptic transmission in the isolated olfactory cortex: receptor identification, Pflugers Arch. 403:141–145.

Reddington, M., et al., 1985, Characterization of adenosine receptors in the hippocampus and other regions of rat brain, In: Purines: Pharmacology and Physiological Roles, ed. T.W. Stone, pp. 17–26, The MacMillan Press, London.

Ribeiro, J.A. and Sebastião, A.M., 1985, On the type of receptor involved in the inhibitory action of adenosine at the neuromuscular junction, Br. J. Pharmac. 84:911–918.

Ribeiro, J.A. and Sebastião, A.M., 1986, Adenosine receptors and calcium: basis for proposing a third (A_3) adenosine receptor, Prog. Neurobiol. 26:179–209.

Ribeiro, J.A. and Sebastião, A.M., 1989, Purinergic modulation of neurotransmitter release in the peripheral and central nervous systems, In: Presynaptic Regulation of Neurotransmitter Release, ed. J. Feigenbaum and M. Hanani, Freund Publishing House, London (in the press).

Schwabe, U., Ukena, D. and Lohse, M.J., 1985, Xanthine derivatives as antagonists at A_1 and A_2 adenosine receptors, Naunyn-Schmiedeberg's Arch. Pharmac. 330:212–221.

Sebastião, A.M., 1989, On the transducing mechanism operated by the adenosine receptor mediating inhibition of transmitter release at the neuromuscular junction, In: <u>Adenosine Receptors in the Nervous System</u>, ed. J.A. Ribeiro, pp. 131-140, Taylor & Francis, London.

Sebastião, A.M. and Ribeiro, J.A., 1988, On the adenosine receptor and adenosine inactivation at the rat diaphragm neuromuscular junction, <u>Br. J. Pharmac.</u> **94**:109-120.

Sebastião, A.M. and Ribeiro, J.A., 1989, 1,3,8- and 1,3,7-substituted xanthines: relative potency as adenosine receptor antagonists at the frog neuromuscular junction, <u>Br. J. Pharmac.</u>, **96**:211-219.

Sebastião, A.M., Ribeiro, J.A. and Stone, 1989, Antagonism of the adenosine receptor at the phrenic nerve-diaphragm and hippocampus of the rat by a xanthine amine congener, <u>Br. J. Pharmac.</u> **96**:265P.

Ukena, D., et al., 1986a, Species differences in structure-activity relationships of adenosine agonists and xanthine antagonists at brain A_1 adenosine receptors, <u>FEBS Letts</u> **209**:122-128.

Ukena, D., et al., 1986b, Functionalized congeners of 1,3-dipropyl-8-phenylxanthine: potent antagonists for adenosine receptors that modulate membrane adenylate cyclase in pheochromocytoma cells, platelets and fatt cells, <u>Life Sci.</u> **38**:797-807.

27
Adenosine Receptors as Drug Targets: Fulfilling the Promise?

M. Williams

Introduction

The purine nucleoside, adenosine, has undergone a research renaissance in the past decade fuelled by a more focussed delineation of the A_1 and A_2 receptor types that mediate its actions in mammalian tissues (Hamprecht and Van Calker, 1985).These receptors have a distinct pharmacology and tissue distribution that offer the potential for the development of selective modulators of adenosine related effects that may have use as therapeutic entities (Williams, 1990)

The physiological and pharmacological actions of adenosine on cardiovascular function were initially reported 60 years ago by Drury and Szent-Gyorgyi (1929) and led to a rapid, if unsuccessful evaluation of the purine nucleoside as an anihypertensive agent (Honey et al.,1930; Jezer et al.,1933).The limited usefulness of the compound in lowering blood pressure even in an era when the β-adrenoceptor antagonists, angiotensin converting enzyme (ACE) inhibitors and other modulators of the renin-angiotensin system were non-existant was predictable based on the metabolic lability of the parent compound. And as more stable analogs became available, their testing as blood pressure lowering agents , reported only anecdotally, was similarly proscribed due to their greater half life, increased potency and lack of receptor selectivity .Thus the comments of Jezer et al. (1933) that adenosine was "not a useful therapeutic preparation for the treatment of heart diseases" became a *sine non qua* that extended to practically all situations involving a potential therapeutic role for the purine. Nonetheless, the physiological characterization of the role of adenosine in a variety of different tissues continued, with intermittant interest from various pharmaceutical companies punctuating a considerable research effort at the basic level. In the 1950s, continued research on the cardiovascular effects (Green and Stoner, 1950) and CNS sedative effects (Feldberg and Sherwoood, 1954), of both adenosine and its nucleotide, ATP were complemented by seminal research in the area of 'non-cholinergic, non-adrenergic (NANC)" neurotransmission (Burnstock, 1976;1978) and studies on the role of adenosine as a metabolic regulator linking coronary function to oxygen supply in the heart and brain (Berne, 1963; Berne et al., 1987) International meetings in Banff in 1978 (Baer and Drummond, 1979) , Charlottesville in 1983 (Berne et al., 1983) and Munich in 1986 (Gerlach and Becker, 1987) further attested to the wealth of new data being generated by an increasing number of researchers.

Efforts in the pharmaceutical industry however, tended to be sporadic primarily because new chemical entities (NCEs) in the area were few and far between. This reflected a lukewarm to almost total disinterest from medicinal chemists who concluded that any NCEs derived from adenosine would prove difficult to patent and would eventually be proven to be antimetabolites. Yet programmes initiated at Merck Darmstatdt, Boehringer Ingelheim,Byk Gulden, Abbott, Takeda, Nova, Parke Davis, Searle, Burroughs Wellcome, Merck,Sharp and Dohme, ICI, CIBA-Geigy, Nelson Research and Smith Kline & French reflected a degree of interest from industrial biologists , primarily in the evaluation of agonists for therapeutic potential in the cardiovascular and CNS areas. A major limitation to therapeutic targeting in such programmes was the ubiquity of effect of the purine nucleoside. Irrespective of the validity of data generated in a given organ system, the major concern was that adenosine agonists would have a plethora of side effects that would make their use as drugs untenable. Interestingly, this perspective remained fashionable as research efforts in the area of peptide neuromodulators became a major focus in a number of therapeutic areas. Like adenosine, many such peptides have a ubiquity of actions and also share with the purine nucleoside a level of bioavailability that can at best be described as mediocre. Yet these considerations, despite the expenditure of many millions of dollars without any significant successes (Williams, 1990), have not discouraged research efforts directed towards therapeutic agents in these areas. In many respects (Williams, 1989), adenosine and the peptides subserve similar neuromodulatory functions offering new approaches to therapeutic intervention via indirect, paracrine effects on tissue function. Given this fact, one must question why the area of purinergic neurotransmission has not been more actively researched in the industrial setting and what can be done to remedy the situation. If not to develop new drugs, at least to objectively evaluate the potential of the area.

Towards a Molecular Target

The seminal observation of Sattin and Rall (1970) that the xanthines, caffeine and theophylline, were adenosine antagonists provided the first direct biochemical evidence that there were adenosine receptors and focussed interest on the area of antagonists as potential drug targets. A concerted chemical effort to improve the selectivity, efficacy and bioavailability of these agents (Daly,1982; 1985; 1989) culminated in several series of 8-phenylxanthines, including CPX (cyclopentylxanthine), CPT (cyclopentyltheophylline), and PD 115,199 (Bruns et al., 1988) and the congeners, XAC (xanthine amine congener) and XCC (xanthine carboxylic acid congener.; Jacobson ,1988) and BW 844U (Linden et al., 1988).

Concurrently, further studies on adenosine modulated adenylate cyclase that had led to the identification of receptor subtypes (*see* Hamprecht and Van Calker,1985) stimulated the development of binding assays for both A_1 and A_2 receptors (Bruns et al.,1980;1986). As with many radioreceptor assays, various groups found evidence for the existence of different affinity forms of the two receptors.Whether such binding sites represented true receptor subtypes or different affinity states of a single receptor (Williams and Sills, 1989) has not been determined. However evidence for different affinity forms of the A_2 receptor (Daly et al., 1983; Bruns et al., 1986) has been substantiated and major species differences in the in vitro pharmacology, especially in regard to antagonists, of both A_1 and A_2 receptors have been documented (Ferkany et al., 1986; Ukena et al.,1986; Stone et al., 1988). To date however, it has not been possible to determine which, if any species mostly closely resembles man. Molecular biological approaches have yet to provide useful information to resolves these issues. However, it has been shown that the A_2 receptor is a molecular species distinct from the A_1 receptor (Barrington et al.,1989)

Implicit in the process of drug discovery has been the discovery of novel compounds showing selectivity for a given receptor. These then foster an iterative process to generate NCEs and thus elucidate the physiological role of the receptor using the NCE as a research tool. While this approach has its limitations (Black, 1986) because of inherent tissue specific differences in receptor pharmacology, the identification of totally new structures for a receptor can exponentially increase understanding of the role of a receptor while maximizing the opportunities to develop new drugs. One example of this approach has come from research related to the 33-amino acid peptide, cholecystokinin (CCK) which has been implicated in disease processes in both the CNS and gastrointestinal tract. As with many peptides, initial interest in CCK was predicated on the development of histochemical and binding assays to document its tissue distribution and that of specific receptors.The majority of research in relation to CCK focussed on its co-localization with dopamine and an anticipated potential to function as a novel antipsychotic agent. However, the only ligands available to assess this potential were peptides, various modified versions of CCK. Their usefulness, like that of adenosine was proscibed by their short half life and poor biovailability. In the course of a random screening program directed towards the pancreatic CCK (CCK-A[alimentary]) receptor, researchers at Merck (Chang et al., 1985) identified a novel, non peptide CCK ligand, asperlicin, from a fermentation broth Chemical modification of this novel structure led to a stable, bioavailable drug candidate, MK 329, an antagonist benzodiazepine related to asperlicin that is now in clinical trials for use as a satiety agent. In addition, the knowledge derived from the discovery and chemical modification of asperlicin has led to a generalized theory of how stable, non-peptidic ligands may be developed for peptide receptors (Williams et al.,1987; Evans et al.,1988).

In the adenosine area, targeted screening has resulted in the identification of a number of series of novel heterocyclic agents which have adenosine antagonist activity (Daly et al., 1988). These include: the triazoloquinazoline, CGS 15943A and related structures (Francis et al., 1988),a series of pyrazolo [4,3d]pyrimidin-7-ones (Hamilton et al., 1987a),a series of triazolo [4,3a] quinoxalinamines (Trivedi and Bruns, 1988) and a series of resorcinol derivatives (Peet et al.,1989). A more novel group of adenosine receptor modulators typified by the 2-amino-3-benzoylthiophene, PD 81,723 (Fig.1) have also been discovered as the result of a focussed screening effort (Bruns and Fergus, 1989). In the agonist area however, beyond selective modifications of the purine nucleoside pharmacophore, there has been very little published in the way of novel i.e. non-ribose, structural entities that are potent and selective agents (Williams et al., 1989).Based on precident as exemplified by the example of asperlicin, a major challenge in the area of adenosine research is to discover novel agonist pharmacophores that are not purine nucleosides.

Therapeutic Potential

Cardiovascular

Agonists

A major emphasis for the potential evaluation of adenosine and its various analogs as therapeutic entities, evolved from the seminal work of Drury and Szent-Gyorgyi (1929). In light of the importance of blood pressure lowering agents , this area has been very attractive from a commercial perspective. However, while adenosine is a potent hypotensive agent (Mullane and Williams, 1990) its transient actions, related to a half life in the order of 10s due to uptake by the coronary endothelium have led to the search for more stable analogs. The latter however , due to their lack of receptor selectivity have had multiple effects mechanisms affecting blood pressure regulation. Three distinct types mechanisms are known: a reduction in cardiac output via activation of A_1 receptors in the atrioventricular node (Bellardinelli et al.,1983); a reduction in coronary and vascular bed resistance resistance vessels via activation of A_2 receptors on smooth muscle (Hamilton et al., 1987b; Oei et al.,1988); and a series of complex, indirect effects involving the renin-angiotensin system and baro-and chemoreceptor mediated autonomic responses. A potent , peripherally active, A_2-selective receptor agonist has long been sought after as the 'ideal' for an effective antihypertensive agent that would directly reduce blood pressure without eliciting a corresponding reduction in peripheral oxygen supply as part of its blood pressure lowering actions. It may be noted however, (Black, 1989) that systemic vasodilation is thought to be less preferable as a means to reduce blood pressure than a reduction in cardiac output. A series of 2-substituted adenosines including CV 1808 (2-phenylamino adenosine; Fig. 1) and CV 1674 (2-(4-Methoxyphenyl)adenosine) with vasodilatatory and antianginal properties developed by Takeda (Kawazoe et al., 1980) were the earliest A_2-selective agents described (Bruns et al.,1986).Further efforts at Parke Davis (Bridges et al.,1987) and CIBA-Geigy (Hutchison et al. 1989) led to the identification of CI-936 (N^6-2,2-diphenylethyl adenosine) and CGS 21680 (2-(2-p-carboxyethyl) phenethylamino adenosine

: Fig.1), respectively, both compounds being potent and selective agonists at the brain A_2 receptor. CI 936 is the most potent with a Ki value of 6 nM but is only 4-fold selective. Thus the compound retains potent activity at A_1 receptors.The methyl-dimethoxy analog of CI-936, (PD 125,944) has approximately the same activity at the A_2 receptor as the parent compound but is some 40 fold more selective (Trivedi et al., 1988, 1990) In contrast, CGS

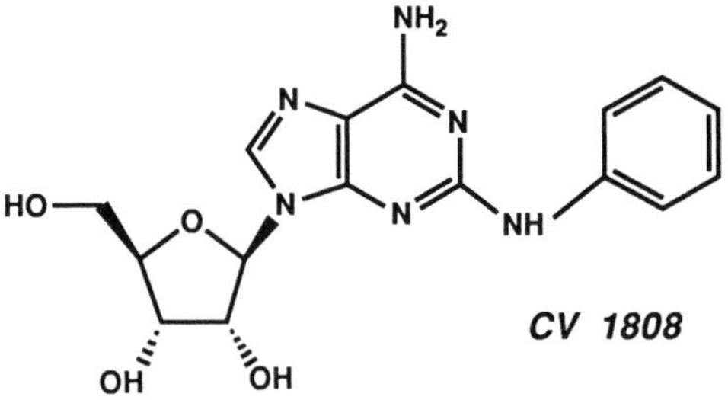

CV 1808

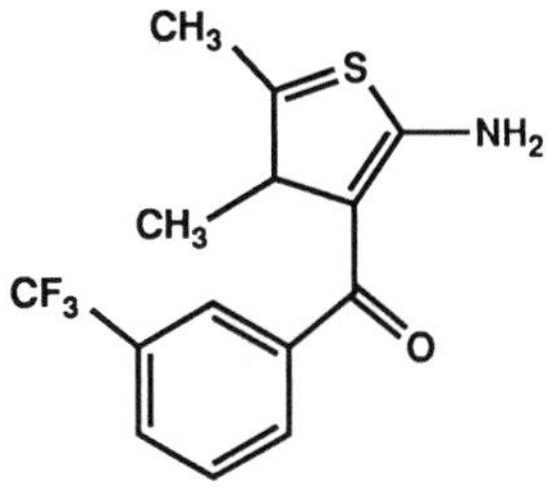

PD 81,723

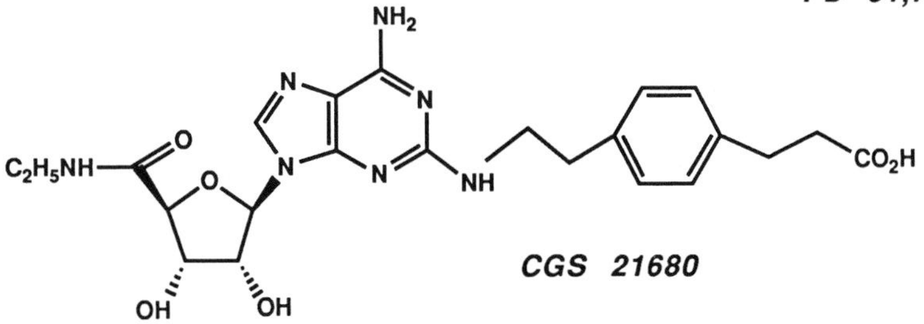

CGS 21680

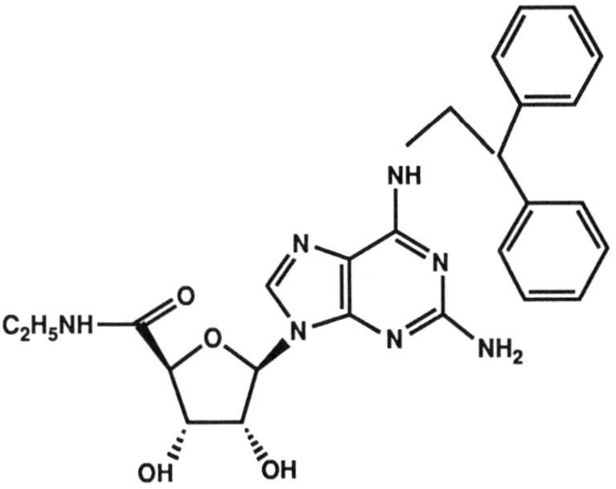

CI-936

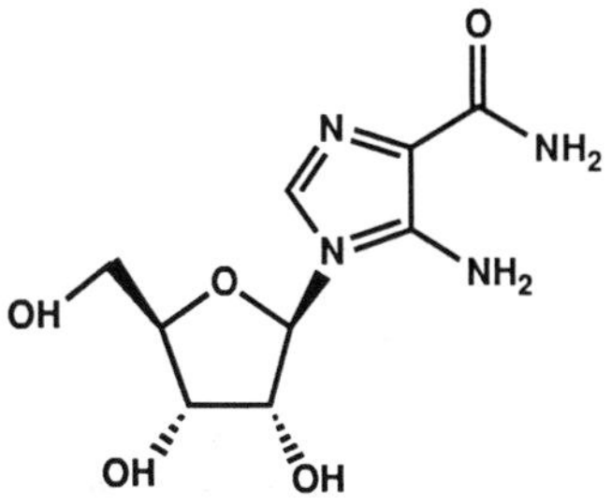

AICA Riboside

Figure 1: Structures of adenosine receptor ligands and modulators

21680 while still relatively active at the A_2 receptor (Ki = 13 nM), is 114-fold selective, its activity at A_1 receptors being greater than 3 μM.As a result, in contrast to other reported agonists, CGS 21680 has no cardiodepressant activity, although a reflex tachycardia due to autonomic related responses has been reported (Hutchison et al.1989). Interestingly, the relatively weak (Ki = 115 nM) and only 5-fold selective compound, CV 1808 has hypotensive activity similar to that seen with CGS 21680 with a corresponding lack of cardiodepressant activity.This profile may be related to both the weak A_1 agonist activity of the compound and its intrinsic nucleoside transport inhibitory properties (Taylor and Williams,1982; Balwiericzak et al., 1990).While these data indicated that a selective A_2 receptor agonist could effectively lower blood pressure without reducing cardiac output, the therapeutic potential of such an agent remains far from proven. The involvement of adenosine in the pathophysiolology of hypertension remains unclear (Ohnishi et al.,1986; Jackson 1987; Mullane and Williams,1990) and in fact does not appear to have been extensively studied. Furthermore, in situations where adenosine or its analogs have been administered there have been various reports of acute side effects. In man, adenosine (Sylven et al.,1986; Watt et al., 1986; Robertson et al., 1988) and stable analogs such as metrifudil (N^6-benzyl adenosine ; Schaumann and Kutscha, 1972) cause unacceptable side effects including flushing, listlessness, panic, angina pectoris-like pain, respiratory dysfunction and headache. Anecdotal evidence for 2-chloroadenosine, CV-1808 and R-phenylisopropyladenosine has suggested similar problems and on this basis, further work will be required to indicate a beneficial effect of adenosine analogs in the treatment of molecular lesions leading to increased blood pressure that may make such agents superior to existing, proven therapies for the treatment of hypertension.

Adenosine has potent actions on renal function (Osswald,1988; Churchill and Bidani, 1990) including inhibition of the renal renin system,a transient regulation of renal blood flow and tubular epithelial function and inhibition of transmitter release from renal nerves. A clear understanding of the physiological consequences of this myriad of actions has yet to emerge because of a lack of data related to the effects of A_2 selective agonists and selective antagonists on kidney function.

Adenosine can inhibit superoxide anion generation in human neutrophils and promote chemotaxis via the activation of A_2 and A_1 receptors, respectively (Cronstein et al., 1987; 1989; Cronstein and Hirschhorn,1990). The purine can also prevent granulocyte activation and adhesion and may thus have potential benefit in maintaining endothelial function during reperfusion injury (Mullane and Williams,1990) . An A_2 receptor agonist that elicits hypotension without bradycardia may have additional benefit in the attenuation of the myocardial necrosis, granulocyte accumulation and endothelial damage associated with reperfusion injury and in addition may function as an anti-atherosclerotic agent. Adenosine-mediated inhibition of platelet aggregation is antithrombogenic and can reduce platelet mediated plaque formation (see Mullane and Williams, 1990) reinforcing a role for the nucleoside in the atherosclerotic process. Unfortunately, current preclinical models appear inadequate to examine the latter hypothesis and necessary human validation to show efficacy may take from 5 -10 years following drug approval. Unless additional benefit for a blood pressure lowering adenosine analog can be shown, it is unlikely to garner commercial interest sufficent to justify the expense of clinical trials However, until clinical evaluation is undertaken, it is unlikely that such data will become available, a peverse 'Catch 22' situation that occurs often in the drug discovery process. One possible solution to this dilemma may be to develop toxicologically 'clean' adenosine agonists for use as adjunct therapy following ballon angioplasty. Adenosine has been successfully used for the treatment of supraventricular tachyarrhythmia (Di Marco et al., 1983) and to produce a controlled hypotension during surgery for cerebral aneurysms (Owall et al.,1987), both indications benefitting from the short half life of the parent nucleoside

Antagonists

The xanthine adenosine antagonists are effective cardiac stimulants, their ability to block the chronotropic and dromotropic effects of endogenous adenosine on cardiac conduction resulting in an increase in cardiac output. Aminophylline has been used clinically for this condition and a series of newer cardiotonics including sulamazole and amrinone and have been reported as adenosine antagonists (Parsons et al.,1988).

In the kidney, adenosine antagonists are effective diuretic agents and may be efficacious in the treatment of acute renal failure (Collis, 1988). Glutamate congeners of XAC have been used as renal selective diuretic prodrugs using the kidney specific enzyme, γ-glutamyl transferase to produce the active xanthine entity (Barone et al.,1989). This prodrug approach may have applicability for agonists as well as antagonists and in addition to potentially circumventing side effects may increase the duration of action of such entities.

Central Nervous System

Agonists

In addition to being an effective CNS depressant, adenosine has been implicated, together with ATP, in the actions of nearly all classes of CNS active drugs (Williams , 1987; Deckert and Gleiter, 1989). This scope of activity may appear bewildering yet more probably indicates the global importance of adenosine in modulating brain function.a factor further reflected by the high densities of A_1 and A_2 receptors in brain tissue and their selective distribution (Jarvis and Williams, 1990).

Adenosine , based on the use of various adenosine analogs and xanthine antagonists has been implicated in the molecular mechanisms of action of hypnotics, anxiolytics, antidepressants, anticonvulsants, antipsychotics and analgesics and is thought to be involved in the central regulation of a variety of peripheral functions including cardiac function and respiration (Williams, 1987). Many of these actions may be explained by the ability of endogenous adenosine to inhibit neurotransmitter release (Fredholm and Dunwiddie, 1988) but also involve direct postsynaptic actions.

The xanthines, caffeine and theophylline are convulsive agents in the CNS and a very attractive hypothesis currently being investigated is that adenosine is an endogenous anticonvulsant, the purine being produced in large amounts as a result of convulsive ischemia (Dragunow, 1988).Adenosine may function in a similar protective manner to prevent the cell death associated with stroke-related ischemia (Evans et al.,1987; Marangos et al.,1990). Central purinergic systems have been implicated in the mechanism of action of the benzodiazepine (BZ) anxiolytics (*see* Phillis and O'Regan,1988). In addition to the antagonist, caffeine being an effective anxiogenic in humans, several BZs have been reported to interact with adenosine A_1 receptors and transport systems. Similarly, adenosine antagonists can antagonize BZ actions in the CNS at both the functional and molecular levels Adenosine and the BZs share a spectrum of related interactions at the anxiety, anticonvulsant, ethanol-modulating and muscle relaxant levels. While there has been no clear outcome from a continuing series of studies that are now over a decade old, neither has there been any convincing evidence that adenosine and the anxiolytic BZs do not share some commonality of action (Williams, 1987, 1990).

The data supporting a role for adenosine in the mechanism(s) of analgesia are complex and contradictory involve both central and peripheral components.The antinociceptive actions of adenosine in the spinal cord appear to be A_2 mediated (DeLander and Hopkins, 1987). Morphine analgesia in the spinal cord occurs as the result of adenosine release from small diameter primary afferent fibers and subsequent activation of postsynaptic receptors mechanisms that suppress nociceptive stimuli (Sawynok et al., 1989).Given the potential interactions with opiates and ethanol, adenosinergic mechanisms may also underlie some aspects of the actions of the drug classes known as substances of abuse. The effects of adenosine on central dopaminergic function are similarly complex (Jarvis and Williams, 1987) and is is unclear, despite some elegant studies on the preclinical antipsychotic effects of adenosine analogs (Heffner et al.,1989), whether the actions of the purine nucleoside are not the direct result of an adenosine receptor-mediated inhibition of striatal dopamine release.The evidence for a role of adenosine in depression is far from convincing and may reflect limitations in the animal models used to characterize such entities

Like many classes of CNS active agents, those whose mechanism of action are thought to involve adenosine are far from ideal in their clinical characteristics. Many anticonvulsants produce their actions via unknown,'black box'-type mechanisms and their side effect profile is accordingly limited. Similarly there is a major need to improve on the side effect profiles of existing therapeutic agents for the treatment of anxiety (dependence liability, alcohol potentiation, sedation) and schizophrenia (tardive dyskinisia) while the field of stroke therapy is still in its early stages of targeting for therapeutic intervention. Accordingly, should CNS selective adenosine agonists prove feasible to develop, it should be easier for such agents to be competitive in the development pipeline than in the more extensively researched area of antihypertensives.

Antagonists

Caffeine, as the prototypic adenosine antagonist is an effective and safe central stimulant that has been in human usage for many thousands of years. The development of more potent, receptor and tissue selective adenosine antagonists has appeared attractive in the major growth area of cognition enhancement. One such entity, HWA 285 is active in humans (Hindmarch and Subhan, 1985). A major limitation however, is that the target geriatric population for a cognition enhancer often suffers from cardiac and renal insufficencies that would be severely compromised by adenosine antagonists with even a modicum of peripheral activity.

Inflammation and Immune Function

Despite the presence of adenosine receptors on lymphocytes, basophils, mast cells, monocytes and neutrophils, the role of the nucleoside in the process of inflammation and its role in immune function have not received as much attention as effects on CNS and cardiovascular function which is to some extent lamentable given the paucity of therapeutic agents in these areas.

Inflammatory mediators such as the cytokines can evoke superoxide release from neutrophils leading to considerable cell damage in a variety of tissues (Mullane, 1988).As noted, adenosine, acting via A_2 receptors can prevent superoxide formation and the deleterious actions of neutrophils in the host response (Cronstein et al.,1989).Adenosine is also effective in inhibiting lymphocyte proliferation , T-cell mediated cell lysis and complement C2 synthesis (*see* Polmar et al.,1988). The effects of adenosine on immune cell function have been related to an immunohomeostatic process that may be important in regulating certain autoimmune responses (Polmar, 1984).The inhibition of mast cell histamine release by adenosine also contributes to the effects of the purine nucleoside on the inflammatory response and may be of potential benefit in the treatment of asthma (Griffith and Holgate, 1990).

The general anti-inflammatory role of adenosine is borne out by a recent report (Krenitsky et al.,1988) on several novel adenosine analogs with antinociceptive activity some 63-fold greater than aspirin and 650-times more potent than acetaminophen. In a classical adjuvant arthritis hyperalgesia model, a compound in this series had an ED_{50} of 0.24 mg/kg p.o. versus 28 mg for aspirin.

Adenosine agonists may prove to be effective in the treatment of various inflammatory disease proceses involving neutrophil and platelet infiltration. Given the paucity of effective entities in this area and the 'black box' mechanism(s) of action of many disease modifying anti-rheumatic drugs (DMARDs), this area may be considered one of the more attractive for evaluation of a purine agonist as a therapeutic agent.

Indirect Approaches to Receptor Modulation

Irrespective of the proposed use for an adenosine agonist, the multiple actions of the nucleoside are inescapable with the majority of ligands currently available. The degree of CNS side effects of an agonist effective in the cardiovascular system has not been adequately resolved anymore than the respective contributions of central and peripheral mechanisms that lower blood pressure. Conversely, any potential CNS active entity will undoubtedly, based on the current knowledge of receptor classification, result in unacceptable blood pressure lowering actions.Given these side effect limitations several research groups have been evaluating agents that can potentiate the actions of endogenous adenosine produced tonically or in response to tissue trauma.

Nucleoside Transport Inhibitors.

Nucleoside transport processes have been extensively studied in relation to the development of antineoplastic agents. Compounds that inhibit adenosine uptake may circumvent some of the lack of specificity of exogenously administered direct agonists. As many as three distinct nucleoside transport systems appear to exist (IJzerman et al., 1989) and while dipyridamole and deoxycoformycin are the best known inhibitors of adenosine uptake, a more potent series of compounds exemplified by mioflazine (Wauquier et al., 1987) have been found to be effective hypnotics and have been rumored to be undergoing evaluation as potential anticonvulsants in man. Structure activity relationships for compounds of the mioflazine type (IJzerman et al. 1989) are suggestive of cooperative effects.

Site and Event Specific Agents

Administration of the purine intermediate, AICA riboside (5-amino-4-imidazole carboxamide-ribose; Fig.1) has been found to lead to a repletion of ATP and an improvement in cardiac function following ischemia (Engler, 1987) While the mechanism responsible for the actions of this compound are controversial (*see* Mullane and Williams, 1990), AICA riboside can effectively increase adenosine levels in situations where adenine nucleotide metabolism is compromised. It may thus be a 'site- and event-specific' potentiator of adenosine with potential use as an adjunct therapy in the attenuation of myocardial reperfusion injury and stroke related CNS cell death.

Allosteric Modulators

Based on a screening approach to novel adenosine ligands, Bruns and Fergus (1989) have described a series of 2-amino-3-benzoylthiophenes including PD 81,723 (Fig.1), which are active as allosteric enhancers of A_1 receptor binding and have provide preliminary evidence for the possible existence of an adenosine receptor complex.

Crucial Issues in Adenosine-Related Drug Therapies

The attractiveness of adenosine as a drug target is confounded by both the broad side effect profile of the purine nucleoside and the existing competition in the marketplace for the various indications cited. The search for a potent antipsychotic based on an adenosine agonist while ignoring the potential for hypotension, or a cognition enhancer while similarly ignoring the renal and cardiotonic actions of an antagonist is naive based on current knowledge in this area. Intellectually, the existence of distinct tissue receptor subtypes would be a major, if not the only key to the development of viable drugs in this area. Ideally, the existence of such entities could be elaborated by the discovery of novel ligands akin to the CCK situation cited above or by the discovery of an adenosine receptor gene family using the same molecular biological techniques that led to the discovery of the steroid receptor superfamily (Evans, 1988). In this task, the recent identification of potent A_2 selective agonists has, for the first time, provided the tools necessary to delineate more accurately the physiological function of these receptors in mammalian tissue function.

Alternatively, given the putative role of adenosine as a paracrine effector agent (Williams,1989) an alternative approach may be to develop indirect modulators of adenosine receptor function. This concept, termed 'isosteric modulation'(Costa , 1989) may represent a more subtle and relatively side effect free approach that will be useful not only for adenosine but also the peptide neuromodulators.For instance in the case of the amino acid neurotrasnmitters, GABA and glutamate, it has not been possible to develop directly acting entities that are either efficacious or side effect free. While the situation with direct allosteric modulators as described by Bruns and Fergus (1989) must await the identification of more potent agents, both the newer nucloside transport inhibitors and AICA riboside are being clinically evaluated.

References

Baer,H.P.and Drummond,G.I. (1979) *Physiological and Regulatory Functions of Adenosine and Adenine Nucleotides,* Raven, New York .

Balwierczak,J.L. et al. (1990) J.Pharmacol. Exp. Ther. submitted.

Barone,S. et al., (1989) J. Pharmacol. Exp. Ther.,250, 79.

Barrington, W.W.et al. (1989) Proc. Natl.Acad. Sci.U.S.A. 86, 6572.

Bellardinelli, L. et al. (1983) In *Regulatory Function of Adenosine*..(Eds.Berne,R.M., Rall, T.W. and Rubio,R.)Nijhoff,Boston, p. 378.

Berne,R.M. (1963) Am.J. Physiol. 204, 317.

Berne,R.M.,et al. (1983) *Regulatory Function of Adenosine*.Nijhoff, Boston.

Berne,R.M. et al. (1987) In *Topics and Perspectives in Adenosine Research,* (Eds.Gerlach E. and Becker,B.F.) Springer Verlag, Berlin.

Black,J.W. (1986) Br.J. Clin. Pharmacol.,22, S5.

Black,J.W. (1989) Science 245,486.

Bridges,A.J.et al. (1987). J .Med. Chem. 30, 1709.

Bruns,R.F.and Fergus,J.H. (1989) In *Adenosine Receptors in the Nervous System* (Ed. Riberio,J.A) Taylor and Francis, London, in press.

Bruns R.F.et al. (1980) Proc.Natl. Acad.Sci. U.S.A. 77, 5547.

Bruns, R.F.et al. (1986) Mol. Pharmacol. 29,331.

Bruns,R.F. et al.(1988) In *Adenosine and Adenine Nucleotides:Physiology and Pharmacology*. (Ed. Paton,D.M.), Taylor and Francis, London,p 39.

Burnstock, G. (1972) Pharmacol. Rev. 24, 509.

Burnstock, G (1978) In *Cell Membrane Receptors for Drugs and Hormones* (Eds. Bolis,L. and Straub,R.N.) Raven, New York, p. 107.

Chang,R.S.L. et al. (1985) Science 239,177.

Churchill, P C.and Bidani, A. (1990) In *Adenosine and Adenosine Receptors* (Ed. Williams,M.), Humana, Clifton, New Jersey, in press.

Collis, M.G. (1988) In *Adenosine and Adenine Nucleotides:Physiology and Pharmacology*. (Ed. Paton,D.M.), Taylor and Francis, London,p 259.

Costa, E. (1989) In *Allosteric Modulation of Amino Acid Receptors: Therapeutic Implications.* (Eds. Barnard,E.A. and Costa,E.) Raven,New York, p. 3.

Cronstein,B.N. and Hirschhorn,R. (1990) In *Adenosine and Adenosine Receptors* (Ed. Williams,M.), Humana, Clifton, New Jersey, in press.

Cronstein,B.N. et al. (1989) J. Clin. Invest. submitted.

Daly,J.W. (1982) J.Med.Chem. 25,197.

Daly,J.W. (1985) In *Purines: Pharmacology and Physiological Roles* (Ed. Stone,T.W.) MacMillan, London, p. 5.

Daly,J.W. (1989) in *Adenosine Receptors in the Nervous System* (Ed. Riberio,J.A.), Taylor and Francis, London, in press.

Daly, J.W. et al. (1983) Cell. Mol. Neurobiol. 3,69.

Daly,J.W.et al. (1986) J.Med. Chem., 29, 1305.

Daly,J.W.et al. (1988) Biochem.Pharmacol. 37, 655.

Deckert,J. and Gleiter,C.H. (1989) Trends Pharmacol. Sci.10,99.

DeLander,G.E. and Hopkins,C.J. (1987) Eur.J. Pharmacol. 139, 215.

Di Marco,J.P. et al. (1983) Circ. 68, 1254.

Dragunow,M.(1988) Prog. Neurobiol. 31,85.

Drury,A. and Szent-Gyorgyi,A. (1929) J. Physiol. (Lond.) 68,213.

Engler,R. (1987) Fed. Proc. 46, 2407.

Evans, B.E. et al. (1988) J.Med. Chem. 31, 2235.

Evans,M.C. et al. (1987) Neurosci. Lett. 83, 287.

Evans,R.E. (1988) Science,240, 889.

Feldberg,W. and Sherwood, S.L. (1954) J.Physiol. (Lond.) 123,148.

Ferkany,J.W. et al. (1986) Drug Dev. Res. 9,85.

Francis,J.E.et al. (1988) J.Med.Chem. 31,1014.

Fredholm,B. and Dunwiddie, T.V. (1988) Trends Pharmacol. Sci. 9, 130.

Gerlach,E. and Becker,B.F. (1987)*Topics and Perspectives in Adenosine Research*, Springer Verlag. Berlin.

Green,H.N. and Stoner, H.B. (1950) *Biological Actions of The Adenosine Nucleotides*, Lewis, London.

Griffiths,T.W.and Holgate,S.T. (1990) in *Adenosine and Adenosine Receptors* (Ed.Williams,M.), Humana, Clifton, NJ, in press.

Hamilton,H.W. et al. (1987a). J.Med.Chem. 30, 91.

Hamilton,H. W. et al. (1987b) Life Sci., 41, 2295.

Hamprecht, B. and Van Calker,D. (1985) Trends Pharmacol. Sci. 6, 153.

Heffner,T.G. et al., (1989) Psychopharmacol. 98, 31.

Hindmarch,I. and Subhan, Z.(1985) Drug. Dev. Res. 5, 379.

Honey,R.M. et al. (1930) Quart. J. Med. 23, 485.

Hutchison,A.J. et al. (1989) J. Pharmacol. Exp. Ther. 251, in press.

IJzerman, A.P. et al. (1989) Eur. J. Pharmacol. 172,273.

Jackson,E.K. (1987) Am. J. Physiol. 253, H909.

Jacobson,K.J. (1988) In *Adenosine Receptors* (Ed.Cooper,D.M.F. and Londos,C.), Liss, New York ,p 1.

Jarvis, M.F. and Williams,M. (1990) in *Adenosine and Adenosine Receptors* (Ed.Williams,M.), Humana, Clifton, NJ, in press.

Jezer,A. et al. (1933). Am. Heart J., 9, 252.

Kawazoe.K. et al. (1980) Arzneim. Forsch. 30, 1083.

Krenitsky,T.A.et al. (1987) European Patent Application 0 260 852.

Linden,J. et al. J.Med. Chem. 31, 745.

Londos,C. and Wolff,J. (1977) Proc. Natl. Acad. Sci.U.S.A. 74, 5482.

Londos,C. et al. (1980) Proc. Natl. Acad. Sci.U.S.A. 77, 2551

Marangos,P.J. et al. (1990) In *Current and Future Trends in Anticonvulsant,Antianxiety and Stroke Therapy* (Eds. Meldrum,B.S.and Williams,M.), Liss, New York, in press.

Mullane,K.M. (1988) In *Human Inflammatory Disease* (Eds. Marone,G. et al.) Decker, Philadelphia, p. 143.

Mullane, K.M. and Williams,M. (1990) in *Adenosine and Adenosine Receptors* (Ed. Williams,M.), Humana, Clifton, NJ, in press.

Oei,H.H. et al. (1988) J. Pharmacol. Exp. Therap.,247, 882.

Ohnishi, A. et al (1986) Hypertension 8, 391.

Osswald,H. (1988) In *Adenosine and Adenine Nucleotides. Physiology and Pharmacology,* (Ed. Paton,D.M.) , Taylor and Francis, London, p 193.

Owall, A. et al. (1987) Anesth. Analg. 66,229.

Parsons,W.J. et al. (1988) Mol. Pharmacol. 33,441.

Peet,N.P.et al. (1988) J.Med. Chem.31, 2034.

Phillis,J.W. and O'Regan,M.H. (1988) Prog. Neuropsychopharmacol. Biol. Psych. 12, 389.

Polmar,S.H. (1984) Surv. Immunol. Res. 3, 274.

Polmar,S.H.et al.(1988)In *Adenosine Receptors* (Ed.Cooper,D.M.F. and Londos,C.), Liss, New York ,p 97.

Robertson,D.et al (1988) In *Adenosine and Adenine Nucleotides. Physiology and Pharmacology,* (Ed. Paton,D.M.) , Taylor and Francis, London, p 241.

Sawynok, J. et al. (1989) Trends Pharmacol. Sci. 10, 186.

Sattin,A. and Rall,T.W. (1970) Mol. Pharmacol. 6, 13.

Schaumann,E. and Kutscha,W. (1972) Arzneim. Forsch. 22, 783.

Stone,G.A. et al. (1988) Drug Dev. Res. 15, 31.

Sylven,C.et al. (1986) Br. Med. J. 293, 227.

Taylor,D.A. and Williams,M. (1982) Eur. J. Pharmacol., 85, 335.

Trivedi,B.K. and Bruns,R.F. (1988) J. Med. Chem. 31, 1011.

Trivedi,B.K. et al., (1988) J.Med. Chem.31,271.

Trivedi,B.K. et al., (1990) in *Adenosine and Adenosine Receptors* (Ed.Williams,M.), Humana, Clifton, NJ, in press.

Ukena, D. et al., (1986) FEBS Letts. 209, 122.

Watt,A.H. et al. (1986) Br. Med. J. 293, 504.

Wauquier, et al (1987) Psychopharmacol. 91, 434.

Williams,M. (1987) Ann. Rev. Pharmacol. Toxicol, 227, 315.

Willliams,M (1989) Neurochem.Inter.,14, 249.

Williams, M. (1990) *Adenosine and Adenosine Receptors* Humana, Clifton, NJ, in press.

Williams,M. and Jarvis,M.J. (1987) Trends Pharmacol. Sci. 8, 330.

Williams,M. and Sills,M.A. (1989) Comp. Med. Chem. 3, in press.

Williams,M. et al. (1987) Chimica Oggi 9/87, 11.

Williams, M.et al. (1989) Curr. Cardiovascul. Patents, 1,599

28
Mechanisms of Adenosine Action

B.B. Fredholm, J. Fastbom, M. Dunér-Engström, P.-S. Hu, I. van der Ploeg, N. Altiok, P. Gerwins, A. Kvanta, and C. Nordstedt

As clearly shown during this symposium adenosine may play an important modulatory role in several cells and tissues (see figure 1). Most of the effects of adenosine are mediated via cell surface receptors which can be grouped into two major categories A_1 and A_2. The pharmacological characterization of these two receptor types has been dealt with extensively elsewhere - the aim of this paper is instead to briefly discuss some aspects of the signalling from these receptors.

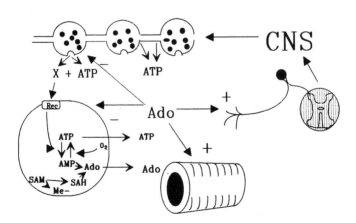

Fig 1. Schematic representation of the role of adenosine in neurotransmission. Adenosine is formed from ATP released either from nerve endings or effector cells or it is formed intracellularly. It can act to depress neurotransmission pre- and postsynaptically, increase blood flow and substrate availability, and stimulate sensory nerve activity.

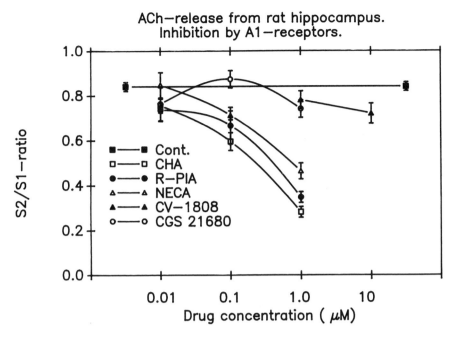

Fig 2. Inhibitory effect of several adenosine analogues on the release of [³H]-ACh from rat hippocampal slices. Note that the two A_2-selective compounds CV-1808 and CGS-21680 are virtually ineffective.

Presynaptic adenosine effects in hippocampus are mediated via A_1-like receptors linked to G-proteins, but not via changes in cAMP.

In the rat hippocampus we have demonstrated inhibition by adenosine and stable adenosine analogues of the release of [³H]-noradrenaline, [³H]-acetylcholine, [³H]-glutamate and [³H]-serotonin whereas the effect of adenosine on the release of GABA is small or non-existent (see Fredholm and Dunwiddie, 1988). One can also show several types of adenosine effects, both pre- and postsynaptic ones, by electrophysiological methods (Dunwiddie, 1985). The pharmacology of these responses is similar to the pharmacology of the A_1- binding sites demonstrated with CHA and R-PIA and to the pharmacology of the adenosine receptors mediating inhibition of cAMP accumulation in hippocampus (Dunwiddie and Fredholm 1989). For example the receptors mediating inhibition of [³H]-acetylcholine release show the typical A_1-receptor agonist profile (figure 2) and are clearly different from the A_2-receptors mediating cAMP accumulation in this preparation.

Changes in cyclic AMP (increases in the case of A_2-receptors or decreases in the case of A_1-receptors) was the first well established mechanism of action of adenosine. However, it was soon realized that adenosine may act independently of cAMP (e.g. Fredholm, 1982) and the terminology A_1 and A_2 (rather than the alternative -Ri and Ra - which refer to changes in adenylate cyclase) was adopted largely because these terms do not assume a particular mechanism of action. The demonstration that transmitter release can be regulated independently of cAMP (see Dunwiddie and Fredholm, 1985) should therefore not be used against the use of the A_1-terminology.

There is good biochemical evidence that A_1-receptors are linked to GTP-binding proteins, as discussed in more detail by others at this meeting. Specifically there is evidence that the presynaptic adenosine receptors, which show A_1-receptor characteristics, are probably linked to GTP-binding G-proteins: 1. In all regions of the brain, including those where presynaptic receptors are quantitatively important (cf. Snyder, 1985), the binding of A_1-receptor ligands is strongly affected by GTP. 2. In some instances the effect of presynaptic adenosine receptor stimulation is blocked by pertussis toxin that irreversibly inhibits some G-proteins (e.g. Dolphin and Prestwich, 1985). 3. N-ethylmaleimide (NEM) which can block G-proteins with some selectivity in hippocampus (Fredholm et al., 1985), antagonizes adenosine receptor mediated effects on [³H]-NA and [³H]-ACh release in an apparently non-competitive manner (Dunér-Engström and Fredholm 1988).

Using quantitative autoradiography we could confirm that essentially all high-affinity binding sites for an adenosine A_1 receptor agonist (CHA) are linked to G-proteins (Fastbom and Fredholm, 1989). Even though there may be interesting regional differences in the coupling between A_1-receptors and G-proteins, with i.a. presynaptic A_1-receptors apparently being somewhat more sensitive to Mg^{2+}, we have not found any A_1-receptors that are independent of G-proteins.

In some experiments male Sprague-Dawley rats (150-250 g) received injections of PTX close to the hippocampus 48-60 h prior to the experiments in order to ADP-ribosylate and inactivate several types of G-proteins of the G_i and G_o families (Gilman, 1987). Following *in vivo* treatment with PTX adenosine lost its ability to block the so-called low calcium bursting that occurs in hippocampal CA_1 neurons following anti-dromic activation and to inhibit cAMP accumulation. By contrast, the inhibitory effect of adenosine on field EPSP's, a largely presynaptic response, was unaffected by PTX treatment in the same animals (Fredholm et al., 1989). The effect of R-PIA on NA and ACh release was also essentially unaffected in the PTX treated animals. The reason for this is not known, but there are two major possibilities: 1) PTX is unable to reach and/or ADP-ribosylate G-proteins in nerve endings and 2) the G-proteins involved in the control of transmitter release are PTX-insensitive. So far we cannot discriminate between those two possibilities.

Involvement of K^+-and Ca^{2+}-channels and of protein kinase C in the presynaptic effects of adenosine

There is excellent evidence that transmitter release from nerve terminals depends on the entry of Ca^{2+}-ions via voltage-sensitive Ca^{2+}-channels. Thus, it is commonly assumed that presynaptically active compounds inhibit transmitter release by limiting Ca^{2+}-entry. Two principally different methods of achieving this end are schematically illustrated in figure 3. In addition, modulation of K^+-channel activity (e.g. Ca^{2+}-activated K^+-channels) could alter the duration of the nerve impulse and in that way the time during which Ca^{2+}-ions may enter the nerve terminal.

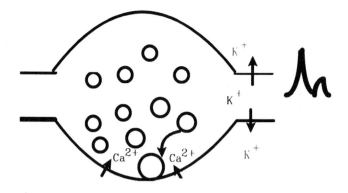

1. Activation of K-channels may decrease the probability that the terminal varicosity is invaded by a nerve impulse and in that way the transmitter release.

2. Inhibition of Ca-channels will decrease the amount of Ca that enters the nerve terminal and in that way the magnitude of the transmitter release.

Fig 3. Schematic illustration of two different targets for presynaptic control of transmitter release.

It has not proven possible to consistently block or reduce the presynaptic adenosine effect by calcium-channel agonists (eg. BayK 8644) or antagonists (eg. nifedipine), that influence L-type calcium channels (e.g. Dunér-Engström and Fredholm, 1988), suggesting that dihydropyridine sensitive Ca^{2+}-channels are not normally involved in electrically evoked transmitter release, but may become important if other channels, eg. N-channels (Miller, 1987), are blocked by adenosine. ω-CgTx, which blocks both N- and L-type channels inhibited both ACh and NA release from hippocampus (Hu and Fredholm, 1989).

Activation of a K^+-conductance could explain why adenosine inhibits transmitter release. We therefore examined if the K^+-channel inhibitor 4-aminopyridine (4-AP) could alter the presynaptic effect of adenosine analogues. For comparison we examined the effect of 4-AP on autoreceptors and on blockade by ω-CgTx. 4-AP in a dose (30-100 μM) that by itself increased transmitter release by 2.5 - 3.5 fold, markedly reduced the effect of autoreceptorstimulation (muscarinic in the case of ACh-release; α_2-receptors in the case of NA-release). By contrast, the effects of R-PIA and Ca^{2+}-blockade was not much affected (Hu and Fredholm, 1989).

Thus, the evidence indicates that adenosine acts to limit transmitter release by acting at receptors that are of the A_1-type and linked to G-proteins. The signal produced may be either a stimulation of a 4-AP insensitive K^+-channel or the inhibition of a Ca^{2+}-channel similar to an N-channel. Although there is evidence

187

that G-proteins may directly interact with N-type Ca^{2+}-channels (e.g. Ewald et al., 1988) it has also been suggested that G-proteins influence Ca^{2+}-channels indirectly via a protein kinase C mechanism. In a study of the role of PKC in the control of NA release from rat hippocampus we found some evidence that PKC may be involved in some of the actions of α_2-adrenoceptors, but not in the effects of presynaptic A_1-receptors (Fredholm and Lindgren 1988). The results presented in table 1 indicate that also with regard to ACh-release PKC plays no major role in mediating the presynaptic inhibitory effect of adenosine.

Table 1. Lack of effect of protein kinase C on the presynaptic effect of the adenosine analogue CHA (1 μM) on [^{3}H]-ACh release from rat hippocampal slices. Protein kinase C was stimulated with phorbol dibutyrate (PDBu) and inhibited by staurosporin (0.1 μM). Mean ± s.e.m. of 4-12 experiments.

Treatment	S_2/S_1-ratio	
(drugs added before S_2)	No CHA	+CHA 1 μM
No drug	0.86 ± 0.02	0.41 ± 0.05
PDBu 0.1 μM	1.30 ± 0.14	0.37 ± 0.04
PDBu 0.1 + Staurosporin	1.00 ± 0.04	-
Staurosporin 0.1 μM	0.88 ± 0.03	0.42 ± 0.02

It is apparent that we may draw some general conclusions from the efforts in several laboratories regarding the mechanism underlying presynaptic inhibition:
1. Adenosine acting at apparently similar receptors in different cells can induce several types of effects - decreased cellular cAMP, decreased Ca^{2+}-entry via Ca^{2+}-channels, increased K^+-channel activity - all of which may lead to inhibition of transmitter release.
2. In a single type of nerve different agents may inhibit transmitter release in somewhat different ways.
3. Depending upon how transmitter release is evoked the mechanism of inhibition could vary.

Hence, it may be fruitless to search for a single explanation of how adenosine limits transmitter release.

Interactions between receptors linked to PIP_2-hydrolysis and adenosine-receptor mediated cAMP accumulation.

It is well known that biogenic amines, such as NA acting on α_1-receptors and histamine acting at H_1-receptors, can stimulate cAMP accumulation in brain slices by enhancing the effects of adenylate cyclase stimulators such as adenosine (see Daly, 1977). This effect of the receptor ligands, which enhance inositol phosphate accumulation, can be mimicked by phorbol esters (Hollingsworth et al., 1985; Nordstedt and Fredholm, 1987). In order to study such interactions in more detail we have shifted our attention from brain slices to cultured cells.

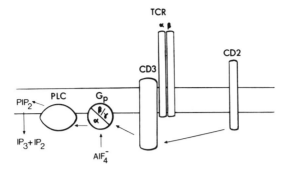

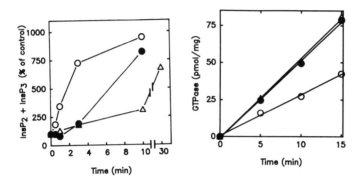

Fig 4. Activation of the CD3 complex by the monoclonal antibody OKT3 or by antibodies (9-1/9-2) to the CD2 receptor leads to an increase in $InsP_3$ accumulation in Jurkat cells and to an increase in GTP'ase activity reflecting activation of a G-protein. The <u>upper part</u> of the figure presents a schematic representation of the signal transduction pathway. <u>Lower left panel</u>: increased accumulation of $InsP_2$ + $InsP_3$ after CD3 stimulation (open circle); CD2 stimulation (closed circle) or aluminium fluoride (triangles). <u>Lower right panel</u>: GTP'ase activity (measured with 0.25 μM [γ -^{32}P]-GTP) in control cells (open circles) or after stimulation of CD3 (closed circles) and CD2 (triangles).

Human T-cells, including the T-cell leukemia cell Jurkat, can be triggered by the T-cell receptor that is associated with the CD3-complex of membrane proteins (see figure 4). In addition, these cells can be activated by an alternative route involving the CD2 receptor, which is a single peptide chain. The results summarized in figure 4 provide the first unequivocal evidence that these two receptor molecules, when triggered with monoclonal antibodies, are able to activate a G-protein. The results also suggest that these receptors and the G-protein are linked to a phospholipase C that hydrolyses PIP_2.

The rise in $InsP_3$ leads to a rise in intracellular Ca^{2+} - via both mobilization of intracellular stores and a Ca^{2+}-dependent Ca^{2+}-entry (Ng et al., 1988). There was also an increase in PKC, as judged by phosphorylation of endogenous PKC substrates and by translocation of the enzyme.

189

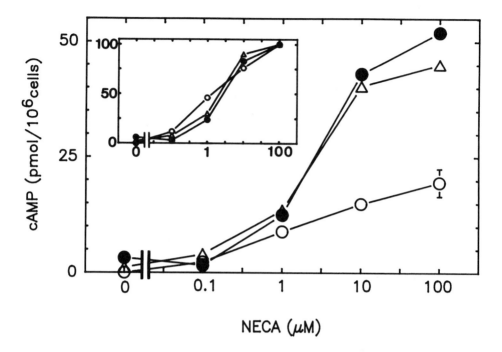

Fig 5. The stimulatory effect of the adenosine analogue NECA in Jurkat-cells (control, unfilled circles) is increased by 0.1 μM 4-β-phorbol-12,13-dibutyrate (triangles) or by the monoclonal antibody OKT3 (corresponding to 1 μg/ml pure antibody; filled circles). The inset shows the data normalized to the respective maximum and illustrates that the enhancement does not involve any change in the potency of NECA.

As seen in figure 5, activation of the CD3 complex by the monoclonal antibody OKT3 and the direct activation of protein kinase C by phorbol esters leads to a marked enhancement of the adenosine-receptor mediated accumulation of cAMP. An increase of essentially equal magnitude was observed if instead cholera toxin was used to stimulate cAMP production (Nordstedt et al., 1989). Thus, the evidence suggests a chain of events from receptor activation of phospholipase C to activation of PKC to activation of a A_2-receptor-mediated cAMP accumulation as schematically illustrated in figure 6.

However, a closer examination reveals that there are complications and that the picture probably is not quite that simple. Thus, it was found that even though PDBu and OKT3 enhanced the accumulation of cAMP to an equal degree OKT3 was far more potent than PDBu in stimulating cholera toxin-induced cAMP accumulation. An even more clear cut difference is apparent when one examines the effect on forskolin-stimulated cAMP. These results clearly show that OKT3 can markedly enhance forskolin-stimulated cAMP formation. This indicates a rather direct effect of CD3 activation on adenylate cyclase. By contrast, phorbol esters have no effect on basal cAMP levels and cause a very minute enhancement of the effect of a high dose of forskolin (figure 7). A further analysis reveals that

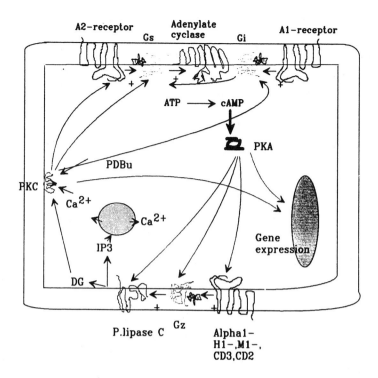

Fig 6. Schematic representation of the interactions between parallel signal transduction pathways.

the effect of OKT3 is extremely Ca^{2+}-dependent, but cannot be blocked by the PKC-inhibitor H-7 (not shown). The effect of OKT3 can, however, be mimicked by combining a Ca^{2+}-ionophore (A 23187) and a phorbol ester.

Thus, the present results clearly suggest that a receptor that is linked to phospholipase C may influence the adenosinereceptor-mediated cAMP accumulation via more than one mechanism. One of those is the PKC that is also activated by phorbol esters already in the absence of added Ca^{2+}. The other mechanism is strongly dependent upon Ca^{2+}. It may represent a subtype of PKC, but it could also represent some completely different form of Ca^{2+}-dependent mechanism.

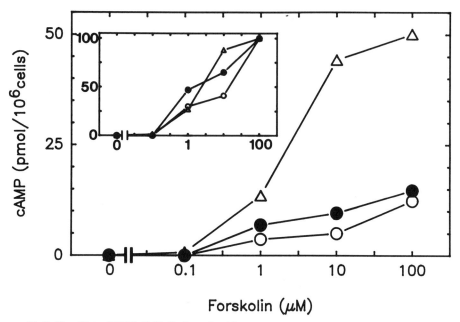

Fig 7. The effect of OKT3 (1:10 dilution; triangle), PDBu (100 nM; filled circle) in enhancing the cAMP accumulation afforded by increasing concentrations of forskolin in Jurkat cells. The inset shows the same data normalized to respective maximum and illustrates that there is no major shift in the potency of forskolin.

Summary

We have briefly summarized the evidence regarding the mechanism(s) underlying the action of adenosine to a) limit the release of neurotransmitters and b) to interact with other receptors that are coupled to phospholipase C.

The release of noradrenaline as well as acetylcholine and probably also excitatory amino acids can be inhibited by adenosine acting at A_1-like receptors that interact with G-protein(s). One major mechanism appears to be a direct inhibition of a N-type Ca^{2+}-channel. However, other mechanisms including inhibition of cAMP, activation of K^+-currents and possibly alterations of intracellular Ca^{2+}-sensitivity also appear to occur - at least under some circumstances.

The adenosine A_2-receptor-mediated stimulation of adenylate cyclase can be amplified by agents that interact with receptors activating phospholipase C. One mechanism appears to be stimulation of a classical phorbol ester activatable protein kinase C, but there are also other ways of producing the interaction.

Thus, it appears possible for adenosine to mediate its effects or to have its effects modulated in more than one way even in a single cell. The existence of multiple, parallel signal transduction pathways leading from the receptor to the final biological response could render the adenosine receptor control of biological phenomena more robust than if it relied on a single mechanism only.

Acknowledgements. This paper summarizes research supported by the Swedish Medical Research Council; the Swedish Cancer Association, Nanna Svartz' foundation, Ostermans Foundation, Gustaf V 80 Years Fund, Hedlunds Foundation and by funds from Karolinska Institutet.

References

Daly, J.W., 1977, Cyclic nucleotides in the nervous system. Plenum Press, New York.

Dolphin, A.C., and Prestwich, S. A., 1985, Pertussis toxin reverses adenosine inhibition of neuronal glutamate release. Nature 316:148-150.

Dunér-Engström, M., and Fredholm, B.B., 1988, Evidence that prejunctional adenosine receptors regulating acetylcholine release from rat hippocampal slices are linked to a N-ethylmaleimide sensitive GTP-binding protein but not to adenylate cyclase or a dihydropyridine-sensitive Ca^{2+}-channel. Acta Physiol. Scand. 134:119-126.

Dunwiddie, T.V., 1985, The physiological role of adenosine in the central nervous system. Int. Rev. Neurobiol. 27: 63-139.

Dunwiddie, T.V., and Fredholm, B.B., 1985, Adenosine modulation of synaptic responses in rat hippocampus: Possible role of inhibition or activation of adenylate cyclase. In Advances in Cyclic Nucleotide Res. Vol 19. DMF. Cooper and K.B. Seamon.(Eds). pp. 259-272.

Dunwiddie, T.V., and Fredholm, B.B., 1989, Adenosine A_1-receptors inhibit adenylate cyclase activity, inhibit neurotransmitter release, and hyperpolarize pyramidal neurons in rat hippocampus. J. Pharmacol. Expt. Therap. 249: 31-37.

Ewald, D.A., Sternweis, P.C., and Miller, R.J., 1988, Guanine nucleotide-binding protein G_o-induced coupling of neuropeptide Y receptors to Ca^{2+} channels in sensory neurons, Proc. Natl. Acad. Sci. 85:3633-3637.

Fastbom, J., and Fredholm, B.B., 1989, Regional differences in the effect of guanine nucleotides on agonist and antagonist binding to adenosine A1-receptors in rat brain, as revealed by autoradiography. Neuroscience. In press.

Fredholm, B.B., 1982, Adenosine receptors. Medical Biol, 60, 289-293.

Fredholm, B.B., and Dunwiddie, T.V., 1988, How does adenosine inhibit transmitter release? Trends in Pharmacological Sciences. 9, 130-134.

Fredholm, B.B., and Lindgren, E., 1988, Protein kinase C activation increases noradrenaline release from rat hippocampus and modifies the inhibitory effect of α_2-adrenoceptor and A_1-receptor agonists. N.S. Arch. Pharmacol. 337:477-483.

Fredholm, B.B., Lindgren, E., and Lindström, K., 1985, Treatment with N-ethylmaleimide selectively reduces adenosine receptor mediated decreases in cyclic AMP accumulation in rat hippocampal slices. Br. J. Pharmacol. 86:509-513.

Fredholm, B.B., Proctor, W., van der Ploeg, I., and Dunwiddie, T.V., 1989, In vivo pertussis toxin treatment attenuates some, but not all, adenosine A1 effects in slices of rat hippocampus. Eur. J. Pharmacol. -Molec.Pharm.Sect. 172:249-262.

Gilman, A.G., 1987, G-proteins: transducers of receptor-generated signals. Ann. Rev. Biochem., 56, 615.

Hollingsworth, E.B., Sears, S.B., and Daly, J.W., 1985, An activator of proteinkinase C (phorbol-12-myristate-13-acetate) augments 2-chloroadenosine-elicited accumulation of cyclic AMP in guinea-pig cerebral cortical preparations. FEBS lett. 184: 339-342.

Hu, P.-S., and Fredholm, B.B., 1989, 4-aminopyridine blocks the inhibition of [^{3}H]-noradrenaline release from rat hippocampus caused by an α_2-adrenoceptor agonist, but not that caused by an adenosine analogue or ω-conotoxin. Acta Physiol. Scand. 136:347-353.

Miller, R.J., 1987, Multiple calcium channels and neuronal function, Science, 235:46-52.

Ng, J., Fredholm, B.B., Jondal, M., and Andersson, T., 1988, Regulation of receptor-mediated calcium influx across the plasma membrane in a human leukemic T-cell line: evidence of its dependence on an initial calcium mobilization from intracellular stores, Biochim. Biophys. Acta 971:207-214.

Nordstedt, C., and Fredholm, B.B., 1987, Phorbol-12,13-dibutyrate enhances the cyclic AMP accumulation in rat hippocampal slices induced by adenosine analogues, N.-S. Arch. Pharmacol. 335:136-142.

Nordstedt, C., Kvanta, A., Van der Ploeg, I., and Fredholm, B.B., 1989, Dual effects of protein kinase-C on receptor-stimulated cAMP accumulation in a human T-cell leukemia line, Eur. J. Pharmacol. 172:51-60.

Snyder, S.H., 1985, Adenosine as a neuromodulator. Ann. Rev. Neurosci. 8:103-124, 1985.

29
Functional Roles of A_1 and A_2 Receptors in Rat Hippocampus and Striatum

T.V. Dunwiddie, C.R. Lupica, W.A. Cass, N.R. Zahniser, and M. Williams

Introduction

We have previously demonstrated that adenosine plays an important role as a modulator of excitatory transmission in the rat hippocampal formation. Although we have provided considerable pharmacological evidence to support our contention that synaptic modulation is mediated via the adenosine A1 receptor subtype (Dunwiddie et al., 1984), there remain a number of unanswered questions. First, although A1 receptors are coupled in an inhibitory fashion to adenylate cyclase (Cooper et al., 1980), we have been unable to demonstrate that inhibition of this enzyme results in inhibition of synaptic transmission. Conversely, co-application of adenosine and cAMP analogs does not result in a reversal of the inhibitory effects of adenosine. Thus, although adenosine can inhibit adenylate cyclase, we hypothesize that another mechanism must underlie the modulatory effects of this agent on synaptic transmission. In particular, we would suggest that activation of a potassium conductance (Greene et al., 1985; Trussell et al., 1985; Trussell et al., 1987), inhibition of a calcium conductance (Dolphin et al., 1985; Macdonald et al., 1986), or perhaps the generation of a second messenger other than cAMP mediates the effects of adenosine.

In addition, adenosine can also increase cyclic AMP levels in brain slices via an interaction with A2 receptors. However, we have been unable to find any electrophysiological correlate to this type of cellular action. Although this might reflect a localization of A2 receptors to non-neuronal cellular elements, the absence of highly selective A2 receptor agonists has made it difficult to characterize A2 effects in the absence of A1 responses. In the present experiments, we have examined the electrophysiological and biochemical actions of a recently developed adenosine A2 receptor-selective agonist (2-[p-(carboxyethyl)-phenylethylamino]-5'N-ethylcarboxamidoadenosine; CGS 21680) in an attempt to characterize A2 receptor-mediated effects in both the hippocampus and the striatum. This compound demonstrates approximately 140-fold selectivity for the A2 versus the A1 receptor (Jarvis et al., 1989a; Jarvis et al., 1989b), and thus constitutes a truly A2 receptor-selective ligand that might prove valuable in characterizing A2 receptor-mediated responses.

Methods & Materials

Male Sprague-Dawley rats (Sasco Animal Laboratories, Omaha, NE) weighing 150-250 grams were used for all experiments. They were housed in groups of four to six under a 12-hr light-dark cycle with food and water available ad libitum. For electrophysiology and biochemistry, subjects were decapitated, and the hippocampus or striatum dissected free of surrounding tissue.

For electrophysiological experiments, coronal sections taken from the middle portion of the hippocampi were prepared (Dunwiddie et al., 1978; Mueller et al., 1981). Slices were not superfused with medium during a 40-60 min recovery period; following recovery, slices were superfused with a constant flow of fresh, oxygenated, preheated medium at a rate of 2 ml/min. For experiments involving antidromic bursting, the superfusion medium was exchanged with one containing 4.0 mM $MgSO_4$ and 0.24 mM $CaCl_2$ at least 30 min prior to recording.

Electrophysiological responses tested were either the synaptic response evoked in the dendrites of the CA1 pyramidal neurons by stimulation of stratum radiatum (field EPSP response), or the repetitive spiking elicited in the cell layer by stimulation of the alveus (low calcium bursting). Drugs were made up in deionized water, and added to the flow of perfusion fluid via a syringe pump. Dose-response curves for adenosine and CGS 21680 were analyzed using an iterative curve-fitting program to determine the maximal response and EC_{50} values. Biochemical data (see below) were analyzed in the same manner.

For cAMP measurements, coronal hippocampal and striatal slices were incubated at 34° C for 60 min in Krebs' buffer. The slices were superfused with aerated Krebs' buffer maintained at 34°C, and then superfused for an additional 15 min (30 min for hippocampal slices) with buffer alone (controls) or with buffer containing various concentrations of CGS 21680 (0.001-10 μM) or 10 μM NECA. Following superfusion the slices were transferred to microfuge tubes, and 100 μl of 10% trichloroacetic acid were added to each tube. The samples were immediately sonicated for 10-20 sec and then placed in a boiling water bath for 5 min. The solubilized cAMP was separated from the precipitated protein by centrifugation (13,600 x g; 8 min) and the supernatant extracted four times with five volumes of water-saturated ether. Cyclic AMP in the final aqueous layer was measured using radioimmunoassay (Biomedical Technologies Inc., Stoughton, MA). The protein in each slice was measured after first solubilizing the precipitated protein with 1 N NaOH (Bradford, 1976).

For the dopamine release studies, striatal slices were prepared and incubated as described above, except that the buffer contained 10 μM nomifensine in order to inhibit the synaptic DA transporter and obtain detectable levels of DA. Stimulation consisted of unipolar square-wave pulses, 1 Hz for 1 min, 2 msec pulse duration, 16-25 mA current flow. The first stimulation period (S1) was always in control buffer, and the second (S2) either in buffer or buffer with drug. Dopamine levels were determined using high-performance liquid chromatography (HPLC) coupled with electrochemical detection (EC) as described previously (Gerhardt et al., 1989). This system has a nominal detection limit of 1.0 pg/injection. Retention times of standards were used to identify peaks, and peak heights were used to calculate absolute amounts of DA (expressed as pg/mg wet weight of tissue/min).

Results and Discussion

In initial experiments, we characterized the effects of CGS 21680 on a variety of electrophysiological responses. Because this compound has been shown to be active as a vasodilator in the nanomolar concentration range (Hutchison et al., 1989), we used concentrations in the 50-1000 nM range in these initial experiments. Insofar as the inhibitory modulation of excitatory transmission was concerned, these concentrations of CGS 21680 were entirely without effect. This analog was also inactive in reducing the repetitive spiking induced in low calcium medium by antidromic stimulation of the pyramidal neurons, a response which is inhibited by adenosine (Haas et al., 1984; Lee et al., 1984) and is also mediated via adenosine A1 receptors (Dunwiddie et al., 1989). CGS 21680 also did not appear to have any effect upon evoked population spike responses, or antidromically evoked responses.

195

TABLE 1
EFFECTS OF ADENOSINE ANALOGS ON DOPAMINE RELEASE

	CGS 100 nM	CGS 1 μM	ADENOSINE 50 μM	CHA 100 nM
Spontaneous DA release (Percent change from control)	4.4$\pm$1.1	7.5$\pm$2.1	-9.6$\pm$18	-4.3$\pm$12
Electrically evoked release (Percent change from control)	1.0$\pm$3.1	3.1$\pm$4.2	-38$\pm$3.2[*]	-33$\pm$6.2[*]

[*] $p < 0.05$

At considerably higher concentrations, CGS 21680 mimicked the effects of A1 receptor-selective agonists. This analog inhibited the evoked field EPSP response, with an EC_{50} value of 24 μM (95% confidence intervals 13-44 μM). Low-calcium bursting was also inhibited at concentrations >5 μM (EC_{50} value estimated to be approximately 100 μM). These effects are not consistent with the low nanomolar affinity of CGS 21680 for the A2 receptor (Kd = 7-20 nM; Jarvis et al., 1989a; Jarvis et al., 1989b) but correspond well with its micromolar affinity for the A1 receptor that mediates both of these responses. Because of the weak responses to CGS 21680, we also considered the possibility that it was a partial agonist insofar as these responses were concerned. However, it did not antagonize any of the effects of adenosine at concentrations of CGS 21680 up to 20 μM, the highest level tested.

The lack of any discernible effect of CGS 21680 on hippocampal electrophysiology in concentrations that would be consistent with actions at an A2 receptor led us to investigate its actions in the striatum, a brain region where A2a receptors are found in relatively high density. In these experiments, we focused on spontaneous and electrically-stimulated release of endogenous dopamine from striatal slices. The ability of adenosine to inhibit the release of DA in vitro and in vivo is well known (Harms et al., 1979; Michaelis, 1979; Meyers et al., 1986; O'Neill, 1986). However, none of these studies have made a clear distinction between A1 or A2 adenosine receptor involvement in this phenomenon, and some results suggest that neither receptor is involved (Ebstein et al., 1982).

In the present studies, the spontaneous overflow of DA was not significantly affected by any adenosine analog (Table 1). When the effects of these agents were tested on electrically-stimulated overflow of DA, CGS-21680 (0.1 and 1 μM) was without effect, while adenosine (50 μM) reduced DA overflow by approximately 40% (Table 1). The effect of adenosine was completely prevented by co-perfusion with 8-PT. CHA (0.1 μM), a selective A1 adenosine receptor agonist, reduced the release of DA to a similar extent as adenosine (Table 1). Thus it appears that in the striatum, as in most other systems that have been studied, it is the adenosine A1 receptor that appears to be linked to the inhibition of neurotransmitter release. Whatever the role of the A2 receptor, its actions do not modify the efflux of dopamine from slices made from this brain region.

Because none of these functional responses to adenosine appeared to be affected by CGS 21680 in concentrations that would be consistent with A2 receptor-mediated actions, we examined its effects on cAMP levels in hippocampal and striatal slices, a parameter that is directly affected by the A2 receptor. As can be seen in figure 1A, concentrations of CGS 21680 ≤ 10 μM failed to significantly promote the formation of cAMP in hippocampal slices, although NECA was quite effective in this regard, as has been previously reported (Dunwiddie et al., 1984). In contrast, in rat striatal slices, both NECA and CGS 21680 increased cAMP levels

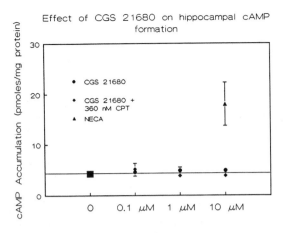

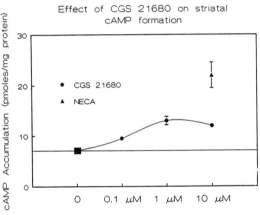

Figure 1: Dose response curves are shown for the production of cAMP by CGS 21680 in hippocampal and striatal slices. Some of the hippocampal studies were also conducted in the presence of 360 nM CPT, a selective A1 antagonist, to block any competing A1 effects. Each point is the mean of at least 3 determinations ± s.e.m.; in some cases the s.e.m. was smaller than the symbol. Mean ± s.e.m. basal levels are indicated by the squares.

(figure 1B); however, the maximal response to NECA was approximately 3-fold higher than the maximal effect of CGS 21680 ($p < 0.05$). As would be expected for an A2 receptor-mediated effect, the increase due to CGS 21680 occurred at relatively low concentrations of agonist (EC_{50} value 110 nM).

Several interpretations of these findings are possible. However, we find the most likely explanation to be that these results reflect the differential distribution of high affinity A2a and low affinity A2b receptors in the nervous system, and that CGS 21680 is an agonist that is highly selective for the A2a subtype of this receptor. High affinity A2a receptors are found only in the striatum and a few other brain regions, whereas the A2b receptor is widespread in its distribution (Jarvis et al., 1989b). If CGS 21680 were selective for the A2a receptor subtype, this would explain not only the lack of effect on cAMP levels in the hippocampus (since the hippocampus lacks the A2a receptor subtype), but also the different maximal responses to NECA and CGS 21680 in striatal slices. Because NECA can activate both receptor subtypes, it can increase cAMP levels in hippocampus via A2b receptors and has a more powerful effect than CGS 21680 in striatum by virtue of the fact that it can act via both A2a and A2b receptors to increase cAMP levels. The > 100-fold greater potency of CGS 21680 in increasing cAMP in the striatum vs. its potency in eliciting electrophysiological actions in the hippocampus, the lack of any effects of this analog on A2b receptors, and the absence of detectable A2a receptors in the hippocampus all make it highly unlikely that the electrophysiological effects of adenosine and adenosine analogs involve A2 receptors of either subtype. If A2 receptors do mediate electrophysiological actions, it would appear that the striatum is the most likely site at which they might be observed.

In sum, our results suggest that CGS 21680 is a highly selective A2 receptor agonist that is selective for the A2a receptor subtype. This is consistent with data from [^{3}H]-CGS 21680 binding showing that this compound interacts with only one high affinity A2 binding site in striatal tissue, and shows little if any specific binding in hippocampal tissue (Jarvis et al., 1989b). Furthermore, our data suggest that this analog is at least as selective for the A2a vs. A2b receptor as it is selective for the A2a vs. A1 receptor. It would seem likely that this analog will prove useful in determining the functional consequences of A2a receptor activation in the brain.

Acknowledgements

This research was supported by grants NS 26851 and DA02702, training grant AA 07464 and the Veterans Administration Medical Research Service. We would like to thank CIBA-GEIGY for making CGS 21680 available to us. Parts of this communication have been presented elsewhere (Lupica et al., 1989).

Abbreviations

8-PT, 8-phenyltheophylline
cAMP, cyclic adenosine-3'-5'-monophosphate
CGS 21680,(2-[p-(carboxyethyl)phenylethylamino]-
     5'-N-ethylcarboxamidoadenosine
CHA, N^6-cyclohexyladenosine
CPT, 8-cyclopentyltheophylline
DA, dopamine
EC$_{50}$, concentration producing a half-maximal response
EPSP, excitatory postsynaptic potential
NECA, 5'-N-ethylcarboxamidoadenosine

References

Bradford, M. M.: A rapid and sensitive method for the quantitation of microgram quantities of protein utilizing the principle of protein-dye binding. *Anal. Biochem.* **72:** 248-254, 1976.
Cooper, D.M.F., Londos, C., and Rodbell, M. Adenosine receptor-mediated inhibition of rat cerebral cortical adenylate cyclase by a GTP-dependent process. *Mol. Pharmacol.* **18:** 598-603, 1980.
Dolphin, A.C., Forda, S.R., and Scott, R.H. Calcium-dependent currents in cultured rat dorsal rood ganglion neurones are inhibited by an adenosine analogue. *J. Physiol.* **373:** 47-65, 1986.
Dunwiddie, T. V. and Fredholm, B. B.: Adenosine receptors mediating inhibitory electrophysiological responses in rat hippocampus are different from receptors mediating cyclic AMP accumulation. *Naunyn Schmiedeberg's Arch. Pharmacol.* **326:** 294-301, 1984.

Dunwiddie, T. V. and Fredholm, B. B.: Adenosine A1 receptors inhibit adenylate cyclase activity and neurotransmitter release and hyperpolarize pyramidal neurons in rat hippocampus. *J. Pharmacol. Exper. Therap.* **249**: 31-37, 1989.

Dunwiddie, T. V. and Lynch G.: Long-term potentiation and depression of synaptic responses in the rat hippocampus: localization and frequency dependency. *J. Physiol.* **276**: 353-367, 1978.

Ebstein, R. P., and Daly, J. W.: Release of norepinephrine and dopamine from brain vesicular preparations: Effects of adenosine analogues. *Cellul. Mol. Neurobiol.* **2**: 193-204, 1982.

Gerhardt, G. A., Dwoskin, L. P. and Zahniser, N. R.: Outflow and overflow of picogram levels of endogenous dopamine and DOPAC from rat striatal slices: improved methodology for studies of stimulus-evoked release and metabolism. *J. Neurosci. Meth.* **26**: 217-227, 1989.

Greene, R.W. and Haas, H.L.: Adenosine actions on CA1 pyramidal neurones in rat hippocampal slices. *J. Physiol.* **366**: 119-127, 1985.

Haas, H. L. and Jefferys, J.G.: Low-calcium field burst discharges of pyramidal neurones in rat hippocampal slices. *J. Physiol.* **354**: 185-201, 1984.

Harms, H. H., Wardeh, G. and Mulder, A. H.: Effects of adenosine on depolarization-induced release of various radiolabelled neurotransmitters from slices of rat corpus striatum. *Neuropharmacol.* **18**: 577-580, 1979.

Hutchison, A. J., Oei, H. H., Ghai, G. R. and Williams, M.: CGS 21680, an A2 selective adenosine receptor agonist with preferential hypotensive activity. *FASEB Abstr.* **3**: 317, 1989.

Jarvis M. F., Schulz, R., Hutchison A. J., Do, U. H., Sills, M. and Williams, M.: [^{3}H]-CGS 21680, a selective A2 adenosine receptor agonist directly labels A2 receptors in rat brain. *J. Pharmacol. Exp. Therap.* In Press, 1989a.

Jarvis, M. F. and Williams, M.: Direct autoradiographic localization of adenosine A2 receptors in the rat brain using the A2 selective agonist, [^{3}H]-CGS 21680. *Eur. J. Pharmacol.* In Press, 1989.

Lee, K. S., Schubert, P. and Heinemann, U.: The anticonvulsant action of adenosine: A postsynaptic, dendritic action by a possible endogenous anticonvulsant. *Brain Res.* **321**: 160-164, 1984.

Lupica, C.R., Cass, W.A., Zahniser, N.R., and Dunwiddie, T.V. Effects of the selective adenosine A2 receptor agonist CGS 21680 on in vitro electrophysiology, cyclic AMP formation and dopamine release in the rat CNS. Submitted for publication, 1989.

Macdonald, R.L., Skerritt, J.H., and Werz, M.A. Adenosine agonists reduce voltage-dependent calcium conductance of mouse sensory neurones in cell culture. *J. Physiol.* **370**: 75-90, 1986.

Michaelis, M. L., Michaelis, E. K. and Meyers, S. L.: Adenosine modulation of synaptosomal dopamine release. *Life Sci.* **24**: 2083-2092, 1979.

Mueller, A. L., Hoffer, B. J. and Dunwiddie, T. V.: Noradrenergic responses in rat hippocampus: Evidence for mediation by alpha and beta receptors in the in vitro slice. *Brain Res.* **214**: 113-126, 1981.

Myers, S. and Pugsley, T. A.: Decrease in rat striatal dopamine synthesis and metabolism in vivo by metabolically stable adenosine receptor agonists. *Brain Res.* **375**: 193-197, 1986.

O'Neill, R. D.: Adenosine modulation of striatal neurotransmitter release monitored in vivo using voltammetry. *Neurosci. Lett.* **63**: 11-16, 1986.

Trussell, L. O. and Jackson, M. B.: Adenosine-activated potassium conductance in cultured striatal neurons. *Proc. Natl. Acad. Sci. U.S.A.* **82**: 4857-4861, 1985.

Trussell, L. O. and Jackson, M. B.: Dependence of an adenosine-activated potassium current on a GTP-binding protein in mammalian central neurons. *J. Neurosci.* **7**: 3306-3416, 1987.

30
Subclassification of Neuronal Adenosine Receptors

L.E. Gustafsson, C.U. Wiklund, N.P. Wiklund, and L. Stelius

Introduction

It has been proposed that adenosine exerts its extracellular biological effects via two distinct receptor subclasses: A_1 which mediate inhibition of adenylate cyclase and A_2 receptors which mediate stimulation of the cyclase. In addition to their opposite effects on adenylate cyclase the receptors have distinct structure activity relationships for certain adenosine analogues. At A_1-receptors the potency order is N^6-R-phenylisopropyladenosine (R-PIA) $\geq$ 5'-N-ethylcarboxamideadenosine (NECA) > S-PIA, whereas at A_2-receptors it is NECA > R-PIA $\geq$ S-PIA (c.f. Stone 1985, Wiklund et al. 1989b). Stimulation or inhibition of adenylate cyclase is no longer regarded as a safe criterion for receptor type (Stone 1985). A further subdivision of A_2 receptors into A_{2a} and A_{2b} was proposed, where A_{2a} is a high affinity receptor (EC_{50} nanomolar range) and A_{2b} is a low affinity receptor (EC_{50} micromolar range)(c.f. Bruns et al. 1987). Studies are at hand, suggesting adenosine receptors with agonist profiles that do not fit the definition of A_1 and A_2 receptors and a new class of adenosine receptors has been suggested: A_3 (Ribeiro & Sebastiao 1986). We have previously reported a difference in structure activity relationships between adenosine A_1 receptors at autonomic nerves versus those characterized e.g. in the CNS (Gustafsson et al. 1985, 1989, Wiklund et al. 1989a). The present study delineates presynaptic P_1 receptors in the autonomic postganglionic nerves as clearly of A_1 type, and probably constituting a novel subclass, A_{1b}.

Methods

General procedure. Guinea pig ileum longitudinal muscle or rat vas deferens preparations were suspended in organ baths. Nerve-induced contractile responses were elicited by transmural stimulation (3 Hz, 0.2 ms, 15 pulses at 1 min intervals) at a load of 2 mN as previously described (Gustafsson et al. 1985). EC_{50}-values for agonist inhibition of nerve-induced contractions and pA_2 values for antagonists versus 2-chloroadenosine (guinea pig ileum) or R-PIA (rat vas deferens) were compared with reported (c.f. legend, Fig. 2) binding affinities in brains from rat and guinea pig. Cholinergic (ileum) or adrenergic (vas deferens) transmitter release was measured as described previously (Wiklund et al. 1989b,1989c).

200

Drugs. N^6-R-phenylisopropyladenosine (R-PIA) was from Boehringer Mannheim GmbH, West Germany. 5'-N-ethylcarboxamideadenosine (NECA) was from Byk-Gulden, Lomberg Chem. Fabr. Konstanz, West Germany, N^6-benzyladenosine (BzADO) and N^6-methyladenosine (MeADO) were from Sigma Chemical Co, St. Louis, USA. 2-phenylaminoadenosine (CV-1808) and 2-(4-methoxyphenyl)-adenosine (CV-1674) were gifts from Takeda Chemical Industries Ltd, Osaka, Japan. Xanthine amine congener (XAC; 8-[4-[[[[(2-aminoethyl)amino]carbonyl]methyl]oxy]phenyl]-1,3-dipropylxanthine), N^6-cyclopentyladenosine (CPA), 8-cyclopentyltheophylline (CPT), 1,3-dipropyl-8-cyclopentylxanthine (DPCPX), 8-(p-sulfophenyl)theophylline (PSØT) and 1,3-dipropyl-8- (p-sulfophenyl)xanthine (DPSPX) were from Research Biochemicals Inc, Natick, Ma, USA). Xanthine carboxyl congener (XCC; 8-[4-[(carboxymethyl)oxy]phenyl]-1,3-dipropylxanthine) was from Dr K. Jacobson, NIH, Bethesda, Md, USA. PD113.297 was from Dr J. Bristol, Warner-Lambert, Ann Arbor, Mi, USA).

Experimental

The capacity of a number of adenosine analogues (R-PIA, NECA, CV-1674, CPA, CV-1808, MeADO and BzADO) and antagonists (PSØT, XAC, XCC, CPT, PD113.297, DPSPX and DPCPX) to modulate adrenergic and cholinergic neurotransmission in rat vas deferens and guinea pig ileum, respectively, was studied. Dose-response curves for all agonists except CV-1674 on the ileum, were displaced to the right in a parallel manner by competitive adenosine receptor antagonists, indicating action at adenosine receptors. Agonist data (EC_{50} values on contractile responses to nerve stimulation) suggested similarity, but not identity, with rat CNS receptors. In guinea pig ileum the inhibitory effect of the adenosine analogues was almost exclusively exerted at prejunctional receptors mediating inhibition of acetylcholine release (c.f. Gustafsson et al. 1985), whereas in rat vas deferens the agonists showed both pre- and postjunctional effects. Thus R-PIA was a preferential prejunctional agonist on [^{3}H]-noradrenaline overflow and this effect was antagonized by A_1 selective antagonists such as CPT and DPCPX. In contrast NECA inhibition of contractile responses was more or less unaffected by CPT or DPCPX at concentrations (10^{-7} M) antagonizing NECA inhibition of noradrenaline release. Inhibition of contractile responses by CV-1674 in the ileum was not antagonized by the adenosine antagonists and it had no effect on acetylcholine release (Fig. 1), indicating that CV-1674 was not an agonist at the postganglionic autonomic adenosine receptors. Instead, compatible with activation of A_{2a} receptors, CV-1674 exerted inhibition of contractile responses in rat vas deferens, an effect antagonized by CPT but only at high concentrations (from 10^{-6} M and above).

R-PIA (rat vas deferens) and 2-chloroadenosine (guinea pig ileum) were used as agonists in antagonist studies with alkylxanthines (PSØT, XAC, XCC, CPT, DPSPX and DPCPX). Regression analysis on guinea pig ileum and rat vas deferens antagonist potencies (pA_2 values) gave an excellent correlation, suggesting identity

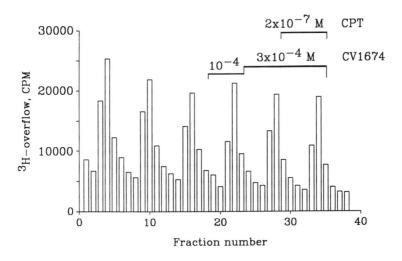

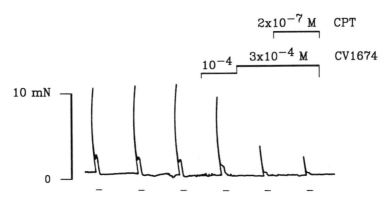

Figure 1. Guinea pig ileum longitudinal muscle. Acetylcholine release estimated as measurement of fractional [3H] overflow. CV-1674 did not affect stimulation-evoked release of [3H] (upper panel) whereas nerve-induced contractile responses were inhibited (lower panel). 8-cyclopentyltheophylline (CPT) (2 x 10-7 M) did not antagonize the effect of CV-1674 (3 x 10-4 M). Horizontal markers (lower panel) indicate periods (60 s) of nerve stimulation.

of receptors (Fig. 2a). When literature data on guinea pig and rat brain antagonist binding was compared they also gave an excellent correlation suggesting identity of receptors (Fig. 2b). When rat brain antagonist binding was compared with antagonist potencies in vas deferens (Fig. 2c) or guinea pig ileum (data not shown), significant correlations were also obtained, but slopes differed from unity. Comparison between the antagonist potencies in rat vas deferens and frog neuromuscular junction (Sebastiao et al. 1989) gave no significant correlation (Fig. 2d).

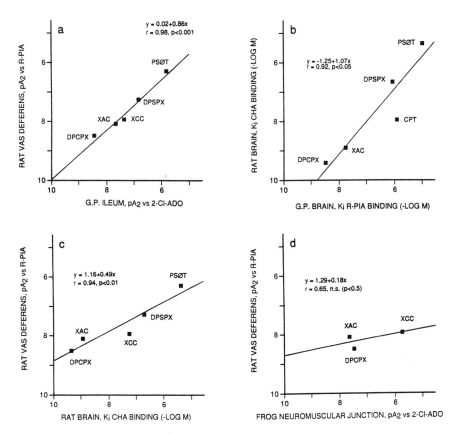

Figure 2. Diagrams comparing reported antagonist affinities (Bruns et al. 1987, Ukena et al. 1986a,1986b, Daly et al. 1985, Jacobson et al. 1985) obtained by binding studies in rat and guinea pig brain, compared with pA2 values in functional studies in rat vas deferens and guinea pig ileum (present study) or frog neuromuscular junction (Sebastiao et al. 1989). Linear regression by least squares, r denotes Pearson correlation coefficient.

Discussion

These findings suggest a near identity between adenosine receptors on peripheral autonomic nerves in rat and guinea pig. The receptors on autonomic nerves seem to closely resemble central receptors, with some noteworthy major differences. Thus, CV-1674 does not seem to be an agonist on the peripheral autonomic nerves, whereas it is active in binding in the CNS (Bruns et al. 1987) and seems to be an agonist on motor nerves (Singh et al. 1986). MeADO is equipotent with CV-1808 at the CNS receptors whereas CV-1808 is much more potent at the postganglionic receptors (Table 1). XAC is much more potent than XCC as antagonist in the brain whereas they are almost equipotent in the periphery. PD113.297 seems to be much less potent at autonomic than at central receptors (Table 1).

Table 1. Suggested criteria for A_{1a} and A_{1b} receptors. Differences in sets of potencies/affinities between functional studies in guinea pig ileum (A_{1b}) and binding in rat brain (A_{1a}), based on sources as in figure 2.

A_{1a} receptors			A_{1b} receptors		

Agonists

R-PIA 1.2 nM	≥	NECA 6.3 nM	R-PIA 17.3 nM	≥	NECA 81.6 nM
R-PIA 1.2 nM	>	S-PIA 49 nM	R-PIA 17.3 nM	>	S-PIA 2 880 nM

CV-1674 agonist: CV-1674 not agonist at 0.3 mM:

CV-1808 560 nM	≥	CV-1674 1300 nM	CV-1808 5 120 nM	>>	CV-1674 > 300 000 nM
MeADO 360 nM	=	CV-1808 560 nM	CV-1808 5 120 nM	>	MeADO > 29 300 nM

Antagonists

XAC 1.2 nM	>	XCC 58 nM	XAC 21.4 nM	≥	XCC 42.7 nM
PD113.297 5.6 nM	>	DPSPX 210 nM	PD113.297 120 nM	=	DPSPX 145 nM

Thus, since the autonomic in comparison with CNS receptors have a much lower affinity for some of the antagonists used, and since CV-1674 was not an agonist on the postganglionic autonomic nerve receptors, we would like to classify them as A_{1b}, whereas an appropriate designation for the CNS receptors is suggested to be A_{1a}. This subclassification agrees well with previous observations on autonomic nerves (Gustafsson et al. 1985) and seems logical from the striking parallels with subclassification characteristics for A_2 receptors (Bruns et al. 1987). The receptors at the neuromuscular junction might be A_3 (Ribeiro et al. 1986), or rather of A_{1a} type, the only support for this being the efficacy of CV-1674 (Singh et al. 1986) and the high potency of XAC (Sebastiao et al. 1989), comparable to binding in guinea pig brain (c.f. Fig. 2). Differences between some A_1-receptors in the autonomic and some in the central nervous system may compare with their different embryological origin, i.e. the central nervous system and motor neurons originate from the neural tube whereas autonomic and dorsal root ganglia originate from the neural crest. Since nerve cells located outside the CNS thus may terminate in the CNS we do not suggest that the A_{1a} receptors be named "central", especially since they might be at hand on cells in the periphery, and vice versa. The differences discussed are not explained only as a species difference since there are substantial within-species differences between A_1-receptors, classified also with *antagonists* (Fig. 2c), a method which should minimize the possibility of comparing the same receptor at different affinity during variable *endogenous* agonist activation. The differences in characteristics are not of a magnitude required for introduction of different main receptor classes, especially since similarities were significant. Instead, subdivision of A1 receptors is suggested, provisional criteria being listed in Table 1, keeping in mind that the separation derives from a comparison of binding data (CNS) with functional studies (vas deferens and ileum, respectively). Final determination would ideally use a similar method (only function or only binding) in both systems. In conclusion, affinities for some *agonists* and *antagonists* are higher in brain of rat and guinea pig than in rat vas or guinea pig ileum. Therefore, certain A1-receptors in rat and guinea pig brain are suggested to be called A$_{1a}$, for high affinity, and A$_1$-receptors in rat vas or guinea pig ileum correspondingly A$_{1b}$, for low affinity. The importance of measurements of transmitter overflow for verification of prejunctional effects, as well as the identification of non-adenosine receptor effects is evidenced.

Supported by the Swedish MRC (proj 7919), the Karolinska Institute, the Swedish Society of Medicine and Torsten and Ragnar Söderberg Foundations.

References

Bruns RF, Lu GH & Pugsley TA (1987) Adenosine Receptor Subtypes: Binding Studies. In: Topics and Perspectives in Adenosine Research. Eds. E Gerlach, BF Becker. Springer-Verlag, Berlin Heidelberg. 59-73.

Daly JW, Padgett W, Shamim MT, Butts-Lamb P & Waters J (1985) 1,3-Dialkyl-8-(p-sulfophenyl)xanthines: Potent water-soluble antagonists for A1- and A2-adenosine receptors. J Med Chem 28:487-492.

Gustafsson LE, Wiklund CU, Wiklund NP & Stelius L (1989) Introduction of A$_{1a}$ and A$_{1b}$ adenosine receptors. Abstract to XXXI International Congress of Physiological Sciences, Helsinki 9-14 July.

Gustafsson LE, Wiklund NP, Lundin J & Hedqvist P (1985). Characterization of pre- and postjunctional adenosine receptors in guinea pig ileum. Acta Physiol Scand 123:195-203.

Jacobson KA, Kirk KL, Padgett WL & Daly JW (1985) Functionalized Congeners of 1,3-Dialkylxanthines: Preparation of Analogues with High Affinity for Adenosine Receptors. J Med Chem 28:1334-1340.

Ribeiro JA & Sebastiao AM (1986) Adenosine receptors and calcium: basis for proposing a third (A$_3$) adenosine receptor. Progr Neurobiol 26: 179-209.

Sebastiao AM & Ribeiro JA (1989) 1,3,8- and 1,3,7-substituted xanthines: relative potency as adenosine receptor antagonists at the frog neuromuscular junction. Br J Pharmacol 96:211-219.

Singh YH, Dryden WF & Chen H (1986) The inhibitory effects of some adenosine analogues on transmitter release at the mammalian neuromuscular junction. Can J Physiol Pharmacol 64:1446-1450.

Stone TW (1985) Summary of a symposium discussion on purine receptor nomenclature. In: Purines: Pharmacology and Physiological Roles. Ed. T Stone. Macmillan, London. 1-4.

Ukena D, Daly JW, Kirk KL & Jacobson KA (1986a) Functionalized congeners of 1,3-dipropyl-8-phenylxanthine: potent antagonists for adenosine receptors that modulate membrane adenylate cyclase in pheochromocytoma cells, platelets and fat cells. Life Sci 38:797-807.

Ukena D, Jacobson KA, Padgett WL, Ayala C, Shamim MT, Kirk KL, Olsson RO & Daly JW (1986b) Species differences in structure-activity relationships of adenosine agonists and xanthine antagonists at brain A$_1$ adenosine receptors. FEBS Lett 209:122-128.

Wiklund CU, Wiklund NP & Gustafsson LE (1989a) Identification of A$_{1b}$ receptors in guinea pig ileum. Abstract presented at: Adenosine Receptors in the Nervous System. Satellite symposium of the 12th ISN congress, Albufeira, Portugal 21-22 April.

Wiklund NP, Cederqvist B, & Gustafsson L (1989b) Adenosine enhancement of adrenergic neuroeffector transmission in guinea pig pulmonary artery. Br J Pharmacol 96:425-433.

Wiklund NP, Wiklund CU, Öhlén A & Gustafsson LE (1989c). Cholinergic neuromodulation by endothelin in guinea pig ileum. Neurosci Lett 101:342-346.

31
Effects of Adenosine on Mast Cells

M.J. Lohse, K.-N. Klotz, K. Maurer, I. Ott, and U. Schwabe

Introduction

Stimulation of mast cells with appropriate antigen results in the rapid release of a large number of mediators of allergy from the cells. The cells possess cell surface receptors for the Fc portion of IgE which serve to transmit the signal of binding of an antigen by the IgE molecule to the interior of the cell. Since monovalent antigens are not capable of inducing mast cell secretion it is believed that crosslinking of IgE molecules by di-or polyvalent antigens represent the appropriate stimuli (Foreman 1980). These stimuli ultimately result in the release of two groups of mediators of allergy: preformed mediators, and secondary or newly formed mediators. Preformed mediators are stored in the mast cell granules and become released upon fusion of the granules with the plasma membrane; they include histamine, serotonin, chemotactic factors, lysosomal enzymes and heparin. Secondary mediators are synthesized only after stimulation of mast cells and then immediately released; they include prostaglandin D_2, leukotrienes and platelet activating factor (Siraganian 1983).

Studies of mast cells are complicated by important intra- and inter-species differences. Within a species, there appear to be at least two groups of mast cells (reviewed by Enerback 1986). The first group is the connective tissue mast cells, which is the most frequently studied type of mast cells and includes, for example, the mast cells obtained by peritoneal lavage. The second group represents the mucosal mast cells, which differ histologically, functionally and in their response to pharmacological agents from those of the first group. Blood basophils are, like mast cells, derived from bone marrow precursors, but their final differentiation in the blood rather than in tissues appears to result in yet another spectrum of properties (Kitamura et al. 1983). This heterogeneity together with differences between species, and often between strains of the same species, or even between the same strains obtained from different suppliers, may explain the discrepant results obtained by different laboratories and should be taken into consideration when generalizing and comparing experimental findings.

206

The first evidence that adenosine can alter mast cell mediator release was reported by Marquardt et al. in 1978. These authors found that adenosine enhanced mast cell release when cells were stimulated with a variety of agents, but that it did not affect release from nonstimulated cells. Since this modulatory effect of adenosine was antagonized by theophylline, it was assumed that it was exerted via classical adenosine receptors. Further supporting the hypothesis of a role for adenosine in the regulation of mast cell mediator release was the observation that antigen-stimulated mast cells generate substantial amounts of adenosine, probably via hydrolysis of ATP (Marquardt et al. 1984). This led to the hypothesis of a positive feedback: stimulation of mast cells by antigen (or other stimuli) leads not only to mediator release, but at the same time to generation of adenosine; adenosine, in turn, enhances mediator release by an action at cell surface receptors.

Two findings challenged this straight-forward concept. First, adenosine may not only enhance, but also inhibit mediator release from basophils and mast cells, and sometimes both enhancement and inhibition were seen in the same models depending on experimental conditions (Marone et al. 1979; Church et al. 1983; Hughes et al. 1984). Second, whereas earlier studies found that methylxanthines antagonized the enhancing effects of adenosine (Marquardt 1978; Church et al. 1983; Hughes et al. 1984), more recent studies failed to observe such antagonism (Vardey and Skidmore, 1985; Hughes and Church, 1986). These observations suggested that adenosine may exert more complex effects on mast cells than thought previously, and that these effects may occur via as yet unknown sites.

Effects of adenosine on peritoneal mast cells.

Our initial studies sought to verify the concept that adenosine receptors exist on the surface of mast cells. We used rat peritoneal mast cells, purified to >95% homogeneity, as a model. In membranes prepared from these cells, adenosine and several of its analogues stimulated the activity of adenylyl cyclase by 30-50% (Lohse et al. 1987). Theophylline and 8-phenyltheophylline completely prevented this stimulation. In intact mast cells, 5'-N-ethylcarbox-amidoadenosine (NECA), caused a two-fold increase in cAMP-levels even in the absence of phosphodiesterase inhibitors. This effect was again antagonized by methylxanthines. These data strongly suggest the presence of adenosine receptors of the A_2 type on rat pertoneal mast cells. No evidence for receptors of the A_1 type was found in experiments investigating adenylyl cyclase activity in membranes, cAMP-levels in intact cells, or in radioligand binding studies (Lohse et al. 1987).

Adenosine analogues enhanced histamine release from the cells, when release was induced by the calcium ionophore A23187 or by the lectin concanavalin A which cross-links IgE receptors. The order of potency was NECA>R-PIA >2-Cl-adenosine with NECA also being somewhat more effective than other adenosine analogues. However, in contrast to adenosine receptor-mediated effects, methylxanthines did not

207

antagonize the enhancement of histamine release by adenosine analogues (Lohse et al. 1987). A similar lack of antagonism by methylxanthines has been observed by other investigators (Vardey and Skidmore, 1985; Leoutsakos and Pearce 1986; Hughes and Church 1986; Peachell et al. 1989).

These results allow two possible conclusions: either the enhancement of histamine release by adenosine is mediated by a novel type of adenosine receptor, or it occurs via an intracellular site. The first conclusion, i.e. the postulate of a new type of adenosine receptor, was made by Hughes and Church (1986), since they observed no effect of the nucleoside transport inhibitor dipyridomole. Our own data, however, support the idea of an intracellular site of action of adenosine (Lohse et al. 1987, 1988). This hypothesis was based on the effects of two more potent nucleoside transport inhibitors, nitrobenzylthioinosine (NBTI) and nitrobenzylthioguanosine (NBTG). First, these compounds prevented the adenosine-induced enhancement of histamine release from rat peritoneal mast cells, when the cells were stimulated either with A23187 or with concanavalin A (Lohse et al. 1987). And second, in the absence of endogenous adenosine but in the presence of strong stimuli such as higher concentrations of A23187, NBTI actually enhances histamine release (Lohse et al. 1988, 1989). Since in the presence of strong stimuli considerable amounts of adenosine are generated by mast cells (Marquardt et al. 1984), we think that the release-enhancing effects of NBTI are due to trapping of the newly generated adenosine inside the mast cells. Hence, it is assumed that adenosine enhances histamine release via an action at an intracellular site.

Synergism between adenosine and calcium.

To further investigate the nature of this presumed intracellular site, we studied the effects of adenosine on the intracellular events that follow antigenic stimulation. Although the exact cascade leading from IgE receptor stimulation to mediator release is still the subject of considerable controversy (see for example Penner 1988), it is generally agreed that the initial event is activation of phospholipase C leading to generation of inositol trisphosphate (which releases calcium from intracellular stores) and diacylglycerol (which activates protein kinase C). Both free cytosolic calcium and protein kinase C are thought to activate mediator release through as yet ill-defined mechanisms.

In our hands, adenosine and its analogues interfered neither with the generation of inositol trisphosphate nor with the elevation of cytosolic calcium after stimulation of the IgE receptors of rat peritoneal mast cells (Lohse et al. 1988). In contrast, Marquardt and Walker (1988, 1989) reported that adenosine increased the cytosolic calcium concentration in antigen-stimulated bone marrow-derived mast cells; it is not clear, however, whether these authors controlled for the artifactual signals due to secretion of fura-2 from mast cells (Almers and Neher 1985).

208

Adenosine did, however, enhance histamine release induced by elevated levels of cytosolic calcium (Lohse et al. 1988). This was demonstrated by varying the free calcium concentration in ATP-permeabilized mast cells. Together with the well-known enhancing effects of adenosine on A23187-induced release this suggests that adenosine acts at a step distal to elevation of cytosolic calcium. At this step there appears to be synergism between calcium and adenosine. Not only does adenosine enhance the effects of intracellular calcium, but in addition elevated intracellular calcium increases the potency of adenosine and its analogues (Lohse et al. 1989).

There is some evidence that adenosine might act via an activation of protein kinase C, since adenosine appears to enhance translocation of the kinase, and also because down-regulation of the kinase reduces the effects of adenosine (Marquardt and Walker 1989). However, whereas phorbol esters, direct activators of protein kinase C, induce some histamine release by themselves, adenosine clearly has no effect when given alone. Furthermore, adenosine does not enhance the effects of submaximal concentrations of phorbol esters (Lohse et al. 1988).

Effects of adenosine on histamine release from human lung tissue.

In order to explore the relevance of these findings to the setting of mediator release from mast cells in the human lung we have more recently studied the effects of adenosine and its analogues on histamine release from human lung fragments obtained during surgery. Histamine release from these fragments was induced by either concanavalin A or A23187. Adenosine and its analogues only slightly stimulated release when given together with these agents. However, in the additional presence of adenosine receptor antagonists, adenosine caused a very marked increase of histamine release. On the other hand, in the presence of nucleoside transport inhibitors adenosine caused an inhibition of histamine release. This suggests that adenosine inhibits histamine release from these fragments via an A_2 receptor (as seen in basophils by Marone et al. 1979, and others) and stimulates the release via an intracellular site. The development of antagonists for this intracellular site may, therefore, result in an inhibition of histamine release in the human lung and lead to the development of novel antiallergic drugs.

References

Almers, W., Neher, E., 1985, The Ca signal from fura-2 loaded mast cells depends strongly on the method of dye-loading. FEBS Lett. **192**, 13-18.

Church, M.K., Holgate, S.T., Hughes, P.J., 1983, Adenosine inhibits and potentiates IgE-dependent histamine release from human basophils by an A_2-receptor mediated mechanism. Br. J. Pharmacol. **80**, 719-726.

Enerback, L., 1986, Mast cell heterogeneity: the evolution of the concept of a specific mucosal mast cell. In: Mast Cell Differentiation and heterogeneity, ed. Befus, A.D. et al., Raven Press, New York, pp.1-26.

Foreman, J., 1980, Receptor-secretion coupling in mast cells. Trends Pharmacol. Sci. 1, 460-462.

Hughes, P.J., Holgate, S.T., Church, M.K., 1984, Adenosine inhibits and potentiates IgE-dependent histamine release from human lung mast cells by an A_2-purinoceptor mediated mechanism. Biochem. Pharmacol. 33, 3847-3852.

Hughes, P.J., Church, M.K., 1986, Separate purinoceptors mediate enhancement by adenosine of concanavalin A-induced mediator release and the cyclic AMP response in rat mast cells. Agents and Actions 18, 81-84.

Leoutsakos, A., Pearce, F.L., 1986, The effect of adenosine and its analogues on cyclic AMP changes and histamine secretion from rat peritoneal mast cells stimulated by various ligands. Biochem. Pharmacol. 35, 1373-1379.

Lohse, M.J., Maurer, K., Gensheimer, H.-P., Schwabe, U., 1987, Dual effects of adenosine on rat peritoneal mast cells. Naunyn-Schmiedeberg's Arch. Pharmacol. 335, 555-560.

Lohse, M.J., Klotz, K.-N., Salzer, M.J., Schwabe, U., 1988, Adenosine regulates the Ca^{++} sensitivity of mast cell mediator release. Proc. Natl. Acad. Sci. USA 85, 8875-8879.

Lohse, M.J., Maurer, K., Klotz, K.-N., Schwabe, U., 1989, Synergistic effects of calcium-mobilizing agents and adenosine on histamine release from rat peritoneal mast cells. Br. J. Pharmacol. (in press).

Marone, G., Findlay, S.R., Lichtenstein, L.M., 1979, Adenosine receptor on human basophils: Modulation of histamine release. J. Immunol. 123, 1473-1477.

Marquardt, D.L., Parker, C.W., Sullivan, T.J., 1978, Potentiation of mast cell mediator release by adenosine. J. Immunol. 120, 871-878.

Marquardt, D.L., Gruber, H.E., Wasserman, S.I., 1984, Adenosine release from stimulated mast cells. Proc. Natl. Acad. Sci. USA 81, 6192-6196.

Marquardt, D.L., Walker, L.L., 1988, Alteration of mast cell responsiveness to adenosine by pertussis toxin. Biochem. Pharmacol. 37, 4019-4025.

Marquardt, D.L., Walker, L.L., 1989, Pretreatment with phorbol esters abrogates mast cell adenosine responsiveness. J. Immunol. 142, 1268-1273.

Peachell, P.T., Columbo, M., Kagey-Sobotka, A., Lichtenstein, L.M., Marone, G., 1988, Adenosine potentiates mediator release from human lung mast cells. Am. Rev. Resp. Dis. 138, 1143-1151.

Penner, R., 1988, Multiple signaling pathways control stimulus-secretion coupling in rat peritoneal mast cells. Proc. Natl. Acad. Sci. USA 85, 9856-9860.

Siraganian, R.P., 1983, Histamine-secretion from mast-cells and basophils. Trends Pharmacol. Sci. 4, 432-437.

Vardey, C.J., Skidmore, I.F., 1985, Characterization of the adenosine receptor responsible for the enhancement of mediator release from rat mast cells. In: Purines: Pharmacology and physiological roles, ed. Stone, T.W. , VCH Verlagsgesellschaft, Weinheim, p 175.

210

Adenosine did, however, enhance histamine release induced by elevated levels of cytosolic calcium (Lohse et al. 1988). This was demonstrated by varying the free calcium concentration in ATP-permeabilized mast cells. Together with the well-known enhancing effects of adenosine on A23187-induced release this suggests that adenosine acts at a step distal to elevation of cytosolic calcium. At this step there appears to be synergism between calcium and adenosine. Not only does adenosine enhance the effects of intracellular calcium, but in addition elevated intracellular calcium increases the potency of adenosine and its analogues (Lohse et al. 1989).

There is some evidence that adenosine might act via an activation of protein kinase C, since adenosine appears to enhance translocation of the kinase, and also because down-regulation of the kinase reduces the effects of adenosine (Marquardt and Walker 1989). However, whereas phorbol esters, direct activators of protein kinase C, induce some histamine release by themselves, adenosine clearly has no effect when given alone. Furthermore, adenosine does not enhance the effects of submaximal concentrations of phorbol esters (Lohse et al. 1988).

Effects of adenosine on histamine release from human lung tissue.

In order to explore the relevance of these findings to the setting of mediator release from mast cells in the human lung we have more recently studied the effects of adenosine and its analogues on histamine release from human lung fragments obtained during surgery. Histamine release from these fragments was induced by either concanavalin A or A23187. Adenosine and its analogues only slightly stimulated release when given together with these agents. However, in the additional presence of adenosine receptor antagonists, adenosine caused a very marked increase of histamine release. On the other hand, in the presence of nucleoside transport inhibitors adenosine caused an inhibition of histamine release. This suggests that adenosine inhibits histamine release from these fragments via an A_2 receptor (as seen in basophils by Marone et al. 1979, and others) and stimulates the release via an intracellular site. The development of antagonists for this intracellular site may, therefore, result in an inhibition of histamine release in the human lung and lead to the development of novel antiallergic drugs.

References

Almers, W., Neher, E., 1985, The Ca signal from fura-2 loaded mast cells depends strongly on the method of dye-loading. FEBS Lett. **192**, 13-18.

Church, M.K., Holgate, S.T., Hughes, P.J., 1983, Adenosine inhibits and potentiates IgE-dependent histamine release from human basophils by an A_2-receptor mediated mechanism. Br. J. Pharmacol. **80**, 719-726.

Enerback, L., 1986, Mast cell heterogeneity: the evolution of the concept of a specific mucosal mast cell. In: Mast Cell Differentiation and heterogeneity, ed. Befus, A.D. et al., Raven Press, New York, pp.1-26.

Foreman, J., 1980, Receptor-secretion coupling in mast cells. Trends Pharmacol. Sci. **1**, 460-462.

Hughes, P.J., Holgate, S.T., Church, M.K., 1984, Adenosine inhibits and potentiates IgE-dependent histamine release from human lung mast cells by an A_2-purinoceptor mediated mechanism. Biochem. Pharmacol. **33**, 3847-3852.

Hughes, P.J., Church, M.K., 1986, Separate purinoceptors mediate enhancement by adenosine of concanavalin A-induced mediator release and the cyclic AMP response in rat mast cells. Agents and Actions **18**, 81-84.

Leoutsakos, A., Pearce, F.L., 1986, The effect of adenosine and its analogues on cyclic AMP changes and histamine secretion from rat peritoneal mast cells stimulated by various ligands. Biochem. Pharmacol. **35**, 1373-1379.

Lohse, M.J., Maurer, K., Gensheimer, H.-P., Schwabe, U., 1987, Dual effects of adenosine on rat peritoneal mast cells. Naunyn-Schmiedeberg's Arch. Pharmacol. **335**, 555-560.

Lohse, M.J., Klotz, K.-N., Salzer, M.J., Schwabe, U., 1988, Adenosine regulates the Ca^{++} sensitivity of mast cell mediator release. Proc. Natl. Acad. Sci. USA **85**, 8875-8879.

Lohse, M.J., Maurer, K., Klotz, K.-N., Schwabe, U., 1989, Synergistic effects of calcium-mobilizing agents and adenosine on histamine release from rat peritoneal mast cells. Br. J. Pharmacol. (in press).

Marone, G., Findlay, S.R., Lichtenstein, L.M., 1979, Adenosine receptor on human basophils: Modulation of histamine release. J. Immunol. **123**, 1473-1477.

Marquardt, D.L., Parker, C.W., Sullivan, T.J., 1978, Potentiation of mast cell mediator release by adenosine. J. Immunol. **120**, 871-878.

Marquardt, D.L., Gruber, H.E., Wasserman, S.I., 1984, Adenosine release from stimulated mast cells. Proc. Natl. Acad. Sci. USA **81**, 6192-6196.

Marquardt, D.L., Walker, L.L., 1988, Alteration of mast cell responsiveness to adenosine by pertussis toxin. Biochem. Pharmacol. **37**, 4019-4025.

Marquardt, D.L., Walker, L.L., 1989, Pretreatment with phorbol esters abrogates mast cell adenosine responsiveness. J. Immunol. **142**, 1268-1273.

Peachell, P.T., Columbo, M., Kagey-Sobotka, A., Lichtenstein, L.M., Marone, G., 1988, Adenosine potentiates mediator release from human lung mast cells. Am. Rev. Resp. Dis. **138**, 1143-1151.

Penner, R., 1988, Multiple signaling pathways control stimulus-secretion coupling in rat peritoneal mast cells. Proc. Natl. Acad. Sci. USA **85**, 9856-9860.

Siraganian, R.P., 1983, Histamine-secretion from mast-cells and basophils. Trends Pharmacol. Sci. **4**, 432-437.

Vardey, C.J., Skidmore, I.F., 1985, Characterization of the adenosine receptor responsible for the enhancement of mediator release from rat mast cells. In: Purines: Pharmacology and physiological roles, ed. Stone, T.W. , VCH Verlagsgesellschaft, Weinheim, p 175.

32
The Role of Adenosine on the Gastric Acid Secretory Response

J.G. Gerber and N.A. Payne

Introduction

Theophylline, an alkylxanthine, has been in extensive clinical use for the treatment of bronchospastic lung disease because the drug causes bronchial smooth muscle relaxation. One side effect associated with its use is dyspepsia, and the drug has been found to stimulate gastric acid production in man (Krasnow *et al.*, 1949; Foster *et al.*, 1979). The mechanism by which theophylline increases gastric acid output has been presumed to be secondary to parietal cell phosphodiesterase inhibition. Biochemically, this hypothesis has a solid basis because histamine stimulates parietal cell acid production by activating adenylate cyclase, resulting in the elevation of cellular cyclic AMP (Wollin *et al.*, 1979). Thus, a phosphodiesterase inhibitor would be expected to enhance the cellular cAMP concentration to histamine stimulation. However, the concentration of theophylline that results in phosphodiesterase inhibition in various tissues is consistently above $500\mu M$ (Butcher *et al.*, 1962; Smellie *et al.*, 1979). The therapeutic plasma concentration of theophylline in man is 50 to $100\mu M$. Thus, it seemed unlikely that the mechanism of enhanced acid secretion to theophylline is secondary to phosphodiesterase inhibition. Since many of the alkylxanthines are also competitive antagonists of the cell surface adenosine receptors, we explored the hypothesis that adenosine is a direct inhibitor of gastric acid output. We chose to explore this hypothesis in the dog model because of the capacity to examine the effect of adenosine both in the *in vivo* gastric fistula dog and in the *in vitro* isolated gastric cell preparation. In addition, the gastric physiology and anatomy of the dog resembles the human.

Materials and Methods

At first we examined the effect of adenosine on gastric acid secretion and gastric blood flow in the anesthetized gastric fistula dog (Gerber *et al.*, 1984). In this preparation the gastro-splenic artery of the dog is isolated, and after performing a splenectomy, the artery supplies exclusively the corpus of the stomach where the parietal cells reside. Two small needles are introduced into the artery, one for the infusion of the secretagogues and the other for the infusion of adenosine. An electromagnetic flow probe is placed around the artery for blood flow measurements. By localizing the infusion of secretagogues and adenosine to the corpus of the stomach, we avoid systemic hemodynamic effects of the compounds. A 10mm cannula is placed in the most dependent portion of the stomach, and the dog is placed in a prone position with gastric secretion collected by drainage from the cannula.

In the first set of experiments, we examined the effect of adenosine (calculated to a concentration of $30\mu M$) on histamine ($0.5\mu g/min$)- and methacholine ($1\mu g/min$)-stimulated acid secretion. Adenosine inhibited histamine-stimulated acid secretion from 197 ± 25 $\mu Eq/15$ min to 68 ± 19 $\mu Eq/15$ min, and methacholine stimulated acid secretion from 280 ± 48 to 81 ± 24 $\mu Eq/15$ min (figure 1). Intravenous infusion of theophylline to a plasma concentration of 94μM resulted in the complete abolition of the adenosine effect on acid secretion. Theophylline demonstrated specificity for adenosine as it did not affect the acid inhibitory response to prostaglandin E₂ (figure 2). In addition, the specific phosphodiesterase inhibitor, RO 20 1724, enhanced the histamine-stimulated acid secretory response, but did not alter the inhibitory effect of adenosine on gastric acid secretion. Infusion of adenosine into the gastrosplenic artery resulted in gastric vasodilation without any change in mean arterial pressure, an effect also inhibited by theophylline.

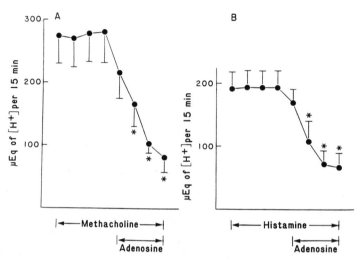

Figure 1. The effect of the gastric artery infusion of adenosine on methacholine-stimulated acid secretion (A) and histamine-stimulated acid secretion (B). The data are expressed as mean ± SEM of 15 min collections. The asterisk signifies that the data points during adenosine infusion are statistically different from the control periods. (From: JG Gerber, S FAdul, NA Payne and AS Nies. Adenosine: A modulator of gastric acid secretion *in vivo*. J Pharmacol Exp Ther 231(1):109-113, 1984. © by American Society for Pharmacology and Experimental Therapeutics)

These data *in vivo* demonstrated that locally infused adenosine can blunt the gastric acid secretory response to both histamine and methacholine. In addition, theophylline at a therapeutic concentration can totally block both the acid inhibitory and the vasodilatory effect of infused adenosine.

In order to localize the site of action of adenosine to the parietal cell, we performed the next set of experiments on parietal cell enriched canine gastric cells (Gerber *et al.*, 1985a). Dispersed gastric cells were isolated from corpus mucosal tissue via sequential EDTA and collagenase digestion. This mixed cell population which contains less than 20% parietal cells was then enriched on a Ficoll density gradient to contain 59±3% parietal cells. The experiments were performed on the parietal cell enriched cells. Acid production by the parietal cells can be estimated by the use of the ¹⁴C aminopyrine uptake technique. Aminopyrine is a weak base that remains unionized at physiologic pH and therefore freely traverses cell membranes. However, at low pH that exists in the secretory canaliculi of the parietal cells, aminopyrine becomes ionized and trapped. The amount of aminopyrine that gets trapped in the parietal cells correlates linearly with the rate of acid production by the parietal cells. Thus, the energy dependent accumulation of ¹⁴C aminopyrine in the gastric cells reflects the amount of acid sequestered intracellularly in the parietal cells.

In the first set of experiments we examined the effect of increasing concentrations of 2-chloroadenosine (2CA) and L-phenylisopropyladenosine (L-PIA) on histamine-stimulated

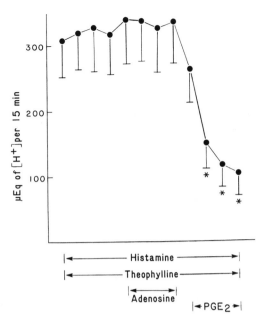

Figure 2. The effect of i.v. infusion of theophylline (94μM) on the inhibition of hista-mine-stimulated gastric acid output by adenosine and prostaglandin E_2. The data are expressed as mean ± SEM of 15 min collections. The asterisk signifies that the data points during PGE_2 infusion are statistically different from the histamine control periods. (From: JG Gerber, S Fadul, NA Payne and AS Nies. Adenosine: A modulator of gastric acid secretion *in vivo*. J Pharmacol Exp Ther 231(1):109-113, 1984. © by American Society for Pharmacology and Experimental Therapeutics)

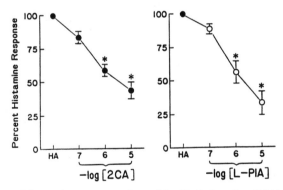

Figure 3. Effect of increasing concentrations of 2-chloroadenosine (2CA) and L-PIA on the cellular accumulation of aminopyrine produced by histamine, 1μM + IBMX, 3μM. Zero percent represents the aminopyrine accumulation with IBMX alone and 100% represents the aminopyrine accumulation in the presence of histamine + IBMX. The asterisk signifies that the accumulation of aminopyrine produced by histamine + IBMX in the presence of 2CA or L-PIA is significantly different from the aminopyrine accumulation with histamine + IBMX alone. The results are expressed as mean ± SEM. (From: JG Gerber, AS Nies and NA Payne. Adenosine receptors on canine parietal cells modulate gastric acid secretion to histamine. J Pharmacol Exp Ther 233(3):623-627, 1985. © by American Society for Pharmacology and Experimental Therapeutics)

215

acid production. All the cells were stimulated with a submaximal histamine concentration, $1\mu M$, in the presence of $3\mu M$ isobutylmethylxanthine (IBMX). In the dog parietal cell preparation, histamine by itself is a weak stimulator of acid production but in the presence of a phosphodiesterase inhibitor, the histamine effect is greatly enhanced. We found that both 2CA and L-PIA inhibited histamine-stimulated acid production with L-PIA being slightly more potent (figure 3). The inhibition appeared concentration-dependent. The inhibition was specific to histamine in that carbachol-stimulated acid production was unaltered by 2CA. In order to demonstrate that the effect of 2CA and L-PIA are adenosine receptor mediated, we coincubated the cells with either $300\mu M$ theophylline or $10\mu M$ 8-phenyltheophylline with 2CA during histamine stimulation. We chose these concentrations of theophylline and 8-phenyltheophylline because in previous experiments we have demonstrated that these concentrations of the drugs had no significant effect on gastric cell phosphodiesterase activity. Both theophylline and 8-phenyltheophylline displaced 2CA from its binding site as indicated by the apparent loss of activity of 2CA in the presence of adenosine receptor antagonists. These data suggested that there are cell surface adenosine receptors on parietal cells involved with inhibition of acid production to histamine.

The effect of adenosine analogs were curiously similar to the effect of PGE_2 on acid production. Soll has previously demonstrated that PGE_2 inhibited histamine-stimulated acid production without affecting carbachol-stimulated acid production. In addition, PGE_2 inhibited histamine-stimulated cAMP levels in the parietal cell enriched gastric cell preparation (Soll, 1980). Therefore, we wanted to be sure that the effect of 2CA on the inhibition of acid production was not secondary to prostaglandin generation. We therefore compared the effect of 2CA in the presence and the absence of $30\mu M$ indomethacin, a drug concentration that causes a high grade inhibition of gastric cell cyclooxygenase activity. We found that indomethacin did not interfere with the activity of 2CA on acid inhibition. We next explored the hypothesis that adenosine analogues inhibited acid production to histamine by modulating the rise in intracellular cAMP. We easily demonstrated that histamine stimulated gastric cell cAMP concentrations three-fold within the first 5 minutes of incubation, but we could not show much inhibition with increasing concentrations of L-PIA. Interestingly, at 100nM, L-PIA consistently decreased histamine-stimulated cAMP levels, but at higher concentrations of L-PIA this inhibitory effect was lost. Since our cell preparation contained on the average 59% parietal cells, it was conceivable that the contaminating cells (chief cells and mucous cells) have adenylate cyclase stimulatory adenosine receptors, making accurate evaluation of the effect of adenosine on cAMP levels in parietal cells imprecise. At low concentrations, L-PIA will preferentially activate A_1 adenosine receptors but at higher concentrations this selectivity is lost. Nonetheless, the acid inhibitory data are suggestive that parietal cells contain predominantly A_1 adenosine receptors.

The final set of experiments was conceived to try to evaluate whether enough endogenous adenosine is generated by gastric cells during an acid secretory stimulus to result in a physiologic response (Gerber et al., 1988). Inclusion of $10\mu M$ 8-phenyltheophylline, an adenosine receptor antagonist, resulted in a $35\pm12\%$ and $31\pm9\%$ increase in parietal cell aminopyrine accumulation at histamine concentrations of $1\mu M$ and $10\mu M$, respectively. 8-Phenyltheophylline had no effect on the basal acid production by the parietal cells. In addition, the effect of 8-phenyltheophylline was specific to histamine in that it did not enhance the acid production to either carbachol or dibutyryl cAMP (figure 4). The inclusion of $1\mu M$ dipyridamole, an inhibitor of adenosine transport, resulted in a $34\pm6\%$ and $31\pm5\%$ decrease in aminopyrine accumulation at histamine concentrations of $1\mu M$ and $10\mu M$. Again, the effect of dipyridamole was specific to histamine in that it did not decrease the acid production to either carbachol or dibutyryl cAMP. Dipyridamole did not depress the basal acid production (figure 5). Finally, the addition of adenosine deaminase, 500mU/ml, to gastric cells resulted in a small but statistically significant enhancement of histamine-stimulated acid production. The effect of the enzyme was specific to histamine-stimulated cells in that neither carbachol- nor dibutyryl cAMP-stimulated cells demonstrated an enhancement in the presence of adenosine deaminase. These data are consistent with the hypothesis of a modulatory role of adenosine during histamine-stimulated acid production.

216

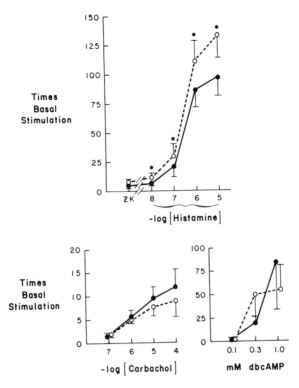

Figure 4. The effect of 8-phenyltheophylline, 10μM, on parietal cell aminopyrine uptake to increasing concentrations of histamine in the presence of the phosphodiesterase inhibitor ZK62711 (rolipram), 10nM (top), carbachol (bottom, left) and dbcAMP (bottom, right). The abscissa represents increasing concentrations of the drug, and the ordinate represents the calculated times basal stimulation derived from dividing the aminopyrine accumulation in the cells in the presence of secretagogue ± 8-phenyltheophylline by the aminopyrine accumulation in the cells without any drugs. ●—●, the dose response curves to the secretagogues in the absence of 8-phenyltheophylline, o--o, the dose-response curves to the secretagogues in the presence of 8-phenyltheophylline. The asterisk signifies that the stimulation of aminopyrine uptake was significantly greater with the addition of 8-phenyltheophylline than without it. (From: JG Gerber and NA Payne. Endogenous adenosine modulates gastric acid secretion to histamine in canine parietal cells. J Pharmacol Exp Ther 244(1):190-194, 1988. © by American Society for Pharmacology and Experimental Therapeutics)

Discussion

We have performed a group of experiments in dogs demonstrating the presence of functional adenosine receptors on parietal cells involved with inhibition of acid secretion. The basis of these experiments was the clinical observation that administration of theophylline to man is associated with an enhanced acid secretory response. Based on our observations, we can now propose a rational hypothesis as to why the administration of methylxanthines like theophylline and caffeine can result in acid hypersecretion. *In vivo*, the infusion of adenosine into the gastric artery in dogs resulted in inhibition of acid secretion to both histamine and methacholine. This effect of adenosine was totally abolished by the infusion of theophylline to a therapeutic plasma concentration of 94μM.

To try to localize the effect of adenosine to the parietal cell, we employed the parietal cell enriched gastric cell preparation from dogs. In these experiments we were able to demonstrate the inhibition of acid production to histamine by stable adenosine analogs L-PIA and 2-chloroadenosine. In the isolated cells, unlike *in vivo*, the cholinergic-stimulated acid secretion was unaffected by adenosine. This observation was not surprising because *in vivo* the cholinergic-stimulated acid secretory response has a histaminergic component. Methacholine-stimulated acid secretion can be inhibited by histamine-2 receptor antagonists *in vivo*, but not in the isolated cell preparation (Gerber *et al.*, 1985b). Since histamine greatly potentiates the cholinergic acid secretory response (Soll, 1982), it is likely that *in vivo* adenosine inhibited the histaminergic component of acid stimulation to methacholine, thereby down-regulating the cholinergic acid secretory response.

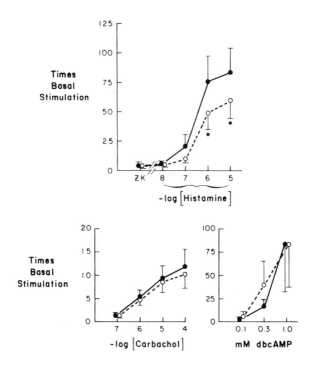

Figure 5. The effect of 1μM dipyridamole on parietal cell aminopyrine uptake to increasing concentrations of histamine in the presence of the phosphodiesterase inhibitor ZK 62711 (rolipram) 10nM (top), carbachol (bottom, left) and dbcAMP (bottom, right). The abscissa represents increasing concentrations of the drug, and the ordinate represents the calculated times basal stimulation derived from dividing the aminopyrine accumulation in the cells in the presence of secretagogues ± dipyridamole by the aminopyrine accumulation in the cells without any drugs. ●—●, the dose response curves to the secretagogues in the absence of dipyridamole; o--o, the dose response curves to the secretagogues in the presence of dipyridamole. Asterisk signifies that the stimulation of aminopyrine uptake was significantly less with the addition of dipyridamole than without it. (From: JG Gerber and NA Payne. Endogenous adenosine modulates gastric acid secretion to histamine in canine parietal cells. J Pharmacol Exp Ther 244(1):190-194, 1988. ©by American Society for Pharmacology and Experimental Therapeutics)

218

We have also demonstrated that the gastric cells generated enough adenosine in cell suspension to down-regulate the parietal cell acid secretory response to histamine. The exact source and stimulus for adenosine release could not be ascertained, but the predominant cell in suspension was the parietal cell. It is quite likely that adenosine, acting as an autocoid, is released from the same cell on which it acts. In addition, stimulation of acid production in parietal cells is associated with increased ATP utilization and turnover. Adenosine release in other cells and tissues is associated with increased ATP turnover (Meghji et al., 1985; Osswald et al., 1980).

We can conclude from our experiments that adenosine, working through plasma membrane adenosine receptors, inhibits histamine-stimulated acid production. The receptor subtype is most likely an A_1 receptor based on the specificity of adenosine towards histamine-stimulated cells. Histamine utilizes cyclic AMP as a second messenger to amplify acid secretory signals. In addition, the gastric cells generate enough adenosine in suspension to modulate the acid secretory response to exogenously added histamine. All these data support the hypothesis that endogenously produced adenosine has an important modulatory role during an acid secretory signal.

References

Butcher, R.W. and Sutherland, E.W., 1962, Adenosine 3',5'-phosphate in biological materials. I. Purification and properties of cyclic 3',5'-nucleotide phosphodiesterases and use of this enzyme to characterize 3',5'-phosphate in human urine, J. Biol. Chem. 237:1244-1250.

Foster, L.J., Trudeau, W.L. and Goldman, A.L, 1979, Bronchodilator effects on gastric acid secretion, JAMA 241:2613-2615.

Gerber, J.G., Fadul, S., Payne, N.A. and Nies, A.S., 1984, Adenosine: A modulator of gastric acid secretion in vivo, J. Pharmacol. Exp. Ther. 231:109-113.

Gerber, J.G., Nies, A.S. and Payne, N.A., 1985a, Adenosine receptors on canine parietal cells modulate gastric acid secretion to histamine, J. Pharmacol. Exp. Ther. 233:623-627.

Gerber, J.G., Payne, N.A., Fadul, S. and Nies, A.S., 1985b, Cholinergic mechanism of acid secretion in the dog: An in vivo and in vitro comparison, Eur. J. Pharmacol. 106:373-380.

Gerber, J.G. and Payne, N.A., 1988, Endogenous adenosine modulates gastric acid secretion to histamine in canine parietal cells, J. Pharmacol. Exp. Ther. 244:190-194.

Krasnow, S. and Grossman, M.I., 1949, Stimulation of gastric secretion in man by theophylline ethylenediamine, Proc. Soc. Exp. Biol. Med. 71:335-336.

Meghji, P., Holmquist, C.A. and Newby, A.C., 1985, Adenosine formation and release from neonatal rat heart cells in culture, Biochem. J. 229:799-805.

Osswald, H., Nabakowski, G. and Hermes, H., 1980, Adenosine as a possible mediator of metabolic control of glomerular filtration rate, Int. J. Biochem. 12:263-267.

Smellie, F.W., Davis, C.M., Daly, J.W. and Wells, J.N., 1979, Aklylxanthines. Inhibition of adenosine elicited accumulation of cyclic AMP in brain slices and of brain phosphodiesterase activity, Life Sci. 24:2475-2482.

Soll, A.H., 1980, Specific inhibition by prostaglandins E_2 and I_2 of histamine-stimulated [^{14}C]aminopyrine accumulation and cyclic adenosine monophosphate generation by isolated canine parietal cells, J. Clin. Invest. 65:1222-1229.

Soll, A.H., 1982, Potentiating interactions of gastric stimulants on [^{14}C]-aminopyrine accumulation by isolated canine parietal cells, Gastroenterology 83:216-223.

Wollin, A., Soll, A.H. and Samloff, I.M., 1979, Actions of histamine, secretin, and PGE_2 on cyclic AMP production by isolated canine fundic mucosal cells, Am. J. Physiol. 237:E437-E443.

33
Adenosine Receptors and Signaling in the Kidney

W.S. Spielman, L.J. Arend, K.-N. Klotz, and U. Schwabe

When adenosine binds to plasma-membrane receptors on a variety of cell types in the kidney, it stimulates functional responses that span the entire spectrum of renal cellular physiology, including alterations in hemodynamics, hormone and neurotransmitter release, and tubular reabsorption (Table 1). This array of diverse responses, appears to represent a means by which the kidney and its constituent cell types can regulate the metabolic demand such that it is maintained at an appropriate level for the prevailing metabolic supply (Figure 1). With the increased recognition of this wide array of renal cellular actions, and the continuing development of relatively specific adenosine receptor agonist and antagonist ligands, investigators have undertaken the task of assigning the different renal actions of adenosine to the known adenosine receptor types, by comparison of relative agonist and antagonist potencies. It is apparent from the inspection of a list of the renal actions of adenosine, that not only does adenosine control a variety of functions but it appears to have a "dual-control" over many aspects of renal function mediated by separate receptors. This approach, while providing useful information on the action and the possible receptor subtypes leaves some questions as to the coupling to second messenger systems, and does not provide molecular information of the subcellular events that may be involved.

With the exception of their ability to respond to adenosine and adenosine analogs, nothing as yet has been described that distinguishes adenosine receptors from the wide variety of receptors that modify adenylate cyclase activity and are therefore likely members of a large class of hormone receptors that, like the visual pigment rhodopsin, are coupled to their intracellular effector systems by guanine nucleotide binding proteins. In some systems, however, it has been impossible to correlate physiological responses to adenosine with changes in levels of cAMP, and therefore, it has been proposed that adenosine may be coupled to other signal transduction systems as well. In the kidney, several of the actions of adenosine associated with activation of the A_1 receptor (i.e. vasoconstriction, renin release inhibition, and inhibition of neurotransmitter release) are effects that have been proposed to be mediated by changes in cytosolic calcium (Churchill and Churchill, 1988). We have recently reported in primary cultures of rabbit cortical collecting tubule cells (Arend et al., 1988) and in an established cell derived from RCCT cells (Arend et al., 1989) that in addition to the classical A_1 and A_2 receptors coupled to the the inhibition and stimulation of adenylate cyclase (Arend et al., 1987), adenosine stimulates the turnover of inositol phosphates and

Table 1. Renal actions of adenosine

Effect	Receptor
Hemodynamic (GFR)........................	
vasoconstriction (preglomerular)	A_1
vasodilation (postglomerular)	A_2
Hormonal/Neurotransmitter................	
Renin release	
Inhibition	A_1
Stimulation	A_2
Erythropoietin..........................	
Inhibition	A_1
Stimulation	A_2
Adrenergic Transmission	
Inhibition (presynaptic)	A_1
Tubular.................................	
Collecting Tubule	
LpA	A_2
Thick Ascending Limb	
T_{NA}	A_1

the elevation of cytosolic free calcium. Furthermore this response is coupled to a pertussis toxin substrate, presumably a G protein, and is inhibited by the highly selective A_1 antagonist, 8-cyclopentyl-1,3-dipropylxanthine (DPCPX). Thus, activation of the renal epithelial A_1 receptor results in the simultaneous acceleration of inositol phosphate production and the inhibition of adenylate cyclase.

The presence of two different mechanisms associated with the adenosine A_1 receptors raises several important questions. The first and most obvious is whether or not two classes of A_1 receptors exist. One possibility is that both the inhibition of adenylate cyclase and the acceleration of inositol polyphosphate production are provoked by a single receptor population via divergent coupling mechanisms. Alternatively, each response may be evoked by independent adenosine receptor populations indistinguishable in their specificity for currently available agonist or antagonist ligands. Although GTP-binding proteins link receptor occupancy to changes in both inhibition of cyclase and the acceleration of inositol phosphate production, the identity of the GTP-binding proteins involved in vivo and the mechanisms are not certain. Finally, is remains to be determined which of the possible signaling events induced by occupancy of receptors linked to the inhibition of adenylate cyclase and/or phospholipase C are causal in mediating a given physiological event, which are permissive, and which are without any functional consequence in a given setting.

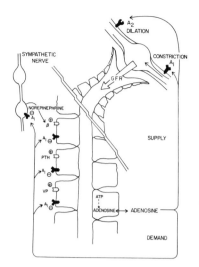

Figure 1. ADENOSINE FEEDBACK HYPOTHESIS
FOR THE KIDNEY. Adenosine, presumably
produced by transporting epithelium,
acts to reduce GFR, via reduction of the
glomerular hydrostatic pressure, by its
vasoconstrictive action on the afferent
arteriole and vasodilatory action on the
efferent arteriole, and thereby,
regulates the supply of delivered solute
to the nephron. The action of adenosine
to inhibit hormone-stimulated cyclic AMP
in various segments of the nephron, both
directly, and indirectly through the
inhibition of neurotransmitter release,
serves to reduce the metabolic demand of
the tubular cells. Together, these
hemodynamic and tubular actions of
adenosine work to return the metabolic
supply and demand ratio toward a level
of transport activity appropriate for
the oxygen and substrate availability of
the tissue.

Radioligand Binding Analysis of Adenosine A_1 Receptors in 28A Cells.

To determine whether or not a single population of A_1 receptors is
coupled to these divergent signaling pathways, we have measured
radioligand binding of [^{3}H]DPCPX to plasma membranes from rabbit renal
medulla and a cell line derived from the rabbit cortical collecting
tubule (RCCT-28A). Saturation binding of [^{3}H]DPCPX in 28A membranes
(Figure 2) analyzed by non-linear curve fitting, gave a one-site model
with an apparent K_D-value of 1.4 nM and a maximum number of binding
sites (B_{MAX}-value) of 64 fmol/mg protein. Scatchard analysis of the
saturation curve gave a linear plot, indicating the presence of only one
homogeneous population of binding sites. The non-specific-binding was
20-30% of the total at the K_D, and saturation of specific binding was
reached with 2 nM [^{3}H]DPCPX.

Competition of several agonists for the [^{3}H]DPCPX binding was measured
to confirm that [^{3}H]DPCPX binds to the A_1 receptor. Competition of
adenosine agonists for [^{3}H]DPCPX binding resulted in biphasic
displacement curves (Table 2) indicating the presence of two affinity
states for the agonists, with approximately one-half of the binding
sites being in the high affinity state and the other half in the low
affinity state. The K_i-values for the various adenosine receptor
agonists exhibit the typical pharmacological profile for A_1 receptors
and the marked stereoselectivity for the PIA enantiomers.

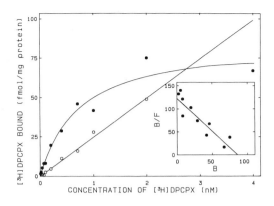

Figure 2. Saturation binding of [3H]DPCPX to RCT-28A cell membranes. Data are given as specific (closed circles) and non-specific binding (open circles). The inset shows the Scatchard plot from the data.

Agonist binding was further characterized by measuring the competition of R-PIA for [^{3}H]DPCPX binding in the presence and absence of GTP (100 uM). In the absence of GTP the competition of [^{3}H]DPCPX by R-PIA resulted in a biphasic displacement curve with an apparent K_D-value of 0.5 nM and B_{MAX}-value of 16.1 pmol/mg protein for the high affinity state and a low-affinity K_D-value of 10.5 nM and B_{MAX}-value of 20.2 fmol/mg protein.

When the competition experiment was carried out in the presence of 100 uM GTP, a monophasic curve was obtained, indicating a single affinity state with a K_D-value of 17.7 nM and a B_{MAX}-value of 54.1 fmol/mg protein. Control binding (100%) increased from 36.3 to 54.1 fmol/mg protein with the addition of 100 uM GTP.

These binding data confirm the previously reported functional data, that cells of the cortical collecting tubule have adenosine A_1 receptors coupled through GTP-binding proteins. Furthermore, these binding data fail to provide any support for the hypothesis that the inhibition of adenylate cyclase and the stimulation of phospholipase C are coupled to two sub-populations of the A_1 receptor, although it is recognized that this conclusion may be a function of the inability of currently available ligands to differentiate between the A_1 receptor subtypes.

Table 2

Pharmacological profile of [^{3}H]DPCPX binding to RCCT-28A membranes.

	K_i (nM)	K_i (nM)
CPPA	0.3	2.7
R-PIA	0.5	7.0
NECA	1.8	47
S-PIA	3.1	275

Desensitization of the adenosine A_1 receptor: Differential effects on adenylate cyclase inhibition and phospholipase C stimulation.

Although the analysis of radioligand binding provided no evidence in support of separate receptor sub-populations of the A_1 receptor mediating the divergent signaling mechanisms, it remained to be determined if activation of A_1 receptors was invariably associated with both inhibition of adenylate cyclase and activation of phospholipase C, or alternatively, could the two signaling pathways be regulated separately, providing for more flexibility in control.

Because prior exposure of cells to agonist ligands is often associated with a desensitization of the response to subsequent addition of agonist, we sought to determine if it was possible to selectively desensitize either the A_1 mediated decrease in adenylate cyclase activity or the A_1 mediated activation of phospholipase C.

To determine if pretreatment of 28A cells with A_1 agonists produced a desensitization of mobilization of intracellular calcium to sebsequent addition of A_1 agonist, 28A cells were treated for 4 hr with increasing concentrations of CHA, an A_1 agonist, ranging from 10^{-8} to 10^{-4} M. During the final hour of exposure to agonist, the cells were loaded with FURA-2, as previously described (Arend et al., 1988), for determination of cytosolic calcium concentration by spectrofluorometry. Cells were then thoroughly washed to remove the extracelluar FURA-2 and A_1 agonist. Without prior exposure to agonist, 1 uM CHA caused a 40% increase in the cytosolic calcium concentration. With prior exposure to agonist, this action of CHA to cause a stimulation of cytosolic calcium concentration is decreased in a concentration dependent manner.

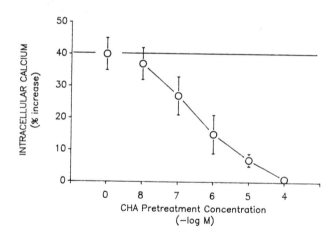

Figure 3. Effect of pretreatment of 28A cells with CHA (4 hr) on CHA induced increase in cytosolic calcium. Values are means ± SEM of 10 experiments.

224

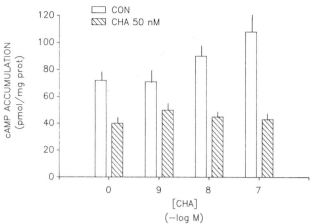

Figure 4. Effect of increasing concentrations of CHA pretreatment (48 hr) on CHA-induced inhibition of cAMP production in 28A cells.

To examine the ability of prior agonist exposure to desensitize the A_1 inhibition of adenylate cyclase activity, RCT-28A cells were pretreated for periods of 4, 12, 24, and 48 hr with vehicle or CHA at concentrations of 1, 10, and 100 nM. The cells were thoroughly washed and then reexposed to 50 nM CHA, the concentration at which we normally see maximal inhibition of cAMP production (Arend et al., 1987) which is approximately 50% (Figure 4). When the cells were pre-exposed to CHA, no alteration in the ability of subsequent addition of 50 nM CHA on cAMP production was observed.

In conclusion, in the absence of evidence of sub-populations of the A_1 receptor, it appears that activation of a single A_1 receptor population results in the inhibition of adenylate cyclase and the mobilization of cytosolic calcium. However, the finding that the divergent signaling mechanisms can be differentially regulated raises the possibility of separate control for the activation of phospholipase C and inhibition of adenylate cyclase by adenosine.

References

Arend, L.J., W.K. Sonnenberg, W. L. Smith, W.S. Spielman. Evidence for A_1 and A_2 adenosine receptors in rabbit cortical collecting tubule cells: modulation of hormone stimulated cyclic AMP. J. Clin. Invest. 79:710-714, 1987.

Arend, L.J., M.A. Burnatowska-Hledin, and W.S. Spielman. Adenosine signal transduction in the rabbit cortical collecting tubule: receptor mediated calcium-mobilization. Am. J. Physiol. 255: F704-F710, 1988.

Arend, L.J., F. Gusovsky, J.H. Daly, J.S. Handler, J.S. Rhim, and W.S. Spielman. Adenosine-sensitive phosphoinositide turnover in a newly established renal cell line. Am. J. Physiol. 256: F1067-F1074, 1989.

Churchill, P.C., and M.C. Churchill. Effects of adenosine on renin secretion. ISI Atlas of Science: Pharmacology. 367-373, 1988.

34
Inhibition of the Function of Guanine Nucleotide Binding Proteins by the New Positive Inotropes

V. Ramkumar and G.L. Stiles

Introduction

Guanine nucleotide binding proteins (G proteins) couple receptors to a variety of effector systems. These proteins are heterotrimers, consisting of α, β and γ subunits. Activation of G proteins by agonist-occupied receptors results in GTP-GDP exchange at the α-subunits, followed by dissociation of the $\beta\gamma$ subunits. In the case of G_s, the α_s-GTP species can then activate adenylate cyclase directly. Inhibition of adenylate cyclase, on the other hand, results from the binding of $\beta\gamma$ released from the dissociation of G_i with free α_s-GTP, thereby preventing activation of adenylate cyclase. Termination of G protein activation is mediated by GTPase activity intrinsic to the α subunits, which cleaves α-GTP to inactive α-GDP. The resulting α-GDP complexes can then reassociate with the $\beta\gamma$ subunits to reform the heterotrimeric proteins (Gilman, 1987).

The normal activity of the G_i protein can be disrupted by several agents. For example, pertussis toxin uncouples receptors from G_i by catalyzing the mono-ADP-ribosylation and inactivation of the α_i subunits. In addition, phorbol esters (Bell et al., 1986) and divalent cations such as Mg^{2+} and Mn^{2+} (Katada et al., 1985) can also inhibit the activity of G_i. In this report, we present data showing that several of the newer positive inotropic drugs with methylxanthine-like ring structures can also inhibit the function of G_i and G_s (to a lesser extent) by interfering with GTP-GDP exchange.

The newer positive inotropic drugs, such as sulmazole, increase cAMP levels in cardiac cells. This is believed to be mediated by inhibition of a low K_m cAMP phosphodiesterase activity (Kariya et al., 1982) characterized in cardiac tissue. However, despite substantial investigation, the relationship between phosphodiesterase inhibition and cardiotonic efficacy remains unclear (Harrison et al., 1986). In addition, these agents are effective inhibitors of the A_1 adenosine receptors (Parsons et al., 1988a), thereby relieving tonic inhibition of adenylate cyclase by endogenous adenosine. Increases in cAMP accumulation correlate well with the inotropic and vasodilatory actions of these drugs (Endoh et al., 1982; 1985), although other mechanisms independent of cAMP might account for these actions (Endoh et al., 1986).

Methods

For in vivo ADP-ribosylation of α_i, pertussis vaccine ($\sim$ 300 opacity units/kg; 0.3-0.5 ml) was administered intraperitoneally to rats three days prior to sacrifice, as described previously (Parsons et al., 1988a). All other methods are as described in the figure legends.

Results

The new cardiotonic agent, sulmazole, stimulated adenylate cyclase activity in a dose-dependent manner, with a greater than 2-fold increase in activity over papaverine at the highest concentration of the drug tested (figure 1A). At a concentration of 100 μM, papaverine fully inhibited the low K_m cAMP phosphodiesterase (IC_{50} for inhibition being 3.8 ± 1.3 μM). Furthermore, the addition of sulmazole to papaverine (100 μM) neither increased nor decreased the maximal inhibitory effect of papaverine on phosphodiesterase activity. Therefore, the additional stimulatory effect of sulmazole on adenylate cyclase must be independent of its inhibition of the low K_m cAMP phosphodiesterase activity (IC_{50} for inhibition being 150 μM). Interestingly, the stimulatory effect of sulmazole is negated by the A_1 adenosine receptor agonist, R-PIA (figure. 1B).

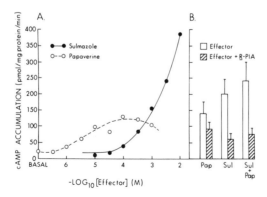

Figure 1. Effector-mediated stimulation of adenylate cyclase activity in rat adipocyte membranes. **A.** Representative dose response curves for sulmazole and papaverine. **B.** Adenylate cyclase activity in presence of sulmazole (3.3 mM) versus a maximally stimulatory concentration of papaverine (100 μM). R-PIA significantly inhibits adenylate cyclase activity in each case to essentially the same level. Pap, papaverine; Sul, sulmazole. Taken from Parsons et al., (1988a).

An active G_i appears essential in order to demonstrate the stimulation of adenylate cyclase by sulmazole. For example, inactivation of G_i in presence of low GTP concentration (10 nM) abolishes the effect of sulmazole. Similarly, ADP-ribosylation of G_i by pertussis toxin also attenuates the stimulatory action of sulmazole (figure 2).

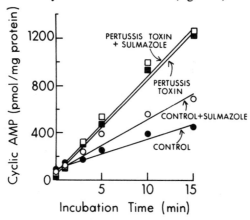

Figure 2. Effect of pertussis toxin on the enhancement of cAMP formation by sulmazole. Adipocyte membranes were incubated with 5 μM GTP in the absence or presence of 2 mM sulmazole. Taken from Ramkumar and Stiles (1988b).

More conclusive demonstration of the importance of an active G_i in mediating the effects of the inotrope was observed by reconstituting purified G_i proteins into membranes previously exposed to pertussis toxin. Reconstituting purified G_i or a mixture of G_i and G_o into these membranes led to the re-establishment of the stimulatory effect of sulmazole in pertussis toxin treated membranes (figure 3). These findings underscore the importance of G_i in the mediation of the stimulatory action of sulmazole.

In addition to enhancing basal adenylate cyclase activity, sulmazole can also increase activity in presence of stimulatory effectors such as isoproterenol, forskolin and fluoride (Ramkumar et al., 1988a). Again, pertussis toxin

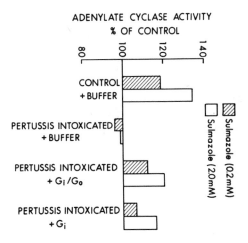

Figure 3. <u>Reconstitution of sulmazole-stimulated adenylate cyclase activity in pertussis-intoxicated rat adipocyte membranes</u>. Adipocyte membranes were prepared from control and pertussis-intoxicated rats and reconstituted with buffer or purified G proteins (either a G_i/G_o mixture or G_i). Data are presented as percentage of stimulation above that obtained in the absence of sulmazole (100%) for each group. Taken from Parsons et al., (1988a).

treatment abolished this stimulatory effect. Interestingly, sulmazole inhibited isoproterenol-stimulated adenylate cyclase activity following inactivation of G_i by performing assays with low GTP concentrations or following pertussis toxin administration. Since the drug does not appear to interact with β-adrenergic receptors, these findings suggest an additional inhibitory effect of the inotrope on α_s.

To determine the mechanism(s) of inhibition of G protein function, the effect of sulmazole on GTP turnover was assessed. The binding of [^{3}H]GTP to G_i and G_s was initiated by R-PIA and isoproterenol, respectively. The release of [^{3}H]GDP from G_i and G_s was initiated by prostaglandin E_1 and isoproterenol, respectively. In all experiments performed, the addition of sulmazole led to the inhibition of agonist-induced [^{3}H]GDP release. For example, sulmazole inhibited release from G_s initiated by 10 μM isoproterenol to 63.9% of control (Ramkumar et al., 1988a). Interestingly, [^{3}H]GDP release initiated by 100 μM was inhibited to only 84.5% of control, suggesting a reversal of the inhibitory action by high concentration of the stimulatory agent. As expected from sulmazole's greater effect on the inhibitory pathway, sulmazole was more effective at inhibiting PGE$_1$-stimulated release of [^{3}H]GDP from G$_i$ (inhibition to 31.9% of control) than inhibiting release from G$_s$ (see above).

Additional experiments were performed to test the effect of the inotrope on agonist-promoted binding of [^{3}H]Gpp(NH)p to G$_i$ protein. These experiments were based on the premise that by stabilizing the α_i-GDP complexes, sulmazole might inhibit the subsequent binding of [^{3}H]Gpp(NH)p. Figure 4 demonstrates that this is exactly the case. The

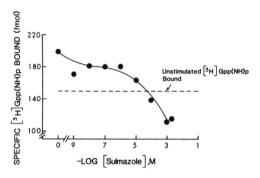

Figure 4. Inhibition of PGE$_1$-stimulated binding of [^{3}H]Gpp(NH)p to G$_i$ by sulmazole. Membranes were incubated with 0.1 μM [^{3}H]Gpp(NH)p in the absence (unstimulated) or presence of 10 μM PGE$_1$. Nonspecific binding was determined as binding remaining in the presence of 100 μM unlabeled GTP and was subtracted from total binding to yield specific [^{3}H]Gpp(NH)p binding. This is a representative experiment of six experiments showing similar results. Taken from Ramkumar et al., 1988a.

inotrope inhibited both basal (unstimulated) and PGE$_1$-stimulated [^{3}H]Gpp(NH)p binding. Stimulated binding was inhibited completely with an IC$_{50}$ about 1 μM. Inhibition of basal guanine nucleotide binding probably reflects the inotrope's inhibitory action at the unstimulated G$_i$ in addition to other GTP binding proteins not activated via PGE$_1$. Taken together, these data suggest that sulmazole's inhibition of G protein function results from inhibition of the GTP-GDP exchange mechanism.

Discussion

This study presents conclusive evidence that the positive inotropic drug sulmazole can interact directly with the G proteins and that this interaction leads to effects on adenylate cyclase activity. Inhibition of G$_i$ relieves a tonic inhibitory influence on adenylate cyclase in adipocyte membranes, resulting in an increase in the basal as well as stimulated enzyme activity. An inhibitory effect of sulmazole on adenylate cyclase activity is only evident following inactivation of G_i by pertussis toxin or in the presence of 10 nM GTP, a concentration of GTP sufficient to activate G_s but not G_i. The concentration of sulmazole required to show direct inhibition at the G protein is several orders of magnitude less that required to stimulate adenylate cyclase. This discrepancy is secondary to a stoichiometric excess of G proteins over catalytic subunits in normal membrane preparations (Gilman, 1987; Rasnas et al., 1988). Thus, a large proportion of G proteins must be inactivated before effects on adenylate cyclase become evident.

The relevance of G protein inhibition in terms of the physiological effects of sulmazole is not clear at present. However, this action would supplement other mechanisms for increasing cAMP levels, such as inhibition of cAMP phosphodiesterase (Kariya et al., 1982) and antagonism of the A_1 adenosine receptors (Parsons et al., 1988a). Thus, sulmazole can elevate cAMP (and probably produce its positive inotropic action) by three independent mechanisms, namely by inhibiting the function of G_i, inhibition of the low K_m cAMP phosphodiesterase and inhibition of A_1 adenosine receptors.

Elevation of adenylate cyclase by inhibition of G_i function is not unique to sulmazole, but is observed with other members of this class of drugs such as amrinone, milrinone and piroximone (Parsons et al., unpublished data). In addition, other agents with the typical xanthine-like ring structure such as isobutylmethylxanthine (Parsons et al., 1988b) and XAC (Ramkumar et al., 1988b) also increase adenylate cyclase activity by inhibiting G_i.

In conclusion, we have provided evidence that the newer positive inotrope sulmazole increases adenylate cyclase activity by inhibiting G_i directly. While the relevance of this action in terms of positive inotropy is not clear, this drug and others of its class might prove useful tools to explore the function of the G proteins.

References

Bell, J.D., and Buxton, L.L., 1986, Enhancement of adenylate cyclase activity in S49 lymphoma cells by phorbol esters: withdrawal of GTP-dependent inhibition, in *J. Biol. Chem.* **261**:12036-12041.

Endoh, M., Yamashita, S., and Taira, N., 1982, Positive inotropic effect of amrinone in relation to cyclic nucleotide metabolism in the canine ventricular muscle, in *J. Pharmacol. Exp. Ther.* **221**:775-783.

Endoh, M., Yanagisawa T., Morita, T., and Taira, N., 1985, Differential effects of sulmazole (AR-L 115 BS) on contractile force and cyclic AMP levels in canine ventricular muscle: comparison with MDL 17,043, in *J. Pharmacol. Exp. Ther.* **234**:267-273 .

Endoh, M., Yanagishawa T., Morita, T., and Blinks, J., 1986, Effects of new inotropic agents on cyclic nucleotide metabolism and calcium transients in canine ventricular muscle, in *Circulation* **73** (Suppl. III):117-133.

Gilman, A.G., 1987, G proteins: transducers of receptor-generated signals, in *Annu. Rev. Biochem.* **56**:615-649.

Harrison, S.A., Reifsnyder, D.H., Gallis, B., Cadd, G., and Beavo, J.A.,1986, Isolation and characterization of bovine cardiac muscle cGMP-inhibited phosphodiesterase: a receptor for new cardiotonic drugs, in *Mol. Pharmacol.* **29**:506-514.

Kariya, T., Willie, L.J., and Dage, R.C., 1982, Biochemical studies on the mechanism of cardiotonic activity of MDL 17,043, in *J. Cardiovasc. Pharmacol.* **4**:509-514.

Katada, T., Gilman, A.G., Watanabe, Y., Baver, S., and Jakobs, K.H., 1985, Protein kinase C phosphorylates the inhibitory guanine-nucleotide binding regulatory component and apparently suppresses its function in hormonal inhibition of adenylate cyclase, in *Eur. J. Biochem.* **151**:431-437.

Parsons, W.J., Ramkumar, V., and Stiles, G.L., 1988a, The new cardiotonic agent sulmazole is an A_1 adenosine receptor antagonist and functionally blocks the inhibitory regulator, G_i, in *Mol. Pharmacol.* **33**:441-448.

Parsons, W.J., Ramkumar, V., and Stiles, G.L., 1988b, Isobutylmethyl-xanthine stimulates adenylate cyclase by blocking the inhibitory regulatory protein, in *Mol. Pharmacol.* **34**:37-41

Ramkumar, V., and Stiles, G.L., 1988a, The new positive inotrope sulmazole inhibits the function of guanine nucleotide regulatory proteins by affecting GTP turnover, in *Mol. Pharmacol.* **34**:761-768.

Ramkumar, V., and Stiles, G.L, 1988b, A novel site of action of a high affinity A_1 receptor antagonist, in *Biochem. Biophys. Res. Commun.* **153**:939-944 .

Rasnas, L.A., and Insel, P.A., 1988, Quantitation of the guanine nucleotide binding regulatory protein G_s in S49 cell membranes using antipeptide antibodies to α_s, in *J. Biol. Chem.* **257**:7485-7490.

35

Blockade of Antigen Activated Signals by Xanthine Analogs in RBL-2H3 Cells: Evidence for a Novel Adenosine Receptor

H. Ali, D. Collado-Escobar, D. O'Reilly, and M. Beaven

Introduction

The two major subclasses of adenosine receptors, A_1 and A_2 which, respectively, inhibit and stimulate adenylate cyclase activity, can be further distinguished by their different affinities for various metabolically stable analogs of adenosine. A rank order of potency or binding affinity of 5'-(N-ethylcarboxamide) adenosine (NECA) > 2 chloro-Adenosine, Adenosine > R-phenylisopropyladenosine (PIA) is characteristic of an interaction with A_2-receptors: the reverse order is typical of A_1-receptors (Daly 1982; Stiles 1986). Although the existence of a third category of adenosine receptors (A_3) that modulate Ca^{2+}-dependent responses independently of adenylate cyclase has been postulated (Ribeiro & Sebastiao 1986), there has been no direct demonstration of coupling of adenosine receptors with an effector system other than adenylate cyclase.

Tissue mast cells and basophils possess receptors for adenosine that can influence secretory responses to immunological and nonimmunological stimuli (Marquardt 1988). Because hydrolysis of membrane inositol phospholipids by phospholipase C, the mobilization of extracellular and intracellular Ca^{2+} and the activation of protein kinase C provide important signals for secretion from antigen-stimulated cells (Beaven & Cunha-Melo 1988), we have investigated the effects of agonists and xanthine antagonists of the adenosine receptors on these signals in a cognate rat mast cell, the RBL-2H3 cell (Barsumian et al., 1981; Seldin et al.,1985). We found that both antigen and the most potent adenosine analog, NECA caused the activation of phospholipase C, but did so through different GTP-binding proteins (G-proteins) as indicated by the responses of cells treated with cholera toxin or pertussis toxin. The responses to the adenosine analogs, however, were transient and insufficient to induce secretion but they synergized antigen-induced signals and thereby markedly enhanced the secretory response to antigen. Interestingly treatment with dexamethasone, down-regulated the systems subserved by antigen but, conversely up-regulated the responses to NECA (Collado-Escobar et al., 1990b). In addition, the xanthine derivatives were either inactive or suppressed the responses to either antigen or NECA in a nonselective manner. These findings point to an action of adenosine at sites other than the currently defined adenosine A_1 or A_2 receptor.

Materials and Methods

Procedures were as described elsewhere (Ali et al., 1989 a,b, & c; Collado-Escobar et al., 1990 a & b). Briefly, cells were sensitized with a dinitrophenol (DNP)-specific IgE (DNP-IgE) and radiolabelled overnight by incubation with [^{3}H]inositol or [^{3}H]5-hydroxytryptamine for the measurement of release of inositol phosphates or secretion. Cells were left intact or permeabilized with streptolysin O (Ali et al, 1989b). Where indicated cells were incubated with pertussis toxin (0.1 μg/ml) or cholera toxin (1 μg/ml) for 3 hr or with dexamethasone for 16 hr before the experiment.

Results and Discussion

<u>Activation of RBL-2H3 Cells by NECA: Comparison with Antigen Stimu-</u>
<u>lation</u>. Addition of NECA, in the absence of Li^+-induced a transient
production of [^{3}H]inositol phosphates (phosphoinositide-response). The
levels of [^{3}H]inositol 1,4,5-trisphosphate reached a maximum within 15
sec. Other [^{3}H]inositol phosphates appeared thereafter, but all of them
had returned to virtually basal levels 10 min after stimulation. In
contrast, antigen (dinitrophenylated bovine serum albumin, DNP-BSA)
induced a sustained response; all metabolites reached elevated steady-
state levels or continued to increase over the course of 20 min.
Simultaneous addition of both stimulants resulted in additive or more
than additive effects on the production of all inositol phosphates
including inositol 1,4,5-trisphosphate (Ali <u>et al.,</u> 1989c).

The marked differences between the two stimulants were also apparent
from the mobilization of Ca^{2+} ions. Unlike antigen, which produced a
sustained increase in $[Ca^{2+}]_i$, NECA induced a transient increase (2 to 3
min) in $[Ca^{2+}]_i$. This increase was associated with a similar transient
influx of $^{45}Ca^{2+}$.   In contrast, after the addition of antigen the
stimulated increase in $[Ca^{2+}]_i$ and an accelerated influx of $^{45}Ca^{2+}$ were
still apparent at 20 min.   Combining the stimulants elicited a prompt
and sustained increase in $[Ca^{2+}]_i$ and an accelerated influx of $^{45}Ca^{2+}$
to produce a profile that suggested a synergistic interaction for 1 to
2 min and an additive response thereafter (Ali <u>et al</u> 1989c).

NECA, by itself, failed to induce secretion from RBL-2H3 cells. The
secretory response to a suboptimal concentration of antigen was,
however, enhanced in a dose-dependent manner by the simultaneous
addition of NECA, 2-chloro-adenosine, adenosine, or <u>R</u>-PIA. The rank
order of potency of nucleosides was NECA > 2-chloro-adenosine,
adenosine > <u>R</u>-PIA with values for EC_{50} (concentrations of nucleosides
required to give 50% of the maximal enhancement) of 0.1 μM, 0.5 μM, 0.5
μM and 50 μM respectively. This order of potency was consistent with the
stimulation of adenylate cyclase via adenosine A_2-receptors. NECA,
however, did not stimulate adenylate cyclase (Ali <u>et al</u> 1989c).

Moreover, cAMP and its analogs failed to stimulate hydrolysis of inositol phospholipids in intact or permeabilized cells and, in fact, at
high concentrations (> 0.5 mM) suppressed the responses to NECA and
antigen (H.M.S.Gonzaga and M.A.Beaven, unpublished data).

<u>Effects of Pertussis Toxin and Cholera Toxin</u>. Studies with the toxins
indicated that NECA and antigen activated phospholipase C through
different G-proteins. For example, treatment of cells with pertussis
toxin blocked the transient production of [^{3}H]inositol phosphates and
the increase in $[Ca^{2+}]_i$ in response to NECA, but did not inhibit the
sustained responses to antigen. In addition, pertussis toxin selectively
inhibited the synergism between NECA and antigen without reducing the
residual stimulatory and secretory responses to antigen. The responses
to antigen and NECA could be further distinguished by treatment with
cholera toxin, which inhibited the stimulatory responses to NECA, but
potentiated those to antigen, primarily by enhancing the influx of Ca^{2+}
(Ali <u>et al</u> 1989c; Narasimhan <u>et al</u> 1988).   The best evidence that an
adenosine receptor was coupled to phospholipase C via one or more G-pro-

teins, however, was that the generation of [^{3}H]inositol phosphates was substantially reduced in cells that had been exposed to the toxins and then permeabilized with streptolysin O (Ali et al 1989c) in which cytosolic proteins, such as lactate dehydrogenase (M.W. 140 kDa), were lost (Ali et al 1989b).

Selective Effects of Dexamethasone. Prolonged exposure (16-20 hr) of RBL-2H3 cells to 0.1 μM dexamethasone resulted in the suppression (by 70-80%) of the generation of [^{3}H]inositol 1,4,5-trisphosphate in response to antigen, whereas this response was enhanced (by 400-500%) when NECA was the stimulant. The effects of dexamethasone were apparent at concentrations as low as 1 nM. The treatment with dexamethasone also prolonged the responses to NECA. Under these conditions the initial increase in [Ca^{2+}]$_i$ was markedly enhanced to give increases as large as those observed in cells that were optimally stimulated with antigen. The phosphoinositide-response, the increase in [Ca^{2+}]$_i$ and an acceler-rated influx of $^{45}Ca^{2+}$ was still evident 20 min after the addition of NECA. These sustained responses to NECA were now accompanied by sub-stantial secretion from the cells. Binding studies indicated a satur-able binding of [^{3}H]NECA to give calculated values of ⁻ 13000 binding sites per cell in dexamethasone-treated cells versus ⁻ 4000 binding sites in untreated cells. The effect of dexamethasone on the kinetics of phosphoinositide-response (an increase in the maximal response and a decrease in value for EC_{50}) were also consistent with an increase in the number of receptors and, possibly, an increased coupling of receptors to phospholipase C (Collado-Escobar et al., 1990a).

Effects of Adenosine Receptor Antagonists. We next examined the effects of antagonists of A_1 and A_2 adenosine receptors on the phosphoinositide responses to NECA (1 μM) or antigen (DNP-BSA, 10 ng/ml). The antagonists 1,3-dipropyl-8-cyclopentylxanthine, (the most potent adenosine receptor antagonist tested) and 1-propargyl-3,7-dimethylxanthine did not inhibit the response to NECA or antigen (Table 1). Some of the other xanthines tested nonselectively inhibited responses to both antigen and NECA but generally only at concentrations that were greater than those expected to inhibit adenosine receptors in other types of cells.

TABLE 1

Inhibitory potencies of xanthine antagonists on antigen (DNP-BSA)- and NECA-induced generation of [^{3}H]inositol phosphates.

	IC_{50}*		K_i**	
		(μM)		
	NECA	ANTIGEN	A_1	A_2
1,3-dipropyl-8-cyclopentylxanthine,	n.i.	n.i.	0.001	0.14
8-phenyltheophylline	> 5	> 5	0.76	0.15
3-Isobutyl-1-methylxanthine,	> 7	> 3	2	5
8-p-sulfophenyltheophylline	> 25	> 25	1	5.5
1-propargyl-3,7-dimethylxanthine,	n.i.	n.i.	45	6
Theophylline,	> 200	> 200	13	14
Enprofylline	> 75	> 100	81	130

n.i. Noninhibitory at concentrations that were at least 10 fold greater than those required to inhibit adenosine receptors.
*Unpublished data; measured against 1 μM NECA and 10 ng/ml DNP-BSA.
** Taken from Daly 1982; Daly et al., 1986; Shamim et al 1989.

Other studies with stimulants of G-proteins suggested that these particular xanthines impaired the coupling of the receptors to G-proteins rather than that of the G-proteins to phospholipase C (unpublished data). We also rejected the possibility that actions of these inhibitory xanthines were a consequence of inhibition of phosphodiesterase and an increase in cAMP, because 8-p-sulfophenyl-theophylline, one of the most potent inhibitors of antigen-induced responses (Table 1), does not penetrate the plasma membrane (Shamim et al 1989). Furthermore, a nonxanthine phosphodiesterase inhibitor such as rolipram (Yamamoto et al, 1984) did not inhibit the responses to either antigen or NECA. The inhibitory effects of the xanthines must, there-fore, be unrelated to actions at A_1 or A_2 adenosine receptors or to an increase in the intracellular concentration of cAMP.

Implications of Our Findings. The studies described here clarify some of the confusion with respect to the actions of adenosine on tissue mast cells that were apparently unrelated to the activation of adenylate cyclase. Although adenosine suppresses mediator-release from blood basophils through adenosine A_2-like receptors, adenylate cyclase and, probably, intracellular P sites (Hughes et al.1987, Peachell et al 1989), its actions on the tissue mast cells are less well defined. For example, adenosine has been reported to inhibit or enhance release of histamine from human lung mast cells, possibly via an adenosine A_2-like receptor (Hughes et al., 1984; Peachell et al., 1988). In contrast, adenosine enhances secretion from rat peritoneal mast cells through receptors other than A_1 or A_2 and probably by mechanisms that do not involve adenylate cyclase (Church et al., 1986; Leoutsakos et al., 1986). Our studies indicated that the enhancement of secretion was the result of the interaction of adenosine with a novel adenosine receptor that transiently activates, via a G-protein, phospholipase C to produce a brief stimulation of hydrolysis of inositol phospholipids and an increased $[Ca^{2+}]_i$. However, these responses are not sufficiently sustained to induce secretion but they do potentiate the responses to low concentrations of antigen.

The studies also point to the importance of the influx of Ca^{2+} in sustaining stimulatory signals and secretion in RBL-2H3 cells. If inositol 1,4,5-trisphosphate causes the mobilization of Ca^{2+} (Berridge 1987) in NECA stimulated cells, a transient influx of Ca^{2+} is still required even for the increase in $[Ca^{2+}]_i$ and the replenishment of intracellular Ca^{2+} stores. The principal difference between antigen and NECA stimulation is that antigen causes a sustained influx of Ca^{2+} and elevation of $[Ca^{2+}]_i$ by a cholera toxin sensitive process. The ability of antigen, but not NECA, to induce secretion is probably a consequence of this sustained influx. In further support of the view that a sus-tained Ca^{2+} influx is required to initiate and maintain stimulatory responses and secretion is that in dexamethasone treatment of cells transforms a transient Ca^{2+} influx into a sustained response; only then does secretion occur.

The opposite effects of dexamethasone on the responses to antigen and NECA suggest that the two stimulants may activate phospholipase C by different mechanisms. Moreover, the diverse effects of the bacterial toxins also indicate that the two stimulants activate phospholipase C through different G-proteins. By analogy with the adenylate cyclase system, the expectation that a G-protein, designated G_p or Np (Gomperts 1986), would activate phospholipase C in response to different stimu-lants in different types of cells has generated conflicting results. For example, stimulation of mast cells with chemical agents, such as compound 48/80, results in hydrolysis of membrane inositol phospholipids and secretion and both of which are inhibited by pertussis toxin (Nakamura & Ui 1985). In contrast, Antigen stimulated responses are resistant to inhibition by this toxin (Saito et al., 1987, Warner et al., 1987), but are inhibited by GDPβS in permeabilized cells (Ali et al., 1989a). The two stimulants thus activate phospholipase C via two different G-proteins in the same cell. Moreover, the responses to adenosine appear to be transduced by yet another G-protein, one that is inactivated by both cholera and pertussis toxins.

In conclusion, our studies provide the first indication that an adenosine receptor stimulates phospholipase C via a G-protein, and thereby modulates a Ca^{2+} dependent response. This receptor can be distinguished from the putative A_3 receptor (Ribeiro & Sebastiao 1986) which, as currently defined, inhibits Ca^{2+} dependent responses in electrically excitable cells. Because some potent xanthine antagonists are inactive, the receptor on RBL-2H3 cells also differs from A_1 or A_2 adenosine receptor. Finally, the xanthines that inhibit NECA- and antigen-induced responses, do so nonselectively by a mechanism that impairs coupling of the receptors to G-proteins.

References

Ali, H., Collado-Escobar, D. M., and Beaven, M. A. (1989a) J.Immunol. 143, 2626-2633
Ali, H., Cunha-Melo, J., and Beaven, M. (1989b) Biochim. Biophys. Acta 1010, 88-99
Ali, H., Cunha-Melo, J. R., Saul, W. F., and Beaven, M. A. (1989c) J. Biol. Chem. in press
Barsumian, E. L., Isersky, C., Petrino, M. G., and Siraganian, R. P. (1981) Eur. J. Immunol. 11, 317-323
Beaven, M. A., and Cunha-Melo, J. R. (1988) Progr. Allergy 42, 123-184
Berridge, M. J. (1987) Annu. Rev. Biochem. 56, 159-193
Church, M. K., Hughes, P. J., and Vardey, C. J.(1986) Br. J. Pharmacol. 87, 233-242
Collado-Escobar, D., Ali, H., and Beaven, M. A. (1990a) J. Immunol. submitted
Collado-Escobar, D., Cunha-Melo, J. R., and Beaven, M. A. (1990b) J. Immunol. in press
Daly, J. W. (1982) J. Med. Chem. 25, 197-207
Daly, J. W., Padgett, W. L., and Shamim, M. T., J. Med. Chem. 29, 1305-1308.
Gomperts, B. D. (1986) Trends in Biochem. Sci. 11, 290-292
Hughes, P. J., Benyon, R. C., and Church, M. K. (1987) J. Pharmacol. Exp. Ther. 242, 1064-1070
Hughes, P. J., Holgate, S. T., and Church, M. K. (1984) Biochem. Pharmacol. 33, 3847-3852
Leoutsakos, A., and Pearce, F. L. (1986) Biochem. Pharmacol. 35, 1373-1379
Marquardt, D. L. (1988) In: "Adenosine Receptors.", D. M. F. Cooper and C. Londos, eds., Alan R. Liss, Inc. New York, NY pp. 87-95
Nakamura, T., and Ui, M. (1985) J. Biol. Chem. 260, 3584-3593
Narasimhan, V., Holowka, D., Fewtrell, C., and Baird, B. (1988) J. Biol.Chem. 263, 19626-19632
Peachell, P. T., Columbo, M., Kagey-Sobotka, A., Lichtenstein, L. M., and Marone, G. (1988) Am. Rev. Resir. Dis. 138, 1143-1151

Peachell, P. T., Lichtenstein, L. M., and Schleimer, R. P.(1989) Biochem. Pharm. 38, 1717-1725

Ribeiro, J. A., and Sebastiao, A. M. (1986) Prog. Neurobiol. 26, 179-209

Saito, H., Okajima, F., Molski, T. F., Sha'Afi, R. I., Ui, M., and Ishizaka, T. (1987) J. Immunol. 138, 3927-3934

Seldin, D. C., Adelman, S., Austen, K. F., Stevens, R. L., Hein, A., Caulfield, J. P., and Woodbury, R. G. (1985) Proc. Natl. Acad. Sci. USA 82, 3871-3875

Shamim, M. T., Ukena, D., Padgett, W. L., and Daly, J. W. (1989) J. Med. Chem. 32, 1231-1237

Stiles, G. L. (1986) Trends Pharmacol. Sci. 7, 486-490

Warner, J. A., Yancey, K. B., and MacGlashan, D. W. (1987) J. Immunol. 139, 161-165

Yamamato, T., Lieberman, F., Osborne, J. C., Manganiello, V. C., Vaughan, M. and Hidaka, H. (1984) Biochem. 3, 670-675

Section 2
Adenine Nucleotide Receptors and Effector Systems

36
Classification and Characterization of Purinoceptors

G. Burnstock

Apart from the well-established roles of purines in cell metabolism and replication, there has been a growing recognition of their importance in cell communication. It has been known for many years that purine nucleotides and nucleosides have widespread and potent extracellular actions on excitable membranes, and that these may indicate a role in physiological regulatory processes (Drury and Szent-Györgyi, 1929; Berne, 1964). Later, it was proposed that ATP was released as the principal neurotransmitter from some non-adrenergic, non-cholinergic ('purinergic') nerves (Burnstock, 1972) or as a cotransmitter with noradrenaline, acetylcholine and other substances (Burnstock, 1976, 1986a, 1988a). In 1978, Burnstock suggested a division of receptors for adenosine and ATP into P_1- and P_2-purinoceptor subtypes respectively. Subsequently, biochemical, pharmacological and receptor-binding studies have led to a proposed subdivision of the P_1-purinoceptor into A_1 and A_2 receptors and P_2-purinoceptors into P_{2X} and P_{2Y} subtypes (see Burnstock and Buckley, 1985; Burnstock and Kennedy, 1985; Daly, 1985; Gordon, 1986; Williams, 1987).

The focus in this article will be on the classification of purinoceptor subtypes, their characterization and distribution and biological roles in a wide variety of cell types.

CLASSIFICATION OF PURINOCEPTORS

P_1- and P_2-Purinoceptors

A basis for distinguishing two types of purinergic receptor was proposed by Burnstock in 1978, based largely on an analysis of the voluminous literature about the actions of purine nucleotides and nucleosides on a wide variety of tissues (Burnstock, 1978). Since that time, many experiments have been carried out that support and extend this proposal (see Burnstock, 1981; Burnstock and Buckley, 1985; Williams, 1987).

The original classification into P_1- and P_2-purinoceptors was based on four criteria: the relative potencies of ATP, ADP, AMP and adenosine; the selective actions of antagonists, particularly methylxanthines; the activation of adenylate cyclase by adenosine, but not by ATP; the induc-

241

tion of prostaglandin synthesis by ATP, but not by adenosine. Thus the following classification was proposed: P_1-purinoceptors are more responsive to adenosine and AMP than to ATP and ADP; methylxanthines such as theophylline and caffeine are selective competitive antagonists; and occupation of these receptors leads to inhibition or activation of an adenylate cyclase system with resultant changes in levels of intracellular cyclic AMP (cAMP). P_2-purinoceptors are more responsive to ATP and ADP than to AMP and adenosine, are not antagonized by methylxanthines, do not act via an adenylate cyclase system, and their occupation may lead to prostaglandin synthesis.

Evaluation and expansion of purinoceptor classification has taken several directions, including studies of the stereoselectivity of P_1- and P_2-purinoceptors, the structural requirements for the actions of purines, chemistry of P_1- and P_2-purinoceptors, analysis of the influence of ectoenzymatic breakdown of nucleotides and uptake of adenosine on measurements of relative agonist potencies, and development of more potent and selective P_1- and P_2-purinoceptor antagonists (see Hogaboom et al., 1980; Satchell, 1984; Burnstock and Buckley, 1985; Williams, 1987; Daly et al., 1988; Dunn and Blakeley, 1988).

Extracellular breakdown of ATP is rapid and involves a number of different enzymes (see Gordon, 1986). This finding means that some of the actions of ATP and ADP might be mediated via P_1-purinoceptors following breakdown to AMP and adenosine (see Moody et al., 1984).

A_1, A_2 and A_3 Subclasses of the P_1-Purinoceptor

The P_1-purinoceptor was subdivided into A_1/R_i and A_2/R_a subtypes according to the relative potencies of a series of adenine analogues and also according to whether they increased or decreased adenylate cyclase activity (Van Calker et al., 1979; Londos et al., 1983). In general, A_1 receptors are preferentially activated by N^6-substituted adenosine analogues, whereas A_2 receptors show preference for 5'-substituted compounds. Thus for A_1 receptors: L-N^6-phenylisopropyladenosine (L-PIA), N^6-cyclohexyladenosine (CHA)$>$2-chloroadenosine (CADO)$>$5'-N-ethyl-carboxamidoadenosine (NECA), D-PIA and adenylate cyclase activity is decreased; while for A_2 receptors: NECA$>$CADO$>$L-PIA, CHA and adenylate cyclase activity is increased.

There have been some problems with this subclassification on the basis largely of inconsistent potency series in different tissues, particularly between central and peripheral tissues, but the recent efforts to develop selective antagonists for A_1 and A_2 subclasses is giving more credibility to this classification (Schwabe et al., 1985; Ukena et al., 1986a, b; Trivedi and Bruns, 1988).

An A_3 subclass of the P_1-purinoceptor has been claimed for an adenosine receptor, present in the heart and nerve endings, that is not coupled to adenylate cyclase (Ribeiro and Sebastião, 1986).

Another type of adenosine recognition site modulating the activity of adenylate cyclase, the intracellular P site, has also been described (Haslam et al., 1978; Londos et al., 1983). This P site is not susceptible to blockade by xanthines.

242

P_{2X}- and P_{2Y}-purinoceptor subclasses have been claimed on the basis of relative potencies of ATP analogues and selective antagonism (Burnstock and Kennedy, 1985). Thus for P_{2X}-purinoceptors: α, β-methylene ATP (α, β-meATP), β, γ-meATP>ATP, 2-methylthio-ATP (2-Me.S.ATP), while arylazidoaminoproprionyl-ATP (ANAPP$_3$) is a selective antagonist and prolonged exposure to α, β-meATP selectively desensitizes this receptor (Kasakov and Burnstock, 1983); P_{2Y}-purinoceptors: 2-Me.S.ATP>>ATP> α, β-meATP, β, γ-meATP, while reactive blue 2, an anthraquinone sulphonic acid derivative, has been claimed to be a selective antagonist, at least over a limited concentration range (Kerr and Krantis, 1979; Manzini et al., 1986; Houston et al., 1987). Studies of the pharmacological actions of isopolar phosphonate analogues of ATP on guinea-pig taenia coli and bladder, have supported the P_{2X}, P_{2Y} subdivision of P_2-purinoceptors in smooth muscle and have also shown that L-adenosine $5'$-(β, γ-methylene)triphosphonate and its analogues are selective agonists of the P_{2X}-purinoceptor (Cusack et al., 1987), while adenosine $5'$-(2-fluorodiphosphate) is a specific agonist for the P_{2Y}-purinoceptor, mediating relaxation of smooth muscle (Hourani et al., 1988).

Since the receptors for ATP on platelets and mast cells (and lymphocytes) do not seem to fit this subclassification, they have been tentatively termed P_{2T}- and P_{2Z}-purinoceptors, respectively (Gordon, 1986).

Table 1 summarizes the purinoceptor classification currently in use.

CHARACTERIZATION OF PURINOCEPTORS

Distribution and Roles

Purinoceptors of various kinds have been identified on a wide variety of cell types (see tables 2 and 3). In general, adenosine is inhibitory in its actions, while ATP is either excitatory or inhibitory.

Nerves and Astrocytes   Adenosine, acting via prejunctional P_1-purinoceptors (usually of the A_1 subtype), is a potent modulator of release of transmitter from terminal varicosities of peripheral adrenergic and cholinergic nerves (see Burnstock, 1986a; Paton, 1987). P_1-purinoceptors are particularly prominent in the brain, where their main role appears to be neuromodulatory (Phillis and Wu, 1981; Williams, 1987). P_2-purinoceptors have been described on cell bodies of sensory neurones in nodose ganglion, spinal cord and brain (Jahr and Jessel, 1983; Fyffe and Perl, 1984; Krishtal et al., 1988) and also on intrinsic ganglionic neurones in heart and bladder (Burnstock et al., 1987). There is recent evidence for P_2- as well as P_1-purinoceptors on astrocytes (Gebicke-Haerter et al., 1988; Pearce et al., 1989).

Muscle ATP has been proposed as a transmitter or cotransmitter in autonomic nerves supplying visceral and vascular organs (see Burnstock, 1976, 1986b, 1988a). Postjunctional receptors for ATP are implicit in the purinergic transmission mechanism: thus it is not surprising that P_2-purinoceptors are present in many smooth muscles. In some muscles,

Table 1. Subtypes of purinoceptors

Purinoceptor	Subclass	Rank Order of Agonist Potency	Antagonists	Adenylate Cyclase Activity	Prosta-glandin Synthesis
P_1	A_1 (R_i)	L-PIA, CHA>CADO >NECA, D-PIA >Adenosine	Non-selective Caffeine Theophylline 8-PT PACPX 8-SPT 9-MeA DPSPX Selective CGS15943A (A_1) DPCPX (A_1) PD116,948 (A_2)	→	–
	A_2 (R_a)	NECA>CADO >L-PIA, CHA >Adenosine		←	–
	A_3 ?	L-PIA, CHA, NECA >CADO		–	–
P_2	P_{2X}	α,β-meATP, β,γ-meATP >ATP=2-Me.S.ATP	ANAPP3 Desensitization by α,β-meATP	–	←
	P_{2Y}	2-Me.S.ATP>>ATP >α,β-meATP, β,γ-meATP	Reactive Blue 2 (an Anthraquinone Sulphonic Acid Derivative)	–	←
	P_{2Z}	ATP4->ATP		–	?
	P_{2T}	2-Me.S.ADP>ADP >α,β-meADP	ATP, AMP, Adenosine (Non-competitively)	→	?

From Burnstock, G., 1989, Purine receptors. In Adenosine Receptors in the Nervous System, edited by J.A. Ribeiro (London: Taylor & Francis), in press.

e.g. those in the intestine and rabbit portal vein, ATP acting via P_{2Y}-purinoceptors is a potent relaxant, while in those in the urinary bladder and vas deferens, and most vascular smooth muscles, ATP acting via P_{2X}-purinoceptors has a potent contractile action (see Burnstock and Kennedy, 1985). P_1-purinoceptors (usually A_2 subtype) are widespread in both vascular and visceral smooth muscle.

Both P_1- and P_2-purinoceptors have been identified in the vertebrate heart (see Burnstock, 1980; Kakei and Noma, 1984). It is proposed that the P_1-purinoceptor present in heart is the A_3 subtype (Ribeiro and

Table 2. Distribution and roles of purinoceptors

Tissue	Purinoceptor	Principal Action
Nerves		
Sympathetic	P_1 (A_1)	Inhibition
Parasympathetic	P_1 (A_1)	Inhibition
Purinergic	P_1 (A_1)	Autoinhibition
Sensory	P_2	Excitation
Astrocytes (CNS)	P_1	Hyperpolarization
	P_2	Accumulation of IP
		Release of Thromboxane A_2
Smooth Muscle	P_{2X}	Contraction
Visceral and Vascular	P_{2Y}	Relaxation
	P_1 (A_2)	Relaxation
Heart Muscle	P_1 $(A_3?)$	Inhibition
	P_2	Excitation
Developing Myotube	P_2	Excitation
Retinal Pericytes	P_2	Contraction
Vascular Endothelial Cells	P_{2Y}	Increase in EDRF
Fibroblasts	P_2	Contraction; Depolarization
Hepatocytes	P_1	Activates Adenylate Cyclase
	P_{2Y}	Glycogenolysis
Adipocytes	P_1	
Spermatozoa and Cilia	P_2	Excitation
Carotid Chemoreceptors	P_1 (A_2)	Excitation
Thyroid Cells	P_1 (A_2)	
	P_2	Increased IP Turnover
Human Amnion Cells	P_{2Y}	Activates Phospholipase C
Chondrocytes	P_2	Increase in Prostaglandins

Sebastião, 1986).   In bullfrog atrial muscle cells, Friel and Bean (1988) observed a biphasic electrical response to ATP. This is consistent with the evidence for release of ATP as a cotransmitter with adrenaline from sympathetic nerves supplying the frog heart (Hoyle and Burnstock, 1986). From a study of the effects of ATP on the papillary and right ventricular muscles of the rat, it was suggested that P_2-purinoceptor activation induces both a positive inotropy and an increase in inositol-lipid metabolism (Legssyer et al., 1988). P_2-purinoceptors have been identified in the developing myotube (Häggblad & Heilbronn, 1988).

Endothelial and Epithelial Cells, Hepatocytes and Pancreatic Secretory Cells Potent actions of ATP on vascular endothelial cells leading to release of endothelium-derived relaxing factor (EDRF) and vasodilatation were first described in 1981 by De Mey and Vanhoutte, and have been described now in many vessels (see Burnstock and Kennedy, 1986; Needham et al., 1987; Sauve et al., 1988). The endothelial ATP receptors have been shown to be of the P_{2Y} subclass (see Burnstock, 1988b).

P_2-purinoceptors have been shown to regulate ion transport in epithelial cells from a variety of different sources, including intestinal epithelial cells and kidney epithelium, where ATP stimulates Cl^- transport and alters Ca^{2+} distribution. ATP also regulates gastric acid secretion and evidence has been presented for involvement of P_{2Y}-purinoceptors in ATP

Table 3. Distribution and roles of purinoceptors

Tissue	Purinoceptor	Principal Action
Mast Cells	P_{2Z}	Degranulation
Immune Cells	P_1 (A_2)	
(Lymphocytes, Granulocytes, Splenocytes, Leucocytes, Basophils, Thymocytes, Macrophages, Neutrophils)	P_{2Z}	Depolarization
Platelets	P_{2T}	Aggregation
Thrombocytes, Megakaryocytes	P_{2T}	Excitation
Erythrocytes	P_1	
	P_{2Y}	
Alveolar Type II Cells	P_{2Y}	Surfactant Secretion
Parotid Acinar Cells	P_{2Y}	Amylase Secretion
Pancreatic B Cells	P_{2Y}	Insulin Secretion
Pancreatic A Cells	P_1 (A_2)	Glucagon Secretion
Intestinal Epithelial Cells	P_2	Ion Fluxes
Renal Epithelioid Cells (MDCK)	P_2	Activation of Renal Tubular Transport
		Activates K^+ Channels
LLC-PK$_1$ Cells	P_2	Increases Intracellular Ca^{2+}
(Renal Epithelial Cell Line)		
Tumours	P_2	Growth Inhibition
Neuroblastoma	P_1 (A_2)	Elevation of cAMP
Transformed Mouse Cell Lines	P_{2Z}	Permeabilization
Ehrlich Ascites Tumour Cells	P_{2Y}	Inhibits Proliferation

regulation of surfactant secretion from type II alveolar epithelial cells (Rice and Singleton, 1987; Gilfillan and Rooney, 1988).

It has been known for some time that ATP has glycogenolytic and hyper-polarizing actions on hepatocytes that are mediated by P_2-purinoceptors (Häussinger et al., 1987), and it has been suggested that the receptor is of the P_{2Y}-purinoceptor subclass (Keppens and De Wulf, 1986). Adenosine, acting via an A_2 receptor, has also been claimed to stimulate hepatic glycogenolysis, but possibly by an indirect mechanism (Buxton et al., 1986).

Pancreatic B cells respond to ATP via P_{2Y}-purinoceptors to increase insulin secretion, while adenosine acts via the A_2 subtype of a P_1-purinoceptor in A cells to increase glucagon secretion (Loubatieres-Mariani and Chapal, 1988).

Mast Cells and Cells of the Immune System   ATP induces calcium-dependent histamine secretion from mast cells (Dahlquist and Diamant, 1974). The agonist form is the tetrabasic acid ATP^{4-} (Cockcroft and Gomperts, 1979; Tatham et al., 1988) and this receptor has therefore been given the separate subclassification of P_{2Z} (Gordon, 1986).

P_1-purinoceptors of the A_2 subtype have been described on various cells of the immune system, including lymphocytes and granulocytes (see Bonnafous et al., 1982). ATP modifies cation fluxes and could thereby deliver the calcium signal for lymphocyte activation (see Cameron, 1984) and extracellular ATP has also been shown to stimulate transmembrane ion fluxes in macrophages, possibly via a P_{2Z}-purinoceptor (Steinberg and Silverstein, 1987).

Platelets and Erythrocytes   ADP causes platelets to rapidly change shape, which leads to platelet aggregation, while P_1-purinoceptors mediate inhibition of ADP-induced platelet aggregation (see Haslam and Cusack, 1981). Since the platelet receptor is unique in being activated by ADP rather than ATP, it has been tentatively classified as a P_{2T}-purinoceptor (Gordon, 1986). P_{2Y}-purinoceptors have been demonstrated in turkey erythrocytes (Berrie et al., 1989; Boyer et al., 1989; Cooper et al., 1989).

Fibroblasts and Other Cell Types   ATP receptors mediating membrane potential changes in fibroblasts have been described (Okada et al., 1984) and the possibility raised that ATP released as a cotransmitter with noradrenaline from sympathetic nerves exerts some control of fibroblast function (Soares-da-Silva and Azevedo, 1985).

Purinoceptors have also been identified on spermatozoa, chemoreceptor cells in the carotid body, and neuroblastoma, adipose, thyroid, salivary acinar and tumour cells.

Transduction Mechanisms

cAMP and P_1-Purinoceptors   cAMP has been claimed to be the primary second messenger associated with A_1 and A_2 (but not A_3) purinoceptor subclasses (see Daly, 1985; Ribeiro and Sebastião, 1986).

The original classification of adenosine receptors into A_1/R_i and A_2/R_a subtypes was based largely on the ability of adenosine and its analogues to stimulate or inhibit the production of cAMP. However, few actions of adenosine have been shown unequivocally to be mediated via changes in the level of cAMP. Although, in many cases, adenosine receptor agonists have been shown to alter the levels of cAMP, the involvement of such changes in the production of the final response is unclear. This underlines the view that a receptor is best conceived as being constructed of two units: a recognition component and a catalytic component.   It is entirely possible that the same recognition component (e.g. A_1 or A_2) could be linked to a variety of catalytic components (e.g. stimulatory or inhibitory regulatory units of adenylate cyclase or Ca^{2+} channels) in the same or different cell types. Hence, it is preferable not to classify adenosine receptors according to their effect on adenylate cyclase, at least until the linkage between receptor occupation and cAMP levels is understood more thoroughly.

Ion Channels and P_2-Purinoceptors The inhibitory actions of ATP acting on P_{2Y}-purinoceptors that lead to hyperpolarization of smooth muscle cells of the intestine appear to be associated with selective opening of K^+ channels (see Hoyle and Burnstock, 1989).

The excitatory actions of ATP acting on P_{2X}-purinoceptors on vascular and visceral smooth muscle cells appear to be associated with the opening of non-selective cation channels, resulting in depolarization and subsequent opening of voltage-dependent Ca^{2+} channels (Nakazawa and Matsuki, 1987; Friel, 1988).   In addition, in some arterial smooth muscles, it has been claimed that increased calcium influx is also the result of direct activation of ATP-gated cation channels without any requirement for depolarization (Benham and Tsien, 1987).

In patch-clamp studies of developing chick skeletal muscle, external ATP has also been shown to activate cation-selective channels (Kolb and Wakelam, 1983). The effects of ATP in neuronal cells are complex, but one direct effect is a rapid depolarization caused by increased cation conductance (Krishtal et al., 1988).

Phosphoinositol Transduction Mechanisms and P_2-Purinoceptors   Extracellular ATP at low (micromolar) concentration stimulates inositol 1,4,5-trisphosphate (IP) production and intracellular Ca^{2+} mobilization in hepatocytes (Charest et al., 1985; Keppens and De Wulf, 1986), adrenal medullary and other vascular endothelial cells (Boeynaems et al., 1988), aortic and ventricular myocytes (Danziger et al., 1988), erythrocytes (Berrie et al., 1989), Ehrlich ascites tumour cells (Dubyak, 1986) and chick myotubes (Häggblad and Heilbronn, 1988). P_{2Y}-purinoceptors coupled to phospholipase C activation and intracellular Ca^{2+} mobilization have also been demonstrated in primary cultures of sheep anterior pituitary cells (van der Merwe et al., 1989) and turkey erythrocyte membranes (Boyer et al., 1989; Cooper et al., 1989).

FUTURE DEVELOPMENTS

The general direction that most receptor studies are taking at present is to clone the receptor following strong ligand binding and then inject the appropriate mRNA into the Xenopus oocyte to express the receptor (see Dascal, 1987). Lotan and collegues have demonstrated a hyperpolarizing response in the oocyte to adenosine and concluded that the adenosine-evoked outward current was carried by K^+; responses to ATP, in contrast, appear to be mediated by Cl^- currents (Lotan et al., 1986).

Adenosine 5'-O-2-thio[^{35}S]diphosphate has been proposed as a radioligand for the P_{2Y}-purinoceptor in purified turkey erythrocyte membranes (Cooper et al., 1989). My own laboratory has recently identified [^{3}H]-α,β-methylene ATP as a strongly binding ligand for the P_2-purinoceptor (Bo and Burnstock, 1989) and we are currently collaborating with molecular biologists to clone this receptor and hopefully to use the Xenopus oocyte to examine the expression of its nucleic acid.

REFERENCES

Benham, C.D. and Tsien, R.W., 1987, A novel receptor-operated Ca^{2+}-permeable channel activated by ATP in smooth muscle. Nature, 328, 275-278.
Berne, R.M., 1964, Regulation of coronary blood flow. Physiological Reviews, 44, 1-29.
Berrie, C.P., Hawkins, P.T., Stephens, L.R., Harden, T.K. and Downes, C.P., 1989, Phosphatidylinositol 4,5-bisphosphate hydrolysis in turkey erythrocytes is regulated by P_{2y} purinoceptors. Molecular Pharmacology, 35, 526-532.
Bo, X. and Burnstock, G., 1989, [^{3}H]-α,β-methylene ATP, a radioligand labelling P_2-purinoceptors. Journal of the Autonomic Nervous

System, in press.

Boeynaems, J.M., Pirotton, S., Van Coevorden, A., Raspe, E., Demolle, D. and Erneux, C., 1988, P_2-purinergic receptors in vascular endothelial cells: from concept to reality. Journal of Receptor Research, 8, 121-132.

Bonnafous, J.C., Dornand, J., Favero, J. and Mani, J.C., 1982, Lymphocyte membrane adenosine receptors coupled to adenylate cyclase: properties and occurrence in various lymphocyte subclasses. Journal of Receptor Research, 2, 347-366.

Boyer, J.L., Downes, C.P. and Harden, T.K., 1989, Kinetics of activation of phospholipase C by P_{2Y} purinergic receptor agonists and guanine nucleotides. Journal of Biological Chemistry, 264, 884-890.

Burnstock, G., 1972, Purinergic nerves. Pharmacological Reviews, 24, 509-581.

Burnstock, G., 1976, Do some nerve cells release more than one transmitter? Neuroscience, 1, 239-248.

Burnstock, G., 1978, A basis for distinguishing two types of purinergic receptor. In Cell Membrane Receptors for Drugs and Hormones: A Multidisciplinary Approach, edited by L. Bolis and R.W. Straub (New York: Raven Press), pp. 107-118.

Burnstock, G., 1980, Purinergic receptors in the heart. In Cardiovascular Receptors: Molecular, Pharmacological and Therapeutic Aspects, edited by P.I. Korner and J.A. Angus. Circulation Research, 46, Suppl. 1, 175-182.

Burnstock, G., editor, 1981, Purinergic Receptors, Receptors and Recognition, Series B, Vol. 12 (London: Chapman & Hall), 365 pp.

Burnstock, G., 1986a, Purines as cotransmitters in adrenergic and cholinergic neurones. In Coexistence of Neuronal Messengers: A New Principle in Chemical Transmission, Progress in Brain Research, Vol. 68, edited by T. Hökfelt, K. Fuxe and B. Pernow (Amsterdam: Elsevier), pp. 193-203.

Burnstock, G., 1986b, The changing face of autonomic neurotransmission. (The First von Euler Lecture in Physiology.) Acta Physiologica Scandinavica, 126, 67-91.

Burnstock, G., 1988a, Sympathetic purinergic transmission in small blood vessels. Trends in Pharmacological Sciences, 9, 116-117.

Burnstock, G., 1988b, Local purinergic regulation of blood pressure. (The First John T. Shepherd Lecture.) In Vasodilatation: Vascular Smooth Muscle, Peptides, Autonomic Nerves and Endothelium, edited by P.M. Vanhoutte (New York: Raven Press), pp. 1-14.

Burnstock, G. and Buckley, N., 1985, The classification of receptors for adenosine and adenine nucleotides. In Methods used in Adenosine Research, Methods in Pharmacology, Vol. 6, edited by D.M. Paton (New York: Plenum Press), pp. 193-212.

Burnstock, G. and Kennedy, C., 1985, Is there a basis for distinguishing two types of P_2-purinoceptor? General Pharmacology, 16, 433-440.

Burnstock, G. and Kennedy, C., 1986, Purinergic receptors in the cardiovascular system. In Receptors in the Cardiovascular System, Progress in Pharmacology, Vol. 6, No. 2, edited by P.A. van Zwieten and E. Schönbaum (Stuttgart: Gustav Fischer), pp. 111-132.

Burnstock, G., Allen, T.G.J., Hassall, C.J.S. and Pittam, B.S., 1987, Properties of intramural neurones cultured from the heart and bladder. Experimental Brain Research, Series 16, 323-328.

Buxton, D.B., Robertson, S.M. and Olson, M.S., 1986, Stimulation of glycogenolysis by adenine nucleotides in the perfused rat liver. Biochemical Journal, 237, 773-780.

Cameron, D.J., 1984, Inhibition of macrophage mediated cytotoxicity by exogenous adenosine 5'-triphosphate. Journal of Clinical and Laboratory Immunology, 15, 215-218.

Charest, R., Prpić, V., Exton, J.H. and Blackmore, P.F., 1985, Stimulation of inositol trisphosphate formation in hepatocytes by vasopressin, adrenaline and angiotensin II and its relationship to changes in cytosolic free Ca^{2+}. Biochemical Journal, 227, 79-90.

Cockcroft, S. and Gomperts, B.D., 1979, Activation and inhibition of calcium-dependent histamine secretion by ATP ions applied to rat mast cells. Journal of Physiology (London), 296, 229-243.

Cooper, C.L., Morris, A.J. and Harden, T.K., 1989, Guanine nucleotide-sensitive interaction of a radiolabeled agonist with a phospholipase C-linked P_{2y}-purinergic receptor. Journal of Biological Chemistry, 264, 6202-6206.

Cusack, N.J., Hourani, S.M.O., Loizou, G.D. and Welford, L.A., 1987, Pharmacological effects of isopolar phosphonate analogues of ATP on P_2-purinoceptors in guinea-pig taenia coli and urinary bladder. British Journal of Pharmacology, 90, 791-795.

Dahlquist, R. and Diamant, B., 1974, Interaction of ATP and calcium on the rat mast cell: effect on histamine release. Acta Pharmacologica et Toxicologica, 34, 368-384.

Daly, J.W., 1985, Adenosine receptors. Advances in Cyclic Nucleotide and Protein Phosphorylation Research, 19, 29-46.

Daly, J.W., Hong, O., Padgett, W.L., Shamim, M.T., Jacobson, K.A. and Ukena, D., 1988, Non-xanthine heterocycles: activity as antagonists of A_1- and A_2-adenosine receptors. Biochemical Pharmacology, 37, 655-664.

Danziger, R.S., Raffaeli, S., Moreno-Sanchez, R., Sakai, M., Capogrossi, M.C., Spurgeon, H.A. and Hansford, R.G., 1988, Extracellular ATP has a potent effect to enhance cystolic calcium and contractility in single ventricular myocytes. Cell Calcium, 9, 193-199.

Dascal, N., 1987, The use of Xenopus oocytes for the study of ion channels. CRC Critical Reviews in Biochemistry, 22, 317-388.

De Mey, J.G. and Vanhoutte, P.M., 1981, Role of the intima in cholinergic and purinergic relaxation of isolated canine femoral arteries. Journal of Physiology (London), 316, 347-355.

Drury, A.N. and Szent-Györgyi, A., 1929, The physiological activity of adenine compounds with special reference to their action upon the mammalian heart. Journal of Physiology (London), 68, 213-237.

Dubyak, G.R., 1986, Extracellular ATP activates polyphosphoinositide breakdown and Ca^{2+} mobilization in Ehrlich ascites tumor cells. Archives of Biochemistry and Biophysics, 245, 84-95.

Dunn, P.M. and Blakeley, A.G.H., 1988, Suramin: a reversible P_2-purinoceptor antagonist in the mouse vas deferens. British Journal of Pharmacology, 93, 243-245.

Friel, D.D., 1988, An ATP-sensitive conductance in single smooth muscle cells from the rat vas deferens. Journal of Physiology (London), 401, 361-380.

Friel, D.D. and Bean, B.P., 1988, Two ATP-activated conductances in bullfrog atrial cells. Journal of General Physiology, 91, 1-27.

Fyffe, R.E.W. and Perl, E.R., 1984, Is ATP a central synaptic mediator for certain primary afferent fibres from mammalian skin? Proceedings of the National Academy of Sciences of the United States of America, 81, 6890-6893.

Gebicke-Haerter, P.J., Wurster, S., Schobert, A. and Hertting, G., 1988, P_2-purinoceptor induced prostaglandin synthesis in primary rat astrocyte cultures. Naunyn-Schmiedeberg's Archives of Pharmacology, 338, 704-707.

Gilfillan, A.M. and Rooney, S.A., 1988, Functional evidence for involvement of P_2 purinoceptors in the ATP stimulation of phosphatidylcholine secretion in type II alveolar epithelial cells. Biochimica et Biophysica Acta, 959, 31-37.

Gordon, J.L., 1986, Extracellular ATP: effects, sources and fate. Biochemical Journal, 233, 309-319.

Häggblad, J. and Heilbronn, E., 1988, P_2-purinoceptor-stimulated phosphoinositide turnover in chick myotubes. Calcium mobilization and the role of guanyl nucleotide-binding proteins. FEBS Letters, 235, 133-136.

Haslam, R.J. and Cusack, N.J., 1981, Blood-platelet receptors for ADP and for adenosine. In Purinergic Receptors, Receptors and Recognition, Series B, Vol. 12, edited by G. Burnstock (London: Chapman & Hall), pp. 221-285.

Haslam, R.J., Davidson, M.M.L. and Desjardins, J.V., 1978, Inhibition of adenylate cyclase by adenosine analogues in preparations of broken and intact human platelets. Evidence for the unidirectional control of platelet function by cyclic AMP. Biochemical Journal, 176, 83-95.

Häussinger, D., Stehle, T., Gerok, W., Tran-Thi, T.-A. and Decker, K., 1987, Hepatocyte heterogeneity in response to extracellular ATP. European Journal of Biochemistry, 169, 645-650.

Hogaboom, G.K., O'Donnell, J.P. and Fedan, J.S., 1980, Purinergic receptors: photoaffinity analog of adenosine triphosphate is a specific adenosine triphosphate antagonist. Science, 208, 1273-1275.

Hourani, S.M.O., Welford, L.A., Loizou, G.D. and Cusack, N.J., 1988, Adenosine 5'-(2-fluorodiphosphate) is a selective agonist at P_2-purinoceptors mediating relaxation of smooth muscle. European Journal of Pharmacology, 147, 131-136.

Houston, D.A., Burnstock, G. and Vanhoutte, P.M., 1987, Different P_2-purinergic receptor subtypes of endothelium and smooth muscle in canine blood vessels. Journal of Pharmacology and Experimental Therapeutics, 241, 501-506.

Hoyle, C.H.V. and Burnstock, G., 1986, Evidence that ATP is a neurotransmitter in the frog heart. European Journal of Pharmacology, 124, 285-289.

Hoyle, C.H.V. and Burnstock, G., 1989, Neuromuscular transmission in the gastrointestinal tract. In Handbook of Physiology, Section 6: The Gastrointestinal System, Vol. I: Motility and Circulation, edited by J.D. Wood (Bethesda: American Physiological Society), pp. 435-464.

Jahr, C.E. and Jessel, T.M., 1983, ATP excites a subpopulation of rat dorsal horn neurones. Nature, 304, 730-733.

Kakei, M. and Noma, A., 1984, Adenosine-5'-triphosphate-sensitive single potassium channel in the atrioventricular node cell of the rabbit heart. Journal of Physiology (London), 352, 265-284.

Kasakov, L. and Burnstock, G., 1983, The use of the slowly degradable analog, α, β -methylene ATP, to produce desensitisation of the P_2-purinoceptor: effect on non-adrenergic, non-cholinergic responses of the guinea-pig urinary bladder. European Journal of Pharmacology, 86, 291-294.

Keppens, S. and De Wulf, H., 1986, Characterization of the liver P_2-purinoceptor involved in the activation of glycogen phosphorylase. Biochemical Journal, 240, 367-371.

Kerr, D.I.B. and Krantis, A., 1979, A new class of ATP antagonist. Proceedings of the Australian Physiological and Pharmacological Society, 10, 156P.

Kolb, H.-A. and Wakelam, M.J.O., 1983, Transmitter-like action of ATP on patched membranes of cultured myoblasts and myotubes. Nature, 303, 621-623.

Krishtal, O.A., Marchenko, S.M. and Obukhov, A.G., 1988, Cationic channels activated by extracellular ATP in rat sensory neurons. Neuroscience, 27, 995-1000.

Legssyer, A., Poggioli, J., Renard, D. and Vassort, G., 1988, ATP and other adenine compounds increase mechanical activity and inositol trisphosphate production in rat heart. Journal of Physiology (London), 401, 185-199.

Londos, C., Wolff, J. and Cooper, D.M.F., 1983, Adenosine receptors and adenylate cyclase interactions. In Regulatory Function of Adenosine, edited by R.M. Berne, T.W. Rall and R. Rubio (Boston: Martinus Nijhoff), pp. 17-32.

Lotan, I., Dascal, N., Cohen, S. and Lass, Y., 1986, ATP-evoked membrane responses in Xenopus oocytes. Pflügers Archiv, 406, 158-162.

Loubatieres-Mariani, M.M. and Chapal, J., 1988, Purinergic receptors involved in the stimulation of insulin and glucagon secretion. Diabete et Metabolisme, 14, 119-126.

Manzini, S., Hoyle, C.H.V. and Burnstock, G., 1986, An electrophysiological analysis of the effect of reactive blue 2, a putative P_2-purinoceptor antagonist, on inhibitory junctional potentials of rat caecum. European Journal of Pharmacology, 127, 197-204.

Moody, C.J., Meghji, P. and Burnstock, G., 1984, Stimulation of P_1-purinoceptors by ATP depends partly on its conversion to AMP and adenosine and partly on direct action. European Journal of Pharmacology, 97, 47-54.

Nakazawa, K. and Matsuki, N., 1987, Adenosine triphosphate-activated inward current in isolated smooth muscle cells from rat vas deferens. Pflügers Archiv, 409, 644-646.

Needham, L., Cusack, N.J., Pearson, J.D. and Gordon, J.L., 1987, Characteristics of the P_2 purinoceptor that mediates prostacyclin production by pig aortic endothelial cells. European Journal of Pharmacology, 134, 199-209.

Okada, Y., Yada, T., Ohno-Shosaku, T., Oiki, S., Ueda, S. and Machida, K., 1984, Exogenous ATP induces electrical membrane responses in fibroblasts. Experimental Cell Research, 152, 552-557.

Paton, D.M., 1987, Presynaptic inhibitory actions of adenosine on peripheral adrenergic and cholinergic neurotransmission. In Pharmacology, Proceedings of the Xth International Congress of Pharmacology, edited by M.J. Rand and C. Raper (Amsterdam: Excerpta Medica), pp. 267-270.

Pearce, B., Murphy, S., Jeremy, J., Morrow, C. and Dandona, P., 1989, ATP-evoked Ca^{2+} mobilisation and prostanoid release from astrocytes: P_2-purinergic receptors linked to phosphoinositide hydrolysis. Journal of Neurochemistry, 52, 971-977.

Phillis, J.W. and Wu, P.H., 1981, The role of adenosine and its nucleotides in central synaptic transmission. Progress in Neurobiology, 16, 187-239.

Ribeiro, J.A. and Sebastião, A.M., 1986, Adenosine receptors and calcium: basis for proposing a third (A_3) adenosine receptor. Progress in Neurobiology, 26, 179-209.

Rice, W.R. and Singleton, F.M., 1987, P_{2y}-purinoceptor regulation of surfactant secretion from rat isolated alveolar type II cells is associated with mobilization of intracellular calcium. British Journal of Pharmacology, 91, 833-838.

Satchell, D., 1984, Purine receptors: classification and properties. Trends in Pharmacological Sciences, 5, 340-344.

Sauve, R., Parent, L., Simoneau, C. and Roy, G., 1988, External ATP triggers a biphasic activation process of a calcium-dependent K^+ channel in cultured bovine aortic endothelial cells. Pflügers Archiv, 412, 469-481.

Schwabe, U., Ukena, D. and Lohse, M.J., 1985, Xanthine derivatives as antagonists at A_1 and A_2 adenosine receptors. Naunyn-Schmiedeberg's Archives of Pharmacology, 330, 212-221.

252

Soares-da-Silva, P. and Azevedo, I., 1985, Fibroblasts and sympathetic innervation of blood vessels. Blood Vessels, 22, 278-285.

Steinberg, T.H. and Silverstein, S.C., 1987, Extracellular ATP^{4-} promotes cation fluxes in the J774 mouse macrophage cell line. Journal of Biological Chemistry, 262, 3118-3122.

Tatham, P.E.R., Cusack, N.J. and Gomperts, B.D., 1988, Characterisation of the ATP^{4-} receptor that mediates permeabilisation of rat mast cells. European Journal of Pharmacology, 147, 13-21.

Trivedi, B.K. and Bruns, R.F., 1988, [1,2,4]Triazolo-[4,3-a]quinoxalin-4-amines: a new class of A_1 receptor selective adenosine antagonists. Journal of Medicinal Chemistry, 31, 1011-1014.

Ukena, D., Daly, J.W., Kirk, K.L. and Jacobson, K.A., 1986a, Functionalized congeners of 1,3-dipropyl-8-phenylxanthine: potent antagonists for adenosine receptors that modulate membrane adenylate cyclase in pheochromocytoma cells, platelets and fat cells. Life Sciences, 38, 797-807.

Ukena, D., Shamim, M.T., Padgett, W. and Daly, J.W., 1986b, Analogs of caffeine: antagonists with selectivity for A_2 adenosine receptors. Life Sciences, 39, 743-750.

Van Calker, D., Müller, M. and Hamprecht, B., 1979, Adenosine regulates via two different types of receptors, the accumulation of cyclic AMP in cultured brain cells. Journal of Neurochemistry, 33, 999-1005.

van der Merwe, P.A., Wakefield, I.K., Fine, J., Millar, R.P. and Davidson, J.S., 1989, Extracellular adenosine triphosphate activates phospholipase C and mobilizes intracellular calcium in primary cultures of sheep anterior pituitary cells. FEBS Letters, 243, 333-336.

Williams, M., 1987, Purinergic receptors and central nervous system function. In Psychopharmacology: The Third Generation of Progress, edited by H.Y. Meltzer (New York: Raven Press), pp. 289-301.

37

Structure Activity Relationships for Adenine Nucleotide Receptors on Mast Cells, Human Platelets, and Smooth Muscle

N.J. Cusack and S.M.O. Hourani

Extracellular adenine nucleotides, adenosine 5'-monophosphate (AMP), adenosine 5'-diphosphate (ADP) and adenosine 5'-triphosphate (ATP) are active on mast cells, platelets and smooth muscle. These P_2 purinoceptor-mediated actions have very different structure-activity relationships, which has lead to a subdivision into P_{2Z} mast cell receptors, P_{2T} platelet receptors, and P_{2Y} inhibitory and P_{2X} excitatory receptors on smooth muscle (Burnstock et al., 1985; Gordon, 1986).

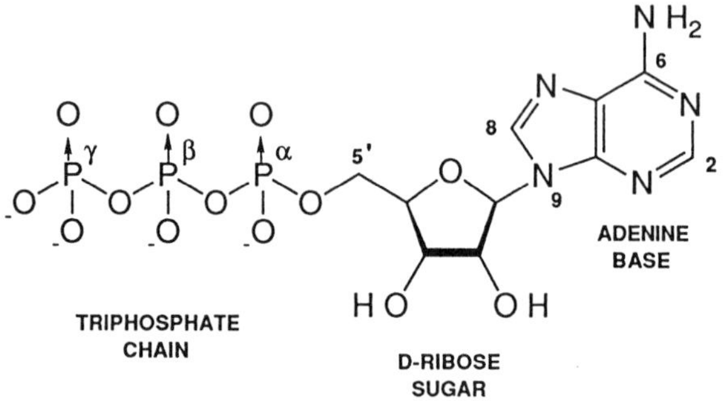

Figure 1: Structure of ATP

Mast Cells

The P_2 purinoceptor on mast cells that mediates permeabilization of the plasma membrane, as measured by uptake of ethidium, is extremely sensitive to alterations to the ATP molecule. Of the naturally occurring nucleotides, only ATP itself induces permeabilization, with maximal activity occurring at 10 μM concentrations, while AMP, ADP, GTP, CTP and UTP are inactive (Cockcroft et al., 1980). Substitutions on the N^6 or C-8 position of the adenine base lead to inactive molecules, but C-2 analogues including 2-chloro-ATP, 2-methylthio-ATP and 2-ethylthio-ATP are as active as ATP at inducing permeabilization. The β-D-ribofuranose sugar is an absolute requirement, as the unnatural β-L-ribofuranose derived enantiomer of ATP, L-ATP, is completely inactive, as are 2-chloro-L-ATP and 2-methylthio-L-ATP (Tatham et al., 1988). The 5'-triphosphate chain can be converted to a phosphorothioate by replacement of an ionized oxygen on the innermost phosphate, ATPα-S, or middle phosphate, ATPβ-S, and leads to enhanced permeabilizing activity relative to ATP, but the receptor does not distinguish between the R_p and S_p diastereoisomers of these potent analogues (Tatham et al., 1988). Other alterations to the triphosphate chain, such as replacement of the innermost, middle or outermost bridging oxygens with methylene, as in homo-ATP, α,β-methylene-ATP and β,γ-methylene-ATP, respectively, generates inactive molecules, and no competitive antagonists of the mast cell P_2 purinoceptor are known, although 2-methylthio-L-ATP at high (100 μM) concentrations inhibited by 50% ATP-induced permeabilization, and may give a lead to more potent inhibitors (Tatham et al., 1988)

Platelets

The human platelet ADP receptor is unique among P_2 purinoceptors in that only ADP itself is an agonist, whereas AMP and ATP are specific competitive antagonists of ADP-induced platelet activation (Macfarlane et al., 1975). ADP released from damaged tissues activates human blood platelets to change their morphology from biconvex discs to spiny spheroids and, in the presence of adequate extracellular calcium ions and fibrinogen, to adhere to each other (aggregate) and to damaged tissue to effect their important physiological role in haemostasis (Born et al., 1984; McClure et al., 1988). Activation of platelets is associated with a potent inhibition of stimulated adenylate cyclase, and calcium mobilization from intracellular stores as well as an influx of extracellular calcium, but the mechanism of the stimulus-response coupling is still unclear. Structure-activity studies have shown that human platelet ADP receptor is very sensitive to modifications of the ADP molecule. The adenine base is an absolute requirement, with the other naturally-occurring nucleoside 5'-diphosphates GDP, CDP, and UDP being inactive, and substitutions on the C-8 or N^6 position lead to a severe reduction or loss of agonist potency, but a variety of substituents may be placed at the C-2 position of adenine with no loss of potency and indeed some C-2 analogues, including 2-azido-ADP, 2-chloro-ADP, and 2-methylthio-ADP, are up to 10-fold more potent than ADP at inducing platelet aggregation and up to 300-fold more potent than ADP at inhibiting stimulated adenylate cyclase (Cusack et al., 1982c). Since very large substituents are tolerated at the C-2 position, this region of the adenine base is probably orientated away from the ADP receptor. The β-D-ribofuranose sugar is essential for agonist activity, and cannot be replaced by other pentose or hexose sugars. The ADP receptor exhibits absolute stereoselectivity for the naturally occurring enantiomer of ADP as the unnatural β-L-ribofuranose derived enantiomer, L-ADP, is completely inactive as agonist or antagonist (Cusack et al., 1979). The unmodified 5-diphosphate chain of ADP is required for full agonist potency. Replacement of an ionized oxygen of the inner or outer phosphate by ionized sulphur produces ADP-α-S and ADP-β-S, respectively, and both analogues are less potent than ADP at inducing platelet aggregation, and are partial agonists with intrinsic activities of 0.75. ADP-α-S exists as a pair of diastereoisomers, which are distinguished by the ADP receptor, since S_p-ADP-α-S is 5-fold more potent than R_p-ADP-α-S at inducing platelet aggregation. ADP-β-S also inhibits stimulated adenylate cyclase as a partial agonist with an intrinsic activity of 0.5 (Cusack et al., 1981a), but ADP-α-S does not and instead antagonizes competitively this action of ADP, and S_p-ADP-α-S is 5-fold more potent than R_p-ADP-α-S (Cusack et al., 1981b). Replacement of the pyrophosphate oxygen linking the inner and outer phosphates by imido produces α,β-imido-ADP, which is 100-fold less potent than ADP at inducing platelet aggregation and it does not inhibit stimulated adenylate cyclase (Cusack et al., 1988a), and replacement of the oxygen joining the ribose to the diphosphate chain by methylene to produce homo-ADP, or a terminal ionized oxygen by fluorine to produce ADP-β-F, generates virtually inactive analogues (Cusack et al., unpublished).

The structure-activity relationships for antagonists at the human platelet ADP receptor are much more relaxed than for agonists, and a wide variety of analogues of AMP, ADP and ATP are competitive inhibitors of both ADP-induced aggregation and ADP-induced inhibition of stimulated adenylate cyclase. Analogues of AMP and of ATP, shown by Schild analysis to be competitive inhibitors of both these actions of ADP, were used to show that one type of ADP receptor, rather than two different subtypes of ADP receptor, mediated both effects of ADP, since a good correlation was found between the apparent dissociation constants for inhibition of each effect of ADP (Cusack et al., 1982b). Isosteric replacement of the oxygen linking the inner and outer phosphates by methylene, difluoromethylene or by dichloromethylene to give α,β-methylene-ADP, α,β-difluoromethylene-ADP, and α,β-dichloromethylene-ADP, generates antagonists, but Schild analysis showed that only α,β-dichloromethylene-ADP had competitive kinetics for inhibition of aggregation, and only the two "isopolar" analogues, α,β-difluoromethylene-ADP and dichloromethylene-ADP antagonized inhibition by ADP of stimulated adenylate cyclase (Cusack et al., 1988a). Substitution of ADP at the C-8 position by bromine to give 8-Br-ADP, generated a specific inhibitor of ADP induced aggregation (Jefferson et al., 1988). ADP receptor antagonism by 2-alkylthio analogues of AMP and of ATP is anomalous since although they do not inhibit aggregation induced by other agents including 5-HT, PAF, norepinephrine, $11\alpha,9\alpha$-epoxymethano-PGH_2 or arachidonic acid, they antagonize completely ADP-induced inhibition of stimulated adenylate cyclase, but only partially ADP-induced aggregation (Cusack et al., 1982a; 1985). One such analogue, 2-methylthio-β,γ-methylene-ATP, inhibits aggregation induced by ADP-β-S, which antagonizes stimulated adenylate cyclase, but not aggregation induced by ADP-α-S, which does not antagonize adenylate cyclase, which suggests that 2-alkylthio analogues are selective for one component of aggregation (Hourani et al., 1986). Radioligand binding of the potent agonist β-[^{32}P]-2-methylthio-ADP to intact platelets in plasma detected one class of binding site, with 400 to 1200 ADP receptors per platelet, an affinity constant of 5-20 nM , and displacements at binding by agonists and by antagonists in accord with their pharmacology (Macfarlane et al., 1983). Radioligand binding of 2-[^{3}H]-ADP to formaldehyde-fixed intact platelets detected both high affinity and low affinity ADP binding sites, with dissociation constants of 0.35 μM and 7.9 μM respectively (Jefferson et al., 1988), and displacement by agonists and antagonists was in harmony with their pharmacology also (Agrawal et al., 1989).

Smooth Muscle Inhibitory P_{2Y} Receptor

ATP induces rapid relaxation of the guinea-pig taenia coli, and this tissue has been used extensively in structure-activity relationships for the inhibitory P_{2Y} purinoceptor (Cusack et al., 1988b). The adenine base is required for maximal activity, ADP is equipotent with ATP, and GTP, CTP, and UTP are also active, but AMP, GDP, CDP and UDP are much less active than ATP at inducing relaxation (Satchell et al., 1975); Cusack et al., unpublished). Substitutions on the adenine base at N^6 abolishes activity, those at C-8 are approximately as potent as ATP, while C-2 substitution enhances potency up to 200-fold that of ATP (Gough et al., 1973; Burnstock et al., 1983). Replacement of the ribose by other sugars generates analogues less potent than ATP (Satchell et al., 1982), and the unnatural β-L-ribofuranose enantiomer, L-ATP, is 3- to 6-fold less potent than ATP (Cusack et al., 1979). Stereoselectivity towards the enantiomers greatly increases with the C-2 substituted analogues, and 2-methylthio-ATP is 700-fold more potent than 2-methylthio-L-ATP (Cusack et al., 1979; Burnstock et al., 1983). The effect of modifications of the 5'-triphosphate chain depends on where they take place, so that homo-ATP is 70-fold more potent

256

than ATP, α,β-methylene-ATP is equipotent with ATP, and β,γ-methylene-ATP is about 10-fold less potent than ATP (Satchell et al., 1975). The isopolar versions, β,γ-difluoromethylene-ATP and β,γ-dichloromethylene-ATP are more potent than β,γ-methylene-ATP but still less potent than ATP itself, but 2-methylthio-β,γ-difluoromethylene-ATP was twice as potent as ATP though again much less potent than 2-methylthio-ATP itself (Cusack et al., 1987). Of the phosphorothioate analogues of ATP, stereoselectivity is exhibited only towards the diastereoisomers of ATP-α-S, the R_p isomer being 50-fold and the S_p isomer 9-fold more potent than ATP, while ATP-β-S and ATP-γ-S are about as potent as ATP (Burnstock et al., 1984). No competitive antagonists of the inhibitory P_{2Y} purinoceptor are known, but adenosine 5'-[2-fluorodiphosphate] (ADP-β-F) is a specific agonist for P_{2Y} purinoceptors, with no agonist or antagonist activity at P_{2X} excitatory receptors, platelet ADP receptors or mast cell ATP receptors (Hourani et al., 1988).

Smooth Muscle P_{2X} Excitatory Receptor

ATP induces rapid contraction of the guinea-pig urinary bladder, and this excitatory P_{2X} purinoceptor has the most promiscuity in its structure-activity relationships, since virtually any triphosphate can cause contraction of this tissue preparation (Cusack et al., 1988b). The adenine base may be replaced by other bases so that GTP, CTP and UTP are nearly as active as ATP and ADP, but GDP, CDP and AMP are inactive (Lukacsko et al., 1982). Substitutions on the adenine base at N^6 as in N^6-phenyl-ATP abolishes activity, at C-8 as in 8-Bromo-ATP do not affect activity, and in contrast to P_{2Y} receptors, C-2 substitutions do not improve potency relative to ATP (Burnstock et al., 1983; Welford et al., 1987). Again in contrast to P_{2Y} receptors, the unnatural ribose enantiomer of ATP, L-ATP is equipotent with ATP and this utter lack of stereoselectivity by the P_{2X} receptor is not improved by C-2 substitution (Burnstock et al., 1983; Welford et al., 1987). The effects of modifications of the 5'-triphosphate chain depends on where they take place, and so homo-ATP is equipotent with ATP, but α-β-methylene-ATP and β,γ-methylene-ATP are much more potent than ATP (Lukacsko et al., 1982); Cusack et al., 1986; Welford et al., 1987). Again unlike the P_{2Y} receptor, the isopolar analogues β,γ-difluoromethylene-ATP and β,γ-dichloromethylene-ATP are no more potent than β,γ-methylene-ATP itself (Cusack et al., 1987). Of the phosphorothioate analogues of ATP, no stereoselectivity is shown towards the diastereoisomers, and ATP-β-S is more potent than ATP, while ATP-α-S is considerably less potent than ATP (Burnstock et al., 1984), again in contrast to the P_{2Y} receptor. No competitive antagonists of the excitatory P_{2X} purinoceptor are known, but the unnatural enantiomer of β,γ-methylene-ATP, β,γ-methylene-L-ATP (L-AMP-PCP) is a very potent specific agonist for P_{2X}-purinoceptors, with no agonist or antagonist activity at P_{2Y} inhibitory receptors, platelet ADP receptors, or mast cell ATP receptors (Cusack et al., 1984; Hourani et al., 1985; Hourani, 1986; Hourani et al., 1988; Tatham et al., 1988) L-AMP-PCP provides a framework for the design of other potent specific agonists for P_{2X} receptors, including the isopolar β,γ-dihalomethylene versions, L-AMP-PCF$_2$P and L-AMP-PCCl$_2$P (Cusack et al., 1987; 1988b), all of which, like L-AMP-PCP itself, are resistant to dephosphorylation by the ectonucleotidases present on smooth muscle (Welford et al., 1986; 1987; Cusack et al., 1988c).

References

Agrawal, A.K., Tandon, N.N., Greco, N.J., Cusack, N.J., and Jamieson, G.A., 1989, Evaluation of binding to fixed platelets of agonists and antagonists of ADP-induced aggregation, *Thrombos. Haemastas.* in press.

Born, G.V.R., and Kratzer, M.A.A., 1984, Source and concentration of extracellular adenosine triphosphate during haemostasis in rats, rabbits and man, *J. Physiol.* **354**:419-429.

Burnstock, G., Cusack, N.J., Hills, J.M., Mackenzie, I., and Meghji, P., 1983, Studies on the stereoselectivity of the P_2-purinoceptor, *Br. J. Pharmacol.* **79**:907-913.

Burnstock, G., Cusack, N.J., and Meldrum, L.A., 1984, Effects of phosphorothioate analogues of ATP, ADP and AMP on guinea-pig taenia coli and urinary bladder, *Br. J. Pharmacol.* **82**:369-374.

Burnstock, G., and Kennedy, C., 1985, Is there a basis for distinguishing two types of P_2-purinoceptor? *Gen. Pharmacol.* **17**:433-440.

Cockcroft, S., and Gomperts, B.D., 1980, The ATP^{4-} receptor of rat mast cells, *Biochem. J.* **188**:789-798.

Macfarlane, D.E., and Mills, D.C.B., 1975, The effects of ATP on platelets: evidence against the central role of released ADP in primary aggregation, *Blood* **46**:309-320.

Cusack, N.J., Hickman, M.E., and Born, G.V.R., 1979, Effects of D- and L- enantiomers of adenosine, AMP and ADP and their 2-chloro-and 2-azido- analogues on human platelets, *Proc. Roy. Soc. Lond. B.* **206**:139-144.

Cusack, N.J., and Hourani, S.M.O., 1981a, Partial agonist behaviour of adenosine 5'-0-(2-thiodiphosphate), *Br. J. Pharmacol.* **73**:405-408.

Cusack, N.J., and Hourani, S.M.O., 1981b, Effects of Rp and Sp diastereoisomers of adenosine 5'-0-(1-thiodiphosphate) on human platelets, *Br. J. Pharmacol.* **73**:409-412.

Cusack, N.J., and Hourani, S.M.O., 1982a, Specific but noncompetitive inhibition by 2-alkylthio analogues of adenosine 5'-monophosphate and adenosine 5'-triphosphate of human platelet aggregation by adenosine 5'-diphosphate, *Br. J. Pharmacol.* **75**:297-400.

Cusack, N.J., and Hourani, S.M.O., 1982b, Adenosine 5'-diphosphate antagonists and human platelets: no evidence that aggregation and inhibition of stimulated adenylate cyclase are mediated by different receptors, *Br. J. Pharmacol.* **76**:221-227.

Cusack, N.J., and Hourani, S.M.O., 1982c, Competitive inhibition by adenosine 5'-triphosphate of the actions on human platelets of 2-chloroadenosine 5'-diphosphate, 2-azidoadenosine 5'-diphosphate, and 2-meththioadenosine 5'-diphosphate, *Br. J. Pharmacol.*, **77**:329-333.

Cusack, N.J., and Hourani, S.M.O., 1984, Some pharmacological and biochemical interactions of the enantiomers of adenylyl 5´-(β,γ-methylene)-diphosponate with the guinea-pig urinary bladder, *Br. J. Pharmacol.* **82**:155-159.

Cusack, N.J., Hourani, S.M.O., Loizou, G.D., and Welford, L.A., 1987, Effects of isopolar phosphonate analogues of ATP on guinea-pig taenia coli and urinary bladder, *Br. J. Pharmacol.* **90**:791-795.

Cusack, N.J., Hourani, S.M.O., and Welford, L.A., 1985, Characterisation of ADP receptors, Advances in Expt. Med. Biol. **192**:29-39.

Cusack, N.J., Hourani, S.M.O., and Welford, L.A., 1988c, The role of ectonucleotidases in pharmacological responses to adenine nucleotide analogues. In: *Adenosine and adenine nucleotides: physiology and pharmacology* (ed. Paton, D.M.) London, Taylor & Francis pp 93-100.

Cusack, N.J. and Pettey, C.J., 1988a, Effects of isopolar isosteric phosphonate analogues of adenosine 5'-diphosphate (ADP) on human platelets, In: *Adenosine and adenine nucleotides: physiology and pharmacology* (ed. Paton, D.M.) London, Taylor & Francis pp 287.

Cusack, N.J. and Planker, M., 1979, Relaxation of isolated taenia coli of guinea-pig by enantiomers of 2-azido analogues of adenosine and adenine nucleotides, *Br. J. Pharmacol.* **67**:153-158.

Cusack, N.J., Welford, L.A., and Hourani, S.M.O., 1988b, Studies on the P_2-purinoceptor using adenine nucleotide analogues. In: *Adenosine and adenine nucleotides: physiology and pharmacology* (ed. Paton, D.M.) London, Taylor & Francis pp 73-84.

Hourani, S.M.O., 1986, Desensitization of the guinea-pig urinary bladder by the enantiomers of adenylyl 5'-(β,γ-methylene)diphosphonate and by substance P, *Br. J. Pharmacol.* **82**:161-164.

Hourani, S.M.O., Cusack, N.J., and Welford, L.A., 1985, L-AMP-PCP, an ATP receptor agonist in guinea-pig bladder, is inactive on taenia coli, *Eur. J. Pharmacol.* **108**:197-200.

Hourani, S.M.O., Loizou, G.D., and Cusack, N.J., 1986, Pharmacological effects of L-AMP-PCP on ATP receptors in smooth muscle, *Eur. J. Pharmacol.* **131**:99-103.

Hourani, S.M.O., Welford, L.A., and Cusack, N.J., 1986, 2-MeS-AMP-PCP and human platelets: implications for the role of adenylate cyclase in ADP-induced aggregation?, *Br. J. Pharmacol.* **87**:84P.

Hourani, S.M.O., Welford, L.A., Loizou, G.D., and Cusack, N.J., 1988, Adenosine 5'-(2-fluorodiphosphate) is a selective agonist at P_2-purinoceptors mediating relaxation of smooth muscle, *Eur. J. Pharmacol.* **147**:131-136.

Gordon, J.L., 1986, Extracellular ATP: effects, sources and fate, *Biochem.* **2233**:309-319.

Gough, G.R., Maguire, M.H., and Satchell, D.G., 1973, Three new adenosine triphosphate analogues: synthesis and effects on isolated gut, *J. Med. Chem.* **16**:1188-1190.

Jefferson, J.R., Harmon, J.T., and Jamieson, G.A., 1988, Identification of high affinity (K_d 0.35 μmol/L) and low affinity (K_d 7.9 μmol/L) platelet binding sites for ADP and competition by ADP analogues, *Blood* **71**:110-116.

Lukacsko, P., and Krell, R.D., 1982, Responses of the guinea-pig urinary bladder to purine and pyrimidine nucleotides, *Eur. J. Pharmacol.* **80**:401-406.

Macfarlane, D.E., Srivastava, P.C., and Mills, D.C.B., 1983, 2-Methylthioadenosine [β-^{32}P] diphosphate. An agonist and radioligand for the receptor that inhibits the accumulation of cyclic AMP in intact platelets, *J. Clin. Invest.* **71**:420-428.

McClure, M.O., Kakker, A., Cusack, N.J., and Born, G.V.R., 1988, Evidence for dependence of arterial haemostasis on ADP, *Proc. Roy. Soc. Lond. B.* **234**:255-262.

Satchell, D.E., and Maguire, M.H., 1975, Inhibitory effects of adenine nucleotide analogs on the isolated guinea-pig taenia coli, *J. Pharmacol. Exptl. Ther.* **195**:540-548.

Satchell, D.G., and Maguire, M.H., 1982, Evidence for separate receptors for ATP and adenosine in the guinea-pig taenia coli, *Eur. J. Pharmacol.* **81**:669-672.

Tatham, P.E.R., Cusack, N.J., and Gomperts, B.D., 1988, Characterization of the ATP^{4-} receptor that mediates permeabilization of rat mast cells, *Eur. J. Pharmacol* **147**:13-21.

Welford, L.A., Cusack. N.J., and Hourani, S.M.O., 1986, ATP analogues and the guinea-pig taenia coli: a comparison of the structure-activity relationships of ectonucleotidases with those of the P_2-purinoceptor, *Eur. J. Pharmacol.* **129**:217-224.

Welford, L.A., Cusack, N.J., and Hourani, S.M.O., 1987, The structure-activity relationships of ectonucleotidases and of excitatory P_2 purinoceptors: evidence that the dephosphorylation of ATP analogues reduces pharmacological potency, *Br. J. Pharmacol.* **141**:123:130.

38

Modulation of Norepinephrine Release by ATP and Adenosine

D.P. Westfall, K. Shinozuka, and R.A. Bjur

Introduction

The evidence is substantial that adenyl purines such as ATP and adenosine can reduce the release of norepinephrine from adrenergic nerves (Paton, 1979; Fredholm and Hedqvist, 1980; Su, 1983). The general notion has been that the effects of adenine nucleosides and nucleotides are mediated by prejunctional P_1-purinoceptors (Burnstock, 1985). Because nucleotides are generally found to be poor P_1-receptor agonists, the prejunctional actions of ATP and other nucleotides are suggested to occur because of metabolism to adenosine which then acts via P_1-receptors.

If the effect of ATP were due to its metabolism to adenosine, one might expect ATP or other nucleotides to be less potent than adenosine since the metabolism is not likely to be instantaneous nor complete. We found it curious therefore that in those situations where the effects of both adenosine and ATP on the release of ^{3}H-norepinephrine have been compared, the agents are equipotent (Clanachan et al., 1977; Enero and Saidman, 1977; Verhaeghe et al., 1977; Moylan and Westfall, 1979; Khan and Malik, 1980). This suggested to us that perhaps both P_2-receptors and P_1-receptors mediate an inhibition of norepinephrine release or that the prejunctional purinoceptor is not a pure P_1- or P_2-receptor but rather possesses some characteristics of both, i.e., a hybrid purinoceptor. These possibilities were tested for by examining the effects of a number of putative P_1- and P_2-receptor agents on release of endogenous norepinephrine from the electrically stimulated rat caudal artery. We review here previous findings from our laboratory (Shinozuka et al., 1988) and discuss additional data that indicates that the prejunctional purinoceptors are distinct from known P_1- and P_2-receptors.

Methods

Experiments were conducted with isolated caudal arteries from male Fisher-344 rats. Arteries were placed in 3.5 ml organ baths at 37°C in a modified Krebs solution. After an equilibration period of 1 hr the tissues were subjected to electrical field stimulation (EFS; 0.5 msec pulses at 1 Hz for 3 min) via platinum electrodes. The EFS was repeated 4 times with 30 min between stimulations. Before and after each stimulation the bathing solution was removed and analyzed for norepinephrine content by high performance liquid chromotography with electrochemical detection. Additional aspects of the methods are described in detail by Shinozuka et al. (1988).

Results and Discussion

As shown by Shinozuka et al. (1988) both P_1-receptor agonists and P_2-receptor agonists reduce the nerve-mediated release of norepinephrine; the relative order of potency being 2-chloroadenosine > beta, gamma methylene ATP > ATP $\geq$ adenosine. This order of potency seems incompatible with the idea that the nucleotides need to be metabolized to adenosine to inhibit release, that is, ATP and adenosine were roughly equipotent and beta, gamma methylene ATP, a compound that is less rapidly hydrolyzed to adenosine than is ATP (Moody and Burnstock, 1982), was more potent than ATP.

Another line of reasoning that suggests that nucleotides do not need to be metabolized to adenosine in order to exert prejunctional effects derives from studies with S-p-nitrobenzl-6-thioguanosine (NBTG), an agent that inhibits adenosine uptake (Kenakin and Leighton, 1985). NBTG potentiates the effects of adenosine but not those of the nucleotides. This is shown in figure 1 where it can be seen that the response to a submaximal concentration of beta, gamma methylene ATP is not enhanced by NBTG. Figure 1 also demonstrates the concentration-response nature of the inhibition of norepinephrine release by beta, gamma methylene ATP and its antagonism by 8-(p-sulfophenyl) theophylline (8 SPT), a known P_1-receptor antagonist.

A curious feature of our findings is that the inhibition of norepinephrine release by both P_1-receptor agonists, such as adenosine and 2-chloroadenosine, and P_2-receptor agonists, such as ATP and beta, gamma methylene ATP, is antagonized by

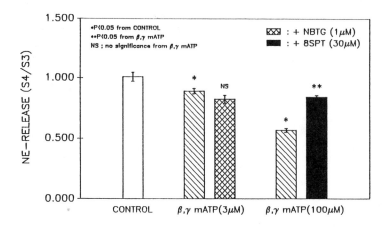

FIGURE 1. Effects of NBTG and 8SPT on the inhibition of norepinephrine release by β,γ—methylene ATP

P_1-receptor antagonists. We sought evidence to determine if the reverse was true, that is, do P_2-receptor antagonists (or receptor desentizing agents) also reduce the response to both P_1- and P_2-agonists. Shown in table 1 are results with alpha, beta methylene ATP, an agent known to desensitize P_2-receptors. Alpha, beta methylene ATP did not affect the release of norepinephrine evoked by 1 Hz EFS. However, pretreatment with this agent effectively reversed the ability of 2-chloroadenosine and beta, gamma methylene ATP to reduce the release of norepinephrine.

TABLE 1.   Effect of α,β—methylene ATP on the inhibition of norepinephrine release by 2—chloroadenosine and β,γ—methylene ATP in rat caudal artery

| Agonist (μM) | Pretreatment (15 min) | |
	None	α,β—methylene ATP (30 μM)
Control	1.038 ± 0.071	1.022 ± 0.103
β,γ—methylene ATP (10)	0.685 ± 0.036 [*]	0.965 ± 0.014 [**]
2—chloroadenosine (10)	0.605 ± 0.061 [*]	1.171 ± 0.161 [**]

[*]   significant difference from control, $P<0.05$.

[**] significant difference from each none treatment, $P<0.05$.

All values represent the norepinephrine release as S4/S3 ratio.
(mean and the standard error of 4—9 experiments)

Some of the features of the prejunctional purinoceptors, as defined by agonists and antagonists, are summarized in table 2. Shown is the order of potency of agonists and the idea that both P_1- and P_2-antagonists antagonize responses. Also indicated is that blockade of adenosine uptake with NBTG, while potentiating the response to adenosine, does not potentiate the response to nucleotides. Taken together, these results are not easily reconciled with the notion that prejunctional purinoceptors are either P_1 or P_2. Thus we propose that the prejunctional purinoceptors may represent a distinct class, which we have termed P_3-purinoceptors (Shinozuka et al., 1988).

TABLE 2. Summary of the actions of adenyl purines on the release of norepinephrine from electrically stimulated rat caudal artery

PURINES	ACTION	EC $_{50}$	8SPT	α,β mATP	NBTG
P1–AGONISTS					
2 CA	Inhibition	0.29	Antagonized	Antagonized	Unaffected
Adenosine	Inhibition	10.86	Antagonized	not examine	Potentiated
P2–AGONISTS					
β,γ–mATP	Inhibition	5.05	Antagonized	Antagonized	Unaffected
ATP	Inhibition	9.44	Antagonized	not examine	Unaffected

2 CA   : 2–Chloroadenosine , β,γ mATP : β,γ–methylene ATP
8–SPT : 8–(p–sulfophenyl)theophylline ; P1–antagonist
α,β mATP : α,β–methylene ATP ; P2–antagonist
NBTG : S–p–nitrobenzyl–6–thioguanosine ; adenosine uptake inhibitor

Features of the three classes of purinoceptors are summarized in table 3. Whereas the P_1-receptor can be viewed primarily as a nucleoside receptor and the P_2-receptor as a nucleotide receptor, the P_3 receptor recognizes the structure of both nucleosides and nucleotides. Indeed adenosine and ATP seem to be nearly equipotent at the P_3-receptor. Agents such as 8SPT which antagonize responses mediated by P_1-receptors, but not P_2-receptors, also antagonize P_3-mediated responses. Alpha, beta methylene ATP, an agent known to cause desentization of P_2-, but not P_1-receptors, and thereby antagonize P_2-receptor mediated responses, also antagonizes P_3-receptor mediated actions. An interesting feature of the P_3-receptor, in distinction from the P_2-receptor, is that alpha, beta methylene ATP does not exhibit agonistic properties prior to blocking responses. Another curious feature of the prejunctional receptors is their sensitivity to light. Shinozuka et al. (1988) have shown that exposure of the rat tail artery to light antagonized the actions of purinoceptor agonists. This is in distinction to the light-evoked activation of P_2-purinoceptors as reported by Burnstock and Wong (1978).

Table 3.   CHARACTERISTICS OF PURINOCEPTORS

	P_1	P_2	P_3
Agonists	Nucleosides	Nucleotides	Both nucleosides and nucleotides
Potency	ADO>>>ATP	ATP>>>ADO	ATP $\approx$ ADO
Influence of block of adenosine uptake on nucleotide potency	Potentiate	No Effect	No Effect
P_1 antagonism (8-SPT)	Antagonism	No Effect	Antagonism
P_2 antagonism (APCPP)	No Effect	Antagonism	Antagonism
Light	No Effect	Agonism	Antagonism
Subtypes	A_1/A_2	P_{2X}/P_{2Y}	–

Summary

Studies with known purinoceptor agonists and antagonists reveal that prejunctional purinoceptors on adrenergic nerves exhibit a number of characteristics that distinguish them from well defined P_1- and P_2-receptors. These receptors should probably be thought of as representing a separate class of receptors; the P_3-purinoceptors.

Acknowledgement

Work cited from the authors' laboratory was supported by a grant from the American Heart Association and NIH grant HL 38126.

References

Burnstock, G., 1985, Nervous control of smooth muscle by transmitters, cotransmitters and neuromodulators, Experientia 41: 869-874.

Burnstock, G., and Wong, H., 1978, Comparison of the effects of ultraviolet light and purinergic nerve stimulation on guinea-pig taenia coli, Brit. J. Pharmacol. **62**: 293-302.

Enero, M.A., and Saidman, B.Q., 1977, Possible feedback inhibition of noradrenaline release by purine compounds, Naunyn-Schmiedeberg's Arch. Pharmacol. **297**: 39-46.

Fredholm, B.B., and Hedqvist, P., 1980, Modulation of neurotransmission by purine nucleotides and nucleosides, Biochem. Pharmacol. **29**: 1635-1643.

Kenakin, T.P., and Leighton, H.J., 1985, The pharmacologic estimation of potencies of agonists and antagonists in the classification of adenosine receptors, In: Paton, D.M. (ed) Methods in pharmacology, vol. 6, Methods used in adenosine research, Plenum Press, New York, pp 213-237.

Khan, M.T., and Malik, K.U., 1980, Inhibitory effect of adenosine and adenine nucleotides on potassium-evoked efflux of [^{3}H]-noradrenaline from the rat isolated heart: lack of relationship to prostaglandins, J. Pharmacol. Exp. Ther. **68**: 551-561.

Moody, C.J., and Burnstock, G., 1982, Evidence for the presence of P_1-purinoceptors on cholinergic nerve terminals in the guinea-pig ileum, Eur. J. Pharmacol. **77**: 1-9.

Moylan, R.D., and Westfall, T.C., 1979, Effect of adenosine on adrenergic neurotransmission in the superfused rat portal vein, Blood Vessels **10**: 302-310.

Paton, D.M., 1979, Presynaptic inhibition of adrenergic neurotransmission by adenine nucleotides and adenosine, In: Baer, H., and Drummond, G.I. (eds) Physiological and regulatory functions of adenosine and adenine nucleotides, Raven Press, New York, pp 69-77.

Shinozuka, K., Bjur, R.A., and Westfall, D.P., 1988, Characterization of prejunctional purinoceptors on adrenergic nerves of the rat caudal artery, Naunyn-Schmiedeberg's Arch. Pharmacol. **338**: 221-227.

Su, C., 1983, Purinergic neurotransmission and neuromodulation, Pharmacol. Rev. **23**: 397-411.

Verhaeghe, R.H., Vanhoutte, P.M., and Shepherd, J.T., 1977, Inhibition of sympathetic neurotransmission in canine blood vessels by adenosine and adenine nucleotides, Circ. Res. **40**: 208-215.

39

The Phospholipase C Linked P_{2y}-Purinergic Receptor

J.L. Boyer, C.L. Cooper, M.W. Martin, G.L. Waldo, A.J. Morris, R.A. Jeffs, H.A. Brown, and T.K. Harden

Introduction

The physiological effects of extracellular ATP and ADP have been broadly described for both the central nervous system and peripheral tissues (Burnstock, 1978; Burnstock and Kennedy, 1986; Gordon, 1986; Williams, 1987; Fleetwood and Gordon, 1987). In contrast to the clearly defined subtypes of P_1-purinergic receptors (A_1- and A_2-receptors) responsible for the physiological effects of extracellular adenosine (Londos and Wolff, 1977; Van Calker, et al., 1979; Stiles, 1986). Subclassification of the ATP- and ADP-regulated P_2-purinergic receptors has proven more difficult. No good antagonists of P_2-purinergic receptors are available, and cell surface hydrolases metabolize ATP and ADP to adenosine, which could produce variable tissue-dependent effects through P_1-purinergic receptors. In spite of these problems some progress in subclassification of P_2-purinergic receptors has been made. The observation of differential effects of a large number of ATP and ADP analogs led Burnstock and Kennedy (1985) to propose that two subtypes of P_2-purinergic receptors exist. P_{2x}-purinergic receptors were suggested to exhibit the potency order of $Ap(CH_2)pp$ > $App(CH_2)p$ > ATP > 2-methylthio ATP (2MeSATP) and P_{2y}-purinergic receptors were suggested to exhibit the potency order of 2MeSATP >> ATP > $Ap(CH_2)pp$ = $App(CH_2)p$. Whether these two putative receptor subtypes account for all of the effects of ADP and ATP is an open question. Indeed, data obtained with several cultured cell lines suggest a third subtype of receptor of the P_2-purinergic receptor class (Dubyak and De Young, 1985).

In comparison to receptors for adenosine, much less also is known about the biochemical sequelae of P_2-purinergic receptor activation. However, in several tissues ATP, ADP, and their analogs have been shown to stimulate what apparently is a phospholipase C-catalyzed breakdown of phosphatidylinositol 4,5-bisphosphate ($PtdIns(4,5)P_2$) to inositol 1,4,5-trisphosphate and a consequential increase in intracellular Ca^{2+}. It has been proposed that this effect is through a P_{2y}-purinergic receptor (Charest et al., 1985; Okajima et al., 1987; Pirotton et al., 1987; Boyer et al., 1989a).

Guanine nucleotide regulatory proteins (G-proteins) are known to be obligatory coupling entities in receptor-mediated transmembrane signalling involving a number of effector proteins, e.g. adenylate cyclase, retinal cyclic GMP phosphodiesterase and ion channels (Gilman, 1987). It is now clear that a member(s) of this class of proteins couples hormone and neurotransmitter receptors to phospholipase C (Boyer et al., 1989b). We have found that turkey erythrocyte membranes are very useful for study of G-protein-regulated phospholipase C. Moreover, this signalling system in turkey erythrocytes is under regulation by a P_{2Y}-

purinergic receptor, which makes the turkey erythrocyte membrane an excellent model to study molecular mechanisms involved in P_{2Y}-purinergic receptor action.

P_{2Y}-Purinergic Receptor Regulation of Phospholipase C

Turkey erythrocytes possess a phosphatidylinositol synthase, and thus, [^{3}H]-labelled inositol can be used to label turkey erythrocyte phosphoinositides to very high specific activity. Membranes prepared from [^{3}H]inositol-labelled erythrocytes exhibit little or no basal release of inositol phosphates, but respond to activators of G-proteins, e.g. GTPγS or AlF$_4^-$, with large increases in phospholipase C activity (Harden et al., 1987; Harden et al., 1988). The effects of ATP and analogs of ATP and ADP on phospholipase C activity can be most unambiguously defined under conditions of short-time assays, e.g. < 5 minutes, where PtdIns(4,5)P$_2$ substrate concentrations are not limiting and in the presence of a guanine nucleotide, e.g. GTP or low (< 1 μM) concentrations of GTPγS, that by itself causes little or no stimulation of the enzyme. In this situation, ATP and ADP analogs cause a marked concentration dependent stimulation of enzyme activity with an order of potency of 2MeSATP > ADPβS > ATPγS > ATP > App(NH)p = ADP > Ap(CH$_2$)pp > App(CH$_2$)p. Adenosine and adenosine analogs have no effect on phospholipase C activity (Boyer et al., 1989a). This pharmacological profile is consistent with the agonist receptor specificity described by Burnstock and Kennedy (1985) for a P_{2Y}-purinergic receptor.

P_{2Y}-purinergic receptor-mediated stimulation of the turkey erythrocyte phospholipase C is absolutely dependent on the presence of guanine nucleotides. The maximal effect of agonists on enzyme activity in the presence of GTP is approximately 15 % of that observed with agonists in the presence of a maximally effective concentration of GTPγS. In the absence of a P_{2Y}-purinergic receptor agonist, GTPγS activates the turkey erythrocyte phospholipase C with a $K_{0.5}$ value of approximately 3 μM. Addition of a P_{2Y}-purinergic receptor agonist produces a concentration dependent increase in the effect of a maximally effective concentration of GTPγS, and the GTPγS activation curve is shifted approximately 50-fold to the left in the presence of a maximally effective concentration of agonist.

Inositol phosphate production in the presence of GTPγS occurs with a time course exhibiting a considerable time lag. This lag is decreased significantly by 2MeSATP. Activation by GTPγS follows first order kinetics and the rate of activation is markedly increased in a concentration dependent and saturable manner by P_{2Y}-purinergic receptor agonists. The rate of activation of the enzyme in the presence of a fixed concentration of agonist is independent of the concentration of guanine nucleotide (Boyer et al., 1989a).

GDPβS competitively blocks guanine nucleotide stimulation of the turkey erythrocyte phospholipase C. By adding high concentrations of GDPβS to turkey erythrocyte membranes previously activated by agonist and GTP, Gpp(NH)p, or GTPγS, the rate of inactivation of the activated enzymic species can be determined. The GTP-preactivated enzyme inactivates very rapidly, whereas the enzyme preactivated in the presence of the hydrolysis resistant analogs is resistant to the inactivating effect of GDPβS (Boyer et al., 1989a).

These properties of the turkey erythrocyte phospholipase C are very similar to those of receptor-regulated adenylate cyclase (Ross et al., 1977; Tolkovsky and Levitzki, 1978), and suggest that P_{2Y}-receptor activation promotes guanine nucleotide exchange on the involved G-protein resulting in transformation of the G-protein into its active GTP- (or GTPγS-) liganded state. As with adenylate cyclase, the life time of the GTP-liganded activated state is apparently determined by the activity of a GTPase, which hydrolyzes GTP returning the G-protein activated catalytic species to a state of inactive catalyst and GDP-liganded G-protein .

The Phospholipase C-Associated G-protein

Little progress has been made in the identification of the G-protein involved in activation of phospholipase C. Indirect evidence obtained from studies of activation/deactivation kinetics described above suggest that the phospholipase C-associated G-protein shares functional homology with other members of the G-protein family. Pertussis toxin impairs receptor-mediated activation of phospholipase C in some, but not all (including the turkey erythrocyte) tissues (Martin, 1989). Thus, heterogeneity in the phospholipase C-associated G-protein may exist, but whether this apparent diversity is receptor- or tissue-specific is not yet known. Attempts to modify turkey erythrocyte membrane phospholipase C by the reconstitution of unactivated or GTPγS-activated G-proteins, or by partially purified GTPγS binding fractions from turkey erythrocyte membranes have been unsuccessful.

Based on analogy with other G-protein-involved signal transduction mechanisms, it has been assumed but not proven that a G-protein with a heterotrimeric structure is involved in receptor-mediated activation of phospholipase C. Recently, Moriarty et al. (1989), reported that the injection of $\beta\gamma$ subunits into *Xenopus* oocytes inhibits a receptor-activated Cl^- current, which apparently occurs secundary to an activation of phospholipase C. Additionally, we have found that $\beta\gamma$ subunits purified from several tissues, inhibit AlF_4^--stimulated adenylate cyclase and phospholipase C activities when reconstituted into turkey erythrocyte "acceptor" membranes (Boyer et al., 1989c). In contrast, reconstitution of $\beta\gamma$ subunits results in potentiation of the stimulation of phospholipase C by a P_{2Y}-receptor agonist + GTP. These results can be explained according to models proposed for the actions of G-proteins in signal transduction. That is, reconstitution of $\beta\gamma$ subunits favors by mass action the formation of the inactive heterotrimeric state of the AlF_4^--activated G-protein in the adenylate cyclase and phospholipase C systems. In addition, $\beta\gamma$ subunits could promote the formation of heterotrimeric complexes, that in the presence of a receptor agonist and GTP were capable of supporting phospholipase C activation. Irrespective of whether these explanations are correct or not, the findings suggest the occurrence of association/dissociation of a heterotrimeric G-protein in the catalytic cycle of phospholipase C (Boyer et al.,1989c).

The Phospholipase C-Associated P_{2Y}-Purinergic Receptor

We have developed methodology to directly label the P_{2Y}-purinergic receptor (Cooper et al., 1989). [^{35}S]ADPβS binds to purified turkey erythrocyte membranes with a B_{max} of 3 pmol/mg protein and K_d of 8 nM. The binding of [^{35}S]ADPβS is inhibited by ADP and ATP analogs in a concentration-dependent manner, the K_i values for these analogs are consistent with those predicted for a P_{2Y}-purinergic receptor, and these K_i values are essentially identical to the $K_{0.5}$ values of the same compounds for stimulation of turkey erythrocyte phospholipase C. Thus, [^{35}S]ADPβS apparently binds to the same binding site as that involved in receptor-mediated activation of the phospholipase C. [^{35}S]ADPβS binding is inhibited in a noncompetitive manner by guanine nucleotides with an order of potency (GTPγS > Gpp(NH)p > GTP = GDP > GDPβS >> GMP) consistent with that for a G-protein-linked receptor. As with other G-protein-linked receptors, these effects of guanine nucleotides on [^{35}S]ADPβS binding are likely a consequence of P_{2Y}-

purinergic receptor agonist-induced association of the receptor and the phospholipase C-linked G-protein.

An analog of ATP, 3'-O-(4-benzoyl)benzoyl ATP (BzATP) also has been used as a photoaffinity ligand for the P_{2Y}-purinergic receptor of turkey erythrocytes. BzATP is a full agonist for the stimulation of phospholipase C, and its activity is strictly dependent on the presence of guanine nucleotides. Photolysis of [^{3}H]inositol-labelled membranes in the presence of BzATP, produced an irreversible activation of the P_{2Y}-purinergic receptor that was also strictly dependent on the presence of guanine nucleotides (Boyer, and Harden, 1989). We currently are utilizing [^{32}P]BzATP as a photoaffinity probe with the goal of obtaining direct information on the P_{2Y}-purinergic receptor protein.

Desensitization of P_{2Y}-Purinergic Receptors

The availability of the turkey erythrocyte membrane as a cell-free preparation that retains high responsivity to receptor stimulation has provided an opportunity to begin investigation of molecular aspects of the regulation of P_{2Y}-purinergic receptors. For example, preincubation of turkey erythrocytes with ADPβS results in a marked decrease in the capacity of ADPβS plus GTP to stimulate phospholipase C activity in membranes derived from these cells (Martin and Harden, 1989). The half-time of occurrence of desensitization is 0.5-2.0 minutes and responsiveness reaches a new quasi-steady state level of 40-50 percent of control within 10 minutes of incubation of cells with agonist. Desensitization reverses in agonist-free medium with recovery of agonist plus GTP responsiveness of the membrane phospholipase C occurring with a half time of 10-20 minutes. The loss of response occurs as a decrease in maximal response with no change in the apparent affinity of P_{2Y}-purinergic receptor agonists. Furthermore, desensitization is apparently a P_{2Y}-purinergic receptor-mediated event since 2MeSATP and ADPβS are potent inducers of desensitization, whereas App(CH$_2$)p has no effect. Desensitization occurs with no change in either the rate of activation or the final phospholipase C activity attained in the presence of GTPγS; AlF$_4^-$-stimulated inositol phosphate formation also is not modified in membranes from desensitized cells. In contrast, there is a marked decrease in the capacity of ADPβS to increase the rate of activation of phospholipase C by GTPγS in membranes from agonist-preincubated cells.

The data obtained with membranes from agonist-preincubated turkey erythrocytes are highly analogous to that reported in detail for agonist-induced desensitization of the receptor-regulated adenylate cyclase system (Harden, 1983; Clark, 1986). We propose that incubation of turkey erythrocytes with a P_{2Y}-purinergic receptor agonist results in activation of phospholipase C by a process involving acceleration of the exchange of GTP for GDP on the involved G-protein. With a somewhat slower, but nonetheless relatively rapid time course, a covalent modification of the receptor-G-protein-phospholipase C ensues resulting in a lesion that is stable to cell lysis and membrane incubation at 30 °C. The location of this lesion has not yet been established. However, it probably does not involve the phospholipase C per se or the part of the G-protein involved in communication between the G-protein and the phospholipase C. The signalling system is apparently modified at the level of receptor-G-protein coupling, which could be explained by a modification of the receptor per se or of that part of the G-protein involved in G-protein-receptor interaction. As with the adenylate cyclase system, phosphorylation of a component of the phosphoinositide signalling pathway could be responsible for desensitization. Indeed, preincubation of turkey erythrocytes with phorbol ester activators of protein kinase C decreases responsiveness to P_{2Y}-purinergic receptor agonists (M.W. Martin, unpublished observations). However, it is not yet clear whether these results presage a feed-back regulation by protein kinase C during receptor activation since long-term incubation of turkey erythrocytes with phorbol esters apparently down-regulates protein kinase C, but does not modify agonist-induced desensitization.

Summary and Future Directions

In summary, progress in understanding P_2-purinergic receptors has been slow as a consequence of a number of factors, not the least of which is the lack of availability of model systems suitable for the study of the biochemistry of a pharmacologically-defined receptor and its associated second messenger signalling system. The P_{2Y}-purinergic receptor and its associated G-protein and phospholipase C on turkey erythrocyte membranes offers some relatively unique opportunities for the study of P_2-purinergic receptor mechanisms. In addition to more general questions concerning the P_{2Y}-purinergic receptor-regulated phospholipase C signalling system, we are beginning to study directly the individual components of this system. Not only can $[^{35}S]ADP\beta S$ be utilized to label the membrane-associated P_{2Y}-purinergic receptor, but the receptor can be followed using radioligand subsequent to solubilization of membranes with a nonionic detergent. The receptor is apparently solubilized in association with its G-protein, which perhaps will offer possibilities in G-protein identification. Furthermore, a covalently binding radioligand has been developed, and preliminary data suggest that this probe labels a $\approx$ 53,000 dalton protein that expresses pharmacological properties of the P_{2Y}-purinergic receptor. Thus, there is promise that the P_{2Y}-purinergic receptor protein can eventually be purified in sufficient quantities to begin structure/function studies. Finally, we recently have purified a phospholipase C from turkey erythrocyte, and have shown by reconstitution that this is a G-protein and P_{2Y}-purinergic receptor-regulated protein. A long-term goal of the laboratory is to reconstruct a P_{2Y}-purinergic receptor-regulated phospholipase C from purified components in a model vesicle system.

References

Boyer, J.L., Downes, C.P., and Harden, T.K., 1989a, Kinetics of activation of phospholipase C by P_{2Y}-purinergic receptor agonists and guanine nucleotides, J. Biol. Chem. 264:884-890.

Boyer, J.L., Hepler, J.R., and Harden T.K., 1989b, Hormone and growth factor receptor-mediated regulation of phospholipase C, Trends Pharmacol. Sci. 10:360-364.

Boyer, J.L., Waldo, G.L., Evans, T., Northup, J.K., Downes, C.P., and Harden, T.K., 1989c, Modification of AlF$_4^-$- and receptor-stimulated phospholipase C by G-protein $\beta\gamma$ subunits, J. Biol. Chem. 264:13917-13922.

Boyer, J.L., and Harden, T.K., 1989, Irreversible activation of phospholipase C-coupled P_{2Y}-purinergic receptors by 3'-O-(4-benzoyl)benzoyl ATP, Mol. Pharmacol, In press.

Burnstock, G., 1978, A basis for distinguishing two types of purinergic receptors. In: Straub RW, Bolis L (eds) Cell membrane receptors for drugs and hormones: A multidisciplinary approach, Raven Press,New York, p 107-118.

Burnstock, G., and Kennedy, C., 1985, Is there a basis for distinguishing two types of P_2 purinoceptors?, Gen. Pharmacol. 16:433-440.

Burnstock, G., and Kennedy, C., 1986, A dual function of adenosine 5'-triphosphate in the regulation of vascular tone. Excitatory cotransmitter with noradrenaline from perivascular nerves and locally released inhibitory intravascular agent, Circ. Res. 58:319-330.

Charest, C., Blackmore, P.F., and Exton, J.H., 1985, Characterization of responses of isolated rat hepatocytes to ATP and ADP, J. Biol. Chem. **260**:15789-15794.

Clark, R.B., 1986, Desensitization of hormonal stimuli coupled to regulation of cyclic AMP levels, Adv. Cyclic Nucleotide Protein Phosphorylation Res. **20**:151-209.

Cooper, C.L., Morris, A.J., and Harden, T.K., 1989, Guanine nucleotide-sensitive interaction of of a radiolabeled agonist with a phospholipase C-linked P_{2Y}-purinergic receptor, J. Biol. Chem. **264**:6202-6206.

Dubyak, G.R., and De Young, M.B., 1985, Intracellular Ca^{2+} mobilization activated by extracellular ATP in erlich ascites tumor cells, J. Biol. Chem. **260**:10653-10661.

Fleetwood, G., and Gordon, J.L., 1987, Purinoceptors in the rat heart, Br. J. Pharmacol. **90**:219-227.

Gilman, A.L., 1987, G-proteins: Transducers of receptor-generated signals, Ann. Rev. Biochem. **56**:615-649.

Gordon, J.L., 1986, Extracellular ATP: Effects, sources and fate, Biochem. J. **233**:309-319.

Harden, T.K., 1983, Agonist-induced desensitization of the β-adrenergic receptor-linked adenylate cyclase, Pharmacol. Rev. **35**:5-32.

Harden, T.K., Stephens, L., Hawkins, P.T., and Downes, C.P., 1987, Turkey erythrocyte membranes as a model for regulation of phospholipase C by guanine nucleotides, J. Biol. Chem. **262**:9057-9061.

Harden, T.K., Hawkins, P.T., Stephens, L., Boyer, J.L., and Downes, C.P., 1988, Phosphoinositide hydrolysis by guanosine 5'-[γ-thio]triphosphate-activated phospholipase C of turkey erythrocyte membranes, Biochem. J. **252**:583-593.

Londos, C., and Wolff, J., 1977, Two distinct adenosine-sensitive sites on adenylate cyclase, Proc. Natl. Acad. Sci. U.S.A. **74**:5482-5486.

Martin, M.W., and Harden, T.K., 1989, Agonist-induced desensitization of a P_{2Y}-purinergic receptor- and guanine nucleotide-regulated phospholipase C, J. Biol. Chem. In press.

Martin, T.F.J., 1989, Lipid hydrolysis by phosphoinositidase C: Enzymology and regulation by receptors and guanine nucleotides, In: Michell, R.H., Drummond, A.H., Downes, C.P., (Eds) Inositol lipids in cell signalling, Academic Press, London, p 81-112.

Moriarty, T.M., Gillo, B., Carty, D.J., Premont, R.T., Landau, E.M., and Iyengar, R., 1988, βγ subunits of GTP-binding proteins inhibit muscarinic receptor stimulation of phospholipase C, Proc. Natl. Acad. Sci. U.S.A. **85**:8865-8869.

Okajima, F., Tokumitsu, Y., Kondo, Y., and Ui, M., 1987, P_2-purinergic receptors are coupled to two signal transduction systems leading to inhibition of cAMP generation and to production of inositol phosphates in rat hepatocytes, J. Biol. Chem. **262**:13483-13490.

Pirotton, S., Raspe, E., Demolle, D., Erneux, C., and Boeynaems, J.M. 1987, Involvement of inositol 1,4,5 trisphosphate in the action of adenine nucleotides on aortic endothelial cells, J. Biol. Chem. **262**:17461-17466.

Ross, E.M., Maguire, M.E., Sturgill, T.W., Biltonen, R.L., and Gilman, A.G. 1977, Relationship between the β-adrenergic receptor and adenylate cyclase. Studies of ligand binding and enzyme activity in purified membranes of S49 lymphoma cells, J. Biol. Chem. **252**:5761-5775.

Stiles, G.L., 1986, Adenosine receptors: Structure, function and regulation, Trends Pharmacol. Sci. **7**:486-490.

Tolkovsky, A.M., and Levitzki, A., 1978, Mode of coupling between the β-adrenergic receptor and adenylate cyclase in turkey erythrocytes, Biochemistry. **18**:3795-3810.

Van Calker, D., Muller, M., and Hamprecht, B., 1979, Adenosine regulates via two different types of receptors the accumulation of cyclic AMP in cultured brain cells, J. Neurochem. **33**:999-1005.

Williams, M., 1987, Purine receptors in mammalian tissues: Pharmacological and functional significance, Ann. Rev. Pharmacol. Toxicol. **27**:315-345.

40
Roles of an ADP Receptor on the Platelet Surface in Mediating Aggregation and Fibrinogen Binding

R.W. Colman

ADP was the first discovered agonist for platelets and remains one of the most important for physiologic hemostasis and pathologic thrombosis. ADP plays two major roles in platelet activation. ADP at low concentrations (0.1 - 0.5 μM) stimulates human platelets to change shape from discs to spiculated spheres. Platelet aggregation and granule secretion occur at higher concentrations (2-5 μM) (Born, 1972). ADP also antagonizes the stimulation of adenylate cyclase by PGE_1 or PGI_2 (Haslam, 1973). Elevation of cAMP leads to inhibition of the activation of platelets by most agonists. By inhibiting stimulated adenylate cyclase ADP reduces cyclic AMP, but this event itself does not activate platelets. The effects of ADP on shape change and adenylate cyclase are mediated by different receptors (Mills, et al, 1985). Because of the occurrence of ecto-ADPases on the platelet surface, study of the binding of ADP has yielded minimum information.

A second approach which has been more productive is to examine the effect of changes in the ligand. Among purinergic receptors the platelet stands alone in preferring ADP over ATP and has been classified as P_{2T}. The ADP receptor for platelet shape change and aggregation displays specificity that permits only a few substitutions. On the purine ring, only the 2-position can be modified. The substitution of a chloro or a methylthiol group at this site results in more active agonists (Maguire and Michal, 1968). In contrast, substitutions at the 6 or 8 position of the purine ring or in the ribose ring result in a decrease in activity (Grant and Scrutton, 1980). The pyrophosphate in the 5'-position is required, but ATP serves as a competitive agonist (McFarlane and Mills, 1975). As discussed below, the receptor for aggregation and shape change has been identified as aggregin, but the distinct receptor linked to adenylate cyclase has not yet been defined except functionally.

Over the past decade, we have used the affinity label 5'-p-fluorosulfonylbenzoyl adenosine (FSBA) to label an ADP receptor protein, which we designate aggregin, which is responsible for shape change in intact platelets (Bennett, et al, 1978). FSBA contains an adenosine group linked to an extended side chain through the 5' position of the ribose which allows formation of a reversible ligand-protein complex at an adenine nucleotide binding site (Colman, et al, 1977). Sulfonyl fluoride, the reactive group in the para position, has been shown to react covalently with several nucleophilic amino acids including: cysteine, tyrosine, lysine and histidine with the displacement of fluoride ion (Colman, 1983).

When intact platelets are incubated with FSBA, only aggregin is labeled. FSBA inhibits ADP mediated platelet shape change from a disc to a spiny sphere as measured by increased absorbance (Bennett, et al, 1978; Bennett, et al, 1981), and scanning electron microscopy (Figures, et al, 1987). Intact platelets incubated with ^{3}H-FSBA show concentration and time dependent covalent incorporation of radioactivity into platelets after extensive dialysis. The extent of inhibition of shape change is closely related to the covalent incorporation of the nucleotide analog (Figures, et al, 1981). When the platelet membranes prepared from the washed platelets exposed to ^{3}H-FSBA were subjected to SDS electrophoresis, a single radioactive peak corresponding to aggregin (M_r = 100,000) was found on the gels (Bennett, et al, 1978; Bennett, et al, 1981; Figures, et al, 1981). The presence of ADP or ATP but not adenosine or GDP completely prevented the labeling.

Little fibrinogen binds to washed intact platelets. However, on exposure to ADP, ^{125}I-fibrinogen binds to platelets proportional to the concentration of fibrinogen at low concentrations (less than 100 µg/ml) (Bennett, et al, 1979; Mustard, et al, 1978; Marguerie, et al, 1979). Incorporation of FSBA into a single membrane protein, aggregin, is associated with inhibition of shape change, aggregation and fibrinogen binding.

Superficially, the external platelet membrane protein labeled by FSBA bears certain similarities to GPIIIa. The molecular weights of the two proteins are indistinguishable on non-reduced gels. Furthermore, platelets of patients with thrombasthenia (which are deficient in GPIIIa), fail to aggregate or bind ^{125}I-fibrinogen upon ADP stimulation (Kornecki, et al, 1981), resembling FSBA-treated platelets. The fact that individuals with thrombasthenia and deficiency of glycoproteins IIb-IIIa display normal ADP-induced shape change (Peerschke, 1985) is against the above proposition. Despite the finding that the thrombasthenic platelets from the two patients evaluated contained less than 5% GPIIIa, we found that ^{3}H-SBA incorporation into normal and thrombasthenic platelets were similar both qualitatively and quantitatively. The dose-dependent inhibitory effect of FSBA on the ADP-induced platelet shape change was similar in experiments with normal and thrombasthenic platelets. Murine monoclonal antibodies directed against GPIIb/GPIIIa complex (A2A6), GPIIIa (SSA6), a monospecific rabbit polyclonal antibody to immunopurified GPIIIa and a human antibody directed against an epitope P1A1 present on glycoprotein IIIa all failed to immunoprecipitate ^{3}H-aggregin while ^{125}I-GPIIIa was immunoprecipitated by the same A2A6 antibodies. Finally, the aggregin and glycoprotein IIIa differ in molecular architecture since, on reduction, the electrophoretic migration of GPIIIa decreased due to the cleavage of intrachain disulfide bonds while that of aggregin remained the same. These results confirm our hypothesis that the two components are distinct proteins (Colman, et al, 1986).

The question of whether the receptor for ADP responsible for platelet activation was the same as that for inhibition of adenylate cyclase has also been explored (Mills, et al, 1985). When various concentrations of FSBA were preincubated for 30 min, no effect on the inhibition by ADP was found. Neither the basal nor the stimulated levels of cAMP were affected by FSBA. 2-methylthio ADP (MeSADP), another analog of ADP, is 75-fold more potent as an inhibitor of stimulated adenylate cyclase (K_i = 7nM), than as a stimulator of platelet aggregation. Even when platelets are incubated with 25 times the concentration of FSBA needed to block aggregation, no change in the number of binding sites for MeSADP (1500) and less than a 2-fold change in the dissociation constant are detected, further suggesting distinct sites for ADP-mediated platelet shape change and adenylate cyclase.

Many antagonists of cell function may also serve as partial agonists, mimicking the action of the agonists but requiring much higher concentrations. We investigated whether FSBA could itself stimulate platelets (Figures, et al, 1987). We found that the addition of either ADP (3 µM) or FSBA (190 µM) induced platelet shape change, which was confirmed by direct examination of platelets by scanning electron microscopy. While ADP induces shape change with an EC_{50} of 0.4 µM, FSBA induces platelet shape change with an EC_{50} of 220 µM. In contrast, FSBG (a structural

analog of FSBA differing only in the nature of purine base), had no measurable effect on platelets even at concentrations of 500 μM. ATP, a
competitive antagonist of ADP, was able to inhibit FSBA induced shape
change. The induction of shape change by FSBA occurred at times too
short to produce significant incorporation of ^{3}H-SBA. This result indicates that prior noncovalent binding of FSBA to platelets resulted in
the activation of platelets and that covalent binding can inhibit their
response. The noncovalent and covalent sites on the platelet are
identical.

Phosphorylation of myosin light chain is an early response of platelets
to stimulation by an agonist and correlates with ADP-induced platelet
shape change (Daniel, et al, 1984). High concentrations of FSBA which
induce platelet shape change also stimulate myosin phosphorylation
while the low doses of FSBA that do not induce shape change fail
(Figures, et al, 1987). The phosphorylation was qualitatively similar
to that seen with ADP. At low FSBA concentrations and long times, the
ADP-induced change in myosin phosphorylation was inhibited. The initial
interaction between FSBA and receptor prior to incorporation may be
the trigger for shape change response and for myosin phosphorylation
which only occurs at the high FSBA concentration. FSBA (50-200 μM) has
also been demonstrated to increase platelet intracellular Ca^{++}
confirming that it serves as a weak agonist (Rao, et al, 1987).

We have utilized FSBA to inhibit ADP effects on the cells at the receptor level in an attempt to define the involvement of ADP in platelet
activation by other effectors including prostaglandin endoperoxides,
collagen, epinephrine and thrombin. Since, unlike ADP-depleting enzymes
which have Michaelis constants in the mM range (Traverso-Cori, et al,
1965; Morrison, et al, 1965) and therefore cannot eliminate very low
concentrations of ADP, FSBA renders the platelets unresponsive to ADP at
the receptor level. We studied FSBA effects on platelet stimulation by
stable endoperoxide analogs (Morinelli, et al, 1983). FSBA and apyrase
inhibit aggregation and fibrinogen binding induced by PGH_2 derivatives,
demonstrating the involvement of ADP. They do not, however, inhibit
endoperoxide induced shape change, demonstrating an ADP independent
mechanism involving an independent prostaglandin endoperoxide (thromboxane A_2) receptor.

The activation of platelets by collagen is thought to require both ADP
and prostaglandin synthesis (Hawiger, et al, 1987). When platelets are
incubated with FSBA (40 μM), the rate of aggregation by collagen is
inhibited progressively over time (Colman, et al, 1986). ADP is essential for platelet aggregation and fibrinogen binding stimulated by
collagen, since prostaglandin effects on aggregation are also mediated
by ADP (Morinelli, et al, 1983).

The biochemical basis for the synergism of agonists is unclear.
In vitro, epinephrine by itself is required in supraphysiological concentrations for platelet aggregation (Cryer, 1980) and thus requires
synergistic agonists for its effects. Quantification of the receptors
and their binding constants have been achieved utilizing the reversible
binding of the specific $α_2$-antagonist-(methyl-^{3}H)-yohimbine (Macfarlane,
et al, 1981). When platelets were incubated with FSBA, epinephrine-
induced aggregation and fibrinogen binding were progressively inhibited
over time (Figures, et al, 1986), indicating the dependence on the ADP
receptor. The inhibitory effect of FSBA could not be explained by the
ability of the nucleotide analog to sterically block the accessibility
of epinephrine to the $α_2$-adrenoreceptors on the platelet surface. Yohimbine was able to inhibit completely by the response of gel-filtered

platelets to epinephrine, but no effect on ADP (2 μM) induced aggregation was noted. These results indicate that the mechanism of cooperativity between ADP and epinephrine may not be a true synergism, but an undirectional dependence of epinephrine activation on ADP receptor occupancy.

Epinephrine did not mediate an increase in the number of total ADP receptors on the cell surface. However, epinephrine increased by decreasing the K_d of the receptor for ADP. The shift in the K_d of the ADP receptor for ADP was would be reflected in an increased ability of ADP to protect against ^{3}H-FSBA incorporation. These results indicate that epinephrine increases the avidity of the receptor to bind ADP by approximately 10-fold. We were able to completely reverse this effect of epinephrine with yohimbine strongly indicating that this effect is mediated at the level of the α_2-adrenergic receptor.

Thrombin, unlike collagen, prostaglandin endoperoxides and epinephrine, induces aggregation by an ADP-independent mechanism (Puri, et al, 1989). Thus, the exposure of fibrinogen binding sites and the rate of aggregation are little affected by FSBA when thrombin (0.2 units/ml) is used to stimulate platelets. Aggregin, which is covalently labeled by FSBA, is completely cleaved after exposure of washed intact platelets to thrombin. However, if thrombin is incubated with platelet membranes containing the covalently labeled aggregin, no cleavage occurs, suggesting that thrombin may act by an indirect mechanism (Puri, et al, 1989a). Metabolic inhibitors also prevented the cleavage. The requirement for metabolic energy for thrombin cleavage suggested that some other intracellular enzyme is responsible. Thiol protease inhibitors leupeptin and antipain (20 μM) (which do not inhibit thrombin enzymatic activity at this concentration) completely inhibited platelet aggregation, suggesting involvement of a thiol protease. Since thrombin can raise intracellular calcium and thus activate calcium dependent neutral proteases (calpains), we tested whether purified platelet calpain could be responsible for some of thrombin actions on platelets. Calpain, unlike thrombin, is able to cleave aggregin in the platelet membrane, suggesting that calpain is the enzyme directly responsible for the proteolysis. In intact platelets, calpain also cleaves aggregin, simulating the effect of thrombin. Calpain can aggregate platelets when fibrinogen is added. The effects of thrombin on platelet aggregation and exposure of fibrinogen binding sites appear to be mediated through the activation of calpain but require occupancy of thrombin receptors (Puri, et al, 1989b).

We tested the effect of high molecular weight kininogen (HK), a potent plasma inhibitor of calpain (Schmaier, et al, 1986), which binds to platelets but is not internalized, for its ability to inhibit thrombin-induced platelet activation (Puri, et al, 1987). At the plasma concentration of HK, it can completely inhibit the ability of calpain to aggregate platelets. We also demonstrated that HK inhibits the cleavage of aggregin induced by thrombin. Since HK cannot be internalized by platelets, calpain must be exposed on the surface of platelets stimulated by thrombin. We tested whether HK in plasma could act as a modulator of the action of thrombin on platelets. The difference between washed platelets and washed platelets mixed 1:1 with normal plasma is that about 10-fold more γ-thrombin is required for 50% maximum activation. Platelets suspended in plasma completely deficient in total kininogen more closely resembles washed platelets than platelets in plasma in the response to γ-thrombin. When HK is added to kininogen deficient plasma it behaves virtually the same as normal plasma, indicating that HK is the most critical component of plasma. Thus, HK appears to modulate thrombin action in normal plasma. This inhibition displays specificity since HK inhibits both alpha and γ-thrombin but fails to inhibit aggregation by ADP, collagen, a calcium ionophore, A23187 (without added Ca^{++}), the protein kinase C stimulator, phorbol myristate acetate, a combination of both, or the stable prostaglandin endoperoxide U46619.

We have developed a working model to explain the action of thrombin to produce platelet aggregation. Thrombin binds to its receptor glycoprotein Ib on the platelet external membrane and stimulates phospholipase C. Metabolic energy is required to form PIP_2, the substrate cleaved by phospholipase C to yield diacylglycerol and inositol-3-phosphate. The latter raises intracellular Ca^{++} activating calpain. Calpain then cleaves the ^{3}H-FSBA labeled 100 kDa protein, aggregin. The cleavage (which can be inhibited by leupeptin) then results in the exposure of the glycoprotein IIb-IIIa complex to allow binding of fibrinogen and platelet aggregation. Plasma HK specifically inhibits the cleavage of aggregin directly by inhibiting calpain and thus indirectly blocks thrombin-induced aggregation by preventing cleavage of aggregin. In contrast, ADP, whose binding is inhibited by FSBA is visualized as producing a conformational change in the 100 kDa membrane protein, aggregin, which allows assembly of the fibrinogen receptor.

ACKNOWLEDGEMENTS

This review was supported by a Program Project grant HL36579 from the NHLBI.

REFERENCES

Bennett, J.S., Colman, R.F., and Colman, R.W., 1978, Identification of adenine nucleotide binding proteins in human platelet membranes by affinity labeling with 5'-p-fluorosulfonylbenzoyl adenosine, J. Biol. Chem. 253:7346-7354.
Bennett, J.S., and Vilaire, G., 1979, Exposure of fibrinogen receptors by ADP and epinephrine, J. Clin. Invest. 64:1393-1401.
Bennett, J.S., Vilaire, G., Colman, R.F., and Colman, R.W., 1981, Localization of human platelet membrane associated actomyosin using the affinity label 5'-p-fluorosulfonylbenzoyl adenosine, J. Biol. Chem. 256:1185-1190.
Born, G.V.R., 1972, Platelets: Functional physiology. In Biggs, R.: Human Blood Coagulation. Haemostasis and Thrombosis. London: Blackwell Scientific Publications, pp. 159-175.
Colman, R.F., Pal, P.K., Wyatt, J, 1977, Adenosine derivatives for dehydrogenases and kinases. Methods Enzymol. 42:240.
Colman, R.F., 1983, Affinity labeling of porcine nucleotide sites. Ann. Rev. Biochem. 52:67-91.
Colman, R.W., Figures, W.R., Scearce, L.M., Strimpler, A.M., Zhou, F., and Rao, A.K., 1986, Inhibition of collagen-induced platelet activation by 5'-p-fluorosulfonylbenzoyl adenosine. Evidence for an ADP requirement and a synergistic influence of prostaglandin endoperoxides. Blood 68:565-570.
Colman, R.W., Figures, W.R., Wu, Q., Chung, S.Y., Morinelli, T.A., Tuszynski, G.P., Colman, R.F., and Niewiarowski, S., 1986, Separation of the 100 kDa protein membrane protein mediating ADP-induced platelet shape change and activation from glycoprotein IIIa. Trans. Assoc. Am. Phys. 59:39-45.
Cryer, P.E., 1980, Physiology and pathophysiology of the human sympathodren neuroendocrine system. N. Engl. J. Med. 303:436-444.

Daniel, J.L., Molish, I.R., Rigmaiden, M., Stewart, G., 1984, Evidence for a role of myosin phosphorylation in ADP-induced platelet shape change. J. Biol. Chem. 259:9826.

Figures, W.R., Colman, R.F., Niewiarowski, S., Morinelli, T.A., Wachtfogel, Y., and Colman, R.W., 1981, New evidence for an ADP-independent mechanism of thrombin-induced platelet activation. Thromb. Haemost. 46:94.

Figures, W.R., Niewiarowski, S., Morinelli, T.A., Colman, R.F., and Colman, R.W., 1981, Affinity labeling of a human platelet membrane protein with 5'-p-fluoro-sulfonylbenzoyl adenosine. J. Biol. Chem. 256:7789.

Figures, W.R., Scearce, L.M., Defeo, P., Stewart, G., Zhou, F., Chen, J., Daniel, J., Colman, R.F., and Colman, R.W., 1987, Direct evidence for the interaction of the nucleotide affinity analog 5'-p-fluorosulfonyl-adenosine with a platelet AD receptor. Blood 70:796-803.

Figures, W.R., Scearce, L.M., Wachtfogel, Y., Chen, J., Colman, R.F., Colman, R.W., 1986, Platelet ADP receptor and α_2-adrenoreceptor interaction: Evidence a requirement for epinephrine-induced platelet activation and an influence of epinephrine on ADP binding. J. Biol. Chem. 261:5981-5986.

Grant, J.A., Scrutton, M.C., 1980, Positive interaction between agonists in the aggregation response of human blood platelets: Interaction between ADP, adrenaline and vasopressin. Br. J. Haematol. 44:109-125.

Haslam, R.J., 1973, Interactions of the pharmacological receptors of blood platelets with adenylate cyclase. Ser. Haematol. 6:333-350.

Hawiger, J., Steer, M.L., and Salzman, E.W., 1987, Intracellular regulatory processes in platelets. Hemostasis and Thrombosis: Basic Principles and Clinical Practice. Colman, R.W., Hirsch, J., Marder, V.J., Salzman, E.W. (eds), 2nd Edition. J. W. Lippincott, Philadelphia, Philadelphia, pp. 710-725.

Kornecki, E., Niewiarowski, S., Morinelli, T.A., and Kloczewiak, M., 1981, Exposure of fibrinogen receptors by chymotrypsin and adenosine diphosphate on normal and Glanzmann's thrombasthenic platelets. J. Biol. Chem. 256:5696-5701.

Macfarlane, D.E., Mills, D.C.B., 1975, The effects of ATP on platelets: Evidence against the central role of released ADP in primary aggregation. Blood 46:309-320.

Macfarlane, D.E., Wright, B.L., and Stump, D.C., 1981, Use of [methyl-^{3}H]yohimbine as a radioligand for alpha-2 adrenoreceptors on intact platelets. Comparison with dihydroergocryptine. Thromb. Res. 24:31-43.

Marguerie, G.A., Edgington, T.S., and Plow, E.F., 1979, Human platelets possess an inducible and saturable receptor for fibrinogen. J. Biol. Chem. 255:154-161.

Maguire, M.H., Michal, F., 1968, Powerful new aggregator of blood platelets-2-chloroadenosine-5'-diphosphate. Nature 217:571-572.

Mills, D.C.B., Figures, W.R., Scearce, L.M., Colman, R.F., and Colman, R.W., 1985, Two mechanisms for inhibition of ADP-induced platelet shape change by 5'-p-fluorosulfonylbenzoyl adenosine: Conversion to adenosine and covalent modification at an ADP binding site distinct from that which inhibits adenylate cyclase. J. Biol. Chem. 260:8078-8083.

Morinelli, T.A., Niewiarowski, S., Kornecki, E., Figures, W.R., Wachtfogel, Y.T., and Colman, R.W., 1983, Platelet aggregation and exposure of fibrinogen receptors by prostagandin endoperoxide analogues. Blood 61:41-49.

Morrison, J.F., and James, E., 1965, The mechanism of the reaction catalized by adenosine triphosphate-creatine phosphotransferase. Biochemistry 97:37-52.

Mustard, J.F., Packham, M.A., Kinlough-Rathbone, R.L., Perry, D.W., and Regoeczi, E., 1978, Fibrinogen and ADP-induced platelet aggregation.

Blood 52:453-466.

Peerschke, E.I.B., 1985, The platelet fibrinogen receptor. Sem. Haematol. 22:241-259.

Puri, R.N., Gustafson, E.J., Zhou, F., Bradford, H., Colman, R.F., and Colman, R.W., 1987, Inhibition of thrombin-induced platelet aggregation by high molecular weight kininogen. Clin. Res. 35:652a.

Puri, R.N., Zhou, F., Bradford, H.N., Hu, C.-J., Colman, R.F., and Colman, R.W., 1989a, Thrombin-induced platelet aggregation and proteolytic cleavage of aggregin by calpain. Arch. Biochem. Biophys. 271: 346-358.

Puri, R.N., Zhou, F., Colman, R.F., and Colman R.W., 1989b, Cleavage of 100 kDa membrane protein (aggregin) during thrombin-induced platelet aggregation is mediated by the high affinity thrombin receptors. Biochem. Biophys. Res. 162:1017-1024.

Rao, A.K., and Kowalska, M.A., 1987, ADP-induced platelet shape change and mobilization of cytoplasmic ionized calcium are mediated by distinct sites on platelets: 5'-p-fluorosulfonylbenzoyl adenosine is a weak agonist. Blood 70:751-756.

Schmaier, A.H., Bradford, H., Silver, L.D., Farber, A., Scott, C.F., and Schutsky, D., 1986, High molecular weight kininogen is an inhibitor of platelet calpain. J. Clin. Invest. 77:1565-1573.

Traverso-Cori, A., Chaimovitch, H., and Cori, I., 1965, Kinetic studies and properties of potato apyrase. Arch. Biochem. Biophys. 109:173-184.

Section 3
Phosphodiesterase Enzymes and Inhibitors

41
Structural and Functional Characterization of Cyclic GMP-Stimulated Phosphodiesterases and Their Role in Intracellular Signal Transduction

R.T. Mac Farland, S.D. Stroop, and J.A. Beavo

Summary

The various families of cyclic nucleotide phosphodiesterases (PDEs) are frequently classified according to distinguishing kinetic features and responsiveness to intracellular effectors. One of these families, the cGMP-stimulated PDEs, includes those forms which are activated by cGMP in vitro. Cyclic GMP-stimulated PDEs isolated from both the soluble and particulate fractions of several tissues display many kinetic and biochemical similarities. A soluble cGMP-stimulated isozyme originally isolated in this laboratory from bovine cardiac and adrenal tissue has been studied in terms of those structural features which may relate to its observed kinetic behavior. Primary sequence analysis, cDNA cloning, limited proteolysis, photoaffinity labeling, and cGMP-binding studies support a model of allosteric regulation occuring through the binding of cGMP to a distinct regulatory domain. Kinetic characteristics of binding and activation by cGMP and various cyclic nucleotide derivatives, as well as primary sequence information, reveal differences between this PDE and cyclic GMP-dependent protein kinase, and may help in distinguishing these two routes of cGMP signal transduction. This article provides a brief overview of cGMP-stimulated PDEs, followed by a description of recent studies in this and other laboratories concerned with the structure and function of these enzymes. These findings are considered in the discussion of two physiological pathways, in cardiac myocytes and adrenal glomerulosa cells, in which a role for the cGMP-stimulated PDE in cGMP signal transduction in indicated.

Introduction

Purine 3',5' nucleoside monophosphate phosphodiesterases (PDEs) constitute a family of enzymes responsible for the hydrolysis and inactivation of cyclic AMP (cAMP) and cyclic GMP (cGMP) second messenger molecules. A key regulatory role for PDEs in cellular signal transduction pathways is implicit in the nature of their catalytic function, and is also suggested by the kinetic changes seen for many PDE isozymes in response to known intracellular effectors including calcium/calmodulin and cGMP itself. Among those phosphodiesterase forms exhibiting regulation by cGMP are the isozymes known collectively as cGMP-stimulated phosphodiesterases (for review see Beavo, 1987). First characterized in soluble fractions from rat liver and particulate fractions of several tissues (Beavo et al., 1971), cGMP-stimulated PDE activities have since been observed in a variety of whole tissues including bovine heart, adrenal, lung, liver and adipose tissue (Martins et al., 1982; Mumby et al., 1982; Moss et al., 1977, Klotz et al., 1972), as well as in extracts from thymic lymphocytes (Franks et al., 1971), L-cell fibroblasts (Manganiello et al., 1972), and human platelets (Weishaar, R.E. et al., 1986).

Although the number of different isozymes is not known, several cGMP-stimulated PDEs have been purified to apparent homogeneity, including a soluble form, originally isolated in this laboratory from bovine heart and adrenal gland (Martins et al., 1982). Cyclic GMP-stimulated PDEs have also been isolated from soluble fractions of bovine adrenal (Miot et al., 1985) and liver (Yamamoto et al., 1983), and particulate fractions from rat liver (Pyne et al., 1986), as well as both bovine and rabbit cerebral cortex (Murashima et al., 1988; Whalin et al., 1988). Future studies using molecular cloning, peptide sequencing, and drug selectivity techniques will help determine the number and diversity of cGMP-stimulated PDE isozymes.

In attempting to assign physiologic roles to cGMP-stimulated phosphodiesterases, we may take advantage of our understanding of the biochemistry and kinetics for individual members of this PDE family. Through comparisons drawn between these enzymes and cGMP-dependent protein kinase, we may be able to distinguish these two routes of cGMP action. Though in few cases have specific PDE isozymes, such as those of the cGMP-stimulated family, been localized to individual cell types, in those cases where this has been accomplished, we may begin to assess their potential function in cellular signal transduction pathways.

Overview of cGMP-stimulated Phosphodiesterase Isozymes

In addition to the soluble cGMP-stimulated phosphodiesterases purified from bovine cardiac and adrenal tissues (Martins et al., 1982; Miot et al.,1985) calf liver has provided a third source for the isolation of what appears to be a very similar isozyme (Yamamoto et al., 1983). In each of these cases, purification has been based upon the use of cyclic nucleotide affinity chromatography. Biochemical, kinetic and immunologic comparison of the purified enzymes reveals strong similarities between members of this PDE family. Distinguished by positive cooperative hydrolysis of both cAMP and cGMP, the soluble cGMP-stimulated PDEs hydrolyze both substrates with approximately equal affinities (10-30 μM range) and at a similar maximal rates (Table 1). A slight preference for cGMP versus cAMP as both substrate and stimulatory effector is noted, with either nucleotide stimulating hydrolysis of the other at subsaturating concentrations. With increasing concentrations of cGMP, a sigmoidal activation curve is seen for hydrolysis of cAMP, present in subsaturating concentrations.

Table 1. Physical and kinetic properties of the soluble cGMP-stimulated PDE from bovine heart and adrenal (from Martins et al., 1982).

molecular weight (SDS-PAGE)	105 kDa (native = 240 kDa)	
Stokes radius (A^{O})	64	
Sedimentation coefficient ($S_{20},w \times 10^{-13}s$)	7.4	
	cAMP	cGMP
Km (μM)	30	10
Vmax (μmoles/min/mg)	120	120
Hill coefficient (npp)	1.9	1.3

As concentrations of the allosteric effector nucleotide are increased into the range of its Km for hydrolysis, a decline in substrate hydrolysis is seen, presumably reflecting competition for binding at the catalytic site. A thorough kinetic analysis of the cGS-PDE isolated from calf liver has been conducted by Yamamoto and co-workers (1983).

In agreement with their kinetic likenesses, isolated soluble cGMP-stimulated PDEs exhibit marked physical similarities. Using SDS-PAGE, a single band of Mr=105,000 is seen for the isozymes from bovine heart, adrenal and liver, with a calculated native molecular weight in the 200 kDa range based on non-denaturing electrophoresis and hydrodynamic measurements (Martins et al., 1982; Yamamoto et al., 1983). In the case of the purified bovine cardiac and adrenal forms isolated in this laboratory, immunologic cross-reactivity is seen using monoclonal antibodies prepared against the cardiac isozyme, further attesting to the similarity between the soluble forms isolated from these two sources (Mumby et al., 1982). Recently obtained amino acid sequence information suggests these may, in fact, represent the same isozyme (unpublished data).

Particulate cGMP-stimulated PDEs isolated from rabbit (Whalin et al., 1988) and bovine (Murashima et al., 1988) cerebral cortex exhibit many kinetic similarities to the soluble forms described, with possible differences in subunit structure suggested for the rabbit enzyme. Interestingly, particulate forms appear to represent the major cGMP-stimulated PDE activity in cerebral cortex (Murshima et al., 1988; Beavo et al., 1971). An additional particulate cGMP-stimulated phosphodiesterase has been purified to apparent homogeneity from rat liver membranes by Pyne and Houslay (1988). In addition to a smaller size seen on SDS-PAGE (Mr=66,000), this isozyme displays a substantially lower Vmax than other cGMP-stimulated forms, and appears to be responsive to crude preparations of "insulin mediator", which elicit 3-4 fold increases in activity.

Structural Features of cGMP-stimulated Phosphodiesterase

Limited Proteolysis and Photoaffinity Labeling. The aforementioned kinetic studies support the suggestion for unique allosteric and catalytic sites on the cGMP-stimulated phosphodiesterase, as previously proposed by several laboratories (Miot et al., 1985; Yamamoto et al., 1983, Wada et al., 1987). Recent work in this laboratory has attempted to define the observed enzymatic behavior of the soluble cGMP-stimulated PDE isolated from bovine cardiac tissue in terms of specifically localized structural features of the monomeric protein itself (Stroop et al., 1989). Using cGMP-stimulated PDE purified from the soluble fraction of bovine cardiac tissue, direct photolabeling with [32P]-cGMP was performed. Scatchard analysis of [32P]-incorporation indicates two binding sites, with calculated half labeling concentrations of 1 and 30μM. These labeling affinities for the putative allosteric and catalytic sites agree with affinities of cGMP for cAMP hydrolysis (1μM) and the Km for cGMP and cAMP as a substrates (10-30 μM) (Yamamoto et al., 1983).

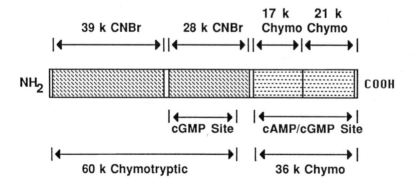

Figure 1. Limited Proteolysis of the cGMP-stimulated PDE showing approximate locations of proposed catalytic and regulatory domains (Stroop et al., 1989).

283

When photolabeling was followed by limited proteolytic digestion using chymotrypsin, two unique labeled fragments of Mr=60,000 and 36,000 were seen on SDS-PAGE. Labeling of the Mr=36,000 fragment was blocked in the presence of 5μM cAMP, suggesting that the low affinity catalytic site may be contained within this region of the protein. Cyclic AMP, as well as 1-methyl, 3-isobutylxanthine (IBMX), increased [^{32}P]-incorporation into the Mr=60,000 band, consistent with the previous reports that cGMP binding to the allosteric site is increased in their presence ((Miot et al., 1985). Time course experiments using chymotrypsin revealed that appearance of the catalytically-active Mr=36,000 fragment correlated with increases in enzyme activity and loss of sensitivity to cGMP-stimulation. Alignment of sequences obtained for these fragments with the complete sequence for the cGS-PDE, recently obtained through direct protein sequencing and a partial cDNA clone, have localized the allosteric and catalytic sites to distinct regions of the protein, and identified the probable chymotryptic site linking these domains (unpublished data) (Fig. 1). Identification of the specific amino acid sequences corresponding to these sites will await the development of labeling agents exhibiting higher degrees of incorporation, or alternatively, may employ site-directed mutagenesis techniques.

Studies Using Cyclic Nucleotide Analogs. Examination of the recently elucidated primary sequence for the cGMP-stimulated PDE has failed to reveal any close homologies with other cGMP-binding proteins, including cGMP-dependent protein kinase, which might also help specifically define the functional domains on this PDE. This apparent lack of homology between the two enzymes presumed responsible for mediation of intracellular cGMP signalling is consistent with differences previously noted in their responsiveness to a variety of cGMP derivatives and cGMP itself.

Erneux and co-workers (1985) have examined the structural requirements for activation of the cGMP-stimulated phosphodiesterase using a number of cyclic nucleotide derivatives. Again using purified soluble enzyme, isolated from bovine adrenal in this case, analogs were compared with respect to their relative abilities to displace bound [^{3}H]-cGMP, and activate phosphodiesterase activity. A perfect correlation was observed between the relative potencies of all the derivatives tested at displacing bound cGMP and activation of hydrolytic activity. Among ribose-substituted compounds, cGMP derivatives were, as predicted, more potent activators than were cAMP derivatives, with substitutions at the 5' position resulting in greater reductions in binding activity than substitutions at the 3' position. Comparison between several base-substituted cyclic nucleotides indicated that activation of the cGMP-stimulated PDE was sensitive to a guanine-type purine structure. Interestingly, 8-substituted analogs, including 8-bromo-cGMP, proved to be very poor activators of phosphodiesterase activity, in contrast to their potent effects on cGMP-dependent protein kinase (Corbin et al., 1986).

A comparison between the two diastereomeric forms of guanosine 3'5'-monophosphorothioate (Rp- and Sp-cGMP) in which one of the two exocyclic oxygens present on the ribose moiety is replaced with an sulfur atom, revealed a lack of regioselectivity for activation of hydrolytic activity by these derivatives (although the potency of the Rp form exceeded that of Sp by approximately 4-fold). This may be contrasted with cAMP-dependent protein kinase, where the corresponding cAMP derivatives differ more dramatically in their actions (Sp-cAMP acting as an activator of kinase activity, while Rp-cAMP does not (Rothermel et al., 1983). The authors interpret this difference as evidence for underlying differences in the patterns of interaction of the ribose moiety with the two enzymes.

Analog studies with cGMP-dependent protein kinase reveal a different pattern in the structural features required for binding and activation by various cyclic nucleotide analogs. Although interactions at the base region are clearly required for the difference in selectivity seen for cAMP- versus cGMP-dependent forms, both kinases exhibit a higher degree of specificity in the ribose cyclic moiety (Corbin et al., 1987), with interactions at the 2', 3', and 5' sites being implied in binding to the regulatory site.

These results suggest fundamental differences exist in the requirements for binding of cyclic nucleotides to these enzymes, and may thus be said to be consistent with the lack of structural homologies seen at the protein level. It may be speculated that the higher concentrations of cGMP required for activation of the cGMP-stimulated phosphodiesterase as compared to cGMP-dependent protein kinase reflect the relative attractive forces resulting from these alternative binding mechanisms.

Recent data from this laboratory has revealed a new class of cGMP analog capable of selectively activating the cGMP-stimulated phosphodiesterase. As shown in Fig. 2, a derivative of cGMP (1-(2-nitrophenylethyl-cGMP) produces activation of antibody-bound, adrenal cGMP-stimulated phosphodiesterase, with maximal activation occurring at concentrations approximately 10-fold higher than those required for cGMP itself. Rp-cGMP, also substituted in the region of the phosphate group, produces a high degree of activation, at concentrations approximately 100-fold greater than are required using cGMP itself.

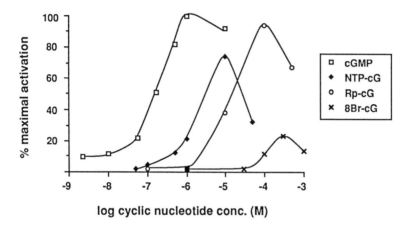

Figure 2. In vitro activation of cGMP-stimulated phosphodiesterase using cGMP derivatives. Abbreviations: NTP-cG (diastereomeric mixture of 1-(2-nitrophenylethyl-cGMP); Rp-cG (Rp isomer of guanosine monophosphorothioate); 8-Br-cG (8-bromo-cGMP).

This derivative does not activate cGMP-dependent protein kinase at these concentrations, in keeping with the apparent greater specificity seen for the ribose cyclic phosphate moiety for cGMP-dependent protein kinase, and may thus be said to define a new class of phosphodiesterase-selective cGMP derivative, substituted at the phosphate group itself. Clearly, in addition to defining structural differences in the requirements for activation of phosphodiesterase versus kinase activity in vitro, identification of selective derivatives will provide tools that will be useful in studying the role of these two pathways of cGMP action in intact cell systems.

Role of Cyclic GMP-stimulated Phosphodiesterase in Signal Transduction

A physiologic role for cGMP-stimulated phosphodiesterase is suggested in cell types which are subject to fluctuations in intracellular cGMP levels, particularly in such cases where cGMP and cAMP are known to act in a physiologically antagonistic manner. In patch clamp studies using isolated frog cardiac myocytes, Hartzell and co-workers (1986) have demonstrated a role for cGMP-stimulated phosphodiesterase in mediating the inhibitory effects of cholinergic agonists on slow inward calcium current (I_{Ca}). Cholinergic agents such as carbachol inhibit ß-adrenergic agonist-induced increases in I_{Ca}, which in turn result from increases in cAMP synthesis and cAMP-dependent phosphorylation of calcium channels (Tsien, 1986). These opposing effects are in keeping with the antagonistic actions of these agents on overall cardiac function, and have been shown to occur partly as a result of carbachol-induced inhibition of adenylate cyclase (Heschler et al., 1986). Cholinergic stimulation results in increases in cGMP, which do not appear to act through cGMP-dependent protein kinase, as evidenced by the failure of 8-bromo-cGMP to mimic these effects on I_{Ca} (Hartzell et al., 1986). A role for cGMP-stimulated phosphodiesterase is indicated by the similarity of the dose-response curves for the effects of cGMP on calcium current and activation of PDE activity. These effects were blocked by high doses of IBMX. Interestingly, approximately 50% of the total cGMP-stimulated PDE activity in these cells was localized to the particulate (20,000 x g) fraction (Simmons et al., 1988).

Cardiac myocytes are known to contain at least one other "low Km", cGMP-inhibited PDE activity, which is believed to be the target for the actions of inotropic drugs such as milrinone, a specific inhibitor of certain members of this PDE class (Harrison et al., 1985). It is tempting to suggest a model in which the cGMP-inhibited and cGMP-stimulated PDEs act in a concerted fashion in such a cell. Under resting conditions, when intracellular cAMP levels are reduced, the cGMP-inhibited PDE would serve to regulate cAMP degradation. Its function under these conditions would seem to fit its role as a target for inotropic drugs, which act to increase contractility under normal resting conditions. Alternatively, under stimulated conditions resulting from ß-adrenergic-induced increases in cAMP, the higher Km cGMP-stimulated phosphodiesterase would assume an increasingly important role in the continued degradation of cAMP, and hence become the relevant target for the effects of cholinergic-induced increases in cGMP.

Interest in cGMP second messenger pathways has grown dramatically with the discovery of atrial natriuretic peptides (ANPs), which are coupled to activation of particulate guanylate cyclase in their major target tissues (Inagami, 1989). In the adrenal cortex, ANP receptors are most heavily concentrated in the *zona glomerulosa* region, the outermost, sub-capsular layer, associated with production of aldosterone (Bianchi et al., 1985). ANP-mediated inhibition of aldosterone synthesis is associated with dramatic increases in cGMP and decreases in intracellular cAMP in isolated glomerulosa cells (Fig. 3). Attempts to mimick these effects using 8-bromo-cGMP have been unsuccessful, thus ruling out a cGMP-dependent protein kinase mechanism in the inhibition of steroidogenesis.

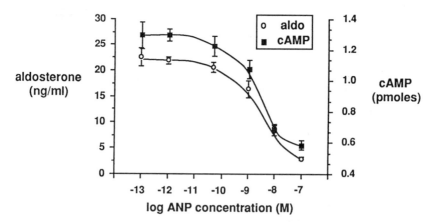

Figure 3. ANF-induced decreases in aldosterone production and intracellular cAMP levels in isolated bovine adrenal glomerulosa cells. Aldosterone concentrations in cell culture media and intracellular cAMP levels were determined following 3 hr. incubation with ANP.

Work in this laboratory has revealed that the majority of cGMP-stimulated phosphodiesterase is localized to the *zona glomerulosa* layer of the adrenal cortex, and represents the highest concentration of this isozyme found to date. Immunoprecipitation experiments using a monoclonal antibody specific for this isozyme indicate that the cGMP-stimulated PDE is the predominant isozyme in this cell layer. These findings suggest that ANP-mediated increases in cGMP may be coupled to the observed inhibition of steroidogenesis through PDE-mediated decreases in cAMP (Fig. 4).

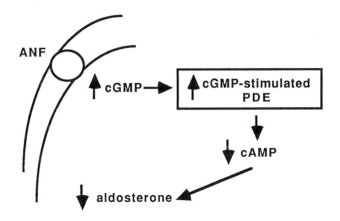

Figure 4. Proposed role of cGMP-stimulated PDE in mediation of ANP effects on aldosterone production.

The significance of cGMP as a second messenger, particularly as concerns the actions of atrial natriuretic peptides, has produced an increased awareness of the importance of cGMP-stimulated PDE as a potential mediator of the actions of this hormone. Our ability to assess the role of this and other PDE isozymes has been greatly enhanced by an understanding of these enzymes at the molecular level. It is by this means that new avenues of therapy, aimed at modifying the activity of individual PDEs as they function in specific intracellular pathways, will be developed.

References

Beavo, J.A., 1988, Multiple Isozymes of Cyclic Nucleotide Phosphodiesterase, *Adv. Second Messenger Phosphoprot. Res.* **22**:1-38.

Beavo, J.A., Hardman, J.G. and Sutherland, E.W., 1971, Stimulation of Adenosine-3',5'-Monophosphate Hydrolysis by Guanosine -3',5'-Monophosphate, *J. Biol. Chem.* **246**:3841-3846.

Bianchi, C., et al., 1985, Radioautographic localization of ^{125}I-atrial natriuretic factor (ANF) in rat tissues, *Histochemistry* **82**:441-452.

Corbin, J.D., et al., 1986, Studies of cGMP Analog Specificity and Function of Two Intrasubunit Binding Sites of cGMP-dependent Protein Kinase, *J. Biol. Chem.* **261**:1208-1214.

Erneux, C., et al, The Binding of Cyclic Nucleotide Analogs to a Purified Cyclic GMP-stimulated Phosphodiesterase From Bovine Adrenal Tissue, *J. Cyc. Nuc. Prot. Phos. Res.* **10**:463-472.

Franks, D. and MacManus, J.P., 1971, Cyclic GMP stimulation and inhibition of cAMP phosphodiesterase from thymic lymphocytes. *Biochem. Biophys. Res. Commun.* **42**:844-849.

Harrison, S.A., et al., 1986, Isolation and Characterization of Bovine Cardiac Muscle cGMP-inhibited Phosphodiesterase: A Receptor for New Cardiotonic Drugs, *Mol. Pharmacol.* **29**:506-514.

Hartzell, H.C. and Fischmeister, R., 1986, Opposite effects of cyclic GMP and cyclic AMP on Ca^{2+} current in single heart cells, *Nature* **323**:273-275.

Heschler, J., Kameyama, M., and Trautwein, W., 1986, On the mechanism of muscarinic inhibition of cardiac calcium current, *Pflugers Arch. Eur. J. Physiol.* **407**:182-189.

Hurwitz, R., et al., 1984, Immunologic Approaches to the Study of Cyclic Nucleotide Phosphodiesterases, *Adv. Cyc. Nuc. Res. and Prot. Phos. Res.* **16**:89-106.

Inagami, T., 1989, Atrial Natriuretic Factor, *J. Biol. Chem.* **264**:3043-3046.

Klotz, W. and Stock, K., 1972, Influence of cyclic guanosine-3',5'-monophosphate on the enzymatic hydrolysis of adenosine-3',5'-monophosphate, *Naunyn-Schmiedeberg's Arch. Pharmacol.* **274**:54-62.

Manganiello, V.C. and Vaughan, M., 1972, Prostaglandin E1 Effects on Adenosine 3-:5'-Cyclic Monophosphate Concentration and Phosphodiesterase Activity in Fibroblasts, *Proc. Natl. Acad. Sci. USA* **69**:269-273.

Martins, T.J., Mumby, M.C. and Beavo, J.A., 1982, Purification and Characterization of a Cyclic GMP-stimulated Cyclic Nucleotide Phosphodiesterase from Bovine Tissues, *J. Biol. Chem.* **257**:1973-1979.

Miot, F., Van Haastert, P.J.M., Erneux, C., 1985, Specificity of cGMP binding to a purified cGMP-stimulated phosphodiesterase from bovine adrenal tissue, *Eur. J. Biochem.* **149**:59-65.

Moss, J., Manganiello, V.C. and Vaughan, M., 1977, Substrate and Effector Specificity of a Guanosine 3':5'-Monophosphate Phosphodiesterase from Rat Liver, *J. Biol. Chem.* **252**:5211-5215.

Mumby, M.C., et al., 1982, Identification of cGMP-stimulated Cyclic Nucleotide Phosphodiesterase in Lung Tissue with Monoclonal Antibodies, *J. Biol. Chem.* **257**:13283-13290.

Murashima, S., et al., 1988, ms. submitted.

Pyne, N.J., Cooper, M.E. and Houslay, M.D., 1986, Identification and characterization of both the cytosolic and particulate forms of cyclic GMP-stimulated cyclic AMP phosphodiesterase from rat liver, *Biochem. J* . **234**:325-334.

Pyne, N.J. and Houslay, M.D., 1988, An Insulin Mediator Preparation Serves to Stimulate the Cyclic GMP Activated Cyclic AMP Phosphodiesterase Rather Than Other Purified Insulin-Activated Cyclic AMP Phosphodiesterases, *Biochem. Biophys. Res. Commun.* **156**:290-296.

Rothermel, J.D. et al., 1983, Inhibition of Glycogenolysis in Isolated Rat Hepatocytes by the Rp Diastereomer of Adenosine Cyclic 3',5'-Phosphorothioate, *J. Biol. Chem.* **258**:12125-12128.

Simmons, M.A. and Hartzell, H.C., 1988, Role of Phosphodiesterase in Regulation of Calcium Current in Isolated Cardiac Myocytes, *Mol. Pharm.* **33**:664-671.

Stroop, S.D., Charbonneau, H., and Beavo, J.A., 1989, Direct Photolabeling of the cGMP-stimulated Cyclic Nucleotide Phosphodiesterase, *J. Biol. Chem.* **264**:13718-13725.

Tsien, R.W., et al., 1986, Mechanisms of calcium channel modulation by beta-adrenergic agents and dihydropyridine calcium agonists, *J. Mol. Cell. Cardiol.* **18**:691-710.

Wada, H., Osborne, J.C., Jr. and Manganiello, V.C., 1987, Effects of Temperature on Allosteric and Catalytic Properties of the cGMP-stimulated Cyclic Nucleotide Phosphodiesterase from Calf Liver, *J. Biol. Chem.* **262**:5139-5144.

Weishaar, R.E., et al., 1986, Multiple Molecular Forms of Cyclic Nucleotide Phosphodiesterase in Cardiac and Smooth Muscle and Platelets, *Biochem. Pharmacol.* **35**:787-800.

Whalin, M.E., Strada, S.J. and Thompson, W.J., 1988, Purification and partial characterization of membrane-associated type II (cGMP-activatable) cyclic nucleotide phosphodiesterase from rabbit brain, *Bioch. Biophys, Acta* **972**:79-94.

Yamamoto, T., Manganiello, V.C. and Vaughan, M., 1983, Purification and Characterization of Cyclic GMP-stimulated Cyclic Nucleotide Phosphodiesterase from Calf Liver, *J. Biol. Chem.* **258**:12526-12533.

42
Phosphodiesterase Genes from Flies to Mammals

R.L. Davis, J. Henkel-Tigges, Y. Qiu, and A. Nighorn

Introduction

The cyclic nucleotide phosphodiesterases (PDEases) comprise a very complex class of enzymes. Numerous forms of the enzyme have been characterized partially using standard biochemical and immunological tools which has led us to a partial understanding of the various forms. The enzyme class has been subgrouped into at least six different families, based on substrate specificity and affinity, molecular mass, and sensitivity to cofactors and drugs (Beavo, 1987). However, the enzymes have been difficult to work with biochemically and the past problems, incomplete nature of the information gathered, and inconsistencies have indicated a need to approach an understanding of the enzymes in a new way. We have adopted the premise that the current insufficiency of information regarding the enzymes can best be filled by isolating the genes and cDNA clones which encode the enzymes. Thus, if every PDEase gene were cloned, we could classify the enzymes according to their gene and/or RNA. Insightful results have already been obtained from the nucleic acids approach.

The first PDEase gene from a higher eucaryote to be cloned was the *Drosophila* dunce[+] gene. This gene has a history separate from its identity as the structural gene for a cAMP-specific PDEase. Dunce mutant flies were identified more than a decade ago on the basis of their failure to perform properly in learning tests (Dudai et al., 1976). Specifically, when normal flies are presented an olfactory cue such as the vapor from certain alcohols or aldehydes simultaneously with the negative reinforcement of electrical shock, they learn this temporal association as evidenced by their avoidance of the shock-associated odorant after training. In contrast, dunce mutants fail in the task, exhibiting a partial deficit in initial learning and a more pronounced deficit in memory (Tully et al., 1985). Thus, the gene became interesting initially because of its role in conditioned behavior. A series of genetic and biochemical studies followed and pointed to the possibility that dunce[+] was the structural gene for a cAMP-specific PDEase (Davis et al., 1984a). Eventually, the gene was cloned through a chromosomal walk (Davis et al., 1984b) and partially sequenced, confirming its identity as a PDEase gene (Chen et al., 1986).

From the cloning and sequence analysis of dunce[+] and several other PDEase genes which have been analyzed subsequently, we have learned a number of things. First, genes which encode the PDEases in higher eucaryotes appear to be remarkably complex. This is illustrated best by dunce[+,] which extends over a minimum of 100 kb of the *Drosophila* genome, encodes at least six distinct RNAs, and exhibits the novel property of having at least two other, and unrelated genes nested within an enormous intron of 80 kb (Davis et al., 1986; Chen et al.,1987). The dunce[+] gene is not alone in this respect, since preliminary information on the size of the bovine cGMP PDEase gene suggests that it also extends over more than 140 kb (Pittler et al., 1989). Second, the open reading frames found in several PDEase cDNA clones predict proteins with a remarkable level of sequence homology. For example, the amino acid sequence identity between the *Drosophila* and rat cAMP PDEase genes is approximately 75% over one extended region (see below). Third, most of the PDEase genes and enzymes which have been sequenced to date are related, indicating that they are encoded by a complex gene family with perhaps one or more members coding for each family of the PDEase enzymes.

Experimental

Rat and *Drosophila* cAMP PDEase Genes Exhibit Exceptional Sequence Conservation

A cDNA clone representing the protein coding region of the *Drosophila* dunce[+] gene was used in genomic blotting experiments to survey the genomes of several other organisms for the presence of homologous genes (Davis et al., 1989). Virtually every organism tested contained one or more restriction fragments which hybridized to the probe for at least one of the three hybridization conditions used. We observed distinct bands using yeast, cockroach, electric ray, chicken, rat, mouse, or human DNAs (Figure 1). This experiment demonstrates that the *Drosophila* dunce[+] gene is conserved in sequence sufficient to detect homologous genes in vertebrate organisms.

The *Drosophila* probe was subsequently used to screen genomic libraries of the rat at a reduced stringency in order to isolate clones representing the homologous sequences. After restriction mapping, a portion of one genomic clone, named ratdnc-1, was used to isolate cDNA clones representing the encoded RNAs. One cDNA clone isolated, named RD1, contained a complete open reading frame sufficient to code for a protein of about 68,000 daltons. A comparison of the predicted amino acid sequence of this open reading frame with the predicted amino acid sequence of the *Drosophila* dunce[+] gene demonstrated a remarkable homology, with approximately 75% of the amino acid residues within the conserved domain of 275 residues being identical (Davis et al., 1989). A dot matrix comparison of the two predicted sequences is shown in Figure 2 to illustrate the region and the extent of homology.

The transcripts homologous to the RD1 cDNA clone have been identified by RNA blotting experiments (Davis et al., 1989). Although they are not as complex as the transcripts homologous to the *Drosophila* dunce[+] gene (Davis et al., 1986), at least two distinct and large transcripts were detected. The most abundant transcript in brain measured about 4.0 kb, while that in lung, heart and testes measured slightly larger at 4.4 kb. The basis for the size difference is not currently known, but in principle, it could be due to the use of multiple promoters, splicing pathways, and/or polyadenylation sites.

291

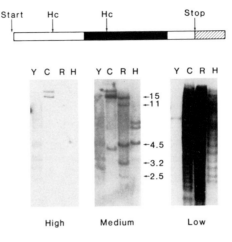

Figure 1, from Davis et al., 1989. Schematic diagram of a *Drosophila* dunce+ cDNA clone showing the predicted start and stop codons, *Hinc* II (Hc) sites, and the "conserved domain" as a filled box. The results of probing yeast (Y), chicken (C), rat (R) and human (H) restriction digests at three stringencies with the rightmost *Hinc* II fragment are shown at the bottom.

Other PDEase cDNA clones, related to RD1, have been isolated from the rat. One, named dunce-related PDEase (DPD), was isolated using a biological assay in yeast (Colicelli et al., 1989). Mutations which elevate cAMP levels in yeast cause the cells to be sensitive to heat. A search for rat cDNA clones which could suppress this heat sensitivity when expressed behind a yeast promoter on an extrachromosomal plasmid produced DPD. In addition, two other cDNA clones related to RD1, were isolated using the *Drosophila* dunce+ gene as an initial probe (Swinnen et al., 1989). Thus, the isolation of the structural gene for cAMP PDEase in *Drosophila* allowed for the isolation of numerous mammalian PDEase genes.

Identity of the PDEase Encoded by the RD1 cDNA Clone

One important question concerns the nature of the PDEase enzyme product of the RD1 and RD1-related cDNA clones. In principle, since the PDEases comprise an enzyme family, a nucleic acid probe representing any one of the enzymes might hybridize to sequences representing a PDEase of a separate family given sufficient sequence homology. However, since the RD1 cDNA clone exhibits extremely high homology with the *Drosophila* cDNA clone, it seemed probable that RD1 would code for an enzyme very similar in properties to that encoded by the *Drosophila* dunce+ gene. As mentioned earlier, this gene codes for low K_m, cyclic AMP-specific PDEase.

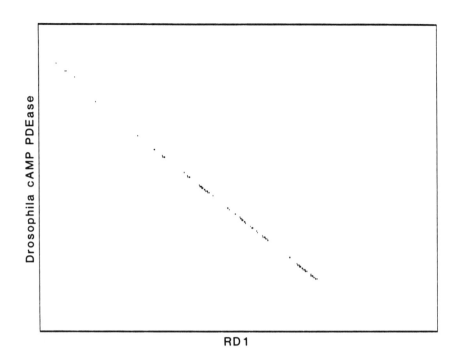

Figure 2. Dot matrix homology diagram comparing the amino acid sequences of the *Drosophila* cAMP-specific PDEase and that predicted from RD1. Each dot represents a match of five residues per window of fifteen residues across each sequence.

One type of low K_m, cAMP PDEase which has been partially characterized biochemically from mammalian tissues is inhibited by the compound, RO 20-1724. A second agent which inhibits the enzyme is the drug, rolipram, an experimental compound which has shown promising antidepressant activity in clinical trials (Horowski et al., 1985; Eckmann et al., 1988).

To determine whether RD1 codes for a PDEase sensitive to RO 20-1724 and rolipram, we expressed the cDNA in yeast and measured the PDEase activity in the presence of the drugs. The RD1 DNA was inserted into the expression vector, pADNS (Colicelli et al., 1989), and used to transform yeast cells carrying mutations in both PDEase genes (*MAT*(α) *leu2 his3 ura3 ade8 pde1:ADE8 pde2:URA3 ras1:HIS3*). Since the pADNS vector carries the selectable marker, *LEU2,* we selected for *LEU2* transformants and used single colonies to innoculate large cultures. After growth, the yeast were harvested, lysed by passage through a French press, and the supernatants collected following centrifugation.

Table 1. Properties of mammalian and *Drosophila* cAMP PDEases.

PDEase		IC_{50} at 1 μM	
	K_m for cAMP (μM)	RO 20-1724	Rolipram
Drosophila	2[a]	>100[b]	>100[b]
RD1 in yeast	4	6	0.6
Rat cerebral cortex cAMP PDEase	2[c]	4[c]	1[c]

a. Taken from Davis et al., 1984a.
b. Measured in crude homogenates with excess cGMP which allows for specific determination of the cAMP-specific PDEase.
c. Taken from Davis, C.W., 1984, and Beavo, J.A., 1987.

PDEase assays (Davis, 1988) on the supernatants revealed the presence of substantial cAMP PDEase activity in the transformants carrying the RD1 insert, but not in cells carrying the expression vector alone. The measured K_m was 4 micromolar (Table 1), which compares favorably to the K_m of 2 micromolar for the *Drosophila* enzyme measured previously. The supernatants showed no detectable cGMP PDEase activity, so that the encoded enzyme can be classed as cAMP-specific. Rolipram and RO 20-1724 both proved to be effective inhibitors of the enzyme activity, with IC50 values similar to that reported for the RO 20-1724/rolipram-sensitive enzyme present in tissue homogenates (Table 1). Interestingly, the *Drosophila* cAMP-specific PDEase is not sensitive to RO 20-1724 or rolipram. Therefore, we conclude that RD1 codes for an enzyme of the RO 20-1724/rolipram-sensitive family of cAMP-specific PDEases.

Discussion

These observations raise a number of important questions. It is intriguing that the *Drosophila* cAMP PDEase is an indispensable part of the biochemistry underlying insect learning/memory processes while the mammalian counterparts may function in the regulation of mood. One question is whether drugs like rolipram affect cognitive abilities. This has not been addressed in a systematic and complete way. The answer to this is of both practical and conceptual importance. On the practical side, if the drugs do affect cognitive abilities, this will likely have impact on the design and use of the drugs for the treatment of depression. A conceptual question posed regards the relationship between the biochemistry underlying learning and memory processes and that underlying the regulation of mood. We believe that the PDEases are likely to be involved in these behaviors by modulating the excitability of neurons intrinsic to the behaviors. If so, it may be difficult to target these PDEases for drugs to treat one behavioral disorder, without altering other types of behavior.

Acknowledgements

We are indebted to M. Wigler and associates for supplying the yeast expression system. The authors' research on the *Drosophila* dunce⁺ gene has been supported by grants from the NIH and the McKnight Foundation . Research on the mammalian cAMP PDEase genes has been supported by grants from the NSF and the McKnight Foundation.

References

Beavo, J.A., 1987, Multiple isozymes of cyclic nucleotide phosphodiesterase, in *Advances in Second Messenger and Phosphoprotein Research* (eds. Robison, G.A. and Greengard, P.) (Raven Press, New York) **22**:1-38.

Chen, C.-N., Denome, S., and Davis, R.L., 1986, Molecular analysis of cDNA clones and the corresponding genomic coding sequences of the *Drosophila* dunce⁺ gene, the structural gene for cAMP phosphodiesterase, *Proc. Natl. Acad. Sci. USA* **83**:9313-9317.

Chen, C.-N., et al., 1987, At least two genes reside within a large intron of the *dunce* gene of *Drosophila*, *Nature (London)* **329**:721-724.

Colicelli, J., et al., 1989, Isolation and characterization of a mammalian gene encoding a high-affinity cAMP phosphodiesterase, *Proc. Natl. Acad. Sci. USA* **86**:3599-3603.

Davis, C.W., 1984, Assessment of selective inhibition of rat cerebral cortical calcium-independent and calcium-dependent phosphodiesterases in crude extracts using deoxycylic AMP and potassium ions, *Biochim. Biophys. Acta* **797**:354-362.

Davis, R.L., and Kauvar, L.M., 1984a, Drosophila cyclic nucleotide phospho-diesterases, *Adv. Cyclic Nuc. Prot. Phosphor. Res.* **16**:393-402.

Davis, R.L., and Davidson, N., 1984b, Isolation of the *Drosophila melanogaster* dunce chromosomal region and recombinational mapping of dunce sequences with restriction site polymorphisms as genetic markers, *Mol. Cell. Biol.* **4**:358-367.

Davis, R.L., and Davidson, N., 1986, The memory gene dunce⁺ encodes a remarkable set of RNAs with internal heterogeneity, *Mol. Cell. Biol.* **6**:1464-1470.

Davis, R.L., 1988, Mutational analysis of phosphodiesterase in *Drosophila*, *Meth. Enzymol.* **159**:786-792.

Davis, R.L., et al., 1989, Cloning and characterization of mammalian homologs of the *Drosophila* dunce⁺ gene, *Proc. Natl. Acad. Sci. USA* **86**:3604-3608.

Dudai, Y., et al., 1976, *dunce*, a mutant of *Drosophila* deficient in learning, *Proc. Natl. Acad. Sci. USA* **73**:1684-1688.

Eckmann, F., et al., 1988, Rolipram in major depression: results of a double-blind comparative study with amitriptyline, *Curr. Therap. Res.* **38**:23-39.

Horowski, R., and Sastre-Y-Hernandez, M., 1985, Clinical effects of the neurotropic selective cAMP phosphodiesterase inhibitor rolipram in depressed patients: global evaluation of the preliminary reports, *Curr. Therap. Res.* **38**:23-39.

Pittler, S.J., et al., 1989, Molecular characterization of human and bovine rod photoreceptor cGMP phosphodiesterase α-subunit and chromosomal localization of the human gene, submitted.

Swinnen, J.V., Joseph, D.R., and Conti, M., 1989, Molecular cloning of rat homologues of the *Drosophila melanogaster* dunce cAMP phosphodiesterase: evidence for a family of genes, *Proc. Natl. Acad. Sci. USA* **86**:5325-5329.

43

Structure and Function of the High Affinity, Rolipram and RO 2-1724 Sensitive, cAMP Phosphodiesterases

J.V. Swinnen, K.E. Tsikalas, and M. Conti

Introduction

The degradation of the second messengers cAMP and cGMP is mediated by phosphodiesterases (PDEs), a complex family of enzymes (Beavo, 1988). Of the diverse forms of PDEs thus far described, the high affinity cAMP PDEs are still poorly understood. Data have accumulated suggesting that the cAMP PDEs comprise two subfamilies on the basis of differences in the sensitivity to cGMP and to specific PDE inhibitors (Beavo, 1988) . There is now general consensus that, in many cells, a form of PDE that is inhibited by cGMP and cilostamide (Weber et al., 1982; Hidaka et al., 1984) coexists with a PDE that is insensitive to cGMP (Weishaar et al., 1985). The latter form is specifically inhibited by rolipram and RO 20-1724 (Weishaar, 1985). The exact relationship between these two subclasses of PDEs has not been defined. It is not known, for instance, if these two forms are the products of two distinct genes or if they are the result of post-transcriptional modifications of a single gene product. Furthermore, even though both forms appear to be under hormonal regulation (Beavo, 1988), the mechanisms mediating this activation are incompletely understood. Here we will report on the characterization of PDEs that belong to the cGMP-insensitive, rolipram and RO 20-1724 inhibited family.

The Sertoli Cell as a Model to Study Phosphodiesterase Regulation

The Sertoli cell of the rat testis is the somatic component of the seminiferous epithelium, performing the function of "nursing" cell for the developing germ cells (Fawcett, 1975). This is similar to the role played by reticulum-endothelial cells in the development of erythrocytes. Some of the features of this cell system make it a very versatile and interesting physiological model for the study of phosphodiesterases and their role in the regulation of cell responsiveness.

The Sertoli cell is a target cell for steroid and protein hormones (Fritz, 1978; Stefanini et al., 1984). The gonadotropin FSH regulates this cell's function via a receptor coupled to adenylate cyclase (Means et al., 1980). One interesting feature of the Sertoli cell is that during testicular development the cell shifts from a stage of high responsiveness to hormone stimulation to a stage of low responsivity (Steinberger et al., 1978; Means et al., 1980). This is associated with terminal differentiation and the production of mature gametes by the seminiferous tubules (Means et al.,1980). The physiological significance and cause of these changes in responsiveness are not well understood. However, together with changes in adenylate cyclase activity (Means et al., 1980), it has been shown that Sertoli cell differentiation is accompanied by changes in the pattern of PDE expression (Geremia et al., 1982). In the undifferentiated Sertoli cell, soluble cAMP hydrolytic activity is represented almost exclusively by the Ca^{++}, calmodulin-dependent PDE (CAM-PDE), while after differentiation, a high affinity cAMP PDE appears in the soluble fraction (Geremia et al., 1982). These changes in PDE activity can be reproduced *in vitro*. Treatment of primary Sertoli cell cultures with FSH, B-adrenergic agonists or cAMP analogs induces the appearance of a high affinity cAMP PDE (Conti et al., 1982). The induction of this form and the consequent changes in the rate of cAMP degradation have been shown to be a cause of the refractory state that follows FSH stimulation (Conti et al., 1983). The fact that phoshodiesterase inhibitors restore the responsiveness of the otherwise refractory Sertoli cell supports this hypothesis (Conti et al., 1983, Conti et al., 1986).

Purification of a cAMP PDE from the Rat Sertoli Cell

In view of the characteristics described above, the Sertoli cell is an optimal source of the high affinity cAMP PDE. Very little cGMP-stimulated or cGMP-inhibited PDE can be recovered in soluble extracts from these cells. The CAM-PDE can be readily separated from the high affinity form during the first chromatography step. And, as previously mentioned, hormone or dibutyryl cAMP treatments cause the appearance of large quantities of a PDE whose characteristics correspond to the cGMP-insensitive, rolipram-inhibited PDE. Thus, the starting material for PDE purification has been the soluble extract of cultured Sertoli cells incubated for 24-48 hours in the presence of 0.5 mM dibutyryl cAMP. Under these conditions the cytosol has a PDE activity of 0.6-1.0 nmoles/min/mg protein, 2-5 times higher than most tissues. Although it has been possible to partially purify the cAMP-PDE using conventional open columns, a purification procedure employing HPLC has several advantages. The entire purification is completed in 48 hours, thus minimizing enzyme degradation. A single symmetrical peak is recovered from the first three columns, but the final hydroxyapatite column yields two peaks of activity (Conti, submitted for publication). These two forms have very similar catalytic properties and sensitivities to inhibitors. The specific activity of the major peak is estimated to be in the order of 2 umoles/min/mg protein. SDS-PAGE and silver staining of fractions containing this predominant form show two major polypetides of 86-88 kDa and 67-68 kDa. The staining intensity of the 67 kDa polypetide closely follows the activity of the major peak, suggesting that this polypeptide contains the catalytic activity. Since the yield has been low, it has not been possible to define the size of the polypetide corresponding to the smaller peak of activity.

The PDE preparations obtained following the above procedure hydrolyze cAMP with a high affinity (K_m 1.6-2.0 uM cAMP) and linear kinetics. Cyclic GMP inhibits cAMP hydrolysis only at concentrations higher than 50 uM, confirming that this form is insensitive to cGMP. The enzyme is sensitive to RO 20-1724 and rolipram but is only marginally inhibited by cilostamide. The molecular weight estimated by gel filtration is 130,000 Da, while as previously mentioned, SDS-PAGE indicates the presence of a 67-68 kDa polypeptide. The reason for this difference has not been established. It can be speculated that the native enzyme is a dimer, but we have no experimental evidence, aside from that reported, to support this hypothesis.

Due to the low yield of this purification, we have not been able up to obtain microsequence of this PDE, nor have we been successful in raising antibodies against the purified polypeptide.

298

Cloning and Characterization of the Sertoli Cell and other Testicular PDEs

To obtain further information about the structure of the hormone regulated cAMP PDE from the rat Sertoli cell, testicular and Sertoli cell libraries were screened using a *Drosophila melanogaster* "dunce" cDNA. This cDNA encodes a cAMP-PDE (Chen et al., 1986). Sequencing of clones derived from these screenings demonstrated the presence of four separate classes of cDNAs (Swinnen et al., 1989a). Since nucleotide sequences were dissimilar even in the most conserved regions, it is concluded that these 4 groups of cDNAs correspond to 4 different genes. These have been named ratPDE1, ratPDE2, ratPDE3, and ratPDE4. Clones corresponding to ratPDE2 and ratPDE4 have also been obtained independently by two other laboratories screening rat brain libraries (Davis et al., 1989; Colicelli et al., 1989).

To understand the biological significance of the observed heterogeneity of the different PDEs, the properties of the proteins encoded by the four genes have been analyzed by using antipetide antibodies and by expressing these cDNAs in prokaryotic and eukaryotic organisms.

Although the data need to be confirmed by further experiments, analysis of the open reading frame indicates that the proteins encoded by these cDNAs have molecular weights between 57 and 68 kDa (Swinnen et al., 1989a; Davis et al., 1989, Colicelli et al., 1989). All four sequences contain a core region that is highly conserved. On the basis of several observations, it is thought that this region corresponds to the catalytic center of the enzyme. This region, in fact, has remarkable similarities to most PDE sequences thus far published (Chen et al., 1986; Charbonneau et al., 1986; Ovchinnikov et al., 1987; Sass et al., 1986). Furthermore, it contains a sequence found also in the cAMP binding domain of the RII regulatory subunit of the cAMP-dependent protein kinase (Scott et al., 1987; Bubis et al., 1988), suggesting that, in the PDEs, this region interacts with the substrate.

Antipeptide antibodies against specific sequences of ratPDE3

Based on the deduced amino acid sequence of ratPDE3, synthetic peptides corresponding to 3 different regions have been synthesized and used to immunize several rabbits. The antisera obtained are still incompletely characterized, but they have been useful in clarifying a number of issues (Swinnen et al., paper in preparation). One of these antisera was used in western blot analysis of a partially purified preparation of cAMP-PDE from the Sertoli cell. This showed that the major immunoreactive species migrated with a MW of 68 kDa. A smaller band of 64 kDa was also observed. The same antibody immunoprecipitated between 80 and 90% of the cAMP PDE activity purified from the Sertoli cell. These data indicate that epitopes of ratPDE3 are shared by the cAMP PDE purified from the Sertoli cell and thus support the hypothesis that ratPDE3 corresponds to the FSH stimulated cAMP PDE. Further studies on the regulation of rat PDE3 confirm this view (see below).

Expression of ratPDE3 and ratPDE4 in bacteria

The cloning data demonstrate that at least two distinct cAMP PDEs (ratPDE3 and ratPDE4) are present in the Sertoli cell (Swinnen et al., 1989a). However the biochemical data available do not allow us to distinguish between these two forms.

To compare the properties of ratPDE3 and ratPDE4, these two cDNAs were expressed in bacteria. This approach has the advantage of providing a preparation which contains only one form of PDE, completely devoid of contamination by similar forms. This is particularly advantageous because, in light of the cloning and purification data, the preparation of cAMP PDEs from mammalian tissues is most likely a mixture of more than one closely related form. Furthermore, large quantities of bacteria can be prepared in a cost-effective manner. It should be pointed out that the information obtained by this expression provides information about the intrinsic activity of these enzymes. Posttranslational modifications occurring in eukaryotic cells are different from those occurring in bacteria. Therefore, the activity expressed by the recombinant PDE might be different from that of the native form. In some instances, this is an advantage, because it provides molecules devoid of post translational modifications. This represents a suitable model to study the effect of post translational modifications, like phosphorylation, on the PDE activity.

To express the entire putative open reading frames of ratPDE3 and ratPDE4, these sequences were inserted downstream from the lambda P_L promoter and a consensus ribosome binding site in the pRC23 plasmid, a vector designed by R. Crowl (Crowl et al., 1985). In the presence of a plasmid (pRK248cIts) encoding a temperature-sensitive lambda cI repressor, transformed E. coli bacteria express the PDE cDNA upon a temperature shift from 30^oC to 42^oC. This control of PDE expression was found necessary because constitutive expression of the PDE appears to be deleterious for bacterial growth. The activity thus far obtained has been characterized in terms of K_m for cAMP, inhibition by cGMP, and sensitivity to different PDE inhibitors. It was found that these two PDEs, apart from a reproducible difference in K_m for cAMP (1.6 uM versus 4 uM), are identical in terms of sensitivity to different inhibitors.

The successful expression in prokaryotic cells indicates that the ratPDE cDNAs contain all the information necessary to encode cAMP PDE activity. It also supports the concept that neither eukaryotic-specific post-translational modifications nor eukaryotic factors like activators are strictly required for activity.

Expression of the PDEs in eukaryotic cells

To obtain further information about differences between ratPDE3 and ratPDE4, cDNAs corresponding to these two enzymes were expressed in monkey COS cells (Swinnen et al., 1989b). The expression confirmed that indeed a high affinity cAMP is encoded in the cDNAs that we have isolated. Surprisingly, the preliminary studies thus far performed show that, while increased PDE activity was recovered in the supernatant fraction of COS cell homogenates after ratPDE3 transfection, no clear increase in PDE activity was observed after transfection with a vector containing ratPDE4 cDNAs. The cause of this difference is under investigation.

Regulation of the Expression of the cAMP PDEs

A possible significance of the heterogeneity of the cAMP PDEs is that each component of this family of enzymes is differentially regulated, the cAMP degradation being highly specialized in different compartments of the cell. With the cloning of cDNAs encoding cAMP PDEs, it has been possible to study the expression of these enzymes by measuring their mRNA levels. Our studies show that complex regulations control the expression of these PDEs.

Northern blot analysis of different tissues with cDNA probes corresponding to the four PDEs indicates organ-specific differences in expression (Swinnen et al., 1989a; Swinnen et al., 1989b). Transcripts corresponding to ratPDE1 are detected only in the testis and the kidney. Conversely, ratPDE2 mRNA is present in brain, heart and testis, but not in kidney and liver. Transcripts corresponding to ratPDE3 and ratPDE4 are present in all organs studied. Thus, these PDEs are differentially expressed in various organs. A further level of complexity is provided by the differences in transcript sizes. For instance, ratPDE2 is expressed with two different transcripts in the brain and four in the testis. This, together with evidence of alternate splicing, suggests that cAMP-PDEs might be highly heterogeneous at the protein level.

Differences in expression are also present within an organ. In the testis, all the data thus far collected indicate that ratPDE1 and ratPDE2 expression is confined to the germ line, while ratPDE3 and ratPDE4 are present in somatic cells (Swinnen et al., 1989a). It is then possible that different types of cells within the same organ express different kinds of cAMP PDEs.

Finally, taking advantage of the Sertoli cell model, it has been possible to study the hormonal regulation of the mRNAs coding for these PDEs. It was found that very low levels of ratPDE3 are present in the immature Sertoli cell in culture. Treatment of the cells with FSH or dibutyryl cAMP produces a more than 100-fold increase in ratPDE3 mRNA (Swinnen et al., 1989b). Similar findings were obtained using two other cell cultures, C6 glioma cells and FRTL-5 thyroid cell line. Thus, a hormone-dependent increase in the intracellular cAMP levels leads to increased expression of this PDE. This is a ubiquitous intracellular regulatory feedback mechanism, in which cAMP is regulating the expression of its own degrading enzymes.

References

Beavo, J.A., 1988, *Adv Second Messenger and Phosphoprotein Res* **22**:1-38.
Bubis, J., et al., 1988, *J. Biol. Chem.***263**:9668-9673.
Charbonneau, H., et al., 1986, *Proc. Natl. Acad. Sci. USA* **83**:9308-9312.
Chen, C.N., Denome, S., and Davis, R.L., 1986, *Proc. Natl. Acad. Sci. USA* **83**:9313-9317.
Colicelli, J., et al., 1989, *Proc. Natl. Acad. Sci USA* **86**:3599-3603.
Conti, M., et al., 1982, *Endocrinology* **110**:1189-1196.
Conti, M., et al., 1983, *Endocrinology* **113**:1845-1853.
Conti, M., et al., 1986, *Endocrinology* **118**:901-908.
Crowl, R., et al., 1985, *Gene* **38**:31-38.
Davis, R.L., et al., 1989, *Proc. Natl. Acad. Sci. USA* **86**:3604-3608.
Fawcett, D.W., 1975, in Greep, R.O., and Astwood, E.B. (eds) *Handbook of Physiology*, Baltimore, Williams and Wilkins p21.
Fritz, I.B., 1978, In G. Litwach (ed) *Biochemical actions of hormones*, New York, Academic Press, pp. 249-281.
Geremia, R., 1982, *Molec. Cell. Endocrinol.* **28**:37-53.
Hidaka, H., and Endo, T., 1984, *Adv. Cyclic Nucleotide and*
Means, A.R., et al., 1980, *Annu. Rev. Physiol.***42**:59.
Ovchinnikov, Yu.E.A., et al., 1987, *FEBS Lett.* **223**:169-173.
*Protein phosphorylation Res.***16**:245-259.
Sass, P., et al., 1986, *Proc. Natl Acad. Sci. USA* **83**:9303-9307.
Scott, J.D., et al., 1987, *Proc. Natl. Acad. sci. USA* **84**:5192-5196.

Stefanini, M., et al., 1984, In Metz and Monroy, (eds) *Biology of Fertilization*, Academic Press, pp. 3-46.

Steinberger, A., Hintz, M. and Heindel, J.J., 1978, *Biol. Reprod.*190:556-572.

Swinnen, J.V., Joseph, D.R., and Conti, M., 1989, *Proc. Natl. Acad. Sci. USA* 86:5325-5328.

Swinnen, J.V., Joseph, D.R., and Conti, M., 1989, *Proc. Natl. Acad. Sci. USA* **86** in press.

Weber, H.W., and Appleman, M.M., 1982, *J. Biol. Chem.* **257**:5339-5341.

Weishaar, R.E., Cain, M.H., and Bristol, J.A., 1985, *J. Med. Chem.*28:537-545.

44

Correlation of 3H-Rolipram Binding to a Cerebral cAMP Phosphodiesterase with Antidepressant Activity

H.H. Schneider and H. Wachtel

Introduction

Rolipram, which has a proven effectiveness as an anti-depressant in the clinic (Eckmann et al., 1988) indirectly enhances noradrenergic transmission. The phosphodiesterase (PDE) inhibitory action of rolipram has been known for several years, yet the high selectivity and potency of cAMP PDE inhibition which had been suggested from behavioural experiments (Wachtel, 1982) has been confirmed only recently after partial purification of a rolipram-sensitive PDE isozyme (RsPDE) by affinity chromatography (Schneider et al., 1988). ^{3}H-rolipram binds specifically to brain protein constituents (Schneider et al., 1986), which may allow an alternative access to the target site of rolipram's action.

Animal models for potential antidepressant agents are related to the monoamine hypothesis of depression. This is because the classical antidepressant drugs are of the type of the tricyclic monoamine uptake inhibitors, or of the monoamine oxidase inhibitors. Rolipram does not exert its pharmacological activity at the level of monoamine receptors but presumably through inhibition of a specific cAMP PDE isozyme (Wachtel, 1983). The cAMP system has gained considerable interest in recent years since chronic treatment of rats with drugs effective in antidepressant therapy rather consistently produce subsensitivity of the β-adrenergic cAMP generating system (Charney et al., 1981). This has also been demonstrated for rolipram (Schultz et al., 1986).

We employed reserpine-induced hypothermia in mice and the induction of head twitches in rats as neuropharmacological tests in antidepressant drug finding routines. The present investigation compares a number of compounds in their potency to inhibit cAMP PDE enzymes, to compete with ^{3}H-rolipram at brain binding sites in vivo and in vitro, and to affect the two pharmacological parameters. The compounds selected belong to variants of rolipram as well as to some PDE inhibitors from different structural classes.

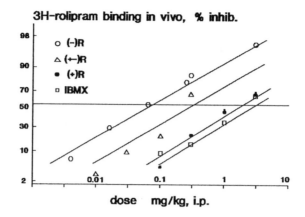

Figure 1 ^{3}H-rolipram binding in vivo to mouse forebrain membranes. Compounds were administered IP 30 min before sacrifice

Materials and Methods

Rolipram (+-)-R and its enantiomers (+)-R,(-)-R, and 3 structural derivatives thereof (A,B,C) were synthesized by Drs. R. Schmiechen and G. Kirsch, ^{3}H-rolipram by Dr. B. Acksteiner, Schering AG. IBMX was from Aldrich Chemicals, reserpine as ampoule formulation from Ciba, Switzerland. We thank the respective companies for the gift of Ro 20-1724 (Hoffmann-La Roche, Switzerland), ICI 63.197 (Imperial Chemical Industries, U.K.), TVX 2706 (Tropon-Werke, F.R.G.)

Experimental

Biochemical Tests. ^{3}H-rolipram binding in vitro using pig brain cortex supernatant (Schneider et al., 1986) and PDE assay at 0.5 μM ^{3}H-cAMP (Marks et al., 1974) were performed as described. Two PDE preparations were used: total PDE activity of the pig brain supernatant (S-PDE), and an affinity purified fraction of rolipram-sensitive PDE (RsPDE; Schneider et al., 1988). For reasons of comparability, the assays were performed at 30 °C. ^{3}H-rolipram binding in vivo was determined 30 min after IP administration of drugs, dissolved in 10 % w/v Cremofor EL (BASF, F.R.G.) and, in the rat, 2 min following the injection of ^{3}H-rolipram (540 kBq in 0.5 ml; 799 GBq/mmol) into the tail vein. The animals were decapitated, the brain dissected immediately and homogenized in ice-cold buffer, 25 mM sodium phosphate, pH 7.4. Three 2-ml aliquots of the homogenate were filtered immediately through Whatman GF/B glass fiber discs soaked with 0.3 % polyethylenimine and washed by three 3-ml portions of cold buffer. The schedule for mice differed only in that the tracer (185 kBq in 0.25 ml) was given 20 min before sacrifice. The inhibitory potency of the drugs on the binding of ^{3}H-rolipram in vivo was determined

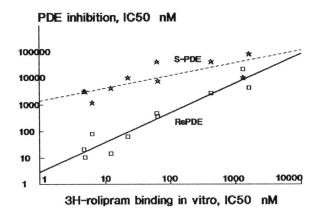

PDE inhibition, IC50 nM

S-PDE

RsPDE

3H–rolipram binding in vitro, IC50 nM

Figure 2   Correlation of PDE inhibitory potency for S-PDE and RsPDE, respectively, and ^{3}H-rolipram binding affinity in vitro for the set of 10 compounds

using 3 - 4 groups of 4 - 5 animals per dose. Basal and non-specific binding were defined by administration of vehicle and (-)-rolipram, 3 mg/kg, respectively. Non-specific binding at a level of 3 - 10 % of basal binding was subtracted from all binding values. ED_{50} values were evaluated graphically from logit % inhi- bition of specific binding vs. log dose plots (figure 1).

All the compounds could totally inhibit ^{3}H-rolipram binding in vitro, and the RsPDE activity. The inhibition of S-PDE revealed very shallow curves, only 60 % inhibition being observed at the highest concentration of 100 μM, with the exception of IBMX. The potencies of the compounds in inhibiting the two PDE preparations were correlated with their affinities to ^{3}H-rolipram binding sites. While the correlation was highly significant between inhibition of RsPDE and ^{3}H-rolipram binding in vitro (r=0.956, p=0.0001), this was much less so when S-PDE was used (r=0.795, p=0.0060, see figure 2).

<u>Pharmacological Tests.</u> Antagonism of reserpine-induced hypothermia was determined immediately following reserpine administration, 10 mg/kg IP (Wachtel et al., 1986). Individual mice (NMRI, 25 g) were placed in perspex cages, 4 hours later the animals were given various doses of the test compound or vehicle and rectal body temperature was registered. For determination of head twitches individual rats (Wistar, 100 g) were placed into plastic cages. Fifteen min following drug or vehicle administration, the incidence of head twitches was counted for 60 min by an experienced observer unaware of the treatment. Each treatment group comprised 8 animals. The minimal effective dose (MED) was defined as the lowest dose differing from the corresponding vehicle treated control group (Dunnett test) at the probability level of p < 0.05. The MEDs for the compounds in antagonizing the reserpine induced hypothermia were correlated with their potencies in inhibiting

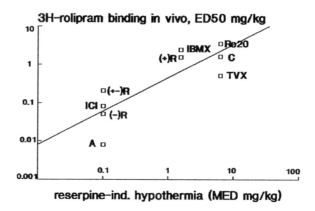

3H-rolipram binding in vivo, ED50 mg/kg

reserpine-ind. hypothermia (MED mg/kg)

Figure 3   Comparison of in vivo ^{3}H-rolipram displacement
activity from mouse forebrain (ED$_{50}$, mg/kg) and antagonism
of reserpine-induced hypothermia in mice (MED, mg/kg) 30 min
after IP administration of test compounds. Correlation
coefficient r = 0.832, p = 0.0054

^{3}H-rolipram binding in vivo (figure 3). For rats, the in
vivo ^{3}H-rolipram binding affinities were correlated with the
induction of head twitches (figure 4). In both cases, the
correlation was highly significant (p=0.0054, r=0.832, and
p=0.0001, r=0.932, respectively).

Discussion

A small number of PDE inhibitors share the ability of
rolipram to inhibit potently and stereoselectively a cAMP
PDE isozyme as well as to compete with ^{3}H-rolipram for
specific binding sites in brain tissue. The significant
correlation observed between these activities in vitro
prompted us to demonstrate a correlation between RsPDE
inhibition in vivo and pharmacological activity that points
to the envisaged therapeutic goal. The in vivo measurement
of a PDE isozyme activity is not possible experimentally. As
an indirect measure of the drug-target interaction of
rolipram with the RsPDE we investigated in vivo the ^{3}H-
rolipram binding site affinity of the compounds. Our study
correlates the in vivo affinity of a number of RsPDE
inhibitors to the ^{3}H-rolipram binding site in rat and mouse
forebrain with pharmacological measures indicative for
antidepressant activity.

Two independent paradigms for the assessment of
antidepressant drug activity were adopted. Antagonism of
reserpine-induced hypothermia (Askew, 1963; Ross et al.,
1967) is regarded as a classic test model for prediction of
possible clinical antidepressant activity. Induction of head
twitches, besides other behavioural phenomena, has
previously been demonstrated to reflect the enhanced
availability of cerebral cAMP induced by specific cAMP PDE

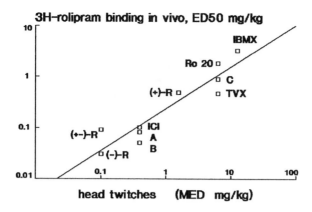

head twitches (MED mg/kg)

Figure 4   Comparison of in vivo ^{3}H-rolipram displacement activity from rat forebrain (ED$_{50}$, mg/kg) and induction of head twitches in rats (MED, mg/kg) 30 min after IP administration of test compounds. Correlation coefficient r = 0.932, p = 0.0001

inhibitors in rats in vivo; it was the first evidence to suggest that rolipram and related compounds act on a distinct subtype of cerebral PDEs (Wachtel, 1982; Schneider, 1984). This type of PDE inhibitor has subsequently been shown in animal experiments to be active in tests predicting antidepressant activity (Wachtel, 1983) which for rolipram has been confirmed in clinical trials (Eckmann et al., 1988).

The correlations were based on data derived from independent experiments using the same species, mode of administration and time of measurement. The correlations reveal a significant interaction, though the evaluation of effectiveness in the pharmacological tests makes use of MEDs and in the binding tests of ED$_{50}$s. This may have caused the shift of the regression lines towards a higher potency of the compounds in the in vivo binding test (figures 3 and 4). Alternatively, this may be interpreted to mean that a high occupation rate of the ^{3}H-rolipram binding sites, i.e. a profound inhibition of the RsPDE, has to be achieved to produce a significant pharmacological effect. The absolute increase in brain cAMP at these doses may be rather low though a significant cAMP increase in rat cerebrum (30 %) has been detected following 0.3 mg/kg rolipram (2 min after IV administration; Schneider, 1984). Depending on the brain structure the effect on cAMP levels may be more pronounced, given the heterogeneous distribution of ^{3}H-rolipram binding sites in the rat brain as visualized by in vitro autoradiography (Kaulen et al., 1989).

In summary, the lines of evidence - from in vitro data of PDE inhibition to the in vivo pharmacological effects - strongly support the notion that the antidepressant activity of rolipram is based on an interaction with a distinct cAMP PDE isozyme, the RsPDE.

Acknowledgements

The authors gratefully acknowledge the competent technical assistence of S. Fritz, J. Seidler, T. Wegner (biochemical studies), P. Böttcher, M. Kunow, P. Pietzuch, K.-J. Rettig, A. Seltz and V. Schulze (pharmacological studies).

References

Askew BM, 1963, A simple procedure for imipramine-like antidepressant agents, Life Sci. **2**: 725-730.

Charney DS, Menkes DB and Heninger GR, 1981, Receptor sensitivity and the mechanism of action of antidepressant treatment, Arch. Gen. Psychiatry **38**: 1160-1180.

Eckmann F, Fichte K, Meya U and Sastre-y-Hernandez M, 1988, Rolipram in major depression: Results of a double-blind comparative study with amitryptiline, Curr. Ther. Res. **43**: 291-295.

Kaulen P et al., 1989, Autoradiographic mapping of a selective cAMP phosphodiesterase in rat brain with the antidepressant ^{3}H-rolipram, Brain Res., in press

Marks F and Raab I, 1974, The second messenger system of mouse epidermis, Bioch. Biophys. Acta **334**: 368-377.

Ross SB and Renyi AL, 1967, Inhibition of the uptake of tritiated catecholamines by antidepressant and related agents, Eur. J. Pharmacol. **2**: 181-186.

Schneider HH, 1984, Brain cAMP response to phosphodiesterase inhibitors in rats killed by microwave irradiation or decapitation, Biochem. Pharmacol. **33**:1690-1693.

Schneider HH, Schmiechen R, Brezinski M and Seidler J, 1986, Stereospecific binding of the antidepressant rolipram to brain protein structures, Eur. J. Pharmacol. **127**: 105-115.

Schneider HH, Pahlke G and Schmiechen R, 1988, Sensitivity of a cAMP PDE to rolipram in different organs. In: Heilmeyer LMG (ed) Signal transduction and protein phosphorylation, NATO ASI Series A, Vol **135**. Plenum Press, New York, pp 81-85.

Schultz JE and Schmidt BH, 1986, Rolipram, a stereospecific inhibitor of calmodulin-independent phosphodiesterase, causes β-receptor subsensitivity in rat cerebral cortex, Naunyn Schmiedeberg's Arch. Pharmacol. **333**:23-30.

Wachtel H, 1982, Characteristic behavioural alterations in rats induced by rolipram and other selective adenosine cyclic 3',5'-monophosphate phosphodiesterase inhibitors, Psychopharmacology **77**: 309-316.

Wachtel H, 1983, Potential antidepressant activity of rolipram and other selective cyclic adenosine 3',5'-monophosphate phosphodiesterase inhibitors, Neuropharmacology **22**: 267-272.

Wachtel H and Schneider HH, 1986, Rolipram, a novel antidepressant drug, reverses the hypothermia and hypokinesia of monoamine-depleted mice by an action beyond postsynaptic monoamine receptors, Neuropharmacology **25**: 1119-1126.

45

Mechanism for Dual Regulation of the Particulate cGMP-Inhibited cAMP Phosphodiesterase in Rat Adipocytes by Isoproterenol and Insulin

C.J. Smith, E. Degerman, V. Vasta, H. Tornqvist, P. Belfrage, and V.C. Manganiello

Introduction

Lipolysis is dually regulated in the rat adipocyte: it is promoted by agents which increase cellular cAMP and antagonized by agents which decrease the synthesis and/or increase degradation of cAMP. The receptor-dependent activations and inhibitions of adenylate cyclase are mediated by the respective GTP-binding transductional proteins, G_s and G_i (Gilman, 1987). Thus, catecholamines, ACTH, and glucagon activate adenylate cyclase and effect the breakdown of triglyceride to glycerol and free fatty acids, via cAMP-dependent protein kinase- (A-kinase) mediated phosphorylation of a hormone-sensitive lipase (Fain, 1980; figure 1). Conversely, basal adenylate cyclase in the rat adipocyte appears to be under tonic inhibition by adenosine, an antilipolytic agent, since incubation of cells with adenosine deaminase (ADA; Schwabe et al., 1975; Honnor et al., 1985) or with pertussis toxin (which inactivates G_i; Gilman, 1987) promotes increases in cellular cAMP, A-kinase and/or lipolysis (Olansky et al., 1983; Smith et al., 1989).

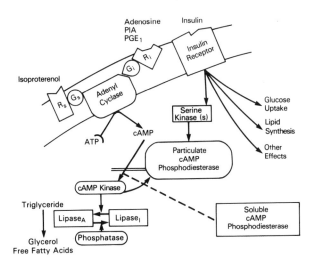

Figure 1.

Insulin is an antilipolytic hormone which can modulate cAMP levels. Since insulin-dependent inhibition of adenylate cyclase has proven to be controversial, activation of a microsomal low K_m cAMP phosphodiesterase (PDE) may be more important in the antilipolytic action of insulin (Elks and Manganiello, 1985; Beebe et al., 1985; Lönnroth and Smith, 1986). Although the detailed mechanisms whereby insulin activates cAMP PDE are unknown, insulin is known to regulate the phosphorylation state of a number of enzymes important in carbohydrate and lipid metabolism (Avruch et al., 1982), and PDE might be the target of an insulin-sensitive serine kinase (Czech et al., 1988). Recent studies also suggest that insulin activation of PDE may regulated by protein similar to G_i since guanine nucleotides activate a similar insulin-sensitive PDE in liver membranes (Heyworth et al., 1983), and incubation of cells with pertussis toxin blocks insulin-activation of PDE in 3T3-L1 adipocytes (Elks et al., 1983) and rat hepatocytes (Heyworth et al., 1986); pertussis toxin also inhibits the antilipolytic action of insulin in rat (Schwabe et al., 1975; Mills and Fain, 1985; Goren et al., 1985) and 3T3-L1 adipocytes (Elks et al., 1983).

Role of the particulate PDE in regulation of A-kinase in rat adipocytes

With the ultimate goal of elucidating the molecular mechanisms involved in hormonal regulation cAMP PDE, our recent studies of particulate PDE, A-kinase and lipolysis were carried out under identical conditions as a function of adenylate cyclase activation or inhibition (Manganiello et al., 1987; Smith and Manganiello, 1989). Studies of the insulin-sensitive PDE are complicated by the fact that the particulate PDE in rat adipocytes is also activated by lipolytic (cAMP-elevating) agents (cf. Francis and Kono, 1982; Gettys et al., 1987), so that it was important to control adenosine concentrations in basal cell incubations by the use of ADA in the absence or presence of the ADA-resistant analog N^6-phenylisopropyladenosine (PIA; Honnor et al., 1985). Adipocytes prepared in this manner exhibited low levels of basal A-kinase and almost undetectable lipolysis.

Incubation of cells with the ß-agonist isoproterenol (ISO, 100 nM) effected a rapid increase (<2 min) in particulate PDE (~2-fold) and A-kinase (~8-10-fold) activities, both of which were stable over the next 25 minutes; glycerol production increased 4-6-fold. The concentration-effect relations for ISO-dependent activation of PDE and lipolysis were similar (Manganiello et al., 1987) and both effects were maximal at an A-kinase activation ratio of 0.5 (Smith and Manganiello, 1989). Similar effects on PDE, A-kinase, and lipolysis were observed when adipocytes were incubated with ADA (Manganiello et al., 1987; Smith et al., 1989) or cAMP analogues (Gettys et al., 1987). The kinetics of activation of A-kinase and particulate PDE suggest that A-kinase-mediated phosphorylation may be responsible for activation of the particulate PDE. Further, considering that PDE activation was observed at sub-maximal activation of A-kinase, it would appear A-kinase mediates rapid feedback inhibition by which the cAMP signal is dampened (Gettys et al., 1987) and/or the turnover of cAMP (and purine nucleotides) is increased (cf. Kather, this volume).

In intact rat adipocytes, the onset of stimulatory effects of ADA were as rapid as observed with ISO; however, ADA-dependent increases in A-kinase and particulate PDE activities were somewhat less stable over the same time period (Smith et al., 1989), and the effects of ADA were more sensitive to adenylate cyclase inhibition. Whereas the stimulatory effects of ADA on PDE and lipolysis were completely inhibited by PIA (or other adenylate cyclase inhibitors), or by the ADA-inhibitor EHNA (Smith et al., 1989), the maximal ISO-activations of PDE and lipolysis were unchanged. However, PIA decreased the sensitivity to ISO for both activation of

PDE and lipolysis (Smith and Manganiello, 1989). Stimulatory effects of adenosine or PIA on basal PDE activity in rat adipocytes were reported by Wong et al. (1985); however, these effects were seen with ~300-fold higher concentrations of PIA (10 μM) than those used to maximally inhibit ADA-dependent PDE activity (Smith et al., 1989).

In the absence lipolytic agents and presence of adenosine, insulin activation of particulate PDE (~50%) was slower (maximal in 10-15 min) than that seen with ISO or ADA, and was not associated with a change in basal A-kinase or lipolysis (Smith and Manganiello, 1989). However, in adipocytes incubated with both ISO and insulin, there was a transient and synergistic activation of particulate PDE activity within 9 min, which then decreased to the level observed with ISO alone after ~20 minutes. The synergistic activation of PDE was associated with a reduction in ISO-stimulated A-kinase and inhibition of lipolysis (Smith and Manganiello, 1989). Similar, but more rapid effects on PDE, A-kinase and lipolysis were observed when adipocytes were incubated with insulin and ADA. However, in the presence of ISO, nanomolar PIA was required for effective insulin regulation of the synergistic activation of PDE, inhibition of A-kinase and antilipolysis (Smith et al., 1989). The differences in kinetics, cAMP-dependence and the synergistic effects on PDE activity clearly suggest distinct PDE activation mechanisms for insulin and lipolytic agents (Smith and Manganiello, 1989; Francis and Kono, 1982).

In the rat adipocyte, G_i protein regulation of PDE activity may be more important for the cAMP-dependent and perhaps "basal" activation state of PDE. Incubation of adipocytes with pertussis toxin (PTX; 1 μg/ml for 2 hours) evoked the same effects as acute incubations with ADA, except that PIA inhibitions of A-kinase, particulate PDE and lipolysis were blocked after PTX. Since insulin lowered A-kinase and inhibited lipolysis stimulated by ADA or PTX (Smith et al., 1989), it appears that adenosine-mediated inhibition of adenylate cyclase is not required for the antilipolytic effect of insulin. Consistent with these results, Weber et al. (1987) reported that PTX blocked ISO- rather than insulin-dependent PDE activation. (In the absence of PIA, ISO plus ADA was only slightly greater than the effect of ADA alone; Smith et al., 1989). These results clearly differed from the effects of pertussis toxin on insulin-sensitive PDE in the 3T3-L1 adipocyte (Elks et al., 1983), which may be related to the unique insulin-mimetic (and pertussis toxin-sensitive) effect of PIA on basal PDE seen in 3T3-L1 adipocyte (Elks et al., 1987).

Immunoprecipitation of cAMP PDE from intact rat adipocytes

Given the importance of the hormone-sensitive cAMP PDE in adipocyte metabolism, we have recently focused our attention on molecular mechanisms of this dual hormonal regulation. To this end, the cGMP- and cilostamide-inhibited particulate low K_m cAMP PDE from rat (Degerman et al., 1987) and bovine (Degerman et al., 1988) adipose tissues was purified 65-100,000-fold by using cilostamide (a potent and selective PDE inhibitor) as an affinity column ligand. On silver-stained SDS reducing gels, the bovine PDE appeared as a 77 kDa polypeptide with a 61-63 kDa doublet. Polyclonal antibodies raised against the pure bovine PDE specifically recognized these same polypeptides on Western immunoblots, and were cross-reactive with the 62/66 kDa PDE purified from the rat (Degerman et al., 1988).

The purified PDEs were substrates in vitro for cAMP-dependent protein kinase, and the ^{32}P-labeling of either the rat or bovine PDE was blocked by the addition of the heat-stable A-kinase inhibitor (Degerman, 1988). Once it was established that the antibody immunoprecipitated either of these ^{32}P-labeled tracers, in the presence or absence of unlabeled rat adipocyte membranes, in situ PDE phosphorylation studies were commenced with intact rat adipocytes (Degerman et al., 1989).

Intact rat adipocytes were prelabeled with ^{32}P-orthophosphate for 2 hours, followed by incubation in the absence or presence of ISO (300 nM, for 3 min) or insulin (1 nM, 12 min). Cells were then homogenized and detergent-solubilized membrane fractions were prepared for immunoprecipitation. Immunoprecipitates from hormone-treated cells contained two phosphoproteins -- a major one at 135 kDa and minor and more variable 44 kDa protein, the labeling of which were at least 10-fold greater than that observed in material from basal cells. This pattern was unexpected since the pure rat PDE is 62/66 kDa; however, both phosphoproteins appeared to be immunologically related to PDE since the immunoprecipitation was prevented if the antibody was preincubated with excess pure bovine (Degerman et al., 1989) or rat PDE.

In order to further characterize the immunologic identity of the phosphoproteins isolated from intact rat adipocytes, solubilized membranes from basal and hormone-activated cells were carried through partial purification on a DEAE-Sephacel column that was eluted with steps of increasing NaBr. DEAE fractions were then precipitated with TCA and prepared for Western immunoblotting onto PVDF membranes (Degerman et al., 1989). With the anti-bovine PDE, an immunoreactive 135 kDa protein was observed in the solubilized membrane prior to DEAE, it was not detectable in the pass-through or low salt wash (in which >90% of the initial protein and <13% of PDE were recovered), and it was greatly enriched per mg protein in the 225 mM salt fraction. About 60% of PDE activity was eluted in this same fraction, with five times less activity recovered in either the 100 mM or 450 mM salt eluates. Less reactive bands of ~90 and ~44 kDa were also detectable in the 225 mM NaBr fraction when 30 μg of protein was present on the PVDF blot. When similar DEAE experiments were carried out with material from hormone-treated ^{32}P-labeled cells, the immunoprecipitated 135/44 kDa phosphoproteins cofractionated with PDE activity; however, a 90 kDa phosphoprotein was never isolated (Degerman et al., 1989).

To further identify the phosphoproteins isolated from the intact adipocytes, membranes from ^{32}P-labeled cells were carried through a two-step partial purification which should be selective for the hormone-sensitive low K_m cAMP PDE. PDE activity which eluted from DEAE-Sephacel was applied to a cilostamide affinity column. In preparations from hormone-treated but not basal cells, the 130 kDa and to a variable extent, 44 kDa phosphoproteins were recovered (Degerman et al., 1989) from eluates of the affinity column. These results suggest, in analogy with cAMP-dependent phosphorylation of 110 kDa PDE in intact human platelets (Grant et al., 1988; MacPhee et al., 1988), that the PDE subunits isolated during the two-week purification may represent proteolytic fragments of the native 135 kDa species.

Activation of cAMP PDE and phosphorylation of the 135 kDa particulate protein

ISO- or insulin-dependent [^{32}P]-serine phosphorylations of the 135 kDa protein (Degerman et al., 1989) were half-maximal at ~3 nM ISO and ~1 pM insulin (Smith et al., 1989), which were 3-4-fold lower K_{act} values than those observed for activation of PDE (Smith and Manganiello, 1989). The time courses for maximal hormone-dependent phosphorylation of the 135 kDa particulate protein in intact cells, however, were in closer agreement with the kinetics of PDE activation (Smith and Manganiello, 1989; Smith et al., 1989): ISO- or ADA-induced labeling was maximal in 1-2 min, which was ~3 times more rapid than that in the presence of insulin; ISO- or insulin-induced labeling was more stable over 25 min as compared to that with ADA. (Smith et al., 1989). The greater sensitivity of phosphorylation may be related to the larger signal for labeling (4-25-fold) versus that for PDE activation (<3-fold). ADA-induced increases in both 135 kDa phosphorylation and PDE activation in intact adipocytes were prevented or reversed with PIA in a dose-dependent manner (K_i ~0.1 nM); similar results were obtained for propranolol inhibition of the effects of ISO (Smith et al., 1989). These data strongly support the idea that the phosphorylation state of the native particulate PDE (135 kDa protein) is closely related to (if not responsible for) enzyme activation.

Summary

The hormone-sensitive particulate low K_m cAMP PDE appears to be important in the regulation of A-kinase and hence triglyceride metabolism in the rat adipocyte. Consistent with recent studies in human platelets (Grant et al., 1988; MacPhee et al., 1988), the ISO- or ADA-activated adipocyte PDE is probably activated by A-kinase-mediated phosphorylation (Gettys et al., 1988). Insulin-induced activation and phosphorylation of PDE is associated with the recruitment of a distinct, and as yet, unidentified serine kinase (Degerman et al., 1989; figure 1). This insulin-dependent activation of PDE does not require, but may be enhanced by adenosine and elevated cAMP (Smith and Manganiello, 1989; Smith et al., 1989). Further studies will involve the identification of the PDE-specific insulin-activated kinase(s) and which particular serine residue(s) are required for the dual hormonal regulation of PDE.

References

Avruch, J., et al., 1985, Protein phosphorylations as a mode of insulin action, *Molecular Basis of Insulin Action* (M. Czech, ed.), Plenum Press, New York, 293-296.

Beebe, S. J., et al., 1985, Discriminative insulin antagonism of stimulatory effects of various cAMP analogs on adipocyte lipolysis and hepatocyte glycogenolysis, *J. Biol. Chem.* **260**:15781-15788.

Czech, M., et al., 1988, Insulin receptor signalling: activation of multiple serine kinases, *J. Biol. Chem.* **263**: 11017-11020.

Degerman, E., et al., 1987, Purification of the putative hormone-sensitive cyclic AMP phosphodiesterase from rat adipose tissue using a derivative of cilostamide as a novel affinity ligand, *J. Biol. Chem.* **262**:5797-5807.

Degerman, E., et al., 1988, Purification, properties and polyclonal antibodies for the particulate, low K_m cAMP-phosphodiesterase from bovine adipose tissue, *Second Mess. and Phosphoprot.* **12**: 171-182.

Degerman, E., 1988, Purification, characterization and regulation of the hormone-sensitive cAMP-phosphodiesterase from adipose tissue, *Ph.D. thesis, Univ. of Lund, Sweden.*

Degerman, E., et al., 1989, Evidence that insulin and isoprenaline activate the cGMP-inhibited low K_m cAMP-phosphodiesterase in rat fat cells by phosphorylation, *Proc. Nat. Acad. Sci. (USA)* (in press).

Elks, M. L., et al., 1983, Selective regulation by pertussis toxin of insulin-induced activation of particulate cAMP phosphodiesterase activity in 3T3-L1 adipocytes, *Biochem. Biophys. Res. Commun.* **116**:593-598.

Elks, M. L., et al., 1987, Effect of N^6-(L-2-phenylisopropyl)-adenosine and insulin on cAMP metabolism in 3T3-L1 adipocytes, *Am. J. Physiol.* **252** *(Cell Physiol.* **21**): C342-C348.

Fain, J. N., 1980, Hormonal regulation of lipid mobilization from adipose tissue, *Biochemical Actions of Hormones* **VII**:119-204.

Francis, S. H., and Kono, T., 1982, Hormone-sensitive cAMP phosphodiesterase in liver and fat cells, *Mol. Cell. Biochem.* **42**:109-116.

Gettys, T. W., et al., 1987, Short-term feedback regulation of cAMP by accelerated degradation in rat tissues, *J. Biol. Chem.* **262**:333-339.

Gettys, T. W., et al., 1988, Activation of particulate low K_m phosphodiesterase activity of adipocytes by addition of cAMP-dependent protein kinase, *J. Biol. Chem.* **263**: 10359-10363.

Gilman, A. G., 1987, G proteins: transducers of receptor-generated signals, *Ann. Rev. Biochem.* **56**:615-649.

Goren, H. J., Northup, J. K., and Hollenberg, M., 1985, Actions of insulin modulated by pertussis toxin in rat adipocytes, *Can. J. Physiol. Pharmacol.* **63**: 1017-1022.

Grant, P. G., et al., 1988, Cyclic AMP-mediated phosphorylation of the low K_m cAMP phosphodiesterase markedly stimulates its catalytic activity, *Proc. Nat. Acad. Sci. (USA)* **85**:9071-9075.

Heyworth, C. M., et al., 1983, Guanine nucleotides can activate the insulin-stimulated phosphodiesterase in liver plasma membranes, *FEBS Lett.* **154**: 87-91.

Heyworth, C. M., et al., 1986, The action of islet activating protein (pertussis toxin) on insulin's ability to inhibit adenylate cyclase and activate cyclic AMP phosphodiesterases in hepatocytes, *Biochem. J.* **235**: 145-149.

Honnor, R. C., et al., 1985, cAMP-dependent protein kinase and lipolysis in rat adipocytes. I. Cell preparation, manipulation and predictability in behavior, *J. Biol. Chem.* **260**:15122-15129.

Lönnroth, P., and Smith, U., 1986, The antilipolytic effect of insulin in human adipocytes requires activation of the phosphodiesterase, *Biochem. Biophys. Res. Comm.* **141**:1157-1161.

Macphee, C. H., et al., 1988, Phosphorylation results in activation of a cAMP phosphodiesterase in human platelets, *J. Biol. Chem.* **263**:10353-10358.

Manganiello, V. C., et al., 1987, Hormonal regulation of adipocyte particulate "low K_m" cAMP phosphodiesterase, *J. Cyc. Nuc. Prot. Phosph. Res.* **11**:497-511.

Mills, I., and Fain, J. N., 1985, Pertussis toxin reversal of the antilipolytic action of insulin in rat adipocytes in the presence of forskolin does not involve cyclic AMP, *Biochem. Biophys. Res. Comm.* **130**:1059-1065.

Olansky, L., et al., 1983, Promotion of lipolysis in rat adipocytes by pertussis toxin: reversal of endogenous inhibition, *Proc. Nat. Acad. Sci. (USA)* **80**: 6547-6551.

Schwabe, U., et al., 1975, Adenosine release from fat cells: Effects on cyclic AMP levels and hormone actions, *Adv. Cyclic Nucleo. Res.* **5**:569-584.

Smith, C. J., and Manganiello, V. C., 1989, Role of hormone-sensitive low K_m cAMP phosphodiesterase in regulation of cAMP-dependent protein kinase and lipolysis in rat adipocytes, *Molec. Pharmacol.* **35**:381-386.

Smith, C. J., et al., 1989, Hormone-sensitive low K_m cAMP phosphodiesterase in rat adipocytes: G_i protein regulation of insulin- and cAMP-dependent phosphorylation, (manuscript in preparation).

Weber, H., et al., 1986, Insulin stimulation of cAMP phosphodiesterase is independent from G-protein pathways involved in adenylate cyclase regulation, *J. Cyc. Nuc. Res. Prot. Phos. Res.* **11**: 345-354.

Wong, E. H.-A., et al., 1985, The action of adenosine in relation to that of insulin on the low-K_m cyclic AMP phosphodiesterase in rat adipocytes, *Biochem. J.* **227**: 815-821.

46
Expression of Calmodulin-Dependent Cyclic Nucleotide Phosphodiesterase in Developing and Adult Rat Brain

M.L. Billingsley, J.W. Polli, C.D. Balaban, and R.L. Kincaid

Introduction

High concentrations of calmodulin (CaM) and CaM-binding proteins such as CaM-dependent cyclic nucleotide phosphodiesterase (CaM-PDE) are expressed in adult neural tissue, and play important roles in calcium-dependent neuronal functions (Caceres, et al., 1983). By characterizing age-related and developmental changes in CaM-PDE, it may be possible to determine potential mechanisms which govern cyclic nucleotide metabolism in developing and aging brain.

In adult brain, CaM-PDE (Kincaid, et al., 1987), calcineurin (Goto, et al., 1986) and CaM kinase II (Ouimet, et al., 1984) have been immunolocalized in neurons. CaM-PDE has been localized to somatodendritic regions of neurons in specific brain regions; such localization is consistent with a role in postsynaptic catabolism of cyclic nucleotides. CaM-PDE is composed of either 60 or 63 KDa subunits which can exist as either homo- or heterodimers (Beavo, 1988). These holoenzymes exhibit different substrate and kinetic properties towards the hydrolysis of cAMP and cGMP; the 63 KDa homodimer has a greater Vmax against cGMP, while the heterodimer and homodimer of the 60 KDa subunit has a greater Vmax for cAMP. Multiple forms of other PDEs have been observed, and effort has been spent determining tissue specificity, kinetic differences, and sensitivity to inhibitors.

The metabolism of cAMP has been studied in embryonic and early postnatal rat brain (Salinas, et al., 1982). Adenylate cyclase and total phosphodiesterase activity were present at low levels during embryonic development; their activities gradually rose, achieving adult levels by postnatal day 30. However, only total PDE activity was determined. We have determined developmental changes in CaM-PDE activity and immunoreactivity, with an ultimate goal of determining mechanisms which regulate its expression in neurons.

315

Preliminary data suggests that CaM-PDE is regulated, in part, by synaptic input (Balaban, et al., 1989). When climbing fiber innervation to cerebellar Purkinje fibers was ablated using 3-acetylpyridine lesioning techniques, we observed a specific loss of CaM-PDE only in cerebellar Purkinje cells, but not in other neurons. Other CaM-dependent enzymes in cerebellum such as calcineurin and CaM kinase II were not affected by such treatment. One hypothesis which emerged from this study was that CaM-PDE may be under transsynaptic control in specific brain regions, and that such control mechanisms may play important roles during synaptogenesis and development. In this report, we hope to summarize several observations concerning the developmental patterns of CaM-PDE expression, to indicate that an isoform of CaM-PDE may be present in developing brain, and to suggest possible implications resulting from the temporal pattern of CaM-PDE expression.

Materials and Methods

Phosphodiesterase Assay:

Tissue fractions from brain were dissected, pooled from 3-5 animals, homogenized, and separated into cytosolic and crude synaptosomal fractions. Incubation conditions were as previously described (Kincaid and Manganiello, 1988), except that 50 μM cAMP was used as a substrate. CaM-dependent activity was determined in the presence of 2 mM calcium and 1 μg purified bovine brain CaM; basal activity was determined in the presence of 2 mM EGTA. Results were expressed as pmol 5'AMP formed/min/μg protein.

Immunocytochemistry and Immunoblotting Procedures:

Sprague-Dawley rats were mated based on the phase of estrus as determined by vaginal lavage. Timed-pregnant females were used for all studies. When necessary, dams were anesthetized with sodium pentobarbital (50 mg/kg; i.p.) and fetuses removed via caesarian section. Immunocytochemistry was performed as previously described, using 4% paraformaldehyde as a fixative. Neonatal rats were perfused transcardially using a 10 cc syringe. Tissue sections were prepared using a vibratome, and all antibody procedures were performed as previously described (Balaban, et al., 1989). The antisera against CaM-PDE was affinity-purified, and demonstrated to have minimal cross-reactivity against other brain proteins (Kincaid, 1988).

For immunoblotting, equal amounts of protein were prepared from developing tissues, and electrophoresed using 10% sodium dodecylsulfate polyacrylamide gels (SDS-PAGE). Resolved proteins were transferred to nitrocellulose sheets, and blots incubated with affinity-purified antisera against CaM-PDE. Immune complexes were visualized using alkaline-phosphatase conjugated secondary antibody.

Chromatographic Procedures:

In order to isolate potential CaM-PDE isoforms present only in embryonic rat brain, a series of chromatographic procedures were employed.

Embryonic brain cytosolic fractions were prepared from brains of 60 embryonic day 15 (E15) rats and 11 postnatal day 0 rats (PND0) using centrifugation, and chromatographed on a mixed-bed column of Biogel P-10 (5 x 5 cm) and DEAE-Trisacryl (5 x 10 cm). The column was eluted with buffer A consisting of 10 mM Hepes, pH 7.4, 250 mM NaCl, 1 mM of EGTA and EDTA, 100 μM leupeptin and 100 μM phenymethysulfonylfluoride. This fraction was dialysed against buffer A without NaCl, and the dialysate was charged on a second DEAE-Trisacryl column; this column was developed with a linear gradient of NaCl (0- 0.3 M). Alternate fractions were subjected to SDS-PAGE, transferred to nitrocellulose and blots were incubated with affinity-purified antisera against CaM-PDE and detected using alkaline phosphatase-conjugated secondary antibody. Fractions enriched in CaM-PDE immunoreactivity were pooled, concentrated using ultrafiltration, and loaded onto a molecular sizing column (TSK-HW50S; 5 x 90 cm). Fractions were collected and subjected to SDS-PAGE and immunoblotting procedures. The sizing column was calibrated using a series of gel filtration standards (BioRad).

Experimental

<u>Differential Localization of CaM-PDE in Adult Brain:</u>

One feature of CaM-PDE distribution in adult rat brain is that considerable regional enrichment exists in the levels of immunoreactivity. This is illustrated in figure 1, which demonstrates that CaM-PDE immunoreactivity is enriched in amygdaloid regions in a pattern almost opposite of the calmodulin-dependent phosphatase calcineurin. This pattern of localization illustrates that there is considerable regional enrichment of CaM-PDE, and that there is no evidence to suggest that CaM-PDE and CN are regulated in tandem fashion.

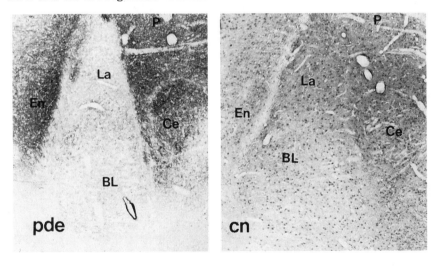

Figure 1: CaM-PDE and Calcineurin (CN) Immunoreactivity in Amygdala. Abbreviations are En-entorhinal cortex; BL-basolateral amygdala; La-lateral amygdala; Ce-central amygdala; P-piriform cortex.

Developmental Pattern of CaM-PDE Activity in Midbrain:

Developmental changes in CaM-PDE activity determined for midbrain cytosol and synaptosomes and is shown in Figure 2. CaM-PDE activity was low during embryonic development, with activity increasing between PND5-20. Activity of CaM-PDE in cytosolic fractions was considerably greater than that in crude synaptosomes. The marked increase in CaM-PDE activity during PND5-20 corresponds to a time of active synaptogenesis, suggesting that the establishment of synaptic connections may modulate expression of this enzyme.

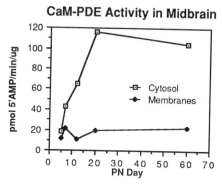

Figure 2: CaM-PDE Activity in Developing Midbrain

Immunoblot Analysis of CaM-PDE in Midbrain:

In order to determine whether levels of immunoreactive CaM-PDE were in concordance with the appearance of enzyme activity, an immunoblot analysis of developing midbrain was conducted. As shown in Figure 3, levels of immunoreactive CaM-PDE increased in both cytosolic and synaptosomal fractions during PND5-20. Immunoreactivity was associated with a 61 KDa peptide in cytosolic fractions, and a 63 KDa peptide in P2 fractions. This suggests the possibility that different isoforms of CaM-PDE reside in different cellular compartments; the mechanisms for such attachment and the consequences of such localization are not known.

Expression of a Potential Isoform of CaM-PDE in Embryonic Brain:

Several developmental schemes concerning the concerted actions of calcium and cyclic nucleotides during axonal growth cone migration have assumed that CaM activation of CaM-PDE may regulate levels of cAMP during development. A combination of immunoblot and activity measurements suggests that 61 and 63 KDa isoforms of CaM-PDE are below limits of detection in early embryonic brain, making it unlikely that these isoforms play dominant roles in early neuronal development. However, tissue-specific isoforms of CaM-PDE have been described (Rossi, et al., 1988); thus, we attempted to determine whether any immunoreactive isoforms of CaM-PDE were present in embryonic brain.

Studies from human SMS-KCNR neuroblastoma cells suggested that a 75 KDa peptide was immunoreactive with the affinity-purified CaM-PDE

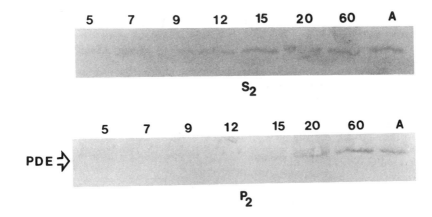

S₂

PDE ⇒

P₂

Figure 3: Immunoblot Analysis of CaM-PDE in Developing Midbrain Cytosol (S2) and Synaptosomes (P2). Numbers above each lanne refer to postnatal day of development.

antisera. Furthermore, induction of differentiation in these cells resulted in a shift of immunoreactive species from the 75 KDa peptide to a doublet at 61-63 KDa. One possible interpretation of such data was that the 75 KDa peptide was an isoform of CaM-PDE only expressed in relatively immature neuronal tissue. Figure 4 shows the result of immunoblot analysis on E15 and E17 rat brain cytosol. A peptide of 75 KDa, termed p75-PDE, was observed in E15 tissues; no evidence of this potential isoform has been seen in adult tisses (data not presented).

E15 E17

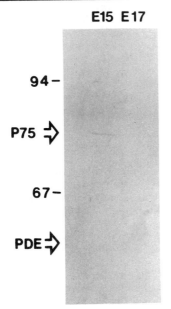

94 –

P75 ⇒

67 –

PDE ⇒

Figure 4: Immunoreactive p75-PDE in Embryonic Rat Brain. A total of 100 ug of cytosolic brain tissue from E15 and E17 fetuses was electrophoresed and subjected to immunoblot analysis using affinity-purified CaM-PDE antisera. The p75 arrow indicates the immunoreactive protein seen in E15 tissue, while the PDE arrow shows the normal position of the 61-63 KDa adult CaM-PDE isoforms.

In an attempt to characterize this isoform, we performed a series of chromatographic procedures. After initial DEAE-Trisacryl chromatography, a second DEAE column was developed using a linear NaCl gradient; p75-PDE immunoreactivity was eluted from the column at 0.15 M NaCl. Immunoblot analysis confirmed that the immunoreactivity was associated primarily with a 75 KDa peptide. Fractions containing immunoreactive p75-PDE were concentrated using ultrafiltration, and chromatographed on a TSK-HW50S gel permeation column. Results of this fractionation are shown in Figure 5, and indicated that p75-PDE immunoreactivity eluted between 150-170 KDa. Immunoblot analysis of these fractions shows that prominent CaM-PDE immunoreactivity was associated a 75 KDa peptide, with lesser immunoreactivity associated with peptides of 61 and 42 KDa. It is not known whether these additional immunoreactive peptides represent proteolytic products of p75-PDE. Thus, it is possible that a potential developmental isoform of CaM-PDE exists in fetal brain.

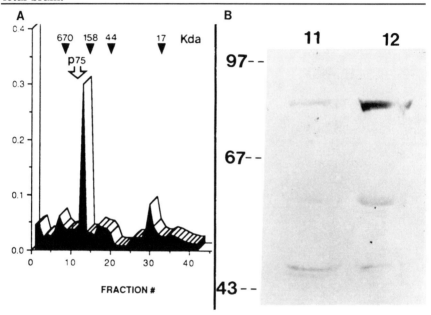

Figure 5: Partial purification of p75-PDE. Panel A shows the elution profile of p75-PDE from a TSK sizing column; the y-axis depicts OD@ 280 nm and the elution pattern of gel filtration standards is shown on the top/ Panel B- Immunoblot analysis of p75-PDE after gel filtration. Fractions 11-12 showed strong p75-PDE immunoreactivity.

Discussion

Considerable efforts have been directed at purifying and characterizing CaM-PDE from adult rat brain. Much less is known concerning the appearance and roles that this enzyme plays in normal development, nor of the mechanisms that govern the selective expression of CaM-PDE. We have demonstrated that CaM-PDE is differentially expressed in adult brain, and that expression is developmentally regulated, with both immunoreactive levels and enzyme activity rising during periods of peak synaptogenesis in the postnatal animal. In the adult rat, the 61 KDa isoform appears to be associated with cytosolic compartments while the 63 KDa isoform is associated with synaptic membranes. We have performed more complete subcellular fractionations, and these results confirm that the 61 and 63 KDa isoforms of CaM-PDE are compartmentalized differentially (Billingsley, et al., submitted for publication). Such subcellular localization may have implications for overall function of each isoform. Further analysis awaits the development of isoform-specific antibodies.

One striking observation consistent throughout brain regions was the relative absence of 61-63 KDa CaM-PDE during embryonic brain development. Although cAMP and calcium have been implicated as important second messengers in the early development of the nervous system (Hyman and Pfenninger, 1985) levels of known isoforms of CaM-PDE were below limits of detection. We attempted to determine whether other potential isoforms may exist based on cross-immunreactivity with the affinity-purified antisera against the 61-63 KDa adult forms. Initial results presented in this study suggest that there is a 75 kDa peptide which is immunoreactive with the affinity purified antisera. Initial biochemical characterization suggests that the p75-PDE exists as a dimer of 150 KDa, and elutes from DEAE colums at 0.15 M NaCl. However, several strong caveats must be considered concerning the role of this protein in cyclic nucleotide hydrolysis. First, we have not demonstrated that this protein can hydrolyse cyclic nucleotides in a CaM-dependent manner. Second, some degree of structural analysis such as limited proteolytic mapping or protein sequencing needs to be performed in order to demonstrate the potential role for this protein in the CaM-dependent hydrolysis of cyclic nucleotides in the developing nervous system.

Studies are in progress to explore transsynaptic mechanisms which may control CaM-PDE expression during synapse formation and to explore the structural and enzymologic features of embryonic forms of CaM-PDE. By approaching the question of CaM-PDE regulation in the developing nervous system, we hope to understand some of the mechanisms which regulate cyclic nucleotide metabolism in the brain.

References

Balaban, C.D., Billingsley, M.L., and Kincaid, R.L. (1989) Evidence for trans-synaptic regulation of calmodulin-dependent cyclic nucleotide phosphodiesterase in cerebellar Purkinje cells. J. Neurosci 9:2374-2381.

Beavo, J.A. (1988) Multiple isozymes of cyclic nucleotide phosphodiesterase. Advances in Second Messenger and Phosphoprotein Research, 22:1-37.

Caceres, A., Bender, P., Snavely, L., Rebhun, L.I., and Steward, O. (1983) Distribution and subcellular localization of calmodulin in adult and developing brain tissue. Neuroscience, 10:449-461.

Goto, S., Matsukado, Y., Mihara, Y., Inoue, N., and Miyamoto, E. (1986). The distribution of calcineurin in rat brain by light and electron microscopic immunohistochemistry and enzyme immunoassay. Brain Res. 397:161-172.

Hyman, C. and Pfenninger, K.H. (1985). Intracellular regulators of neuronal sprouting: calmodulin-binding proteins of nerve growth cones. J. Cell. Biol. 101:1153-1160.

Kincaid, R.L., Balaban, C.D., and Billingsley, M.L. (1987) Differential localization of calmodulin-dependent enzymes in rat brain: evidence for selective expression of cyclic nucleotide phosphodiesterase in specific neurons. Proc. Natl. Acad. Sci. USA. 84:1118-1122.

Kincaid, R.L. (1988) Preparation, characterization, and properties of affinity-purified antibodies to calmodulin-dependent cyclic nucleotide phosphodiesterase and the protein phosphatase calcineurin. Meth. Enzymol. 159:626-636.

Kincaid, R.L. and Manganiello, V.C. (1988) Assay of cyclic nucleotide phosphodiesterase using radiolabelled and fluorescent substrates. Meth. Enzymol. 159:457-470.

Salinas, M., Galan, A., and Herrara, E. (1982) Fetal and early postnatal development of adenylate cyclase and cyclic AMP phosphodiesterase activities in rat brain. Biol. Neonate. 41:94-100.

47

Regulation of cAMP Metabolism in PC12 Cells by Type II (cGMP-Activatable) Cyclic Nucleotide Phosphodiesterase

M.W. Whalin, S.J. Strada, J.G. Scammell, and W.J. Thompson

Introduction

The cyclic nucleotide phosphodiesterase (CN PDE) enzyme system consists of multiple molecular forms. These activities can be distinguished by substrate and immunological specificities, physico-chemical, kinetic and regulatory properties (Strada et al., 1984; Beavo, 1988). A number of drugs have been developed that exhibit pharmacological specificity for one or more isoforms. These drugs can be used to selectively inhibit these enzymes and, therefore can be used to study their role in the intact cell.

One isoform that is abundant in brain (Whalin et al., 1988) and adrenal (Beavo, 1988) tissues is termed the Type II or cGMP-activatable CN PDE. A novel characteristic of this enzyme is that cGMP is able to enhance the hydrolysis of cAMP provided that neither is at saturating levels (Beavo, 1988). This isoform may represent a potential regulatory pathway by which cGMP can modulate cAMP metabolism in cells.

The PC12 rat adrenal medullary pheochromocytoma cell line is a well-characterized cell strain which has been utilized as a model system for the study of neurotransmitter release (Green and Tischler, 1976). These cells have been shown to contain functional adenosine (ADO) receptors coupled to cAMP synthesis (Noronha-blob et al., 1986) and atrial natriuretic factor (ANF) receptors (Rathinavelu and Isom, 1988) coupled to cGMP synthesis (Fiscus et al., 1987). Evidence suggests the involvement of both cAMP and cGMP in the release of neurotransmitters in this cell line. For example, Rabe and McGee (1983) reported that elevation of cAMP levels in response to adenosine potentiated depolarization-dependent norepinephrine release from PC12 cells. On the other hand, ANF has been shown to inhibit neurotransmitter release in these cells (Drewett et al., 1988). The mechanism by which ANF inhibits release is unknown, but its effects are correlated with an increase in cGMP levels (Drewett et al., 1989). The results from the above studies suggest that there may be an interaction between the two cyclic nucleotide systems.

To date, there have been no reports related directly to the regulation of cyclic nucleotide metabolism in this cell line. The goals of the present study were: (1) to characterize the CN PDE activities present in PC12 cells; and (2) to investigate the possible role of Type II (cGMP-activatable) CN PDE activity in the regulation of cAMP metabolism in the intact cell.

Characterization of Cyclic Nucleotide Phosphodiesterase Activity in Subcellular Fractions from PC12 Cells

Centrifugation of PC12 cell sonicates showed an equal distribution of total CN PDE activity between the cytosolic and particulate compartments when measured at saturating cAMP or cGMP substrate concentrations. Activity analysis under conditions designed to optimize the contribution of the various isoforms, including the use of isozyme-selective inhibitors in _in vitro_ assays, showed that the Type II (cGMP-activatable) CN PDE was the dominant isoform in both the cytosolic and particulate fractions. No Type I (calcium/calmodulin-activatable) CN PDE or Type IV (cGMP-inhibitable) low K_m PDE activity were detectable. A small amount of a rolipram-sensitive Type IV (cGMP-insensitive) cAMP PDE activity was found to be enriched in particulate fractions.

Fractionation of PC12 cell 20,000 X g supernatants by DEAE-cellulose ion-exchange chromatography demonstrated a major peak of CN PDE activity in which cAMP, cGMP-stimulated cAMP, and cGMP hydrolysis coeluted (figure 1). This peak of activity was not inhibited by several isozyme-specific inhibitors (e.g., CGS 9343B, RO 20-1724, rolipram, LY 195115, or RS 82856) but was very sensitive to inhibition by the isoquinoline compound HL-725 (Trequisin).

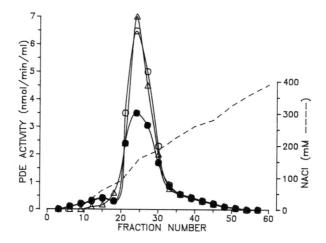

Figure 1. DEAE-cellulose chromatographic profile of CN PDE activity present in 20,000 X g supernatant from PC12 cells. Recovery was 80% of the applied activity. • 5 μM cAMP, o 5 μM + 2 μM cGMP, Δ 25 μM cGMP.

The major peak of CN PDE activity obtained after DEAE chromatography was purified further using cGMP-Sepharose affinity chromatography. The purified enzyme demonstrated a major protein band at 102,000 daltons with 7.5% SDS gel electrophoresis. Non-denaturing gel conditions showed that cGMP-stimulated cAMP hydrolysis comigrated with the major protein band. Gel electrophoresis (7.5% SDS-PAGE in the second dimension) confirmed that phosphodiesterase activity correlated with an identifiable protein of M_r 102,000 daltons. The kinetic and regulatory properties of the purified Type II CN PDE from PC12 cells are similar to those of enzymes purified from bovine heart, adrenal and liver (Martins et al., 1982; Yamamoto et al., 1983) and rabbit brain (Whalin et al., 1988).

Cyclic AMP Accumulation in Intact PC12 Cells

Cyclic AMP accumulation was studied in the intact cell using an adenine prelabeling technique (Kebabian et al., 1979). This method has been used successfully in brain slices (Smellie et al., 1979; Whalin et al., 1989) and S49 lymphoma cells (Barber et al., 1988) to investigate the role of CN PDE activity in intact cells.

PC12 cells, in which ATP stores had been prelabeled with $[^3H]$-adenine, rapidly accumulated cAMP in response to treatment with ADO. Steady state levels were reached within 5 min and remained constant for 10 min in the continued presence of agonist. For the cAMP accumulation experiments described below, cells were incubated for 5 min with a maximally effective concentration of ADO (50 µM). When CN PDE inhibitors were used, cells were preincubated with drug for 10 min prior to the addition of ADO. Cyclic GMP analogs or cGMP elevating agents such as SNP or ANF were added simultaneously with ADO.

Selective inhibitors of Type I (calcium/calmodulin-activatable) CN PDE [CGS 9343B], Type III (cGMP-specific) PDE [M&B 22948], Type IV (cGMP-inhibitable) low K_m PDE [LY 195115] and Type IV (cGMP-insensitive) cAMP PDE [rolipram, RO 20 1724] had no effect on ADO-stimulated cAMP accumulation in PC12 cells (figure 2). In contrast, the isoquinoline compounds, HL-725 (Trequisin) and papaverine markedly potentiated the ability of ADO to increase cAMP accumulation.

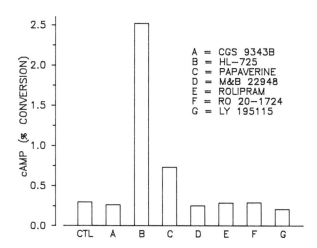

Figure 2. Effect of selective CN PDE inhibitors on ADO-stimulated cAMP accumulation in PC12 cells. Inhibitors (100 µM) were added 10 minutes prior to ADO (50 µM). The ordinate represents the ratio of $[^3H]$cAMP to total cellular tritium which is >90% $[^3H]$ATP.

The results of these experiments suggest that a form of CN PDE sensitive to inhibition by the isoquinoline derivatives, HL-725 and papaverine is involved in regulating agonist-stimulated cAMP accumulation in PC12 cells.

Treatment of PC12 cells with SNP (500 µM) or ANF (1 µM) alone resulted in a 5-8 fold increase in cGMP accumulation with cGMP levels increasing rapidly to a new steady state level within 2 min and remaining constant for 10 minutes (data not shown). Figure 3 shows the effect of either ANF, SNP or two cGMP analogs, (8-bromo-cGMP or N^2, $O^{2'}$-dibutryl-cGMP) on ADO-stimulated cAMP accumulation. Treatment of PC12 cells with ANF or SNP resulted in a 50% attenuation of cAMP accumulation elicited by ADO, whereas cGMP analogs (which failed to activate cAMP hydrolysis <u>in vitro</u>) had no effect. HL-725 abolished the SNP or ANF attenuation of ADO-stimulated cAMP accumulation (data not shown). These results indicate that the attenuation of the cAMP response is specific for cGMP since neither 8-bromo- nor dibutyrl-cGMP reproduced the effects of SNP or ANF. The results also suggest that the HL-725 sensitive enzyme and the mediator of cGMP action may be the same protein i.e., the Type II (cGMP-activatable) CN PDE.

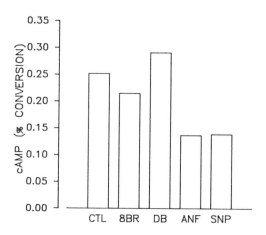

Figure 3. Effect of 1 mM 8-bromo-cGMP (8Br), 1 mM N^2, $O^{2'}$-dibutryl-cGMP (DB), 1 μM atrial natriuretic factor (ANF) or 500 μM sodium nitroprusside (SNP) on ADO-stimulated cAMP accumulation.

Cyclic AMP Decay in Intact PC12 Cells

The results of the cAMP accumulation studies suggest that the inhibition of Type II CN PDE by HL-725 results in a potentiation of the ADO-stimulated cAMP response and that ANF- or SNP-induced elevation of cGMP and subsequent activation of the Type II CN PDE may mediate the attenuation of the ADO-stimulated cAMP response. An alternative explanation for these observations may be that the activity of adenylate cyclase is altered by these agents.

To address this issue further, cAMP decay experiments in intact cells were performed using published procedures (Barber et al., 1988; Whalin et al., 1989). When the action of ADO is terminated by the addition of adenosine deaminase (ADA) the resulting decay of cAMP is a function of CN PDE activity. In the cAMP decay experiments, ANF and/or HL-725 were added simultaneously with the addition of ADA. Figure 4 shows that the cAMP decay in the absence of ANF or HL-725 was very rapid reaching basal, non-stimulated levels within 2 min. Cyclic AMP decay in the presence of ANF was accelerated as compared to control, whereas addition of HL-725 retarded cAMP decay. HL-725 abolished the effects of ANF on cAMP decay.

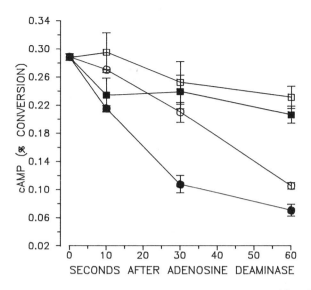

Figure 4. Effect of ANF and HL-725 on cAMP decay in PC12 cells.
o Control, ● ANF 1 μM, □ HL-725 (100 μM); ■ ANF 1 μM + HL-725 (100 μM).

Conclusions

The results of these studies show that the Type II (cGMP-activatable) CN PDE represents the major cAMP hydrolytic activity present in PC12 cells. The properties of this enzyme are similar, if not identical, to those of purified enzymes from other sources (Martins et al., 1982; Yamamoto et al., 1983; Whalin et al., 1988). Experiments in intact PC12 cells indicate that the alteration of the catalytic status of the Type II (cGMP-activatable) CN PDE is manifested in changes in cAMP metabolism. That is, inhibition of the enzyme by HL-725 markedly potentiates the cAMP response elicited by ADO and retards the rate of cAMP decay. Furthermore, the activation of the enzyme by an increase in cGMP induced by ANF or SNP can attenuate cAMP accumulation and increase the rate of cAMP decay. Future studies will address the involvement of this enzyme in the ANF inhibition of neurotransmitter release.

The results of these studies show that PC12 cells may provide a model system for the study of the coupling of adenosine receptors to cAMP accumulation and the role of Type II CN PDE in regulating cyclic nucleotide metabolism and cyclic nucleotide mediated cellular events.

Acknowledgements

This work was supported by USPHS Grants GM33538 and DK40593, USAF (44920-85-K-0014), a Faculty Development Award from the Pharmaceutical Manufacturers Association, and a Predoctoral Fellowship from Eli Lilly & Co. The authors thank Ms. Judi Naylor for her secretarial assistance in preparing this manuscript.

References

Barber, R., Goka, T.J., and Butcher, R.W., 1988, Role of high affinity cAMP phosphodiesterase activities in the response of S49 cells to agonists, Mol. Pharmacol. 32:753-759.

Beavo, J.A., 1988, Multiple isozymes of cyclic nucleotide phosphodiesterase, Adv. Second Messengers and Phosphoprotein Res. 22:1-38.

Drewett, J., et al., 1988, Atrial natriuretic factor inhibits norepinephrine release in an adrenergic clonal cell line (PC12), Eur. J. Pharmacol. 150:175-179.

Drewett, J., et al., 1989, Cyclic guanosine 3',5'-monophosphate mediates the inhibitory effect of atrial natriuretic factor in adrenergic, neuronal pheochromocytoma cells, J. Pharmacol. Exp. Ther. 250:428-432.

Fiscus, R.R., et al., 1987, Atrial natriuretic factors stimulate accumulation and efflux of cyclic GMP in C6-2B rat glioma and PC12 rat pheochromocytoma cell cultures, J. Neurochem. 48:522-528.

Greene, L.A., and Tischler, A.S., 1976, Establishment of a noradrenergic clonal line of rat adrenal pheochromocytoma cells which respond to nerve growth factor, Proc. Natl. Acad. Sci. USA 73:2424-2428.

Kebabian, J.W., Kuo, J.-F., and Greengard, P., 1972, Determination of relative levels of cyclic AMP in tissues or cells prelabeled with radioactive adenine, Adv. Cyclic Nucleotide Res. 2:131-136.

Martins, T.J., Mumby, M.C, and Beavo, J.A., 1982, Purification and characterization of a cyclic GMP-stimulated cyclic nucleotide phosphodiesterase from bovine tissues, J. Biol. Chem. 257:1973-1979.

Noronha-blob, L., et al., 1986, Pharmacological profile of adenosine A_2 receptors in PC12 cells, Life Sci. 39:1059-1067.

Rabe, C.S., and McGee, R., Jr., 1983, Regulation of depolarization dependent release of neurotransmitters by adenosine: Cyclic AMP-dependent enhancement of release from PC12 cells, J. Neurochem. 41:1623-1634.

Rathinavelu, A., and Isom, G.E., 1988, High affinity receptors for atrial natriuretic factor in PC12 cells, Biochem. Biophys. Res. Commun. 156:78-85.

Smellie, F.W., Daly, J.W., and Wells, J.N., 1979, 1-isoamyl-3-isobutyl-xanthine: A remarkedly potent agent for the potentiation of norepinephrine, histamine, and adenosine-elicited accumulations of cAMP in brain slices, Life Sci. 25:1917-1924.

Strada, S.J., Martins, M.W., and Thompson, W.J., 1984, General properties of multiple molecular forms of cyclic nucleotide phosphodiesterase in the nervous system, Adv. Cyclic Nucleotide and Phosphoprotein Res. 16:13-29.

Whalin, M.E., Strada, S.J., and Thompson, W.J., 1988, Purification and partial characterization of membrane-associated Type II (cGMP-activatable) cyclic nucleotide phosphodiesterase from rabbit brain, Biochim. Biophys. Acta 972:79-94.

Whalin, M.E., et al., 1989, Correlation of cell-free brain cyclic nucleotide phosphodiesterase activities to cyclic AMP decay in intact brain slices, Second Messengers and Phosphoproteins 12:(in press).

Yamamoto, T., Manganiello, V.C., and Vaughan, M., 1983, Purification and characterization of cGMP-stimulated cyclic nucleotide phosphodiesterase from calf liver, J. Biol. Chem. 258:12526-12533.

48
cGMP-Binding Phosphodiesterase: Novel Effects of cGMP Binding on the Interaction Between Functional Domains in the Enzyme

S.H. Francis, M.K. Thomas, and J.D. Corbin

INTRODUCTION

Historically, the mechanism whereby changes in intracellular cGMP elicit physiological effects has been assumed to parallel the more thoroughly studied cAMP system, i.e., that the increase in cGMP levels causes increased activity of the cGMP-dependent protein kinase (cGK) which then mediates the physiological response. However, in addition to the cGK there are numerous proteins in the cell that bind cGMP with high specificities and must be included when considering the mechanism of cGMP action (Lincoln and Corbin, 1983). These include a cGMP-dependent cation channel protein (Cook and Kaupp, 1987) as well as a family of cGMP binding phosphodiesterases. Some representatives of the latter group are the cGMP-stimulated phosphodiesterase (Martins et al., 1982), the retinal cGMP-binding cGMP-phosphodiesterases (Yamazaki et al., 1980; Gillespie and Beavo, 1988) and a cGMP-binding cGMP-specific phosphodiesterase (cG-BPDE) from lung and platelets (Hamet and Coquil, 1978; Francis, et al., 1980).

The cG-BPDE was first studied in rat lung, and its resolution from the cGK on DEAE-cellulose is shown in Figure 1. A major peak of cGMP phosphodiesterase activity elutes at low salt concentrations, but there is virtually no cAMP phosphodiesterase activity in these fractions (data not shown). The cGMP hydrolytic activity is unaffected by calcium, calmodulin or trifluoperazine. Two peaks of cGMP-specific binding activity can be resolved on this column. In the absence of 3-isobutyl-1-methylxanthine (IBMX), there is one major peak of cGMP binding activity corresponding to the peak of cGK activity. However, in the presence of IBMX, another peak of cGMP binding activity is observed, and this peak of binding corresponds to the peak of cGMP phosphodiesterase activity. The IBMX stimulation of cGMP binding is not due to

329

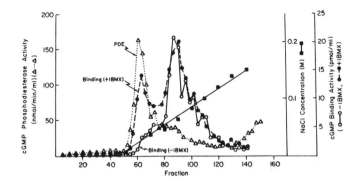

Figure 1. DEAE-cellulose chromatography of rat lung supernatant.

substrate protection since the binding assay is conducted at 0-4° in the
presence of 1 mM EDTA which blocks any phosphodiesterase activity.

CHARACTERISTICS OF THE cG-BPDE

The cG-BPDE has been purified to homogeneity from rat (Francis and
Corbin, 1988) and bovine lung (Thomas, et al., 1990a), and the properties of the enzyme are summarized in Table I. The enzyme exhibits a 100
fold specificity for cGMP over cAMP at both its hydrolytic and catalytic
sites. The cyclic nucleotide analog specificities of the binding site
and the catalytic site are markedly different since the binding site
stringently excludes almost all cyclic nucleotides other than cGMP. The
K_m for cGMP hydrolysis is 5μM, and under optimal conditions the K_d for
cGMP binding to the binding site is 0.2μM. The affinity for cGMP
binding varies significantly with the conditions in the binding assay.
The catalytic activity of the enzyme is insensitive

Table I

PROPERTIES OF THE PURIFIED BOVINE cG-BPDE

Hydrolytic site K_m	5 μM
V_{max}	10 μmol/min/mg
Binding site K_d	0.2 μM
cGMP Bound	2 moles/mol enzyme
Metal Specificity	Mg^{2+} or Mn^{2+}
Native MW	180,000
Subunit MW	93,000
Inhibitor Potencies	zaprinast>dipyridamole>papaverine=IBMX=8MeOxMeMIX>cilostamide>theophylline>rolipram

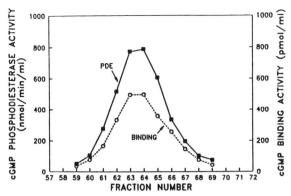

Figure 2. Gel filtration chromatography of purified bovine lung cG-BPDE.

to calcium and calmodulin, and catalysis requires the presence of a divalent cation, either magnesium or manganese. The cGMP binding activity and the cGMP hydrolytic activity co-migrate in all steps of the purification (Figure 2), and the activities correspond to a 93 kDa protein on SDS-PAGE. Based on the molecular weight determination for the native enzyme of ~180,000, this supports the interpretation that the enzyme is comprised of two 93 kDa subunits. Two moles of cGMP are bound per mole of enzyme, and catalysis proceeds with a V_{max} of ~10 µmol/min/mg. Competitive inhibitors of cGMP hydrolysis, such as IBMX, increase cGMP binding to the purified enzyme, and photoaffinity labeling of the enzyme with [^{32}P]cGMP in the presence of IBMX increases the labeling of the 93 kDa protein on SDS-PAGE. cAMP has little effect on the labeling pattern whereas cGMP abolishes the labeling. When a hydrolytic site-specific analog such as N^2-hexyl-cGMP is included in the photoaffinity labeling experiment, the labeling at the binding site is also increased.

INHIBITOR POTENCIES AND EFFECTS

The hydrolytic activity of the cG-BPDE is inhibited by a variety of phosphodiesterase inhibitors as shown in Figure 3. The most potent inhibitors are zaprinast and dipyridamole with K_i values of ~300 and 800 nM, respectively. IBMX, 8-methoxymethyl-3-isobutyl-1-methylxanthine (MeOxMeMIX) and papaverine have very similar K_i values of 6-7 µM, and theophylline and cilostamide are much less potent. Rolipram and RO-1724 are even weaker inhibitors. Thus, this enzyme exhibits an inhibition pattern which is relatively distinct from other cGMP phosphodiesterases. For instance, the inhibition of the hydrolytic activity of the cG-BPDE by zaprinast is at least an order of magnitude more potent than that

331

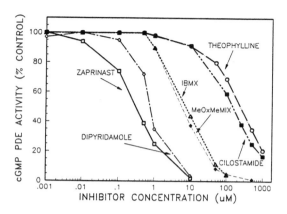

Figure 3. Inhibition of cGMP phosphodiesterase hydrolytic activity by selected inhibitors.

reported for zaprinast inhibition of cGMP hydrolysis by the calcium/cal-modulin stimulated phosphodiesterase. Comparison of the effects of a variety of cyclic nucleotide analogs on the hydrolytic sites of these two phosphodiesterases reveals striking similarities (Francis, et al., 1989). Thus, many of the structural determinants within the catalytic sites of these two enzymes are likely to possess similar features although the substrate specificity for cGMP versus cAMP is far greater for the cG-BPDE than for the calmodulin-stimulated phosphodiesterase. Such similarities must be considered when cyclic nucleotide analogs or phosphodiesterase inhibitors are used to study metabolic processes in vivo since most agents currently in use would have significant overlap between these two enzymes, and perhaps others as well.

All of the phosphodiesterase inhibitors which have been tested increase the affinity for cGMP binding at the binding site, thus emphasizing the communication of a structural change in the protein upon occupation of the catalytic site. By the principle of reciprocity, one would predict that occupation of the binding site would effect changes at the catalytic site. Although such a change has not been demon-strated, cGMP binding to the binding site of the enzyme produces a conformational change in the protein which has been observed as increased electronegativity when the enzyme is chromatographed on DEAE-cellulose.

DOMAIN STRUCTURE OF THE cG-BPDE

Partial chymotryptic digestion of the cG-BPDE cleaves the enzyme into two functional domains, a cGMP-binding domain and a hydrolytic domain (Thomas, et al., 1990a) The hydrolytic function is correlated with a 55 kDa fragment on SDS-PAGE, and the binding function can be correlated with two fragments (45 and 35 kDa) that represent different degrees of proteolysis of the same region of the molecule. The 45 and 35 kDa fragments have identical amino terminal amino acid sequences and by analysis of the molecular weight of the fragment retaining the binding activity (106 kDa), these fragments appear to be dimerized. This suggests that the cGMP binding domain and the dimerization domain reside within the sequence of the 35 kDa fragment. The partial proteolysis of the enzyme greatly decreases the phosphodiesterase activity of the enzyme, but the binding function is markedly increased. The limited proteolysis of the cG-BPDE appears to remove an inhibitory constraint on the binding function to allow its full expression. This reinforces the interpretation that the binding site of this enzyme is unavailable for cGMP binding under basal conditions and becomes accessible through events which alter the conformation of this region of the molecule. Thus, the cGMP binding site on the cG-BPDE is somewhat unique among cGMP binding proteins since a) the site is unavailable for interaction with cGMP in the basal state of the enzyme and requires the induction of a conformational change to make it accessible and b) the binding site exhibits an unusually high stringency for cGMP as compared to other sites which interact with cGMP, including the hydrolytic site of the cG-BPDE, cGMP binding sites of cGK, the cGMP binding sites of the cation channel protein, and the cGMP binding site of the cGMP-stimulated phosphodiesterase.

cGMP-DEPENDENT PHOSPHORYLATION OF THE cG-BPDE BY KINASES

One important function of the cGMP binding site in the cG-BPDE has recently been discovered (Thomas, et al. 1990b). In the presence of cGMP, the cG-BPDE is phosphorylated at a specific serine residue by the catalytic subunit of the cAMP-dependent protein kinase (Thomas, et al., 1990b). The catalytic subunit activity is unaffected by cGMP, so the dependency of the phosphorylation upon cGMP must be due to cGMP interaction with the cG-BPDE. The cG-BPDE is also phosphorylated by the cGK at a rate at least 10 times greater than that observed with the catalytic subunit. Purified protein kinase C and the purified calcium/calmodulin kinase type II do not phosphorylate the cG-BPDE in a cGMP-dependent manner. The phosphorylated peptide from a tryptic digest of the cG-BPDE has been purified and sequenced; it exhibits a typical consensus sequence (KISASEFDRPLR-) for potential phosphorylation sites

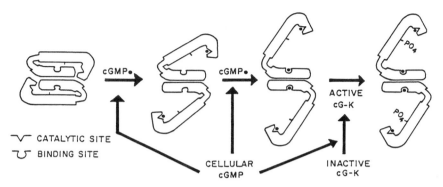

CATALYTIC SITE

BINDING SITE

cGMP

cGMP

ACTIVE
cG-K

CELLULAR
cGMP

INACTIVE
cG-K

Figure 4.   Working model for modulation of cG-BPDE structure upon
interaction with cGMP.

for cyclic nucleotide dependent protein kinases with the first serine in
the sequence being phosphorylated. When the phosphorylation reaction is
conducted in the presence of analogs specific for the hydrolytic site of
the cG-BPDE, no phosphorylation occurs. However, in the presence of
cGMP or a combination of cGMP and IBMX the phosphorylation proceeds
rapidly. Thus, cGMP must occupy the binding site in order for the
phosphorylation site to be available as a substrate for the kinases
(Figure 4). Whether simultaneous interaction of cGMP with the hydrolytic site is required is unknown. However, the dual regulation of a
kinase as well as its substrate is a novel mechanism for second
messenger action. When cGMP concentrations in tissues are increased,
cGMP would bind to the cGK to activate the catalytic function of that
enzyme. Concomitantly, cGMP would interact with the hydrolytic site of
the cG-BPDE which would result in a conformational change allowing
interaction of the cGMP with the binding site of the enzyme. Only at
this stage would the phosphorylation site on the cG-BPDE be available
for modification by the cGK. The site would only be accessible to the
cAMP-dependent protein kinase when cAMP and cGMP were simultaneously
increased. The importance of the phosphorylation event in the functions
of the enzyme and the reasons for such rigid control of access to this
site remain important topics for consideration in our research.

REFERENCES

Cook, N.J., and Kaupp, U.B. (1986): The cGMP-dependent cation channel from
    vertebrate photoreceptors: purification and function reconstitution.
    Photobiochem. Photobiophys. 13, 331-343.

Francis, S.H., Lincoln, T.M. and Corbin, J.D. (1980): Characterization of a novel cGMP binding protein from rat lung. J. Biol. Chem. 255, 620-626.

Francis, S.H. and Corbin, J.D. (1988): Purification of cGMP-binding protein phosphodiesterase from rat lung. Methods Enzymol. 159, 722-729.

Francis, S.H., Thomas, M.K., and Corbin, J.D. (1989): cGMP-Binding cGMP-specific phosphodiesterase from lung. in Cyclic Nucleotide Phosphodiesterases. (J.A. Beavo and M.D. Houslay, eds.) John Wiley & Sons Ltd., Chichester, Eng.

Gillespie, P.G. and Beavo, J.A. (1988): Characterization of a bovine cone photoreceptor phosphodiesterase purified by cyclic GMP-Sepharose chromatography. J. Biol. Chem. 263, 8133-8141.

Hamet, P., and Coquil, J.-F. (1978): Cyclic GMP binding and cyclic GMP phosphodiesterase in platelets. J. Cyclic Nucleotide Res. 4, 281-290.

Lincoln, T.M. and Corbin, J.D. (1983): Characterization and biological role of the cGMP-dependent protein kinase. Adv. Cyclic Nucleotide Res. 15, 139-192.

Martins, T.J., Mumby, M.C. and Beavo, J.A. (1982): Purification and characterization of a cyclic GMP-stimulated cyclic nucleotide phosphodiesterase from bovine tissues. J. Biol. Chem. 257, 1973-1979.

Thomas, M.K., Francis, S.H. and Corbin, J.D. (1990a): Purification and characterization of a bovine lung cGMP binding phosphodiesterase. (submitted for publication).

Thomas, M.K., Francis, S.H. and Corbin, J.D. (1990b): Phosphorylation of a cGMP binding phosphodiesterase. (submitted for publication).

Yamazaki, A., Sen, I., Bitensky, M.W., Casnellie, J.E., and Greengard, P. (1980): Cyclic GMP-specific, high affinity, noncatalytic binding sites on light activated phosphodiesterase. J. Biol. Chem. 255, 11619-11624.

49
Clinical Investigations with Phosphodiesterase Inhibitors

R.E. Weishaar

Introduction

The involvement of the second messengers cyclic AMP and cyclic GMP in regulating cardiac and vascular muscle contractile function, lipolysis, secretion, and other key cellular processes is well known (Hidaka and Endoh, 1984; Elks and Manganiello, 1985; Weishaar et al, 1985a). Efforts to discover therapeutic agents which influence intracellular levels of cyclic AMP and cyclic GMP have focused on two approaches; agents which stimulate cyclic nucleotide synthesis, e.g., beta-receptor agonists, adenylate and guanylate cyclase activators, and on agents which inhibit cyclic nucleotide degradation, e.g., phosphodiesterase inhibitors. Interest in this latter approach has increased in recent years with the discovery of multiple molecular forms of phosphodiesterase, with different substrate specificities, intracellular responsibilities and tissue distributions (Weishaar, 1987). In addition, several new chemical classes have been identified which exert selective inhibitory effects on the different forms of phosphodiesterase (Weishaar, 1987). In the paragraphs to follow a number of these "second generation" phosphodiesterase inhibitors will be described, and the status of studies aimed at evaluating their therapeutic utility will be discussed.

Selective Phosphodiesterase Inhibitors

The different molecular forms of phosphodiesterase can be broadly divided into three distinct classes based on their order of elution from anion-exchange resin (Thompson, et al, 1979). The first class (Type I) is often referred to as the calmodulin-stimulated or cyclic GMP-specific phosphodiesterase. Recently, subclasses of the Type I phosphodiesterase have been identified which vary in their sensitivity to calmodulin and their substrate specificity (Lugnier, et al, 1986; Weishaar, et al, in press). The second class (Type II) is a high K_m enzyme which hydrolyzes both cyclic AMP and cyclic GMP. This enzyme if often referred to as the cyclic GMP-stimulated phosphodiesterase since its ability to hydrolyze cyclic AMP can be enhanced by low concentrations of cyclic GMP (Beavo, et al, 1981). The third class (Type III) of phosphodiesterase is a low K_m enzyme which selectively hydrolyzes cyclic AMP. The activity of this enzyme is potently inhibited by low concentrations of cyclic GMP, and thus this enzyme is often referred to as the cyclic GMP-inhibited phosphodiesterase. Two subclasses of this enzyme have been identified which vary in their sensitivity to cyclic GMP and to different selective Type III phosphodiesterase inhibitors. These two subclasses have been referred to as Type IIIB and IIIC by some investigators (Yamamoto, et al, 1984), and Type III and IV phosphodiesterase by other investigators (Reeves, et al, 1987).

The biochemical characteristics of the different molecular forms of phosphodiesterase are summarized in Table 1.

TABLE 1. Biochemical Characteristics of the Different Molecular Forms of Phosphodiesterase and Their Subclasses

Type of Phosphodiesterase	Subclasses	Substrate	K_m (μM)	Response to Calmodulin	Response to Cyclic GMP
	A	Cyclic AMP, Cyclic GMP	~1	+++	NR
I	B	Cyclic GMP	~5	+	NR
	C	Cyclic GMP	~1	NR	NR
II	None Identified	Cyclic AMP, Cyclic GMP	20-40	NR	+
III	B	Cyclic AMP	~1	NR	-
	C	Cyclic AMP	~1	NR	---

+ = Stimulates enzyme activity

- = Inhibits enzyme activity

NR = No response

Most of the "first generation" phosphodiesterase inhibitors, e.g., theophylline, isobutyl methylxanthine, and papaverine, exert comparable inhibitory effects on all molecular forms of phosphodiesterase (Weishaar, et al, 1985a). In contrast, second generation inhibitors exert a selective inhibitory effect on a particular type of phosphodiesterase. Recently, selective inhibitors of various subclasses of phosphodiesterase, e.g., TCV-3B, M&B 22948, imazodan, and rolipram, have also been identified (Hidaka and Endoh, 1984; Weishaar, 1987; Weishaar, et al, in press). A partial list of selective inhibitors of the different molecular forms of phosphodiesterase is provided in Table 2.

TABLE 2. Selective Inhibitors of the Different Molecular
Forms of Phosphodiesterase and Their Subclasses

Type I			Type II	Type III		
A	B	C		B	C	
None Identified	TCV-3B	M&B 22948	Dipyridamole AR-L 57	Rolipram Ro 20-1724	Amrinone Milrinone Imazodan CI-930 EG-626 Enoximone	Amipizone Y-590 Cilostamide OPC 13,013 OPC 13,135 Piroximone

Clinical Studies with Selective Phosphodiesterase Inhibitors

To date, relatively selective inhibitors of every form of phosphodiesterase except the putative Type IA enzyme have been evaluated in
clinical studies. In most cases, however, these studies do not provide
sufficient information to assess the overall therapeutic utility of these
agents. For several inhibitors, most notably milrinone, rolipram, and
dipyridamole, the findings are somewhat clearer. These three agents have
undergone extensive clinical evaluation for the treatment of congestive
heart failure, depression, and vascular occlusive disease, respectively,
and the results of these studies will be summarized below. In addition,
several studies aimed at identifying more speculative therapeutic uses
for selective inhibitors will also be noted.

Heart Failure

Depending on the severity, therapy for congestive heart failure currently
consists of diuretics, afterload reducing agents, and/or cardiotonics
(Meinertz, et al, 1988). The latter class of compounds is administered
to correct the deficit in myocardial contractile function which underlies
the symptomology generally associated with heart failure, e.g., edema,
shortness of breath, kidney and liver failure, etc. Preclinical studies
with selective Type IIIC phosphodiesterase inhibitors such as amrinone
and imazodan have shown that these agents exert a direct positive
inotropic effect on the heart, presumably due to an increase in
ventricular cyclic AMP levels (Endoh, et al, 1982; Weishaar, et al,
1985b; Silver, 1989). These agents have also been shown to increase
cardiac output and ejection fraction in patients with heart
failure (Benotti, et al, 1978; Jafri, et al, 1986). The first such
inhibitor to be studied in detail was amrinone. The adverse side effects
associated with amrinone administration, e.g., nausea, thrombocytopenia,
and liver function abnormalities, made long-term studies with this agent
difficult to evaluate (Massie, et al, 1985).

338

The selective Type IIIC inhibitors milrinone, which is a more potent and better tolerated analogue of amrinone, and enoximone, which is structurally unrelated to amrinone, have both undergone extensive investigation in more than 40 clinical studies. In a randomized, double-blind crossover study, White et al (1985), demonstrated that administration of milrinone produced an immediate improvement in exercise tolerance of heart failure patients. Acute administration of enoximone was also associated with an improvement in hemodynamic responses in patients with Class III or IV heart failure (Uretsky, et al, 1983). Both agents were effective following intravenous or oral administration.

Long-term uncontrolled clinical trials of 2-12 weeks duration have provided equivocal results regarding the hemodynamic effects of milrinone and enoximone. Most studies with milrinone have reported sustained hemodynamic improvement following long-term administration (Simonton, et al, 1985; LeJemtel, et al, 1986; Hood, 1989). These studies, however, typically noted that overall morality was unaffected or increased with repeated administration of milrinone (Hood, 1989). Mortality was also high in most long-term studies with enoximone, and in five uncontrolled trials only one concluded that enoximone produced a long-term beneficial effect (Uretsky, 1985; Wood and Hess, 1989). Rubin and Tabak (1985) reported that withdrawal of enoximone after five weeks administration did not produce any change in symptoms. Subsequent readministration of higher doses failed to elicit an improvement in hemodynamics. The contribution of a decline in ventricular performance in these patients, not associated with drug administration, to the development of such tolerance cannot be excluded.

DiBianco and coworkers (1989) recently reported the results of a 12-week, double-blind clinical trial in which heart failure patients were treated with milrinone, digoxin, or a combination of the two drugs. The authors demonstrated that milrinone significantly increased exercise tolerance and reduced the frequency of worsened heart failure. However, neither milrinone nor the combination of milrinone plus digoxin offered any advantage over treatment with digoxin alone. There was also a suggestion that milrinone may aggravate ventricular arrhythmias. The authors concluded that no clinical benefit could be derived from substituting milrinone for digoxin in treating heart failure patients, or from adding milrinone to digoxin treatment.

In summary, although the selective Type IIIC inhibitors provide acute benefit in patients with congestive heart failure, long-term clinical trials with these agents have generally been disheartening (Massie, et al, 1985; Uretsky, 1986; Hood, 1989; DiBianco, et al, 1989). The basis for this finding has not been established, but may be related to the adverse effect of increasing contractile activity and calcium influx in severely damaged myocardial tissues. The possibility still exists that these agents might provide useful therapy when administered to patients at earlier stages of the disease, when ventricular performance was less compromised.

Many investigators have suggested that deficiencies in monoaminergic neurotransmission in the central nervous system play an important role in the etiology of endogenous depression (Schildkraut, 1965; Coppen, 1967; Davis, 1970). Such an involvement is supported by the observation that most conventional therapies for depression influence the turnover of norepinephrine in the synaptic nerve terminal, either by inhibiting norepinephrine degradation, e.g., monoamine oxidase inhibitors, or by blocking reuptake of norepinephrine, e.g., tricyclic antidepressants (Glassman and Perel, 1973; Schildkraut, 1973).

Increasing cyclic AMP levels in the nerve terminal represents a novel approach to influencing monoamine neurotransmission, since cyclic AMP has been shown to increase the presynaptic release of norepinephrine, and to potentiate the postsynaptic response to norepinephrine (Wachtel, 1983; Kehr, et al, 1985). In addition, deficiencies of cyclic nucleotide metabolism have been reported for patients with depression (Extein, et al, 1979; Pandley, et al, 1979). Wachtel and coworkers (1989, in press) suggested that decreased cyclic AMP availability coupled with diminished cellular responsiveness to monoaminergic neurotransmitters may represent an essential feature in the etiology of depression.

In preclinical studies the selective Type IIIB phosphodiesterase inhibitors rolipram and Ro 20-1724 produce actions comparable to those observed with the reference antidepressant imipramine (Wachtel, 1983; Przegalinski and Brigajska, 1983; Wachtel and Schneider, 1986). These actions are thought to represent an effect of rolipram and Ro 20-1724 on pre- and postsynaptic levels of cyclic AMP, leading to increased norepinephrine release and enhanced responsiveness (Wachtel, 1983). Subsequent experiments have demonstrated that rolipram does not inhibit monoamine oxidase or reuptake of monoamines, or possess anticholinergic activity (Kehr, et al, 1985; Wachtel, et al, 1988).

When administered to healthy volunteers, rolipram had no effect on pulse rate, blood pressure, or respiratory rate, and did not produce any extrapyramidal motor side effects (Horowski and Sastre-y-Hernandez, 1985). Emesis was observed, but only at high doses of the drug. Uncontrolled studies in patients with depression demonstrate that rolipram possesses antidepressant activity (Zeller, et al, 1984; Horowski and Sastre-y-Hernandez, 1985). Horowski and Sastre-y-Hernandez (1985) reported that the antidepressant response to rolipram was observed within the first ten days of administration, and that the drug was well tolerated. Two double-blind studies in which rolipram was administered for four weeks to patients with depression have recently been reported. In one study, the efficacy of rolipram was compared with the reference antidepressant amitryptyline (Eckmann, et al, 1988), and in the other study rolipram was compared with the reference agent desipramine (Bobon, et al, 1988). Both studies demonstrated that rolipram is an effective agent for treating depression with efficacy similar to that of the two reference agents. Comparable rates of onset were observed for all three drugs. Anticholinergic and hypotensive side effects were observed more commonly for desipramine than rolipram (Bobon, et al, 1988).

Gastrointestinal complaints were the most frequent side effect associated with rolipram administration.

The results of clinical trials conducted thus far with the selective Type IIIB phosphodiesterase inhibitor rolipram demonstrate that altering cyclic AMP metabolism in the central nervous system represents a novel approach to treating depression. The observation that rolipram produces minimal side effects, in contrast to nonselective phosphodiesterase inhibitors such as theophylline, demonstrates the importance of designing agents which exert inhibitory effects on specific forms of the enzyme.

Inhibition of Platelet Aggregation

Dipyridamole (Persantin) is a relatively selective Type II phosphodiesterase inhibitor which has been shown to possess ex vivo antiplatelet activity (Rajak, et al, 1977). This activity is thought to result from the ability of dipyridamole to increase cyclic AMP levels in platelets, leading to inhibition of thromboxane A_2 synthesis and prevention of platelet aggregation (Rivey, et al, 1984). Inhibition of adenosine uptake may also contribute to the effect of dipyridamole on platelet functions (Rivey, et al, 1984).

The involvement of cyclic AMP in modulating arachidonic acid metabolism in platelets prompted Moncada and Korbut (1978) to speculate that the combination of a phosphodiesterase inhibitor with a cyclooxygenase inhibitor could restore hemostatic balance in certain pathological conditions. A number of clinical studies have subsequently evaluated the efficacy of combined dipyridamole and aspirin therapy in treating various conditions associated with altered platelet activity, e.g., coronary and lower extremity bypass grafts, prosthetic valve replacement, and disorders of the peripheral circulation. In addition, this combination has also been evaluated in the treatment of stroke and renal disease, and prevention of a second myocardial infarction.

Harajola, et al (1981), were the first to show that coadministration of aspirin and dipyridamole prevented reocclusion following arterial reconstructive surgery. This finding has subsequently been confirmed in a number of clinical trials (for a complete summary see Rivey, et al, 1984). Several investigators, however, have failed to demonstrate efficacy using this regimen (Pantely, et al, 1979; Sharma, et al, 1983). In these latter trials therapy with aspirin and dipyridamole was initiated postoperatively, in some cases three to five days following surgery. Chesebro and Fuster (1986) have recommended that maximum efficacy is obtained when dipyridamole therapy is begun preoperatively, followed by postoperative addition of aspirin.

The combination of aspirin and dipyridamole has also been found effective in preventing systemic embolism in patients receiving prosthetic heart valves (Hirst, 1981), and in treating patients with peripheral vascular disease (Negus, et al, 1980; Steele, 1980; Hess, 1985). In the latter patients, combination therapy with aspirin and dipyridamole significantly attenuated the natural course of occlusive arterial disease. The beneficial effect observed was also significantly greater than that produced by aspirin alone.

A limited number of clinical studies have also evaluated the efficacy of aspirin and dipyridamole in treating stroke and myocardial infarction. In a randomized double-blind trial of patients undergoing carotid endarterectomy, Findlay and coworkers (1985) reported that perioperative use of aspirin plus dipyridamole may reduce the risk of operative stroke and the long-term risk of repeat carotid stenosis. In the European Stroke Prevention Study (1987), survival curves for patients with a clinical diagnosis of a recent cerebrovascular event of atherothrombotic origin, e.g., transient ischemic attack, showed a significant benefit in favor of patients treated with aspirin and dipyridamole.

Two clinical trials have shown that the combination of aspirin and dipyridamole may decrease the risk of a second myocardial infarction. In the Persantin-Aspirin Reinfarction Study (1980), combination therapy produced a trend toward improved survival rates in patients who had recovered from a documented acute myocardial infarction. A similar finding was reported by Klimt and coworkers (1986). In this latter study, the combination of aspirin and dipyridamole significantly reduced the incidence of nonfatal myocardial infarction, or death due to an acute cardiac event in patients recovered from a previous infarction.

Taken together, these results demonstrate that in those pathological conditions for which alterations in platelet activity are suspected to play an important role the Type II phosphodiesterase inhibitor dipyridamole, in combination with the cyclooxygenase inhibitor aspirin, produces a clinically beneficial effect.

Other Potential Indications for Selective Phosphodiesterase Inhibitors

Mopidamole (RA-233; Rapenton), a derivative of dipyridamole, has been shown to reduce the progression of malignancy in certain experimental animal models (Li and Hellman, 1983; Maniglia, et al, 1982), and in a pilot study in humans (Gastpar, 1982). Zacharski and coworkers (1988) recently compared the efficacy of RA-233 plus chemotherapy with chemotherapy alone in a five-year double-blind trial involving patients with advanced carcinomas of the lung and colon. RA-233 treatment was associated with a statistically significant prolongation of survival in patients with nonsmall cell lung cancer limited to one hemithorax, but not in patients with other types of tumors. This beneficial effect appeared to result directly from RA-233 therapy. The combination of dipyridamole and methotrexate was also associated with clinical improvement in patients with advanced solid tumors (Higano and Livingston, 1989).

The selective Type IC phosphodiesterase inhibitor M&B 22948 (Zaprinast) has been evaluated in several clinical studies for treating asthma. In a double-blind placebo-controlled cross-over trial Rudd and coworkers (1983) showed that M&B 22948 significantly inhibited the exercise-induced fall in forced expiratory volume in asthmatic patients. M&B 22948 did not, however, prevent the fall in forced expiratory volume in response to inhaled histamine. In a subsequent study in asthmatic children, Reiser, et al (1986), showed that M&B 22948 was only effective in preventing exercise-induced asthma in children whose asthma was not steroid dependent.

Topical application of the nonselective phosphodiesterase inhibitor papaverine produces a modest but significant improvement of psoriatic lesions (Van Scott, et al, 1971). Stawiski and coworkers (1979) reported that in two double-blind clinical trials the selective Type IIIB phosphodiesterase inhibitor Ro 20-1724 also improved psoriatic lesions. Although Ro 20-1724 was less effective than intensive occlusive therapy using triamcinolone acetonide, no adverse systemic or cutaneous effects were noted.

Conclusions

In less than two decades the role of phosphodiesterase in regulating cyclic nucleotide metabolism has evolved from the simplistic concept of a single enzyme with ubiquitous distribution which hydrolyzes both cyclic AMP and cyclic GMP, to the present concept of perhaps seven different enzymes, which are differentially distributed in various tissues and cells, each with unique properties and intracellular responsibilities. The therapeutic potential of phosphodiesterase inhibitors has likewise evolved, as a new generation of potent and selective inhibitors has begun replacing the previous generation of nonselective inhibitors such as theophylline and papaverine. Although not all the clinical trials with these new selective inhibitors have produced the anticipated results, studies to date have demonstrated that these agents are effective in treating a number of pathological conditions, including acute heart failure, depression, occlusive vascular disease, and certain forms of cancer, and can also improve recovery following vascular bypass grafting, valve replacement, and myocardial infarction.

References

Beavo, J.A., Hardman, J.G., and Sutherland, E.W., 1971, J. Biol. Chem. 246:3041-3046.

Benotti, J.R., et al., 1978, N. Engl. J. Med. 299:1373-1377.

Bobon, D., et al., 1988, Eur. Arch. Psychiatr. Neurol. Sci. 238:2-6.

Chesebro, J.H., and Fuster, V., 1986, Cardiology 73:292-305.

Coppen, A., 1967, Brit. J. Psychiat. 113:1237-1264.

Davis, J.M., 1970, Intl. Rev. Neurobiol. 12:145-175.

DiBianco, R., et al., 1989, N. Engl. J. Med. 320:677-683.

Eckmann, F., et al., 1988, Curr. Ther. Res. 43:291-295.

Elks, M.L., and Manganiello, V.C., 1985, J. Cell. Physiol. 124:191-198.

Endoh, M., Yamashita, S., and Taira, N., 1982, J. Pharmacol. Exp. Therap. 221:775-783.

The European Stroke Prevention Study, 1987, Lancet 2:1351-1354.

Extein, I., et al., 1979, Psychiatr. Res. 1:191-197.

Findley, J.M., et al., 1985, J. Neurosurg. 63:693-698.

Gastpar, H., 1982, Ann. Chin. Gynaecol. 71:142-150.

Glassman, A.H., and Perel, J.M., 1973, Arch. Gen. Psychiat. 28:649-653.

Harajola, P.T., Meurala, H., and Frick, M.H., 1981, J. Cardiovasc. Surg. 22:141-144.

Hess, H., Mietaschk, A., and Deichsel, G., 1985, Lancet, 1:415-419.

Hidaka, H., and Endo, T., 1984, Adv. Cyclic Nucleotide Res. 16:245-259.

Higano, C.S., and Livingston, R.B., 1989, Cancer Chemo. Pharmacol. 23:259-262.

Hirsh, J., 1981, Arch. Intern. Med. 141:311-315.

Hood, W.B., 1989, Am. J. Cardiol 63(Suppl.):46A-53A.

Horowski, R., and Sastre-y-Hernandez, M., 1985, Curr. Ther. Res. 38:23-29.

Jafri, S.M., et al, 1986, Am. J. Cardiol. 57:254-259.

Kehr, W., Debus, G., and Neumeister, R., 1985, J. Neural. Trans. 63:1-12.

Klimt, C.R., et al., 1986, J. Am. Coll. Cardiol. 7:251-269.

LeJemtel, T.H., et al., 1986, Circulation 73(Suppl. III):III-213-III-218.

Li, X.-T., and Hellman, K., 1983, Clin. Exp. Metastasis 1:51-59.

Lugnier, C., et al., 1986, Biochem. Pharmacol. 35:1743-1751.

Maniglia, C.A., Tudor, G., and Gomez, J., 1982, Cancer Lett. 16:253-260.

Massie, B., et al., 1985, Circulation 71:963-971.

Meinertz, T., Drexler, H., and Just, H., Cardiovasc. Drugs Therap.
2:413-418.

Moncada, S., and Korbut, R., 1978, Lancet, 1:1286-1289.

Negus, D., et al., 1980, Lancet 1:891. Pandley, G., Dysken, M.W., and
Garver, D.L., 1979, Am. J. Psychiat.
136:675-678.

Pantely, G.A., et al., 1979, N. Engl. J. Med. 301:962-966.

The Persantin-Aspirin Reinfarction Study, Circulation 62:V85-V88.

Przegalinski, E., and Bigajska, K., 1983, Pol. J. Pharmacol. Pharm.
35:33-40.

Rajak, S.M., et al., 1977, Br. J. Clin. Pharmacol. 4:129-133.

Reeves, M.L., Leigh, B.K., and England, P.J., 1987, Biochem. J.
241:535-541.

Reiser, J., Yeang, Y., and Warner, J.O., 1986, Br. J. Dis. Chest
80:157-163.

Rivey, M.P., Alexander, M.R., and Taylor, J.W., 1984, Drug Select.
Perspect. 18:869-880.

Rubin, S.A., and Tabak, L., 1985, J. Am. Coll. Cardiol. 5:1422-1427.

Rudd, R.M., et al., 1983, Br. J. Dis. Chest 77:78-86.

Schildkraut, J.J., 1965, Am. J. Psychiat. 122:509-522.

Schildkraut, J.J., 1973, Ann. Rev. Pharmacol. 13:427-454.

Sharma, G.V.R.K., et al., 1983, Circulation 68 (Suppl. II):II-218-II-221.

Silver, P.J., 1989, Am. J. Cardiol. 63(Suppl.):2A-8A.

Simonton, C.A., et al., 1985, J. Am. Coll. Cardiol. 6:453-459.

Stawiski, M.A., et al., 1979, J. Invest. Dermatol. 73:261-263.

Steele, P., 1980, Lancet 2:1320-1329.

Thompson, W.J., et al., 1979, Adv. Cyclic Nucleotide Res. 10:69-92.

Uretsky, B.F., et al., 1983, Circulation 67:823-828.

Uretsky, B.F., et al., 1985, J. Am. Coll. Cardiol. 5:1414-1421.

Uretsky, B.F., 1986, Postgrad. Med. J. 62:585. Van Scott, E.J., and
Farber, E.M. In: Dermatology in General Medicine.
Fitzpatrick, T.B., Arndt, K.A., Clark, W.H., Eisen, A.C.,
Van Scott, E.J., Vaughan, J.E., eds. New York, McGraw-Hill, 1971,
p. 219.

Wachtel, H., 1983, Neuropharmacol. 22:267-272.

Wachtel, H., and Schneider, H.H., 1986, Neuropharmacol. 25:1119-1126.

Wachtel, H., Loschmann, P.-A., and Pietzuch, P., 1988, Pharmacopsychiatr.
21:218-221.

Wachtel, H. In: New Concepts in Depression. Briley, M., Fillion, G., eds. London: McMillian Press (in press).

Weishaar, R.E., Cain, M.H., and Bristol, J.A., 1985a, J. Med. Chem. 28:537-545.

Weishaar, R.E., et al., 1985b, J. Cyclic Nucleotide Protein Phos. Res. 10:551-564.

Weishaar, R.E., 1987, J. Cyclic Nucleotide Protein Phos. Res. 11:463-472.

Weishaar, R.E., et al., Hypertension (in press).

White, H.D., et al., 1985, Am. J. Cardiol. 56:93-98, 1985.

Wood, M.A., and Hess, M.L., 1989, Am. J. Med. Sci. 297:105-113.

Yamamoto, T., et al., 1984, Biochemistry 23:670-675.

Zacharski, L.R., Moritz, T.E., and Baczek, L.A., 1988, J. Natl. Cancer Instit. 80:90-96.

Zeller, E., et al., 1984, Pharmacopsychiat. 17:188-190.

50

Chemical Probes of Type IV Phosphodiesterase: Design, Synthesis, and Characterization of (₃H)-2Y186126 as a Radioligand for Sites Within Cardiac Membranes

D.W. Robertson and R.F. Kauffman

Introduction

During the past decade, there has been an enormous research effort directed towards congestive heart failure (CHF). The etiologies, pathophysiology and natural history of the disease have been studied, and using this knowledge, there have been multiple attempts to design and develop new drugs for its treatment. Fortunately, these efforts have met with some success, and angiotensin converting enzyme inhibitors and vasodilators can now be used to not only ameliorate the symptoms of CHF patients, but increase their duration of life as well. (CONSENSUS Trial Study Group, 1987).

Development of new types of inotropes, which are structurally and mechanistically distinct from the cardiac glycosides and catecholamines, has been the target of many drug development programs, and this effort was stimulated, in large part, by the discovery of amrinone (Alousi, et al., 1986). Although this drug is fraught with many difficulties such as impotency and lack of biochemical specificity, it spawned a host of increasingly more potent and biochemically more specific progeny, including milrinone, isomazole, pimobendan, indolidan, and imazodan (Robertson, et al., 1988). The structures of some of these compounds are depicted in figure 1. In patients with severe CHF, these drugs increase cardiac output by increasing the inotropic state of the heart and by peripheral vasodilation. However, their clinical use has been plagued with considerable controversy, and there is speculation that the acute symptomatic relief brought about by these drugs may be accompanied by long-term decrements in life-span (Packer, 1989).

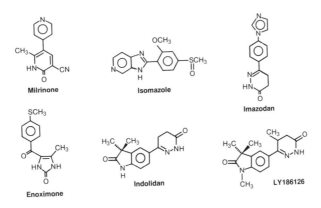

Figure 1. Structures of some PDE-inhibitor cardiotonics.

Although the clinical applications of these drugs will not be delineated fully for some time, basic scientists have clearly benefited from these drug development efforts, and some exceedingly potent and isozymically selective PDE inhibitors are now widely available for biochemical and pharmacological studies. It is now generally accepted that these drugs inhibit the cGMP-sensitive Type IV (cAMP specific) PDE (Strada et al., 1984). Moreover, in cardiac tissue, the Type IV PDE which mediates the inotropic effects of these drugs appears to be present in high density in the sarcoplasmic reticulum (SR) (Kauffman et al., 1986; Kithas et al., 1988). To permit further biochemical and molecular pharmacology studies regarding this class of drugs, we designed a radioligand that could be employed as a label for Type IV PDE, and selected LY186126 (figure 1), an analogue of indolidan, as a suitable candidate for radiolabeling.

Materials and Methods

The synthesis of the precursor to [^{3}H]-LY186126 was recently published (Robertson et al., 1989). In the synthesis of the labeled material, an indolone desmethyl precursor to [^{3}H]-LY186126 was deprotonated with sodium hydride, alkylated with tritium-labed iodomethane, and then purified to produce [^{3}H]-LY186126 (1,3-dihydro-3,3-dimethyl-1-[^{3}H$_3$] methyl-5-(1,4,5,6-tetrahydro-4-methyl-6-oxo-3-pyridazinyl)-2H-indol-2-one).

Purified SR vesicles were prepared as described (Jones et al., 1981), and the conditions for [^{3}H]-LY186126 binding to these vesicles has been described recently (Kauffman et al., 1989). Briefly, the binding studies were conducted using varying concentrations of [^{3}H]-LY186126 in a medium containing 50 mM Tris-Cl,5 mM MgCl$_2$, 0.01% BSA, pH 7.4. After 60 minutes, the membrane suspension was filtered over Whatman GF/C filters pretreated with 50 mM Tris-Cl (pH 7.4) and 0.05% BSA, and the radioactivity retained by the filter was determined by scintillation counting. Specific binding was quantitated as the portion of total bound radioligand displaced by 100 µM indolidan.

Results

The alkylation of the precursor to form [^{3}H]-LY186126 proceeded well. In control experiments it was evident that the regiochemistry of the alkylation was almost exclusively on the indolone nitrogen as judged by nOe NMR experiments. However, after initial alkylation at this nitrogen, some ancillary alkylation occurred on the dihydropyridazinone nitrogen, but this dimethylated material could be easily removed by HPLC. In the actual synthesis of the radioactive material, tritium-labeled iodomethane (85 Ci/mmol) was reacted with the precursor, and after HPLC purification [^{3}H]-LY186126 was obtained with a radiochemical purity of 98% and a specific activity of 79.2 Ci/mmol.

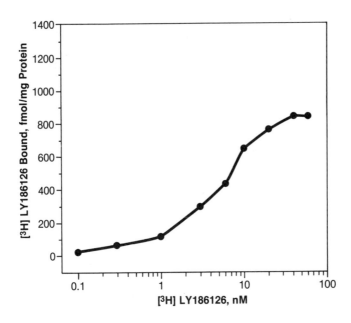

Figure 2. A representative saturation isotherm for [^{3}H]-LY186126 binding to canine cardiac SR vesicles. Binding was determined at 25°C in the presence of the indicated concentrations of [^{3}H]-LY186126.

[^{3}H]-LY186126 binds with high affinity and selectivity to canine cardiac SR vesicles, and results from a typical experiment are shown in figure 2. In this preparation, the B_{max} value was 780 ± 100 fmol/mg protein and the K_d value was 3.8 ± 1.3 nM. Scatchard replot of the binding data was linear, suggesting a homogeneous population of binding sites.

Although binding of [^{3}H]-LY186126 could be displaced with unlabeled indolidan, its specificity was probed further using additional drugs, and results are depicted in figure 3. The PDE-inhibitor cardiotonics milrinone and unlabeled LY186126 were able to virtually obliterate [^{3}H]-LY186126 binding when included at 1 μM, whereas other cardiovascular drugs such as nifedipine, propranolol, or prazosin were without effect.

The drug also binds to enriched SR vesicles prepared from sheep or rabbit ventricular myocardium, and a summary of data from different species is given in table 1. Moreover, specific binding could be obtained in myocardial vesicles, a readily obtained microsomal fraction, prepared from the dog, monkey, rat and guinea pig (data not shown).

Table 1: Binding of [^{3}H]-LY186126 to SR Vesicles From Different Species

Species	B_{max} (fmol/mg protein)	K_d (nM)	Reference
Dog	780 ± 100	3.8 ± 1.3	Kauffman et al., 1989
Rabbit	714 ± 77	6.2 ± 1.4	Artman et al., 1989
Sheep	944 ± 115	8.5 ± 2.3	Artman et al., 1989

To determine the pharmacological relevance of the specific binding, competition experiments were conducted with a variety of structurally distinct PDE inhibitors. The experiments were conducted using 10 nM [3H]-LY186126, K_d values for the PDE inhibitors were determined, and results are shown in table 2. Also shown are the abilities of these compounds to inhibit SR-PDE and to produce positive inotropic effects in pentobarbital-anesthetized dogs (ED_{50}'s).

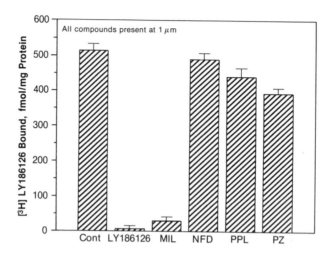

Figure 3. Displacement of [3H]-LY186126 binding with various cardiovascular drugs. Binding experiments were conducted at 10 nM [3H]-LY186126 in the presence or absence of the indicated drugs at a concentration of 1 μM. Data are mean ± SEMs. Cont, control; MIL, milrinone; NFD, nifedipine; PPL, propranolol; PZ, prazosin.

Discussion

LY186126 was chosen as a radiolabeling candidate since the indolone methyl substituent could be used as a carrier for the ready incorporation of three tritia into the molecule. The second methyl substituent in the dihydropyridazinone ring was required to attenuate the decrease in potency brought about by the indolone methyl group. The resulting compound is a close chemical, biochemical and pharmacological mimic of indolidan; inotropic ED_{50} values after i.v. administration of the two drugs to anesthetized dogs were 4.4 and 6.8 μg/kg, respectively. [3H]-LY186126 appears to be an excellent biochemical tool in that it is readily prepared and purified, binds with high affinity and selectivity to its biochemical target, and is very stable in high specific activity, labeled form; material stored at -20°C in ethanol for one year experienced little radiolytic decomposition.

We have shown that [3H]-LY186126 binds specifically and with high affinity to SR vesicles and myocardial vesicles derived from a variety of species. This binding appears to be pharmacologically relevant in that it is displaced selectively by structurally distinct PDE inhibitor cardiotonics, but not by several mechanistically distinct cardiovascular drugs. Moreover, there were strong correlations between the abilities of a compound to inhibit [3H]-LY186126 binding, to inhibit SR-PDE mediated cAMP hydrolysis, and to produce positive inotropic effects in dogs. These data strongly suggest that [3H]-LY186126 labels the inotropic receptor for this class of cardiotonics and that the inotropic receptor is coupled, in some fashion, to the Type IV PDE located in SR. However, whether this inotropic receptor is in fact Type IV PDE awaits further biochemical scrutiny.

Table 2: Biological Effects of Several PDE-inhibitor Cardiotonics

Compound	-log ED$_{50}$ [a] (mg/kg,iv)	-log IC$_{50}$ [b] (M)	-log K$_d$ [c] (M)
Indolidan	2.30	6.85	8.26
Isomazole	1.52	4.66	6.40
Milrinone	1.43	5.96	7.93
Imazodan	1.35	6.34	7.13
CI-930	1.89	6.96	7.41
Enoximone	0.55	5.57	6.70
Piroximone	0.46	4.93	---
Amrinone	0.41	5.32	6.18
LY195114	2.48	7.24	8.60
LY197055	2.49	7.86	9.26
LY195562	2.28	7.54	8.29
LY181512	1.62	6.64	7.11
LY195640	1.06	6.47	---
LY186232	1.15	6.19	---
LY197096	0.57	5.64	4.21
LY198419	2.0	6.42	---
LY217567	0.59	5.70	---
LY197140	2.0	5.57	---
LY198518	1.28	6.12	7.23
LY197540	2.32	6.42	---
LY197913	0.57	5.85	---
Caffeine	---	3.22	~3.00
Theophylline	---	3.66	4.56
Rolipram	---	3.00	~3.00

[a]ED$_{50}$ values for increasing contractility after iv administration to pentobarbital-anesthetized dogs. Data are from the published sources indicated in Kauffman et al., 1986. [b]IC$_{50}$ values for inhibition of SR-PDE. Data are from Kauffman et al., 1986, or were generated according to methods described therein. [c]K$_d$ values for inhibition of [^{3}H]-LY186126 binding to canine myocardial vesicles. Data are from Kauffman et al., 1989b.

Acknowledgements

We thank Pamela J. Edmonds for her expert assistance in preparation of this paper. Moreover, we are indebted to Drs. Michael Artman and W. Joseph Thompson of the University of South Alabama College of Medicine for stimulating discussions and careful collaborative efforts.

Abbreviations

CHF - congestive heart failure; PDE - phosphodiesterase; SR - sarcoplasmic reticulum; nOe - nuclear Overhauser effect.

References

Alousi, A. A., 1986, Pharmacology of the bipyridines: amrinone and milrinone, *Circulation* **73 (3 Pt 2)**:III 10-24.
Artman, M. et al., 1989, Analysis of the binding sites for the cardiotonic phosphodiesterase inhibitor [^{3}H]-LY186126 in ventricular myocardium, *Mol. Pharmacol.*, In Press.
CONSENSUS Trial Study Group, 1987, Effects of enalapril on mortality in severe congestive heart failure. Results of the Cooperative North Scandinavian Enalapril Survival Study (CONSENSUS), *N. Engl. J. Med.* **316**:1429-1435.
Jones, L.R., et al., 1981, Biochemical evidence for functional heterogeneity of cardiac sarcoplasmic reticulum vesicles, *J. Biol. Chem.* **256**:11809-11818.

Kauffman, R.F. et al., 1986, LY195115: A potent, selective inhibitor of cyclic nucleotide phosphodiesterase located in the sarcoplasmic reticulum, *Mol. Pharmacol.* **30**:609-616.

Kauffman, R.F. et al., 1989a, Specific binding of [^{3}H]-LY186126, an analogue of indolidan (LY195115) to cardiac membranes enriched in sarcoplasmic reticulum vesicles, *Circ. Res.* **64**:1037-1040.

Kauffman, R.F. et al., 1989b, Characterization and pharmacological relevance of high affinity binding sites for [^{3}H]-LY186126, a cardiotonic phosphodiesterase inhibitor, in canine cardiac membranes, *Circ. Res.* **65**:154-163.

Kithas, P.A. et al., 1988, Subcellular distribution of high affinity type IV cyclic AMP phosphodiesterase activity in rabbit ventricular myocardium: Relations to effects of cardiotonic drugs, *Circ. Res.* **62**:782-789.

Packer M., 1989, Effect of phosphodiesterase inhibitors on survival of patients with chronic congestive heart failure, *Am. J. Cardiol.* **63**:41A-45A.

Robertson, D.W. et al., 1988, Positive inotropic agents in management of congestive heart failure, *ISI Atlas Sci.: Pharmacology* **2**:129-135.

Robertson, D.W. et al., 1989, Synthesis of a tritium-labeled indolidan analog and its use as a radioligand for phosphodiesterase inhibitor cardiotonic binding sites, *J. Med. Chem.* **32**:1476-1480.

Strada, S. J. et al., 1984, Cyclic nucleotide phosphodiesterases, in Strada S. J., Thompson W.J. (eds): Advances in cyclic nucleotide and protein phosphorylation research. New York, Raven Press Publishers, 1984, Vol 16, p. vi.

51
Purification of a cGMP-inhibited cAMP Phosphodiesterase from Vascular Smooth Muscle

A. Rascon, P. Belfrage, E. Degerman, S. Lindgren, K.E. Andersson, A. Newman, V.C. Manganiello, and L. Stavenow

Introduction

Cyclic nucleotide phosphodiesterases (PDEs) constitute a complex group of enzymes, multiple forms of which are found in various amounts and proportions in most mammalian cells. These enzymes differ in subcellular localization, substrate affinities, kinetic characteristics, physiochemical properties, responsiveness to various effectors or drugs and mechanisms of regulation (Wells et al., 1977; Beavo et al., 1982; Manganiello et al., 1987). At least 6-7 distinct major classes of PDEs have been isolated, purified and characterized. One class, the cGMP inhibited low K_m PDE (hereafter referred as cGI-PDE) is very sensitive to inhibition by a number of inotropic and antithrombotic agents including cilostamide and other OPC derivatives, imazodan, milrinone, fenoxamine, LY 195115, Y 590, anegrelide. Inhibition of cGI-PDEs from heart and platelets by therapeutically effective concentrations of these drugs correlates with physiological effects.

A cGI-PDE has also been proposed to be involved in regulation of vascular smooth muscle tone since specific inhibitors of this enzyme class caused vasorelaxation of isolated human mesenteric arteries and veins (Lindgren et al., 1989), as well as PDE inhibition in partially purified fractions from several vascular (aortic) smooth muscle sources (Silver et al., 1988; Weishaar et al., 1986).

To examine this possibility and to be able to study the mechanism of regulation of this enzyme, identification and purification of the enzyme protein is required. cGI-PDEs have previously been purified from cardiac tissue (Harrison et al., 1986), platelets (Macphee et al., 1986; Grant et al., 1984), adipose tissue (Degerman et al., 1987; 1988) and liver (Pyne et al., 1987; Boyes et al., 1988) but not from vascular smooth muscle. In this study we have used cilostamide agarose affinity chromatography (CIT-agarose) to develop a purification procedure for the cGI-PDE from aortic smooth muscle. Some of the properties of the purified enzyme are described.

Materials and Methods

Phosphodiesterase activity was assayed by the method previously described by Manganiello et al. (1973) using 0.5 μM cAMP concentration as substrate. cGI-PDE activity (referred to as OPC-inhibited activity)

353

was defined as the difference of (activity in absence of OPC-3911)−(activity in presence of 3 μM OPC-3911). For inhibition studies the assay was performed in presence of different concentrations of OPC-3911, CI-930, milrinone, cGMP or RO 20-1724.

Bovine aortic smooth muscle tissue (200 g) was knife-homogenized in a solution containing 20 mM Tris, pH 7.5, 1 mM EDTA, 250 mM sucrose, 10 μg/ml pepstatin. The homogenate was centrifuged at low velocity and the resulting supernatant centrifuged at 50,000 xg for 1 h. The 50,000 xg supernatant was applied to a DEAE-Sephacel column equilibrated with 50 mM sodium acetate buffer pH 7.0 containing 10% glycerol, 1 mM EDTA, 1 mM DTE, 1 μg/ml leupeptin, 1 μg/ml pepstatin and 10 μg/ml antipain (buffer A). The elution was performed with a continuous sodium acetate gradient from 0.05 M to 1.2 M, pH 7.0. The fractions containing OPC-3911-inhibited PDE activity were pooled, dialyzed and subjected to CIT-agarose affinity chromatography, slab gel electrophoresis under denaturing (SDS-PAGE) and non-denaturing conditions, and Western blot analysis as described by Degerman et al. (1987,1988).

Experimental

The 50,000 xg supernatants used as starting material contained > 90% of total smooth muscle cAMP-PDE activity; about 40% of the enzyme activity in this fraction was inhibited by OPC-3911 and by the specific antibody raised against the bovine adipose tissue cGI-PDE. PDE activity in the supernatants could be separated into two peaks by DEAE-Sephacel chromatography. PDE activity of the second peak was partially inhibited by OPC-3911, indicating the presence of other PDEs in this fraction. The cGI-PDE was purified to apparent homogeneity from the pool of the OPC-3911 inhibited fractions using CIT-agarose chromatography, whereas a RO 20-1724 sensitive cAMP-PDE was found in the non-binding fraction.

Several polypeptides were identified after SDS-PAGE and silver staining of the purified cGI-PDE (105, 77, 74, 53 and 30 kDa); all, except the smallest, crossreacted with a monoclonal antibody raised against the bovine heart cGI-PDE, indicating that extensive proteolytic nicking took place during the purification. Gel electrophoresis under non-denaturing conditions demonstrated that more than 90% of the protein-stained material co-migrated with enzyme activity. The bovine aortic smooth muscle cGI-PDE was very sensitive to inhibition by OPC-3911, milrinone and CI-930 but not by RO 20-1724 (IC$_{50}$, (μM), OPC 3911 < CI 930 ≈ cGMP < Milrinone <<< RO 20-1724) and hydrolyzed cAMP and cGMP with linear Michaelis-Menten kinetics. The apparent K$_m$ for cAMP was 0.2 μM with an estimated V$_{max}$ of 3 μmol/min/mg at 30°C. Corresponding values for cGMP hydrolysis were 0.1 μM and 0.3 μmol/min/mg, respectively.

Incubation of purified cGI-PDE with the catalytic subunit of the cAMP-dependent protein kinase resulted in phosphorylation of the 105 and 77/74 kDa polypeptides, with minor phosphorylation in the 68/62 kDa area.

Discussion

A cGI-PDE from vascular smooth muscle has been purified and identified for the first time. The physiological importance of such an enzyme was recently indicated (Tanaka et al., 1988; Lindgren et al., 1989) from studies where specific inhibitors of this enzyme class at therapeutic doses resulted in vascular smooth muscle relaxation and well as in the inhibition of crude enzyme preparations (Weishaar et al., 1986; Lugnier et al., 1986; Silver et al., 1988). The purified smooth muscle enzyme had kinetic and inhibitory properties similar to the cGI-PDEs purified from other tissues (e.g. adipose tissue, platelets, heart, liver) including an exquisite sensitivity to proteolytic nicking during the purification.

The enzyme was phosphorylated by the catalytic subunit of the cAMP-dependent protein kinase as found with the adipose tissue, cardiac tissue and platelet enzymes. This suggests that cGI-PDE activity can be regulated by phosphorylation/dephosphorylation and may be involved in the vascular relaxation process observed in response to β-agonist and the "new" generation of cardiotonic drugs, by regulating cAMP levels. Characterization of the "in vitro" phosphorylation of the cGI-PDE (e.g. stoichiometry, sites phosphorylated) and intact cell experiments are needed to verify the physiological significance of changes in the phosphorylation state and the activity of the enzyme. The availability of the purified cGI-PDE opens the possibility of study in more detail the role of this enzyme in regulation of aortic smooth muscle tone.

Acknowledgements

We wish to thank Dr. J. Beavo for supplying us with the monoclonal antibody directed against the bovine cardiac cGI-PDE. Financial support was obtained from Swedish MRC grant No. 3362, the Medical Faculty, University of Lund and A. Pahlssons Foundation, Malmo, Sweden.

Table 1. Purification of low K_m-PDE from bovine aorta smooth muscle

Fraction	Total Protein (mg)	Total Activity nmol/min	OPC-inhibited Activity nmol/min	Specific Activity nmol/min/mg protein	Purification fold	Yield %
50,000 g supernatant	1870	1865	858	0.459	1	100
DEAE-Seph (OPC-inhibited)	498	803	401	0.805	1.75	47
Dialysis, CIT-agarose DEAE-Sephacel	0.037	86	86	2324	5063	10

a Substrate concentration 0.5 μM [^{3}H]cAMP
b Part of total activity that could be inhibited by 3 μM OPC-3911, was calculated as the difference of (activity in absence of OPC-3911)- (activity in presence of OPC-3911) and reflects cGI-PDE
c Based on the OPC-inhibited activity

Table 2. Effect of Inhibitors on the Eluate and the Non-binding Fraction From CIT-agarose Affinity Chromatography

Inhibitors	Non-binding Fraction	cGI PDE
Milrinone	ND[1]	0.40
OPC-3911	> 30	0.054
cGMP	> 5	0.25
RO 20-1724	3	> 30
CI 930	ND[1]	0.25

PDE preparations were assayed with 0.5 μM [^{3}H]cAMP with and without the indicated inhibitors. IC_{50} values were determined graphically.

[1]Not determined

References

Beavo, J. A. et al., 1982, Identification and properties of cyclic nucleotide phosphodiesterases, Mol. Cell. Endocr. 28:387-410.

Boyes, S. and Loten, G. E., 1988, Purification of an insulin-sensitive cyclic AMP phosphodiesterase from rat liver, Eur. J. Biochem. 174:303-309.

Degerman, E. et al., 1987, Purification of the putative hormone-sensitive cyclic AMP phosphodiesterase from rat adipose tissue using a derivative of cilostamide as a novel affinity ligand, J. Biol. Chem. 262:5797-5807.

Degerman, E. et al., 1988, Purification, properties and polyclonal antibodies for the particulate, low K_m cAMP phosphodiesterase from bovine adipose tissue, Second Messengers and Phosphoproteins 12:171-182.

Grant, P. G. and Colman, R. W., 1984, Purification and characterization of a human platelet cyclic nucleotide phosphodiesterase, Biochemistry 23:1801-1807.

Harrison, S. A. et al., 1986, Isolation and characterization of bovine cardiac muscle cGMP-inhibited phosphodiesterase: A receptor for new cardiotonic drugs, Mol. Pharmacol. 29:506-514.

Lindgren, S. et al., 1989, Relaxant effects of the selective phosphodiesterase inhibitors Milrinone and OPC 3911 on isolated human mesenteric vessels, Pharmacol. Toxicol. 64:440-445.

Lugnier, C. et al., 1986, Selective inhibition of cyclic nucleotide phosphodiesterases of human, bovine and rat aorta, Biochem. Pharmacol. 35:1743-1751.

Macphee, C. H. et al., 1988, Phosphorylation results in activation of a cAMP-PDE phosphodiesterase in human platelets, J. Biol. Chem. 263:10353-10358.

Manganiello, V. C. and Elks, M. L., 1986, Regulation of particulate cAMP phosphodiesterase activity in 3T3-L1 adipocytes: The role of particulate phosphodiesterase in the antilipolytic action of insulin, in Mechanisms of Insulin Action (Belfrage, P., Donner, J. and Stralfors, P., eds.) Elsevier Publishing Co., Inc., New York, pp. 147-166.

Pyne, N. J. et al., 1987, Specific antibodies and the selective inhibitor ICI 118233 demonstrate that the hormonally stimulated "dense-vesicle" and the peripheral-plasma-membrane cyclic AMP phosphodiesterases display distinct tissue distributions in the rat, Biochem. J. 248:897-901.

Silver, P. L. et al., 1988, Differential pharmacologic sensitivity of cyclic nucleotide phosphodiesterase isozymes isolated from cardiac muscle, arterial and airway smooth muscle, Europ. J. Pharmacol. 150:85-94.

Tanaka, T. et al., 1988, Effects of cilostazol, a selective cAMP phosphodiesterase inhibitor on the contraction of vascular smooth muscle, Pharmacology 36:313-320.

Weishaar, B. E. et al., 1986, Multiple molecular forms of cyclic nucleotide phosphodiesterase in cardiac and smooth muscle and in platelets, Biochem. Pharmacol. 35:787-800.

Wells, J. N. and Hardman, J. G., 1977, Cyclic nucleotide phosphodiesterases, Adv. Cyclic Nucleotide Res. 8:119-143.

52

Low K_mcAMP Phosphodiesterase Isozymes and Modulation of Tone in Vascular, Airway and Gastrointestinal Smooth Muscle

P.J. Silver, A.L. Harris, R.A. Buchholz, M.S. Miller, R.J. Gordon, R.L. Dundore, and E.D. Pagani

INTRODUCTION

Both cAMP and cGMP are second messengers which mediate a reduction in tone for a variety of smooth muscle relaxants in vascular, airway and gastrointestinal smooth muscle. In recent years, the development of better biochemical purification methods, as well as the identification/discovery of selective inhibitors of cyclic nucleotide phosphodiesterase (PDE) isozymes, have led to a better understanding of the presence and role of PDE isozymes in various tissues. In particular, recent evidence suggests that there are two distinct PDE isozymes which hydrolyze cAMP with a high affinity. One isozyme is sensitive to inhibition by submicromolar concentrations of cGMP and by various cardiotonic agents (referred to as "cGMP inhibitable" or cGi-PDE). The other isozyme is not inhibited by cGMP but is sensitive to inhibition by rolipram and Ro 20-1724 (referred to as Ro-PDE).

We have initially sought to examine these low K_m cAMP PDEs in vascular (aortic), airway and gastrointestinal smooth muscle and to evaluate the role that they may have in modulating the regulation of tone. In these studies, we have utilized some recently developed selective low K_m cAMP PDE inhibtors in biochemical and physiological experiments. The results of these studies indicate that these two pharmacologically distinct low K_m cAMP PDE isozymes are present in these smooth muscles, that the relative amount of these isozymes within a given smooth muscle type is species-related, and that these isozymes may have differential roles in modulating tone in different smooth muscles.

METHODS

PDE isozymes from the indicated smooth muscles were isolated via DEAE-Sephacel chromatography and PDE assays were performed as previously described (Silver et al., 1988). Changes in isometric force development for aortic and tracheal smooth muscle in isolated tissue baths were quantitated according to established procedures (Harris et al., 1989a,b). Effects of PDE inhibitors on histamine-induced bronchoconstriction in guinea pigs (Harris et al., 1989b) and on gastrointestinal motility in rats (Miller et al., 1981) were as previously detailed.

RESULTS and DISCUSSION

The selectivity for various low K_m cAMP PDE inhibitors was determined by quantitating inhibition of canine aortic peak III (from a DEAE-Sephacel column) low K_m cAMP PDE (which is primarily cGi-PDE) and inhibition of canine tracheal peak IV (similarly from a DEAE-Sephacel column) low K_m cAMP PDE (which is mostly Ro-PDE). Determination of the ratio of the IC_{50} values for cGi-PDE/Ro-PDE (Fig. 1) demonstrates that the pyridazinone PDE inhibitors CI-930 and imazodan are the most selective for cGi-PDE and are at least 10-50 fold more selective than the bipyridine PDE inhibitors amrinone and milrinone. The methylxanthines (IBMX and theophylline) and papaverine are relatively non-selective inhibitors, while Ro 20-1724 and rolipram are the most selective for inhibiting Ro-PDE. Since CI-930 and rolipram were the most selective inhibitors of the cGi- and Ro-PDEs, respectively, these agents were used to characterize the respective isozymes in smooth muscle.

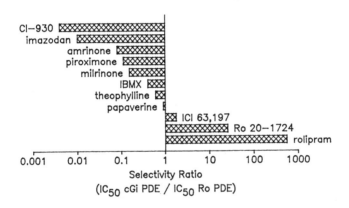

Figure 1. Selectivity ratios for various PDE isozyme inhibitors. Ratios were determined by quantifying the IC_{50} values for inhibition of cGi-PDE (obtained from canine aorta) and dividing by the IC_{50} values for inhibition of Ro-PDE (obtained from canine trachealis). This analysis shows that CI-930 is approximately 300x more selective as an inhibitor of cGi-PDE, whereas rolipram is approximately 500x more selective as an inhibitor of Ro-PDE.

All smooth muscles examined contained a peak of low K_m cAMP PDE activity which eluted from a DEAE-Sepahacel column at approximately 250 mM Na acetate (Table 1). In some cases, there was evidence of a single peak which was pharmacologically "pure", in some cases there was visible evidence of 2 partially-overlapping peaks of activity, and in other cases there was evidence of a single peak which contained two pharmacologically-distinct peaks. The latter case was characterized by biphasic concentration-response curves to CI-930 or rolipram which could be converted to single-phase concentration-response curves by performing assays in the presence of the other inhibitor. Thus, by pharmacological removal of one isozyme with a selective inhibitor, the presence of the other isozyme could be revealed. This combination protocol was also useful for examining the role of these isozymes in intact smooth muscle (see later discussion).

359

In aortic vascular smooth muscle, there was little, if any, Ro-PDE in the rodent, canine or primate sources examined in this study (Table 1), as evidenced by the high IC_{50} values of rolipram. However, combination experiments did reveal the presence of Ro-PDE in rat aortic smooth muscle. The IC_{50} values obtained in most of these species for CI-930 (0.2-0.3 uM) are similiar to those obtained in cardiac muscle (Silver et al., 1988). Interestingly, the rat aortic enzyme is more sensitive to inhibition by this agent, with an IC_{50} value of 0.06 uM. Other cGi-PDE inhibitors are also relatively more potent against this isozyme in this species.

Table 1. Low K_m cAMP PDEs in Smooth Muscle

| | IC_{50}, uM | | | | | | | |
| | Rat | | Guinea Pig | | Dog | | Monkey | |
Smooth Muscle Source	CI	Ro	CI	Ro	CI	Ro	CI	Ro
Aortic (Vascular)	0.06	57*	0.2	195	0.3	389	0.2	ND
Tracheal (Airway)	>10	0.8	2*	18*	74	0.7	1	117
Gastrointestinal (Longitudinal)	35*	13*	>10	0.5	>10	1.3	ND	ND

CI=CI-930; Ro=rolipram
*Evidence of 2 peaks separable via DEAE-Sephacel chromatography or by inhibition of each isozyme by a combination of CI-930 and rolipram.
ND=not determined.

Both airway and gastrointestinal smooth muscle contain cGi-PDE and Ro-PDE (Table 1), as suggested by the low IC_{50} values of both inhibitors. The relative amount of each isozyme varies according to species. For example, the guinea pig trachea appears to contain equal amounts of both isozymes; the dog trachea contains primarily Ro-PDE. Primate (cynomolgus monkey) trachealis appears to contain mostly cGi-PDE. Similarly, guinea pig and dog longitudinal smooth muscle contain mostly Ro-PDE, while that of the rat appears to be mostly mixed.

The correlations between inhibition of cGi-PDE and either vasorelaxation or tracheal relaxation for tissues obtained from guinea pigs is shown in Fig. 2. For aortic smooth muscle, a positive correlation exists (r=0.86, P<0.01) which is similar to correlations obtained for various cardiac muscle models (Silver et al., 1989). Similar correlations are evident for rat aortic smooth muscle obtained from either normotensive or hypertensive animals (SHR r=0.96, p<0.001). Rolipram is not a potent vasorelaxant in these models (IC_{50}> 100 uM).

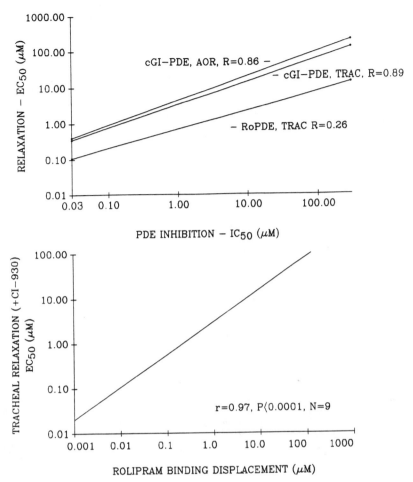

Figure 2. Comparison of correlations between cGi-PDE inhibition and vasorelaxation or tracheal relaxation, between Ro-PDE inhibition and tracheal relaxation, and between displacement of rolipram from a high affinity binding site and tracheal relaxation. IC_{50} values for 8-10 PDE inhibitors were quantitated for cGI-PDE inhibition in guinea pig aorta, guinea pig trachea (obtained in the presence of 3 μM rolipram to block Ro-PDE), or for Ro-PDE inhibition in guinea pig trachea (obtained in the presence of 3 μM CI-930 to block cGi-PDE). K_d values for displacement of rolipram-binding for the same PDE inhibitors used for Ro-PDE inhibition were obtained from the published data of Russo et al. (1987). EC_{50} values (concentrations that produced 50% relaxation) for these PDE inhibitors were obtained versus phenylephrine-contracted guinea pig aortic smooth muscle or versus histamine-contracted guinea pig tracheal smooth muscle following cumulative dosing during contraction. Tracheal force data for cGI-PDE comparison, and for Ro-PDE or rolipram binding displacement comparisons, were obtained in the presence of either 3 μM rolipram, or 3 μM CI-930, respectively.

In vivo, cGi-PDE inhibitors such as CI-930 and milrinone lower blood pressure in a dose-related manner in conscious SHR (Fig. 3), consistent with the biochemical data in aortic smooth muscle. CI-930 also potentiates the hypotensive effect of the adenylate cyclase activator fenoldopam, consistent with inhibition of low K_m cAMP cGi-PDE as a mechanism of action for CI-930 (Dundore et al., 1989). Interestingly, rolipram also lowers blood pressure in conscious SHR (Fig. 3). Whether this is related to a central effect on blood pressure regulation, or reflects a role for Ro-PDE in arteriolar smooth muscle, is yet to be determined.

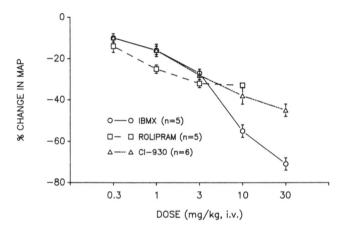

Figure 3. Dose-related reductions in mean arterial blood pressure (MAP) were determined in chronically-cannulated conscious SHR following cumulative intravenous dosing with either CI-930, rolipram, or IBMX. The average MAP prior to dosing was 170 ± 4 mm Hg. Values are the mean ± standard error for percentage change for the number (n) of animals indicated.

In airway smooth muscle in vitro, rolipram and CI-930 exhibit biphasic concentration-response relationships for relaxation of carbachol-, histamine- and LTD_4-contracted guinea pig tracheal smooth muscle. However, as with PDE isozyme inhibition, each agent produced a monophasic (sigmoidal) concentration-response curve when tested in the presence of a fixed concentration (3 μM) of the other. While CI-930 was approximately equipotent in inhibiting cGi-PDE and relaxing rolipram-pretreated trachea, rolipram was substantially more potent (EC_{50} = 20 nM) in relaxing CI-930-pretreated trachea than in inhibiting CI-930-pretreated PDE (Ro-PDE IC_{50} = 2.6 μM). Among a series of PDE inhibitors (Table 2), there was a highly significant correlation (r=0.89, p<0.01) between cGi-PDE inhibition and rolipram-pretreated tracheal relaxation, but not between Ro-PDE inhibition and CI-930-pretreated tracheal relaxation (r=0.26). Nine of the PDE inhibitors used in this study have been reported to displace rolipram from a high affinity binding site in rat brain (Russo et al., 1987). A highly significant correlation between relaxation of CI-930-pretreated trachea and displacement of rolipram binding by these agents was observed (r=0.97, p<0.0001).

362

In vivo, a similar pattern of activity was evident. Significant correlations were observed between bronchodilation (inhibition of histamine-induced bronchoconstriction) and cGi-PDE inhibition (r=0.88, p<0.01), rolipram-displacing potency (r=0.88, p<0.001) but not Ro-PDE inhibition (r=0.21).

Table 2. Correlations between PDE isozyme inhibition or inhibition of rolipram binding and regulation of tone for isozyme inhibitors in smooth muscle

Parameter of Tone	cGi-PDE	Ro-PDE	Ro Binding
Vasorelaxation (Guinea Pig Aorta)	r=0.86 p<0.01	ND	ND
Tracheal relaxation (Guinea Pig Trachea)	r=0.89 p<0.01	r=0.26 ns	r=0.97 p<0.0001
Bronchodilation (Guinea Pig, in vivo)	r=0.88 p<0.01	r=0.21 ns	r=0.88 p<0.001
Inhibition of GI motility (Rat, in vivo)	r=0.53 ns	r=0.14 ns	r=0.94 p<0.01

ND=not determined; ns=not significant

The relationship between inhibition of rat ileal smooth muscle cGi-PDE or Ro-PDE, and inhibition of gastrointestinal motility in vivo was also assessed (Table 2). As with airway smooth muscle, there was no correlation between Ro-PDE inhibition and inhibition of motility, although there was a significant correlation (r=0.94) between displacement of rolipram binding and decreased gastrointestinal transit. In contrast to airway smooth muscle and vascular smooth muscle, there was no correlation between inhibition of cGi-PDE and inhibition of gastrointestinal transit.

In summary, these data indicate that cGi-PDE is present in the vascular, airway and gastrointestinal smooth muscles examined in this study. Functionally, cGi-PDE appears to play a major role in vitro and in vivo in modulating contractility in vascular and airway smooth muscle, but not longitudinal gastrointestinal smooth muscle. Ro-PDE is present predominantly in airway and gastrointestinal smooth muscle, and not in aortic smooth muscle. However, other laboratories have demonstrated the presence of this PDE in other vascular smooth muscles (Prigent et al., 1988). Functionally, there is no correlation between apparent inhibition of Ro-PDE and relaxation of either airway or gastrointestinal smooth muscle. Instead, there is a highly significant correlation between displacement of rolipram from a high affinity binding site and inhibition of either airway or gastrointestinal smooth muscle tone. The identity of this high affinity binding site in these smooth muscles remains to be determined. Possibilities include a second site on Ro-PDE, another PDE isozyme which is present in the Ro-PDE fraction, or a PDE-independent site which is involved in the modulation of smooth muscle tone.

Acknowledgements

The authors acknowledge the technical assistance of Mary Connell, Linda Hamel, Ross Bentley, Annette Wallace, Doriann Van Liew, Edward Ferguson, Phillip Pratt, Wendy Hallenbeck, and Lee Hildebrand in performing these studies. The secretarial assistance of Peggy Branch in complilng this manuscript is also appreciated.

REFERENCES

Dundore, R. L., et al., 1989, The hypotensive response of SHR to the selective low K_m-cAMP phosphodiesterase inhibitor CI-930 is blunted by sodium nitroprusside (SNP), The Pharmacologist 31(3):189.

Harris, A. L., et al., 1989a, Phosphodiesterase isozyme inhibition and the potentiation by zaprinast of endothelium-derived relaxing factor and guanylate cyclase stimulating agents in vascular smooth muscle, J. Pharmacol. Exp. Ther. 249:394-400.

Harris, A. L., et al., 1989b, The role of low K_m cAMP phosphodiesterase isozyme inhibition in tracheal relaxation and bronchodilation in the guinea pig, J. Pharmacol. Exp. Ther. (in press).

Miller, M. S., Galligan, J. J. and Burks, T. F., 1981, Accurate measurement of intestinal transit in the rat, J. Pharmacol. Methods 6:211-217.

Prigent, A. F., et al., 1988, Comparison of cyclic nucleotide phosphodiesterase isoforms from rat heart and bovine aorta. Separation and inhibition by selective reference phosphodiesterase inhibitors, Biochem. Pharmacol. 37:3671-3681.

Russo, L. L., Lebel, L. A., Koe, B. K., 1987, Effects of selected phosphodiesterase inhibitors on calcium-dependent PDE activity and rolipram binding sites of cerebral cortex, Society for Neuroscience Abstracts 13:903, 1987.

Silver, P. J., et al., 1988, Differential pharmacological sensitivity of cyclic nucleotide phosphodiesterase isozymes isolated from cardiac muscle, arterial and airway smooth muscle, Eur. J. Pharmacol. 150:85-94.

Silver, P. J., et al., 1989, Phosphodiesterase isozyme inhibition, activation of the cAMP system, and positive inotropy mediated by milrinone in isolated guinea pig cardiac muscle, J. Cardiovasc. Pharmacol. 15:530-540.

53

Structure Activity Relationships of Xanthines as Inhibitors of Cyclic Nucleotide Phosphodiesterases and as Antagonists of Adenosine Receptors

J.N. Wells, L. Nieves, and T. Katsushima

Antagonism of the actions of endogenous adenosine and inhibition of cyclic nucleotide phosphodiesterase activity are thought to be important mechanisms in the pharmacological actions of methylxanthines. Adenosine exerts its effects by interaction with at least two receptors (Londos et al., 1977; Van Calker et al., 1979), termed A_1- and A_2-receptors. Most xanthines inhibit both A_1-and A_2- adenosine receptors in a manner that appears to be the result of competitive antagonism of agonist binding. Many (but not all) of the effects of adenosine appear to result from the coupling of A_1- and A_2-adenosine receptors through different G-proteins to adenylate cyclase causing inhibition or stimulation, respectively, of adenylate cyclase activity. Thus, adenosine and other agonists of adenosine receptors alter the intracellular levels of cyclic AMP; whether cAMP levels are increased or decreased in a particular cell by agonists of adenosine receptors depends, presumably, upon the type of adenosine receptor or the ratio of the types of adenosine receptors present.

In addition to adenylate cyclase activity, cyclic nucleotide phosphodiesterase activity also regulates cyclic AMP levels in a cell. Multiple forms of cyclic nucleotide phosphodiesterases exist in all cell types examined and many of these forms are discussed elsewhere in this volume. The commonly consumed methylxanthines are relatively non-selective inhibitors of the activity of the various forms of cyclic nucleotide phosphodiesterase (Wells et al., 1975a). Although these methylxanthines (caffeine and theophylline) are less potent as inhibitors of cyclic nucleotide phosphodiesterases than as adenosine receptor antagonists, inhibition of cyclic AMP hydrolysis is, nonetheless, a complicating factor in the interpretation of experiments using the xanthines to elucidate the effects of interactions of agonists with adenosine receptors. For the same reasons it is difficult to determine the mechanism(s) of the multiple pharmacological effects of the xanthines because endogenous adenosine is likely to be present during in vitro or in vivo experiments. Adenosine interaction with A_1-receptors should result in a reduction of cyclic AMP levels. The presence of a xanthine should attenuate the reduction in cyclic AMP levels both by inhibition of the adenosine-receptor interaction and by inhibition of the cyclic AMP phosphodiesterase activity. On the other hand, adenosine interaction with A_2-receptors would result in an increase

[1] Present address is Toyoba Co. LTD. Katata Research Institute 1-1-Kata 2-chome, Ohtsu Shiga, 520-02 Japan.

in cyclic AMP levels that is inhibited by xanthines at the receptor level but is potentiated by inhibition of the cyclic AMP phosphodiesterase activity.

Substituted xanthines represent the most potent class of adenosine receptor antagonists reported to date and also constitute the most general class of potent phosphodiesterase inhibitors available. A knowledge of the structure-activity-relationship (SAR) of the xanthines as inhibitors of these two systems may allow the design of inhibitors that are selective for one or the other site of action. In addition, the results of such studies could provide a rational basis upon which to choose a compound or series of compounds as investigational tools. This brief article presents our attempts to summarize the SAR of xanthines as inhibitors of these two systems.

In our own studies we have determined the effect of adenosine antagonists on concentration-response curves for the NECA-induced stimulation of adenylate cyclase activity of a membrane preparation from human platelets and for the R-PIA-induced decrease in forskolin-stimulated adenylate cyclase activity of membranes from rat fat pads (Martinson et al., 1987). A series of concentration-response curves to an adenosine agonist were constructed, each assayed in the presence of a different concentration of inhibitor or the solvent for the inhibitor (control). K_i values were obtained using the method of Arunlakshana and Schild (Arunlakshana et al., 1959). Since the Affinities of antagonists for adenosine receptors appear to depend upon the species of animal from which the cells were derived (Ukena et al., 1986a), comparison of data from different laboratories may be misleading. Therefore, most of the conclusions summarized in this manuscript are derived from data collected in our laboratory utilizing rat fat cell and human platelet membranes. Although the absolute values of the K_i's and the apparent selectivity of the compounds may differ with other membrane preparations, the relative effects of the structural modifications on the K_i values are likely to be universal. The ability of the compounds to inhibit cyclic nucleotide phosphodiesterase activities were determined using Ca^{2+}·calmodulin-sensitive and cyclic GMP-inhibited cyclic AMP phosphodiesterases separated from each other and partially purified from porcine coronary arteries as described before (Wells et al., 1975b)). Potencies to inhibit these enzyme activities were estimated by determining the I_{50} values (concentration required to inhibit by 50% the hydrolysis of 1 μM substrate).

Structural modification of the 1- and 3-position of the alkylxanthines does not dramatically alter the ability of the xanthines to inhibit either A_1- or A_2-adenosine receptors (Martinson et al., 1987) but can greatly influence the ability of the xanthines to inhibit the Ca^{2+}·calmodulin-sensitive phosphodiesterase or the cyclic GMP-inhibited cyclic AMP phosphodiesterase (Garst et al., 1966; Kramer et al., 1977; Wells et al., 1981). For example, xanthines with any combination of alkyl groups ranging from methyl to isoamyl on the 1- and 3-position inhibit A_1- and A_2-adenosine receptors of rat fat cells and human platelets, respectively with K_i values between 1 and 10 μM (Martinson et al., 1987). On the other hand, while 1-methyl-3-isobutylxanthine is 3-fold more potent as an inhibitor of Ca^{2+}·calmodulin phosphodiesterase than of cGMP-inhibited cyclic AMP phosphodiesterase activity (Wells et al., 1975a), 1-isoamyl-3-isobutylxanthine is 5-fold more potent as an inhibitor of the cGMP-inhibited cyclic AMP phosphodiesterase than of the Ca^{2+}·calmodulin-sensitive phosphodiesterase (Kramer et al., 1977). Likewise, theophylline (1,3-dimethylxanthine) is a relatively weak inhibitor of cyclic nucleotide phosphodiesterases but 1-methyl-3-isobutylxanthine is among the most potent xanthines as an inhibitor of both the

Ca^{2+}·calmodulin-sensitive and the cyclic AMP-specific phosphodiesterase activities; these compounds are not dramatically different in their potencies as antagonists of either A$_1$- or A$_2$-adenosine receptors (Martinson et al., 1987).

Activity of the xanthines as cyclic nucleotide phosphodiesterase inhibitors (Wells et al., 1981) and as antagonists of adenosine receptors (Martinson et al., 1987) are greatly reduced by substitution of highly polar groups such as carboxylalkyl moieties onto the 1- or 3-position, however, binding to the adenosine receptors is much less affected than is inhibition of cyclic nucleotide phosphodiesterase activities by less polar substituents such as carboxylic acid esters. In fact, 1-carbomethoxymethyl-3-isobutylxanthine is equipotent with and 1-methyl-3-(2-acetoxyethyl)xanthine is only slightly less potent than 1-methyl-3-isobutylxanthine as an antagonist of A$_1$- and A$_2$-adenosine receptors. On the other hand these compounds are about one-twentieth as potent as 1-methyl-3-isobutylxanthine as inhibitors of the phosphodiesterase activities.

The most dramatic alteration in potencies of the xanthines as cyclic nucleotide phosphodiesterase inhibitors (Wells et al., 1981) and as antagonists of adenosine receptors (Wu et al., 1982; Bruns et al., 1983; Schwabe et at., 1985; Daly et al., 1985; Hamilton et al., 1985; Ukena et al, 1986; Martinson et al., 1987) results from substitution in the 8-position of this heterocyclic system. Electron withdrawing substituents such as trifluoromethyl reduce the potency of xanthines to inhibit phosphodiesterase activity and the potency as antagonists of adenosine receptors. 8-Alkyl, -amino and -alkoxymethyl groups do not affect the ability of 1-methyl-3-isobutylxanthine to inhibit Ca^{2+}·calmodulin-sensitive phosphodiesterase activity, but consistently reduce the ability of the xanthine to inhibit cyclic AMP-specific phosphodiesterases. Thus 8-methoxymethyl- and 8-methyl substituted 1-methyl-3-isobutylxanthine are equipotent with the parent 1-methyl-3-isobutylxanthine as inhibitors of Ca^{2+}·calmodulin-sensitive phosphodiesterase activity but are one-thirtieth to one-fiftieth as potent, respectively, as the parent compound to inhibit cyclic AMP-specific phosphodiesterase activity (Kramer et al., 1977; Wells et al., 1981). Thus, 8-substituted-1-methyl-3-isobutylxanthines are relatively selective inhibitors of Ca^{2+}·calmodulin-sensitive cyclic nucleotide phosphodiesterase activity by virtue of the cyclic AMP-specific form being apparently unable to accommodate increased bulk in the 8-position of the xanthines. On the other hand, substitution of larger alkyl, cylcoalkyl and aromatic groups on the 8-position of 1-methyl-3-isobutylxanthine or, even more impressively, 1,3-dipropylxanthine generates compounds that are, in general, more potent than the parent, 8-unsubstituted-xanthine as inhibitors of adenosine receptors. This stratagem has led to highly selective A$_1$-antagonists by virtue of a much greater increase in affinity of the 8-substituted xanthines for the A$_1$-receptor subtype than for the A$_2$-subtype. The most striking example is 1,3-dipropyl-8-cyclopentylxanthine (Ukena et al., 1986; Martinson et al., 1987). The parent 1,3-dipropylxanthine inhibits A$_1$- and A$_2$-adenosine receptors with K$_i$ values of 0.94 and 1.9 μM, respectively. 1,3-Dipropyl-8-cyclopentylxanthine has a higher affinity for both receptor subtypes than the parent compound, but is about 150-fold more potent as an antagonist of A$_1$-receptors (K$_i$ = 0.47 nM) than of A$_2$-receptors (K$_i$ = 69 nM).

Recently we have investigated the SAR of the 8-cycloalkyl substituted 1,3-dipropylxanthines in an attempt to determine the characteristics of the domain of the A$_1$-adenosine receptor that are responsible for the rather remarkable affinity of the 8-cycloalkylxanthines for the A_1-receptor subtype. In our hands an aromatic substituent in the 8-position, although imparting increased affinity to the 1,3-dipropylxanthine for both A_1- and A_2-receptors, does not increase the affinity for A_1-receptors as much as, nor give rise to the receptor subtype selectively that is exhibited by, 1,3-dipropyl-8-cyclopentylxanthine. It is also clear, that the 8-cyclopentyl group is unique among the 8-cycloalkyl xanthines because 1,3-dipropyl-8-cyclopentylxanthine is at least 5-fold more potent as an antagonist of A_1-adenosine receptors than any other 8-cycloalkyl analog studied. While no other 8-cycloalkyl analog that we have examined is as potent as the cyclopentyl analog, 1,3-dimethyl-8-(2-methylcyclopropyl)xanthine is highly selective for inhibition of the A_1-adenosine receptor. This compound is about 1000-fold more potent as an inhibitor of A_1- than of A_2-adenosine receptors with K_i values of 11 nM and greater than 10 μM, respectively. We also studied a series of 1,3-dipropyl-8-cyclohexylxanthines that were substituted with amino or carboxylic acid groups in either the cis or trans (with respect to the xanthine nucleus) orientation of the 2-, 3- or 4-position of the cyclohexyl group. In general, such substitution reduces the potency of 1,3-dipropyl-8-cyclohexylxanthine to inhibit A_1- and A_2-receptor subtypes to approximately the same extent, but these analogs could prove to be useful because they represent compounds with a reactive functionality that may allow the design of affinity chromatography matrices or affinity labels for the A_1-adenosine receptor.

Acknowledgements: The authors' work was conducted under the aegis of NIH grant GM21220. The expert technical work of John Wiley and Jaime Nieves is gratefully acknowledged. The authors would like to thank Mary Couey for the excellent assistance and infinite patience in the preparation of this manuscript.

References:

Arunlakshana, O. and Schild, H.O., 1959, Br. J. Pharmacol. **14**:48-59.

Bruns, R.F., Daly, J.W. and Snyder, S.H., 1983, Proc. Natl. Acad. Sci. USA **80**:2077-2080.

Daly, J.W., Padgett, W., Shamin, M.T., Butts-Lamb, P. and Waters, J., 1985, J. Med. Chem. **28**:487-492.

Garst, J.E., Kramer, G.L. and Wells, J.N., 1966, J. Med. Chem. **19**:499-503.

Hamilton, H.W., Ortwine, D.F., Worth, D.F., Badger, E.W., Bristol, J.A., Bruns, R.F., Haleen, S.J. and Steffen, R.P., 1985, J. Med. Chem. **28**:1071-1079.

Kramer, G.L., Garst, J.E. and Wells, J.N., 1977, Biochemistry **16**:3316-3321.

Londos, C. and Wolff, J., 1977, Proc. Natl. Acad. Sci. USA **74**:5482-5486.

Martinson, E.A., Johnson, R.A. and Wells, J.N., 1987, Mol. Pharmacol. **31**:247-252.

Schwabe, U., Ukena, D. and Lohse, M.J., 1985, Naunyn-Schmiedeberg's Arch. Pharmacol. **330**:212-221.

Ukena, D., Jacobson, K.A., Padgett, W.L., Ayala, C., Shamin, M.T., Kirk, K.L., Olsson, R.O. and Daly, J.W., 1986a, FEBS Letters **209**:122-128.

Ukena, D., Daly, J.W., Kirk, K.L. and Jacobson, K.A., 1986b, Life Sciences **38**:797-807.

van Calker, D., Muller, M. and Hamprecht, B., 1979, J. Neurochem. **33**:999-1005.

Wells, J.N., Wu, Y.J., Baird, C.E. and Hardman, J.G., 1975a, Mol. Pharmacol. **11**:775-783.

Wells, J.N., Baird, C.E., Wu, Y.J. and Hardman, J.G., 1975b, Biochim. Biophys. Acta **384**:430-442.

Wells, J.N., Garst, J.E. and Kramer, G.L., 1981, J. Med. Chem. **24**:954-958.

Wu, P.H., Phillis, J.W. and Nye, M.J., 1982, Life Sciences **31**:2857-2867.

54
Mechanisms of Xanthine Actions in Models of Asthma

R.E. Howell, R.F. Gaudette, M.C. Raynor, A.T. Warner, W.T. Muehsam, J.A. Schenden, J. Perumattam, R.N. Hiner, and W.J. Kinnier

Introduction

Theophylline has long been one of the most effective and frequently used drugs for the treatment of asthma. Recently another xanthine, enprofylline, has been found useful for the treatment of asthma (Pauwels, et al, 1985). Unfortunately, the mechanism of action of xanthines in the lung remains unclear. Inhibition of cyclic nucleotide phosphodiesterase (PDE) activity was originally considered the mechanism of theophylline's anti-asthmatic effects (Piafsky and Ogilvie, 1975). However, it has been claimed that at therapeutic plasma levels theophylline does not appreciably inhibit PDE activity and thus PDE inhibition cannot account for the clinical effects of theophylline (Persson, et al, 1986).

Adenosine antagonism has more recently been proposed as the mechanism of theophylline's anti-asthmatic effects (Fredholm, 1980). This hypothesis is supported by observations that adenosine can potentiate the immunologically-induced release of bronchoactive mediator compounds from lung mast cells (Hughes, et al, 1984; Peachell, et al, 1988) and can induce bronchoconstriction in asthmatic patients (Polosa, et al, 1989), effects that are inhibited by theophylline. However, enprofylline and theophylline exhibit similar pulmonary effects despite enprofylline's weaker adenosine antagonist activity, suggesting that adenosine antagonism may be unimportant to the anti-asthmatic effects of xanthines (Pauwels, et al, 1985; Persson, et al, 1986).

We investigated the mechanism of action of xanthine in the lung, using several xanthines exhibiting a range of potencies as PDE inhibitors and as adenosine antagonists. We found that the pulmonary effects of xanthines appear to occur through multiple mechanisms of action.

Materials and Methods

<u>Animal Preparation.</u> Male guinea pigs were anesthetized by urethane and the trachea, carotid artery and jugular vein were cannulated. Animals were ventilated with room air at a constant rate and volume (1 cc per 100 gm, 55 breaths per min). Pulmonary inflation pressure was measured from a side arm of the tracheal cannula connected to a pressure transducer. Arterial blood pressure and heart rate were measured from the carotid artery. Bronchoconstriction was induced by placing a

Table 1. Bronchodilator Activity Of Xanthines

COMPOUND	vs. HISTAMINE ED_{50} (MG/KG)	vs. CARBACHOL ED_{50} (MG/KG)
Theophylline	31.6 ± 5.2(6)	29.2 ± 3.7(4)
Enprophylline	13.1 ± 3.1(5)*	27.4 ± 5.4(3)
8P-Theophylline	> 100(3)a	> 100(3)b
IBMX	2.4 ± 0.4(4)*	7.9 ± 2.7(4)*
DPX	7.4 ± 2.2(4)*	17.9 ± 6.1(4)*
DPHPX	11.8 ± 1.3(7)*	44.7 ± 8.1(4)
DPFX	7.8 ± 1.6(6)*	23.4 ± 6.2(4)

* Significantly different from theophylline (p <0.5).
a 8-Phenyltheophylline caused a 37 ± 8% reversal of histamine-induced bronchoconstriction at 100 mg/kg.
b 8-Phenyltheophylline caused a 45 ± 7% reversal of carbachol-induced bronchoconstriction at 100 mg/kg.

maximum concentration of one of several agents described below into an ultrasonic nebulizer inserted between the respirator and animal. Xanthines were administered intravenously.

Bronchodilator Studies Following a period of stable baseline inflation pressure (10-14 cmH$_2$0), an aerosol of histamine (100 ug/ml) or carbachol (50ug/ml) was administered for approximately 45 min. After the maximum response was attained each xanthine was given in increasing doses to induce cumulative reductions in inflation pressure. An ED_{50} value was determined from each animal and represents the dose required for a 50% reversal of bronchoconstriction.

Prophylactic Studies Following the baseline period, each xanthine was given in a single dose 2 min prior to challenge with an aerosol of: (1) ovalbumin (10mg/ml) to ovalbumin-sensitized animals, (2) LTD$_4$ (50ug/ml) to normal animals or (3) PAF (500ug/ml) to normal animals. The maximum increase in inflation pressure was compared to the maximum response in vehicle-injected controls.

Chemicals The following xanthines were used: theophylline, enprofylline, 8-phenyltheophylline (8-PT), isobutylmethyl xanthine (IBMX), 1,3-diethyl-8-phenyl xanthine (DPX), 1,3-dipropyl-8-hydroxyphenyl xanthine (DPHPX) and 1,3-dipropyl-8-tetrahydrofuranyl xanthine (DPFX).

Results

Aerosols of histamine or carbachol increased inflation pressure approximately 50 cm. Inflation pressure remained constant in the absence of drug administration and was not reduced by vehicle injection. All xanthines except 8-PT reversed bronchoconstriction 70-90%. ED_{50} values for bronchodilator activity are summarized in table 1.

Ovalbumin aerosols increased inflation pressure an average of 36 cm in 8 control sensitized groups. As illustrated in figure 1, all of the xanthines except DPFX partially inhibited antigen-induced bronchoconstriction at 10-20 mg/kg, whereas DPFX inhibited the response only at 40 mg/kg. Similarly, at 20mg/kg all of the xanthines except DPFX inhibited LTD$_4$-induced bronchoconstriction whereas DPFX

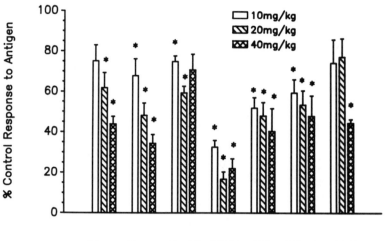

Figure 1 Effects of xanthines on antigen-induced bronchoconstriction, as a percent of control responses. * indicates a
significant difference from control (p<.05). n = 5-7 per group.

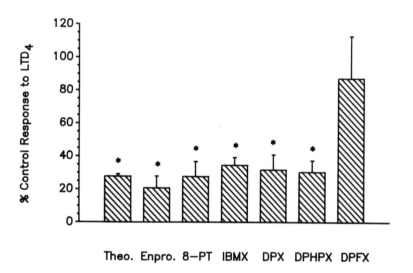

Figure 2 Effects of xanthines (20mg/kg only) on LTD$_4$-induced bronchoconstriction, as a percent of control responses. * indicates a
significant difference from control (p<.05). n = 3-5 per group.

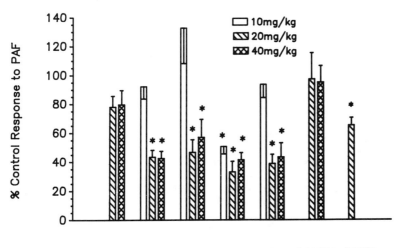

<u>Figure 3</u> Effects of xanthines on PAF-induced bronchoconstriction, as
a percent of control responses. * indicates a significant difference
from control (p<.05). n = 5-7 per group.

was inactive, as illutrated in figure 2. LTD_4 aerosols increased inflation pressure an
average of 20 cm in 3 control groups.

PAF aerosols increased inflation pressure an average of 20 cm in 7 control groups.
As illustrated in figure 3, enprofylline, 8-PT, IBMX, DPX and DPFX inhibited PAF-
induced bronchoconstriction at 20 mg/kg. In contrast, theophylline and DPHPX
failed to inhibit PAF-induced bronchoconstriction at 20 or 40 mg/kg. PAF aerosols
decreased blood pressure approximately 40% in control animals and in animals given
8-PT or DPFX (which were not in themselves hypotensive). The effects of the other
xanthines upon PAF-induced hypotension could not be evaluated due to their
hypotensive effects.

Discussion

The bronchodilator effects of xanthines were examined using the reversal of
bronchoconstriction induced by histamine and carbachol. All of the xanthines except
8-PT were effective bronchodilators against both agonists, indicating that this effect
is relatively nonspecific, although most xanthines were more potent in reversing
histamine-induced bronchoconstriction than in reversing carbachol-induced
bronchoconstriction. The relative potencies of xanthines as bronchodilators against
either agonist (IBMX > DPX > enprofylline ≥ theophylline > 8-PT) correlate very well
with their relative potencies as cAMP PDE inhibitors (Smellie, et al, 1979; Scotini
et al, 1983; Fredholm, et al, 1985; Clemo, et al, 1987; Choi, et al, 1988), although
we are unaware of any reports on the effects of DPHPX or DPFX on PDE activity.

Our observation that xanthine-induced bronchodilation in vivo correlates with cAMP PDE inhibition is supported by previous observations that xanthine-induced tracheal relaxation in vitro correlates with cAMP PDE inhibition (Polson, et al, 1982; Ogawa, et al, 1989). Our results suggest that PDE inhibition is involved in the bronchodilator effects of xanthines.

In addition to causing bronchodilation, theophylline and enprofylline exert anti-allergic and anti-inflammatory effects useful for the treatment of asthma (Pauwels, et al, 1985; Pauwels, 1987; Persson, et al, 1988). We examined the prophylactic effects of xanthines against bronchoconstriction induced by antigen challenge in sensitized animals. Notably, 8-PT inhibited antigen-induced bronchoconstriction at doses causing no bronchodilation while DPFX failed to inhibit antigen-induced bronchoconstriction at doses causing substantial bronchodilation. These results suggest that the anti-allergic effect of xanthines occurs through a different mechanism from the bronchodilator effect. This effect cannot be readily attributed to PDE inhibition or to adenosine A_1 or A_2 receptor antagonism as a single mechanism at action, since (1) at lower doses the weak PDE inhibitor 8-PT was as effective as the more potent PDE inhibitors enprofylline and theophylline and (2) the weak adenosine A_1/A_2 antagonists enprofylline and theophylline were as effective as the much more potent adenosine A_1/A_2 antagonists 8-PT, DPX and DPHPX (Ukena, et al, 1985; Schwabe, et al, 1985; Bruns et al, 1986; Williams, et al, 1986; Noronha-Blob, et al, 1986). Therefore, the anti-allergic effect of xanthines may involve an unknown mechanism.

In order to determine if the inhibition of antigen-induced bronchoconstriction occured only via the inhibition of mediator release from inflammatory cells, we examined the prophylactic effects of xanthines upon bronchoconstriction induced by two important mediators of anaphylaxis - LTD_4 and PAF (Barnes, 1987). Every xanthine that inhibited antigen-induced bronchoconstriction at 20 mg/kg greatly inhibited LTD_4-induced bronchoconstriction at the same dose, whereas the one xanthine (DPFX) that did not inhibit the response to antigen at 20 mg/kg also failed to inhibit the response to LTD_4. These results indicate that inhibition of LTD_4-induced bronchoconstriction can at least partly account for the inhibition of antigen-induced bronchoconstriction. The results with 8-PT and DPFX again suggest that this effect occurs through a different mechanism from the bronchodilator effect.

PAF-induced bronchoconstriction was also inhibited by several xanthines. Interestingly, theophylline and DPHPX did not inhibit the response to PAF despite being as effective as other xanthines against responses to antigen and LTD_4. Furthermore, DPFX inhibited PAF-induced bronchoconstriction at a dose that was ineffective against antigen or LTD_4. Inhibition of PAF-induced bronchoconstriction thus appears to occur by a mechanism different from the inhibition of bronchoconstriction induced by antigen or LTD_4. The inactivity of theophylline and DPHPX together with the activity of 8-PT suggest that this prophylactic effect is also different from the bronchodilator effect and is difficult to attribute to PDE inhibition or to adenosine A_1/A_2 receptor antagonism as a single mechanism of action. In addition, the hypotensive response to PAF was unaltered in animals in which bronchoconstriction was inhibited, suggesting that xanthines are not PAF receptor antagonists. Rather, xanthines probably inhibit bronchoconstriction resulting from the release of one or more mediators from various cells by PAF, such as leukotienes, cyclo-oxygenase products or serotonin (Lefort, et al, 1984; Jancar, et al, 1988; Barnes, et al, 1988). Our results indicate that in guinea pigs LTD_4 is not the primary mediator of PAF-induced bronchoconstriction inhibited by xanthines.

References

Barnes, P.J., 1987, Am. Rev. Resp. Dis. **135**: 526-531.

Barnes, P.J., Chung, K.F. and Page, C.P., 1988, J. Allergy Clin. Immunol. **81**:919-934.

Bruns, R.F., Lu, G.H. and Pugsley, T.A., 1986, Mol. Pharmacol. **29**:331-346.

Clemo, H.F., Bourassa, A., Linden J. and Belardinelli, L., 1987, J. Pharmacol. Exp. Ther. **242**:478-484.

Choi, O.H., Shamin, M.T., Padgett, W.L. and Daly, J.W., 1988, Life Sci. **43**:387-398.

Fredholm, B.B., 1980, Eur. J. Respir. Dis. (Suppl. 109) **61**:29-36.

Fredholm, B.B., Bergman, B., Fastbon, J. and Jonzon, B., 1985, in <u>Anti-Asthma Xanthines and Adenosine</u>, K-E. Andersson and C.G.A. Persson (eds), 291-308, Excerpta Medica, Amsterdam.

Hughes, P.J., Holgate, S.T. and Church, M.K., 1984, Biochem. Pharmacol. **33**: 3847-3852.

Jancar, S., Theriault, P., Provencal, B., Cloutier, S. and Sirois, P., 1988, Can. J. Physiol. Pharmacol. **66**:1187-1191.

Lefort, J., Rotilio, D. and Vargaftig, B.B., 1984, Br. J.Pharmacol. **82**:525-531.

Noronha-Blob, L., Marshall, R.P., Kinnier, W.J. and U'Prichard, D.C., 1986, Life Sci. **39**:1059-1067.

Ogawa, K., Takagi, K. and Satake, T., 1989, Br. J. Pharmacol. **97**:542-546.

Pauwels, R., 1987, Chest **92**:325-375.

Pauwels, R., Van Renterghem, D., Van Der Straeten, M., Johannesson, N. and Persson, C.G.A., 1985, J. Allergy Clin. Immunol. **76**:583-590.

Peachell, P.T., Columbo, M., Kagey - Sobotka, A., Lichtenstein, L.M. and Marone, G., 1988, Am. Rev. Resp. Dis. **138**:1143-1151.

Persson, C.G.A., Andersson, K-E. and Kjellin, G., 1986, Life Sci. **38**:1057-1072.

Piafsky, K.M. and Ogilvie, R.I., 1975, N.E. J. Med. **292**:1218-1222.

Polosa, R., Holgate, S.T. and Church, M.K., 1989, Pulmonary Pharmacol. **2**:21-26.

Polson, J.B., Krzanowski, J.J. and Szentivanyi, A., 1982, Biochem. Pharmacol. **31**:3403-3406.

Schwabe, U., Ukena, D. and Lohse, M.J., 1985, N-S. Arch. Pharmacol. **330**:212-221.

Scotini, E., Carpenedo, F. and Fassina, G., 1983, Pharmacol. Res. Commun. **15**:131-143.

Smellie, F.W., Davis, C.W., Daly, J.W. and Wells, J.M., 1979, Life Sci. **24**:2475-2482.

Ukena, D., Schirren, C.G., Klotz, K.N. and Schwabe, U., 1985, N-S. Arch. Pharmacol. **331**:89-95.

Williams, M., Braunwalder, A. and Erickson, T.J., 1986, N-S. Arch. Pharmacol. **332**:179-183.

Section 4
Abstracts

A1. EXTRACELLULAR AND INTRACELLULAR FORMATION OF ADENOSINE.

JERZY BARANKIEWICZ[1], AMOS COHEN[2], AND HARRY GRUBER[1], [1]Gensia Pharmaceuticals, Inc., 11075 Roselle Street, San Diego, CA 92121, USA [2]Division of Immunology/Rheumatology, Research Institute, The Hospital for Sick Children, Toronto, Ontario M5G 1x8, Canada

Adenosine is an important natural purine nucleoside which regulates a number of functions in cardiovascular, neuro, and immuno-systems. Production of adenosine is a complex process because adenosine can be produced both intracellularly and extracellularly, and only by specific cells. In addition, different substrates can serve as the source of adenosine (ATP, SAH, cAMP), and the amount of adenosine produced depends on the physiological state of cells. To better understand adenosine formation it was studied in human B and T lymphoblasts.

Human B lymphoblasts, in contrast to T lymphoblasts, can produce significant amounts of adenosine from both extracellular and intracellular ATP. Extracellularly, adenosine is formed in a three step pathway catalyzed by ectoATPase, ectoADPase, and ectoAMPase that are present on the surface of B cells, but not T cells. At an initial 500 uM concentration of extracellular ATP, adenosine accumulated in the medium linearly; however, at 5 uM extracellular ATP, adenosine accumulated extracellularly only when adenosine uptake was inhibited by dipyridamole. Adenosine produced extracellularly depends on the amount of ATP released from cells, different physiological conditions, and the presence of ectonucleotidases which are active on B lymphoblasts, B lymphocytes, thrombocytes, endothelial cells, but have little or no activity on T lymphoblasts, T lymphocytes, or erythrocytes.

Intracellular formation of adenosine and its release into the medium seems to be more complicated, because intracellular ATP can be degraded by alternative pathways of which only one includes adenosine formation. The importance of alternative pathways of intracellular ATP degradation seems to be different in various cells. In T lymphoblasts, catabolism of ATP proceeds mainly via AMP deamination because of their relatively low AMP-5'-nucleotidase activity. The relatively high adenosine deaminase and adenosine kinase activities (enzymes utilizing adenosine) do not allow adenosine release from these cells. In contrast, B lymphoblast's ATP is catabolized significantly in part via AMP dephosphorylation; however, because of the low activity of adenosine utilizing enzymes (AK and ADA), adenosine is released into the medium from these cells. These results indicate that B lymphoblasts can produce adenosine from both extracellular and intracellular sources whereas T cells probably cannot.

A2. INACTIVATION OF ADENOSINE AND ADENINE NUCLEOTIDE TRANSMITTERS BY ASTROGLIA.

V. MADELIAN, E. LAVIGNE, D.L. MARTIN, W. SHAIN, Laboratory of Neurotoxicology and Nervous System Disorders, Wadsworth Center and Department of Environmental Health and Toxicology, School of Public Health, SUNYA, Albany, NY 12201-0509

Adenosine and ATP are reported to have neurotransmitter/neuromodulator activities in the brain. We have used two preparations of astroglial cells to show that astroglia can regulate adenosine and ATP neurotransmission in the CNS by various mechanisms.

LRM55 cells are a continuous cell line derived from a rat spinal cord tumor and exhibit a number of astocytic properties. Primary cultures of astrocytes were prepared by dissociation of 1-3 day old rat cerebellar cortices. Adenosine uptake was measured using radiolabelled compounds. Adenine nucleosides and nucleotides were separated by HPLC and measured by liquid scintillation counting.

Both cell preparations can inactivate adenosine by active uptake from the extracellular space. This process results in a net movement of adenosine out of the medium. [^{3}H]-adenosine uptake is (i) sensitive to dipyridamole but not to nitrobenzylthioinosine or nitrobenzylthioguanine, (ii) partially inhibited by AMP but not other adenine nucleosides or nucleotides, and (iii) sensitive to adenosine receptor agonists (cyclopentyladenosine > cyclohexyladenosine = 1-PIA > 2-chloroadenosine >> NECA). Uptake of [^{3}H]-adenosine appears linear for 20 min and proceeds at a rate of 0.53 ± 0.02 nmoles/mg protein/min. After 60 min total label incorporated by cells corresponds to 24.0 ± 0.5 nmoles adenosine/mg protein. Analysis of cell contents indicates that uptake results first in an increase in intracellular adenosine content followed by rapid conversion into ATP. At the end of 1- min the nucleotide pool accounts for 90% of the

379

label while only 2% is present as adenosine. The remaining 8% of the label is present as adenine, inosine, hypoxanthine, cAMP, and unidentified metabolites. Incorporation of [3H]-adenosine continues even after 60 min; however, the relative distribution of label into various compounds shows little change after 10 min.

Astroglial cells can also inactivate adenosine and ATP by metabolizing them extracellularly. This mechanisms of inactivation was studied by following the degradation of [3H]-ATP or [3H]-adenosine by a HEPES-buffered Hanks saline wash collected after a 60 min incubation with cells. This enzyme preparation was centrifuged and incubated with either substrate. Analysis of aliquots removed at various times showed a time-dependent disappearance of both substrates. ATP degradation was accompanied by the concomitant appearance of dephosphorylated products (ADP, AMP, adenosine, adenine, and inosine). Adenosine degradation produced adenine, inosine, and some unidentified products.

These results indicate that astroglial cells can inactivate adenosine and ATP by intra- and extra-cellular mechanisms. Adenosine is both taken up and metabolized intracellularly, as well as, metabolized extracellularly. ATP is degraded extracellularly into adenosine and adenine, both of which can then be removed from the medium by uptake into either neurons or glia.

A3. ASTROGLIAL CELLS RELEASE COMPOUNDS LABELLED WITH [3H]ADENINE WHEN STIMULATED WITH ISOPROTERENOL OR ELEVATED [KCl]$_0$.

D. L. MARTIN, V. MADELIAN, W. SHAIN, Wadsworth Center for Laboratories and Research, NY State Health Dept. and Dept. of Environmental Health and Toxicology, SUNYA, Albany, NY 12201-0509, USA.

Astrocytes have receptors for many neurotransmitters and neuropeptides implying that they receive messages from neurons. We have investigated whether astrocytes can respond to neuronal signals by releasing neuroactive substances, as such processes could have important developmental and regulatory functions in neuronal-glial interactions. The release of compounds labelled with [3H]adenine was studied with primary cultures of cerebellar astrocytes and the LRM55 astroglial cell line. For release studies, astroglial cells were grown to confluence on flexible plastic cell-support film (Bellco). Prior to the experiment cells were incubated with 10 μCi of [2-3H]adenine for 1 h at 35°C to load the cells with radioactivity. The cell support film with adherent cells was then transferred to a cylindrical perfusion chamber and perfused with Hepes buffered Hank's saline. The perfusion medium was changed and drugs were applied without changing the flow of perfusate with a system of in-line valves controlled by a computer. The β-adrenergic agonist isoproterenol (IPR) stimulated the release of adenine-labelled compounds. IPR-stimulated release was dose-dependent (EC$_5$0 = 3 nM) and was reversibly inhibited by the -adrenergic antagonists alprenolol and propranolol (0.1 μM) but not inhibited by the α-adrenergic antagonist phentolamine (100 μM). Elevated [K$^+$]$_0$ (10-50 mM) also stimulated the release of adenine labelled compounds. K$^+$-stimulated release was only observed when [K$^+$]$_0$ was elevated by adding KCl to the medium but not when KCl was isosmotically substituted for NaCl. Release was stimulated to a similar degree by adding equiosmolar concentrations of KCl, NaCl or sucrose, indicating that KCl-stimulated release is related to the elevated osmotic pressure of the medium and is not attributable to depolarization by elevated [K$^+$]$_0$. These results differed from the effects of osmolarity and elevated [K$^+$]$_0$ on taurine release from astroglial cells indicating that the release mechanisms for taurine and adenine-base compounds are different. The released compounds were identified by high performance liquid chromatography on a Waters μBondapak C18 column. Baseline release samples contained ATP, ADP, AMP, adenosine, adenine, inosine and an unidentified metabolite. IPR and high-[KCl]$_0$ stimulated the release of different populations of labelled compounds. IPR stimulated the release of cyclic adenosine 3'-5'monophosphate (cAMP) much more strongly than other compounds. Since cAMP is a small fraction of total adenine-base compounds (<1%), IPR appears to selectively stimulate cAMP release. Elevated [KCl]$_0$ stimulated the release of ATP, ADP and AMP but not cAMP. Adenosine release was reduced below baseline by high [KCl]$_0$.

Supported by contract N1487G0179 from the US Naval Medical Research and Development Command and by grants NS21219 and AA007155 from USPHS/DHHS.

A4. CHARACTERIZATION OF THE ENZYMES INVOLVED IN THE DEGRADATION OF EXOGENOUS ATP AT THE INNERVATED SARTORIUS MUSCLE OF THE FROG. A.M. SEBASTIA~O, R.A. CUNHA, J.A. RIBEIRO.

ATP might contribute to the pool of endogenous adenosine that inhibits neuromuscular transmission (Ribeiro and Sebastia~o, 1987). At the innervated frog sartorius muscle, exogenous ATP can be sequentially degraded into AMP and then either into IMP or into adenosine (Ado) (Cunha et al., 1989). The present work was undertaken to characterize the enzymes involved in the degradation of exogenous ATP in this preparation.

The preparations were mounted in a 2 ml bath at room temperature (22-25 °C). The bathing solution (pH 6.8) contained (mM): NaCl 117, KCl 2.5, $MgCl_2$ 1.2, $CaCl_2$ 1.8, NaH_2PO_4 1, Na_2HPO_4 1. Unless otherwise stated the solutions contained dipyridamole (0.5uM). Throughout the assays air was bubbled into the bath. At different times after the addition of ATP or its metabolites, samples (75ul) were collected from the bath and analysed by reverse-phase HPLC. Kinetic parameters (Km and Vmax) were usually calculated from progress curves (Boeker, 1984), with 7-17uM initial concentrations of substrate. In the case of the ATPase, Km and Vmax values were also calculated from initial velocity measurements (Hanes plot) with initial concentrations of substrate of 5-1200 uM.

ATP was degraded into ADP, AMP, IMP, Ado and inosine (Ino). Km and Vmax values for the degradation of ATP into ADP were 616 ± 48uM and 135_+10nmol/min (progress curves, n=7). Similar Km (754 ± 33uM) and Vmax (147 ± 5.2nmol/min) values were obtained from initial velocity measurements (n=3). Removal of $CaCl_2$ or $MgCl_2$ from the bath did not cause appreciable changes in Km or Vmax values. In an EDTA (1.25mM) medium without both $CaCl_2$ and $MgCl_2$, the Km value was increased by $46 \pm 4\%$ and the Vmax decreased by $38 \pm 3\%$ (n=3). Neither ADP (50 uM) nor p-nitrophenylphosphate (1 mM) modified the ATP degradation kinetics. ADP was degraded into AMP, IMP, Ado and Ino. Km and Vmax values for degradation of ADP into AMP were 1095 ± 393uM and 284 ± 49nmol/min (n=3). AMP (50 uM) did not modify the degradation of ADP. AMP was degraded into IMP, Ado and Ino. AOPCP (200uM) inhibited by 100% Ado formation from AMP. The Km and Vmax values for degradation of AMP into IMP (determined in the presence of 200uM AOPCP) were 192 ± 34uM and 52 ± 9.2nmol/min (n=3). IMP and Ado were both degraded into Ino. Km and Vmax values for degradation of IMP into Ino were 672 ± 167uM and 118 ± 29nmol/min (n=4). Ino (50uM) did not modify degradation of IMP. AOPCP (200uM) inhibited by 100% Ino formation from IMP. Dipyridamole (0.5 uM) inhibited ($47 \pm 13\%$ at 90 min, n=3) Ino formation from Ado. EHNA (5- 25uM), applied in the presence of dipyridamole (0.5uM), did not affect the formation of Ino from Ado. In the absence of exogenously applied nucleotides or nucleosides, only trace amounts of inosine (0.19 ± 0.01uM at 120 min, n=25) were formed. None of the other compounds were formed.

The results suggest the presence of the following ectoenzymes at the innervated frog sartorius muscle: 1)ATPase, which requires either Ca^{2+} or Mg^{2+}, 2) ADPase, 3) 5'-nucleotidase, sensitive to AOPCP and accepting either AMP or IMP as substrates, 4) AMP-deaminase. Deamination of Ado probably occurs intracellularly. This work was supported by a project grant (BAP-0470P(EDB)) from EEC.

Boeker, E.A. (1984). Biochem. J., 223, 15-22.
Cunha, R.A., Sebastia~o, A.M. and Ribeiro, J.A. (1989). In: Adenosine Receptors in the Nervous System, pp.213   ed. J.A. Ribeiro, Taylor & Francis, London.
Ribeiro, J.A. and Sebastia~o, A.M. (1987). J. Physiol., 384, 571-585.

A5. DESIGN AND USE OF INHIBITORS OF CYTOSOLIC 5'-NUCLEOTIDASE TO PREVENT PURINE NUCLEOTIDE DEGRADATION. A.C. SKLADANOWSKI, G.B. SALA, A.C. NEWBY. Department of Cardiology, University of Wales College of Medicine, Cardiff, U.K.

High energy output organs which rapidly lose purine nucleosides during ischaemia might be protected by preventing the dephosphorylation of mononucleotides, thus retaining nucleotides within the cell. This might be achieved by inhibiting the soluble cytoplasmic 5'-nucleotidases responsible for adenosine and inosine formation. A soluble IMP-specific 5'-nucleotidase present in many cells including liver, heart and polymorphonuclear leucocytes was chosen as a target for drug action. The reaction path of this enzyme involves a phosphoenzyme intermediate, hydrolysis of which occurs only after the reversible dissociation of inosine or adenosine (Worku,Y. & Newby, A.C. Biochem. J. 205, 503-510, 1982). Thus nucleoside analogues with a free 5'-hydroxyl group become phosphorylated in the presence of a substrate nucleotide. Nucleoside analogues with a modified 5'-group should cause noncompetitive inhibition.

Among the series of commercially-available or easily-synthetised nucleoside analogues, 5'-deoxy-5'-isobutylthioadenosine (IBTA) and 5'-deoxy-5'-isobutylthioinosine (IBTI) were found to be the most potent inhibitors. Both compounds noncompetitively inhibited the cytoplasmic 5'-nucleotidases from rat liver, heart and polymorphonuclear leucocytes. Inhibition by 50% was achieved at 2-6 mM-IBTA or 4-10 mM-IBTI. IBTI was found also to inhibit the newly discovered AMP-specific cytoplasmic 5'-nucleotidase from pigeon heart by 50% at 6 mM but did not inhibit plasma membrane 5'-nucleotidase.

The ability of these inhibitors to penetrate and act in intact cells was tested with 2-deoxyglucose poisoned polymorphonuclear leucocytes. IBTA (3 mM) and IBTI (7 mM) inhibited the formation of adenosine (by 80% and 70% respectively) and hypoxanthine (35% and 51% respectively). After adding 2-deoxyglucose for 10 min, the percentage of the initial total nucleotide pool remaining as AMP+IMP was increased from 12% in the absence of inhibitors to 20% and 25% in the presence of IBTA and IBTI respectively.

The data suggest that IBTA and IBTI may be model compounds for the development of more potent inhibitors of cytoplasmic 5'-nucleotidase. These might be useful pharmacologically, to maintain nucleotide concentrations in transplanted organs, during cardioplegia or during transient ischaemia.

A6. DOES LOW Km 5'-NUCLEOTIDASE PRODUCE ADENOSINE IN RAT LIVER? SPYCHALA, J., PARK, J.K. AND FOX, I.H. University of Michigan, Departments of Biological Chemistry and Internal Medicine, Ann Arbor, MI 48109.

Adenosine may be produced in the liver by the action of two soluble 5'-nucleotidases, the high Km and low Km form. High Km 5'-nucleotidase is specific for IMP and GMP, is activated by ATP and inhibited by Pi. The activity of this enzyme is very low with AMP. On the other hand, low Km 5'-nucleotidase prefers AMP and has Km of 12 uM. However, this enzyme is strongly inhibited by ATP and ADP. To elucidate whether low Km 5'-nucleotidase is capable of AMP dephosphorylation in vivo, kinetic experiments were designed to examine the kinetic properties of the rat liver 5'-nucleotidase and to simulate physiological conditions under which the enzyme operate in vivo. The pH optimum is at 7.7 to 8.2. ATP and ADP are competitive inhibitors of the enzyme with Ki values of 13 and 3 uM respectively. a,b, methylene ADP, a synthetic analogue of ADP, was the most potent inhibitor with a Ki value of 0.5 uM. ATP was a potent inhibitor at low concentrations (5 - 50 uM) but did not substantially increase inhibition at physiological range of concentrations from 2 to 3 mM. On the other hand, AMP from 1 to 100 uM at fixed ATP (3 mM) and ADP (50 uM) concentrations, linearly increases enzyme activity. The relative enzyme insensitivity to physiological changes of ATP concentrations and the pseudo-first order kinetics with increase of AMP concentrations seems to be compatible with the known mechanisms of adenosine production. Total low Km 5'-nucleotidase activity extracted from 1 g of rat liver and measured at 100 uM AMP was extrapolated, using kinetic data, for different AMP, ADP and ATP concentrations at pH 7.4. These calculations show that at fixed ATP (3 mM) and ADP (50 uM) concentrations the soluble low Km 5'-nucleotidase activity may range from 0.5 nmoles of produced adenosine/min/g wet tissue at 1 uM AMP, to 5 nmoles/min/g at 10 uM AMP and 60 nmoles/min/g at 100 uM AMP. We conclude that there is sufficient soluble low Km 5'-nucleotidase activity to explain adenosine generation in rat liver, even in the presence of ATP and ADP..

A10. ADENOSINE TRANSPORT REGULATION IN CHROMAFFIN CELLS
POSSIBLE INVOLVEMENT OF KINASES A AND C. M.T. MIRAS-PORTUGAL, R.
SEN. D. FIDEU, M. TORRES AND E. DELICADO, Departamento de Bioquimica, Facultad de
Veterinaria. Universidad Complutense de Madrid, 28040 Madrid, Spain.

Adenosine transport has been largely studied in chromaffin cells and a transport with high
affinity and capacity found (1,2). In spite of the increasing importance of adenosine as a
neuromodulator and as a vasoactive substance, there is no work, so far, about the possibility of
adenosine transport regulation by external effectors and/or the corresponding intracellular
signals, as it has been done for the glucose transport even in the same cellular model (3,4).

The adenosine transport was measured in cultured chromaffin cells, at a density of 250.000
cells by well, as described in reference 1.

The nitrobenzylthioinosine (NBTI) binding experiments were done with 10^6 cells in Petri
dishes, and the binding to chromaffin cell membranes with 0.5 mg protein by assay as described
in reference 2.

The effects of protein kinase A stimulation on adenosine transport were studied in several
ways: 1) increasing the intracellular cAMP with forskolin. 2) with cAMP analogues that enter
the cell, as the 8-(4-chlorophenylthio)adenosine-3':5'-monophosphate cyclic (ClPhcAMP). 3)
with inhibitors of the cAMP phosphodiesterase as the 3-isobutyl-1-methylxanthine (IBMX).

Forskolin (0.5 μM), ClPhcAMP (100 μM) and IBMX (1 mM), inhibited the adenosine
transport, the percentage inhibition being 50, 50 and 85% respectively. Only minor changes
were observed on the K_M values.

The effect of protein kinase C on adenosine transport were done in the presence of some
activators of this enzyme as the phorbol esters and other compounds. PMA (0.1 μM), PDBu
(0.1 μM), Dicaproin (0.1 μM) and Tricaprylin (0.1 μM) were inhibitors of the adenosine
transport, the percentage of inhibition being 20, 15, 40 and 45% respectively.

The binding of NBTI to cultured chromaffin cells was modified by the kinases A and C
effectors, but without changes in the K_d values. Both ClPhcAMP and forskolin decreased the
number of high affinity binding sites. The percentage of decrease being close to 20-25% of the
controls. Similar decrease was obtained with PDBu (20%).

External effectors at the plasma membrane as insulin and secretagogues also modified the
transport capacity and the number of binding sites for NBTI in cultured chromaffin cells.

1.- Miras-Portugal et al (1986) J. Biol. Chem. 261, 1712-1719.

2.- Torres et al (1988) Biochim. Biophys. Acta 969, 111-120.

3.- Delicado and Miras-Portugal (1987) Biochem. J. 243, 541-547.

4.- Delicado et al (1988) FEBS Lett. 229, 35-39.

A11. AUTORADIOGRAPHICAL LOCALIZATION OF ADENOSINE RECEPTORS
AND TRANSPORT SITES IN CARDIAC CONDUCTION AND ENDOTHELIAL
CELLS. F.E. PARKINSON and A.S. CLANACHAN, Department of Pharmacology, University
of Alberta, Edmonton, Alberta, CANADA, T6G 2H7

Adenosine has pronounced depressant effects on cardiac function including decreased rate and
force of contractions and reduced atrioventricular conduction velocity. These effects are
mediated by cell membrane-located adenosine A_1 receptors. The effects of adenosine are limited
by its rapid transport into cells by nucleoside transport systems. Inhibitors of nucleoside
transport potentiate and prolong the effects of adenosine.

We investigated the mechanisms involved in the prominent depression of atrioventricular
conduction induced by adenosine in 10 μm sections of guinea pig heart. Atrioventricular
conduction cells were identified by histochemical detection of acetylcholin-esterase. The
density of adenosine A_1 receptors, identified by autoradiography of the binding of the selective
antagonist 8-cyclopentyl-1,3-[^{3}H]dipropylxanthine ([^{3}H]DPCPX), showed that conduction
cells have a higher density of receptors than cardiac myocytes. In contrast, autoradiography of
nucleoside transport sites in these cells, performed with [^{3}H]nitrobenzylthioinosine
([^{3}H]NBMPR), a widely used probe for transport sites, indicated that there was no marked
difference in site density between conduction cells and cardiac myocytes.

383

However, nucleoside transport sites were elevated in cells associated with the coronary vasculature and the endocardial lining. Comparison of the areas of high site density to the radioimmunohistochemical distribution of von Willebrand Factor, an endothelial cell marker, indicated that the two markers were co-localized. Quantitative autoradiography was used to determine the binding constants for [^{3}H]NBMPR binding to endothelial cells and cardiac myocytes. These two cell types had high affinity for [^{3}H]NBMPR; K_D values were 1.4 nM and 4.5 nM for the endothelial cell and cardiac myocyte areas, respectively. The endothelial cell binding component had about twice (1751 fmol/mg protein) the number of [^{3}H]NBMPR binding sites as the myocytes (990 fmol/mg protein).

The results of these studies indicate that the pronounced dromotropic effects of adenosine in guinea pig heart are correlated with a higher density of adenosine A_1 receptors in atrioventricular conduction cells. The transport capacity of these cells, as estimated by [^{3}H]NBMPR binding site density, is apparently not increased in proportion to A_1 receptors. The high density of transport sites associated with endothelial cells points to the key role of these cells in adenosine homeostasis in heart.

During these studies F.E.P. was a recipient of an A.H.F.M.R. Research Studentship. Supported by the Alberta Heart and Stroke Foundation.

A12. AMP-INDUCED AFTERLOAD REDUCTION INCREASES CARDIAC INDEX DURING ACUTE LOW OUTPUT CARDIAC FAILURE IN DOGS.

A.S. CLANACHAN and B.A. FINEGAN, Departments of Pharmacology and Anaesthesia, University of Alberta, Edmonton, AB, CANADA, T6G 2H7.

Adenosine receptor stimulation elicits systemic vasodilation that may be useful in reducing impedance to left ventricular ejection (afterload) during treatment of acute cardiac failure with catecholamines. The objectives of this study were to compare the hemodynamic properties of adenosine monophosphate (AMP) with a clinically useful vasodilator, sodium nitroprusside (SNP) when infused alone or in combination with the positive inotropic agent, dopamine (DA), in normal anesthetized dogs. Also, we have evaluated the effects of afterload reduction with AMP or SNP on the cardiac index (CI) of anesthetized dogs in which acute cardiac failure was induced by the administration of microbeads (50 μm) into the left coronary artery.

In normal anesthetized dogs, both SNP(2 to 25 μg.kg⁻1.min⁻1) and AMP (200 to 2500 μg.kg⁻1.min⁻1) when given alone or with DA

(5 μg.kg⁻1.min⁻1) were effective vasodilators and reduced systemic vascular resistance (SVR) and arterial pressure in a dose-dependent manner. Heart rate and CI were elevated by both agents. When compared at equi-effective hypotensive infusion rates, CI was increased more by AMP than by SNP. Neither vasodilator, at the doses used, depressed A-V nodal conduction as judged by stimulus artefact-R intervals during periods (15s) of right atrial pacing (3Hz). In addition, measurements of renal blood flow by an electromagnetic flow probe placed on the left renal artery indicated that maintained renal vasoconstriction or antagonism of DA-induced renal vasodilation did not occur.

Cardiac failure was induced in a separate group of dogs by the administration of microbeads until left ventricular end diastolic pressure was increased from 5 to 15 mmHg; this was associated with a decrease in dP/dt and CI of 30% and 32%, respectively. AMP (100 to 1000 μg.kg⁻1.min⁻1) or SNP (1 to 10 μg.kg⁻1.min⁻1) reduced SVR, and at doses that caused similar reductions in afterload, AMP increased CI significantly more than SNP. In both groups of animals, AMP-induced cardiac depression was not demonstrable. The mechanism responsible for the higher efficacy of AMP relative to SNP may be related to its more selective arterial vasodilatory activity. SNP, by also causing venorelaxation, reduces cardiac preload that, in turn, limits increases in CI.

We conclude that DA-AMP combination therapy may offer significant advantages to DA-SNP in situations where elevation of CI and reduction of SVR are required. Further investigation of the role of AMP in the clinical management of acute low output cardiac failure is warranted.

Supported by Alberta Heart & Stroke Foundation.

A13. ETHANOL-INDUCED INCREASES IN EXTRACELLULAR ADENOSINE CONCENTRATION DUE TO INHIBITION OF ADENOSINE UPTAKE VIA NUCLEOSIDE TRANSPORTER. LAURA E. NAGY, DAVID J. CASSO, CHRISTOPHER FRANKLIN, IVAN DIAMOND, AND ADRIENNE S. GORDON, University of California, San Francisco and the Ernest Gallo Clinic and Research Center, Building 1, Room 101, San Francisco General Hospital, San Francisco, CA 94110.

Chronic exposure to ethanol results in a heterologous desensitization of receptors coupled to adenylate cyclase via G_s, the stimulatory guanine nucleotide regulatory protein. We have previously reported that accumulation of extracellular adenosine is required for the development of ethanol-induced heterologous desensitization. In order to understand the mechanism underlying ethanol-induced extracellular adenosine accumulation, we examined the interaction of ethanol with the adenosine transport system in S49 lymphoma cells and NG108-15 neuroblastoma cells. We found that ethanol non-competitively inhibited nucleoside uptake without affecting deoxyglucose or isoleucine transport. Ethanol-induced increases in extracellular adenosine concentrations appear to be due to decreased influx of adenosine via the nucleoside transporter as well as decreased intracellular trapping of adenosine. After chronic exposure to ethanol, cells became tolerant to the acute effects of ethanol, i.e. ethanol no longer inhibited uptake and, consequently, no longer increased extracellular adenosine concentration. Moreover, in S49 cells, chronic ethanol exposure also decreased uptake of adenosine in the absence of ethanol. This was most likely the result of an ethanol-induced decrease in the number of transporters since binding studies using a high affinity ligand for the nucleoside transporter, NBMPR, also showed a concomitant decrease in the maximal number of binding sites. These results suggest that acute ethanol inhibition of adenosine uptake leads to the desensitization of receptor-dependent cAMP signal transduction which develops after chronic exposure to ethanol.

A14. EFFECT OF ETHANOL ON BRAIN PURINE METABOLITES IN AWAKE AND ANESTHETIZED RATS. J.R. MENO, E.L. GORDON, J. MIYASHIRO AND H.R. WINN. Department of Neurological Surgery, University of Washington, Seattle, Washington, 98195

The purine nucleoside adenosine (ADO) is an important regulator of cerebral blood flow. In portal circulation, ADO has been implicated to mediate increases in blood flow following ethanol administration (Orrego et al, Am. J. Physiol. 254:G495-G501, 1988). This study was undertaken to determine if brain purine metabolites are altered after ethanol administration.

Adult male Sprague Dawley rats were initially anesthetized with halothane, tracheostomized and immobilized with tubocurarine (1mg/kg, IV). All operative sites were infiltrated with mepivacaine HCl (1%). Ethanol or saline was administered by gavage (2g/kg). The animals were further divided into subgroups. In one group, halothane (0.5%) was continued throughout the experiment and in the other, general anesthetic was discontinued (unanesthetized). Blood samples (1 ml) were collected from the ethanol treated animals for the determination of plasma ethanol levels. Brain tissue was sampled by the freeze-blow technique 30 minutes after ethanol or saline was administered. Mean plasma ethanol levels were 227±48 mg/dl (unanesthetized) and 192±26 mg/dl (anesthetized).

There was a significant increase in total purine metabolite levels (i.e. ADO + inosine + hypoxanthine) in unanesthetized as compared to anesthetized animals. In the anesthetized group, the ethanol-treated animals had decreased purine metabolites compared to saline-treated animals ($p<0.05$). In the unanesthetized group, the levels of these compounds remained unchanged following ethanol administration. We conclude that, unlike portal circulation, increases in brain purine metabolites are not observed in association with ethanol administration.

A16. PLASMA ADENOSINE CONCENTRATION DURING EPISODES OF OBSTRUCTIVE SLEEP APNEA. H.G.E. LLOYD, L. LAKS, R. GRUNSTEIN, H. LAGERCRANTZ, M. MORRISON, G.A.R. JOHNSTON, C.E. SULLIVAN. Departments of Medicine and Pharmacology, Sydney University, N.S.W. 2006, Australia and Department of Neurophysiology, Karolinska Institute, Stockholm, Sweden.

In the sleep apnea syndrome there is repetitive nocturnal hypoxemia of varying severity. The level of hypoxemia may be sufficiently severe to cause increased cellular ATP breakdown and hence increased adenosine production. Findley et al., (1988) have recently reported that adenosine plasma concentration is significantly correlated with two indices of nocturnal hypoxemia. We have examined this further by measuring plasma adenosine in blood taken from an indwelling venous catheter (forearm) during wakefulness and during apneic sleep in 8 patients with severe repetitive obstructive sleep apnea.

The measurement of plasma adenosine was essentially as described by Gewirtz et al.(1987). The ultrafiltrated plasma was assayed using reverse-phase HPLC with isocratic elution (0.02M sodium phosphate buffer, pH 4.0, 9% methanol; 0.8ml/min). The dietary compound theobromine, which acted as an internal standard, was also measured. Arterial oxygen saturation (P_{AO2}) was recorded using an ear oximeter. Values of P_{AO2} are the maximum depressions recorded during the apneic period.

The mean awake plasma adenosine concentration (P_{AO2}: 97%) was $2.7 \pm 0.6 \times 10^{-8}M$ (x $\pm$ sem) compared with 3.6 ± 0.9 (P_{AO2}: 81-90%), $^*6.5 \pm 1.2$ (P_{AO2}: 71-80%), $^*5.7 \pm 1.2$ (P_{AO2}: 55-70%) $\times 10^{-8}M$ ($^*P < 0.01$: Wilcoxon rank test). Regression analysis showed a significant correlation between adenosine plasma concentration and P_{AO2} (correlation coefficient = 0.40; $P < 0.01$) with a regression coefficient equal to 0.0103 ($P < 0.01$). There was no correlation between theobromine plasma concentration and P_{AO2}.

The two-fold increase in plasma adenosine concentration which occurred when P_{AO2} fell below 80% indicates increased rate of cellular ATP breakdown and adenosine production. However, the increase is relatively small compared with the decrease in P_{AO2}, and the data were widely scattered. These findings are likely to reflect the rapid removal of adenosine in plasma by erythrocytes (>95% within 20s), and serve to emphasize the tendency for intra-erythrocytic and plasma adenosine pools to maintain equilibrium.[2]

..In conclusion, plasma adenosine concentration increases during sleep apnea reflecting an increased rate of tissue adenosine production. Elevated adenosine levels may be of considerable importance in this syndrome mediating, for example, the reversible depression of respiratory drive found in some sleep apneic patients. However, measurement of plasma adenosine concentration during cyclical repetitive hypoxemia is unlikely to provide a sensitive indicator of tissue hypoxia.

1. Findley LJ, Boykin M, Fallon T, Belardinelli L (1988) J Appl Physiol 65: 556-561.
2. Gewirtz H, Brown P, Most AS (1987) Proc Soc Exp Biol Med 185: 93-100.

A17. REGULATION OF MYOCARDIAL ADENOSINE FORMATION. A ^{31}P-NMR SPECTROSCOPY STUDY IN ISOLATED RAT HEART. J. P. HEADRICK, R. J. WILLIS. Griffith University, Division of Science and Technology. Nathan, 4111, QLD, Australia.

While it is clear that adenosine may be involved in modulation of cardiac function and coronary flow, little is known concerning the regulation of adenosine formation in the myocardium. Adenosine formation is thought to be catalyzed by 5'-nucleotidase, however, several forms of the enzyme exist in different cellular compartments, displaying different kinetic characteristics. ^{31}P-NMR methodology was employed to non-invasively monitor changes in cytosolic metabolism during graded ischaemia (12, 7.2, 5.3, 3.4, and 1.8 ml/min/g), graded hypoxia (95%, 65%, 50%, 35%, and 5% O_2), and isoproterenol infusion (0.4, 3.0, and 60 nM) in isovolumic rat heart. Adenosine and inosine release into the coronary effluent was simultaneously monitored. Purine release was hyperbolically related to log [ATP]/[ADP][Pi] during hypoxia and stimulation (r=0.83 and 0.82 respectively), and was linearly related during ischaemia (r=0.91). Purine formation correlated significantly with free cytosolic [AMP], linearly during ischaemia (r=0.92), and exponentially during hypoxia and isoproterenol infusion (r=0.87 and 0.91 respectively), indicating that the formation of adenosine may be allosterically activated during hypoxia and metabolic stimulation, but is regulated solely by substrate availability during ischaemia. The observed changes in free cytosolic [ATP], [ADP], and [AMP] (unchelated to Mg^{++}), and changes in [Pi], [Mg^{++}], and [creatine phosphate] are inconsistent with the involvement of the ATP-activated cytosolic 5'-nucleotidase characterized by Itoh or the recently characterized lysosomal enzyme in adenosine formation. Alternatively, the data is fully consistent with the involvement of an enzyme with kinetic characteristics similar to the membrane-bound 5'-nucleotidase, or the ATP-inhibited cytosolic enzyme recently isolated from placental tissue and kidney. It is concluded that adenosine formation in heart is linked to energy metabolism via alterations in cytosolic metabolites and substrate levels ([AMP]), and that the observed efflux of adenosine is catalyzed by an enzyme which possesses kinetic characteristics similar to those for the currently characterized membrane-bound 5'-nucleotidase.

A18. COMPARATIVE EFFECTS OF PROGESTERONE AND OF SYNTHETIC PROGESTOGENS ON cAMP LEVELS IN FEMALE RAT ADIPOCYTES. SUTTER-DUB M.-TH., B.CH.J. SUTTER, Laboratoire d'Endocrinologie, Université de Bordeaux I, Avenue des Facultés, 33405 Talence Cédex, France.

Progesterone is concerned with insulin resistance in the pregnant and the lactating rat, as in the progesterone-treated rat (5 mg/rat/kg). In isolated female rat adipocytes this steroid caused a dose-dependent decrease of glucose oxidation and lipogenesis in a 20-min incubation. This rapid acute action of progesterone appeared to be independent of protein synthesis. Synthetic steroids, agonists (the progestogen, RU-5020; the synthetic glucocorticoid, RU-26988) or antagonist (the antiprogesterone, antiglucocorticoid RU-38486) decrease also glucose oxidation of adipocytes as early as 20-min incubation and was sustained during a 2-h incubation. The early effect was less marked for progesterone agonist RU-5020 than the other three steroids and, when incubation was prolonged for 2h, the lowest inhibitory effect was observed with progesterone antagonist, RU- 38486.
This acute action of steroids seems difficult to reconcile with the classical mode of action of steroids. As second messengers have been shown to interfere in biological responses to steroids in various cells (Xenopus oocytes, human thymocytes and endometrium...), their intervention in the effects of progesterone, RU-5020, and RU-38486 in adipocytes, was investigated. - Glucose[1- ^{14}C]oxidation to CO_2, according to Rodbell (1964), was determined, as an index of glucose metabolism. cAMP levels were measured, after different incubation times, with a sensitive commercial available radioimmunoassay dosage (CEA, France). - In female rat fat cells dibutyryl cAMP, as well as inhibitors of phosphodiesterase (caffeine, theophylline, imipramine, papaverine, dipyridamole), or stimulators of adenylate cyclase (cholera toxin, glucagon), known to increase the intracellular levels of cAMP, produced the same inhibitory effect as the three steroids, on [1-^{14}C]glucose oxidation. Moreover, progesterone increased directly cAMP levels in a dose-dependent manner after a 5-min incubation of fat cells. This effect is not produced by RU-5020, neither by RU-38486, nor by other steroids such as corticosterone, 17ß-estradiol, although all these steroids decrease also [1 ^{14}C]glucose oxidation.
When adenylate cyclase activity was increased by glucagon, or phosphodiesterase activity inhibited by papaverine, the response was not the same as with the steroids tested : effects differ in function of concentrations of the steroid, duration of incubation etc.
So the first steps of the action of the synthetic steroids RU- 5020 and RU-38486, on glucose metabolism in female rat adipocytes, differ from that of the natural steroid, progesterone, and this may be one of the reasons of the differences observed in the acute effects of these two steroids on glucose metabolism in female rat fat cells.

A20. EFFLUX OF CYCLIC AMP IS REGULATED BY OCCUPANCY OF THE ADENOSINE RECEPTOR IN CULTURED ARTERIAL SMOOTH MUSCLE CELLS. L. L. SLAKEY, E. S. DICKINSON, T. F. FEHR, S. J. GOLDMAN. Dept. of Biochem., Univ. of Massachusetts, Amherst, MA 01003

Efflux of cAMP has been demonstrated in several cell types. In some cases it has been shown that the process is unidirectional and energy dependent (Barber and Butcher, Adv. Cyc. Nuc. Res. 15, 119, 1983, Brunton, Meth. in Enz. 159, 83, 1988). Little is known about the cellular regulation of cAMP efflux, although inhibitors have been identified. We have found that when pig aortic smooth muscle cells are treated with adenosine, intracellular cAMP rises, and cAMP accumulates in the incubation medium. No ATP escapes from the cells in the presence or absence of agonist under the conditions of these experiments. Cellular cAMP rises to a maximum level within 4-6 minutes of stimulation and then declines slowly. Efflux persists at a constant fractional rate over 60 minutes of observation as long as agonist is present. When agonist is removed, efflux stops abruptly, while cellular cAMP falls through a range sufficient to support efflux in the presence of agonist.

At cellular cAMP levels below 30 pmol/10^6 cells, efflux displays pseudo first order dependence on cellular cAMP, as judged by linear dependence of the accumulated extracellular cAMP on the time integral of the intracellular cAMP. The apparent first order rate constant for efflux increases in a dose-dependent manner in response to adenosine or 5-(N-ethyl)-carboxamide adenosine (NECA). The EC50 of the rise in the rate constant for cAMP efflux is 12 μM for adenosine and 5 μM for NECA. Treatment of cells with either cholera toxin (1 μg/ml) or forskolin alone (1 μM) also causes efflux together with a rise in cellular cAMP. These data are consistent with the hypothesis that in this cell type, activation of efflux of cAMP is tightly coupled to activation of adenylate cyclase. The kinetic data suggest that cAMP serves only as the substrate for efflux and not as a regulator. The rate constant for efflux is not affected by H8 (N-[2-(methylamino)ethyl]-5-isoquinolinesulfon-amide), an inhibitor of cAMP-dependent protein kinase.

Treating cells with adenosine together with low doses of cholera toxin (100 ng/ml), or 0.1 μM forskolin, or the phosphodiesterase inhibitor RO 20-1724, causes cellular cAMP to rise high enough so that it becomes evident that efflux is a saturable process, as is also the case with adenosine alone in some high responder cell lines. We estimated the kinetic parameters for efflux by numerically fitting the time course of change in extracellular cAMP (product) to the time course of change in intracellular cAMP (substrate), assuming simple Michaelis-Menten kinetics. The apparent Km for efflux is approximately 250 pmol cAMP/10^6 cells at 50 μM extracellular adenosine.

This work was supported by NIH grant number HL 38130.

A21. **RADIOCHEMICAL EVIDENCE FOR FREE INTERSTITIAL 5'-AMP IN ISOLATED PHYSIOLOGICALLY PERFORMING WORKING HEART.** ROLF BÜNGER, DAVID A. HARTMAN, ROBERT T. MALLET, Department of Physiology, Uniformed Services University of the Health Sciences, F. Edward Hébert School of Medicine, Bethesda, Maryland 20814-4799.

We previously reported (J.Mol.Cell.Cardiol. 20 (Suppl.III), 58,1988) in good agreement with data from dog heart in vivo (Berne et al., Am.J.Physiol. 254,H207,1988) that the epicardial transudate adenosine concentration was 75 ± 10 nM in normoxic isolated working guinea-pig heart; perfusion medium was a cell-and colloid-free Krebs-Henseleit buffer (pH ≈ 7.43) containing calcium (1.25 mM) and magnesium (0.65 mM) in physiological free concentrations, supplemented with 5 mM glucose (5 U/L insulin) plus 5 mM pyruvate as energy-yielding substrates. Epicardial fluid was sampled from cardiac apex, while a transverse pericardial sinus drainage between aorta and pulmonary trunk, in front, and atria, behind, was discarded. A marked interstitial radiotracer dilution of arterially infused [8-^{14}C]adenosine (0.4 - 1.4 μM) was also observed when adenosine traversed the ventricular wall. Thus, free adenosine preexisted in the interstitium of the isolated working guinea pig heart. The source of interstitial adenosine formation is not fully elucidated, e.g., coronary endothelium vs. working cardiomyocyte vs. free interstitial adenylates. Since coronary endothelial cells seem to release adenosine only into the luminal, not abluminal space (Nees et al., Basic Res.Cardiol. 80,515,1985) and because the thermodynamics of the free cardiac adenylates require submicromolar cytosolic free 5'-AMP concentrations in the normoxic heart (Bünger et al., Eur. J. Biochem. 159,203,1986), which implies minimal formation of adenosine in the myocyte cytosol, we tested the possibility that free adenylates, especially 5'-AMP existed in the interstitial space. Isolated working hearts were perfused (non-recirculating) with 0.6 - 0.8 μM [8-^{14}C]5'-AMP without and with 10 μM AOPCP (α ß-methylene adenosine 5'-diphosphate), a specific ecto 5'-nucleotidase inhibitor. Concentrations and specific radioactivities of 5'-AMP in arterial, venous, and epicardial transudate samples were measured after separation of 5'-AMP from AOPCP and the natural purine nucleosides and bases using reverse phase and ion-exchange HPLC techniques. The data (Table) show that 5'-AMP was degraded during a single passage through the coronary vasculature, but especially when traversing the ventricular wall. AOPCP completely and reversibly blocked 5'-AMP degradation. Prior to AOPCP the sp. activity of 5'-AMP was decreased about 19 % and 91 % in venous and epicardial samples, respectively. AOPCP increased the labeling of venous and epicardial 5'-AMP, but the relative sp. activities were still significantly lower than unity, especially in the epicardial samples (30.4 %). Consequently, interstitial dilution of exogenously applied tracer 5'-AMP must have occurred despite zero net degradation of 5'-AMP due to AOPCP. From the measured 5'-AMP concentrations and isotope dilution factors the endogenous (unlabeled) 5'-AMP concentration was estimated in presence of AOPCP, 0.071 μM and 0.493 μM in venous effluent and epicardial transudate samples, respectively. Consequently, there was little doubt that free adenylates existed in the interstitium of the isolated working guinea pig heart. Our radiochemical estimates are in reasonable agreement with chemical AMP measurements from non-working Langendorff-perfused guinea-pig heart not receiving exogenous 5'-AMP (Imai et al., Eur.J.Physiol., in press). Two main hypotheses appear plausible: 1) an AOPCP-insensitive adenylate permease near-equilibrates exogenous tracer with endogenous unlabeled 5'-AMP, 2) the myocardium releases intact adenylates (cyclic AMP or ATP) leading to the formation of unlabeled 5'-AMP (ecto phosphodiesterase, ecto nucleotidases) which dilutes arterially infused tracer 5'-AMP. Supported by NIH HL 36067 and USUHS RO7638.

CONDITION	n	$[AMP]_a$	$[AMP]_v$	Rel. sp. activity			AMP uptake (nmol/min /g dry)
				$[AMP]_{e}pi$	AMP_v	$AMP_{e}pi$	
				(nM)	(% of AMP_a)		
CONTROL	15-17	668 ±28	558[a] ±18	210[a] ±38	81.3[a] ±1.9	8.9[a] ±1.2	17.33[d] ±5.34
10 μM AOPCP	15-16	695 ±40	694[e] ±10	681[e] ±65	90.0[a],[e] ±1.3	30.4[a],[e] ±1.3	5.90[ns] ±7.21
POST-AOPCP	8-11	777 ±36	699[b] ±45	288[c] ±53	82.3[a],[c] 2.4	11.6[a],[c] ±0.9	32.41[d] ±10.30

a: p<0.01 vs. $[AMP]_a$, b: p<0.05 vs. $[AMP]_a$, c: p≤0.01 vs. AOPCP, d: p < 0.05 vs. zero, e: p < 0.01 vs. control, ns: not significant vs. zero; a,v, epi: arterial, venous, epicardial

A22. DUAL EFFECTS OF ADENOSINE ON NEUROMUSCULAR TRANSMISSION. SUSAN R. BARRY. Department of Physical Medicine & Rehabilitation, University of Michigan, 1500 East Medical Center Drive, Room UH1D204, Ann Arbor, MI 48109-0042

The extracellular concentration of adenosine at the neuromuscular junction (NMJ) may increase with synaptic activity. Adenosine triphosphate is released with the neurotransmitter, acetylcholine (ACh), from motor nerve terminals and may be broken down to adenosine in the synaptic cleft. In addition, adenosine may be released from both nerve and muscle cells following intracellular ATP degradation. Since the levels of adenosine increase with neuromuscular activity, adenosine may play an important feedback role in neuromuscular transmission.

The effects of adenosine were tested on spontaneous transmitter release from the motoneurons innervating the cutaneous-pectoris muscle of the frog, Rana pipiens. Intracellular microelectrodes were placed in the endplate region of individual muscle fibers in order to record spontaneous miniature endplate potentials (mepps). A mepp results from the spontaneous release of a single quantum of ACh from the non-spiking nerve terminal. A change in the frequency of mepps which is not accompanied by a change in mepp amplitude is interpreted as a change in spontaneous transmitter release.

The most common action of adenosine (1μM to 1mM) was to depress spontaneous transmitter release, an effect which was not accompanied by a change in mepp amplitude or the resting potential of the muscle. This action was probably mediated by an A1-like adenosine receptor

389

since the adenosine analog, L-N6-henylisopropyladenosine (L-PIA), reduced mepp frequency at a threshold concentration of about 1nM and was 12 times more potent than 5-N-ethylcarboxamidoadenosine (NECA) and 237 times more potent than its stereoisomer, D-PIA, at decreasing spontaneous transmitter output. The non-specific adenosine receptor antagonists, theophylline (20μM) and 8-phenyltheophylline (1μM), and the potent A1-antagonist, 1-3-dipropyl-8-cyclopentylxanthine (100nM), reversed the inhibitory effects on mepp frequency of maximal concentrations of adenosine and L-PIA.

Adenosine may also stimulate spontaneous transmitter release. In normal frog Ringer, theophylline, at a concentration of 10μM, produced no consistent effect on mepp frequency. In contrast, 10μM theophylline significantly depressed mepp frequency in hyperosmotic Ringer. This effect may result from inhibition of an excitatory effect of endogenous adenosine because the inhibitory action of theophylline was mimicked by adenosine deaminase, an enzyme which degrades and thus blocks effects of endogenous adenosine.

Additional data support the hypothesis that adenosine may stimulate as well as inhibit spontaneous transmitter output. In 2 out of 8 preparations, adenosine (50μM to 1mM) enhanced mepp frequency. This excitatory effect may be mediated by an A2-receptor since NECA (1 to 10μM), an analog with approximately equal affinity for the A1 and A2 receptors, produced biphasic effects on spontaneous transmitter release.

In summary, adenosine may both stimulate and inhibit spontaneous transmitter release from frog motoneurons. These two opposing effects may be mediated by separate receptors on the motor nerve terminal.

B6. **AFFINITY CHROMATOGRAPHY OF THE A_1 ADENOSINE RECEPTOR FROM RAT BRAIN.** F.J. ZIMMER, K.-N. KLOTZ, R. KEIL, U. SCHWABE, Pharmakologisches Institut der Universität Heidelberg, Im Neuenheimer Feld 366, D-6900 Heidelberg, FRG.

The A_1 adenosine receptor has been solubilized with the zwitterionic detergent CHAPS and purified by affinity chromatography. The adenosine receptor antagonist XAC (xanthine amine congener) was covalently attached to epoxy-activated Sepharose 6B and the resulting XAC agarose gel adsorbed about 50 % of the solubilized receptors using 3 volumes of the gel and 10 volumes extract from rat brain membranes. Desorption was achieved with different antagonists and was most effective with theophylline yielding approximately 50 % of the adsorbed receptors in the eluate. In this way a 100 - 150-fold purification of the A_1 receptor was achieved. Adsorption of the receptors to the gel was inhibited by adenosine receptor agonists and antagonists indicating the specificity of the process. The partially purified receptor preparation was further characterized by radioligand binding with the antagonist 8-cyclopentyl-1,3-[^{3}H]dipropylxanthine ([^{3}H]DPCPX) in the presence of 100 uM GTP. Competition for this radioligand exhibited a pharmacological profile for the affinity-purified receptors which is consistent with an A_1 adenosine receptor.

B7 **CHARACTERIZATION OF ATRIAL ADENOSINE RECEPTORS.** H. TAWFIK-SCHLIEPER, K.-N. KLOTZ, A. WILKEN, V. KREYE[*], U. SCHWABE, Pharmakologisches Institut der Universität Heidelberg, Im Neuenheimer Feld 366, and [*]II. Physiologisches Institut der Universität Heidelberg, Im Neuenheimer Feld 326, D-6900 Heidelberg, FRG.

Adenosine, as reported from electrophysiological studies, causes hyperpolarisation and shortening in the action potential of atrial myocytes. This effect was suggested to be caused by an increase in the potassium conductance. In the present work we studied the pharmacological profile of adenosine receptors in guinea pig left atria by investigating the effect of adenosine analogues on ^{86}Rb$^+$-efflux, on cAMP level in isolated guinea pig atrial myocytes and on the binding of the antagonist radioligand 8-cyclopentyl-1,3-[^{3}H]dipropylxanthine ([^{3}H]DPCPX) in guinea pig atrial membrane preparations. Adenosine receptor agonists increased the basal

$^{86}Rb^+$-efflux rate by about 2.5 times. The EC_{50}-values of 2-chloro-N^6-cyclopentyladenosine (CCPA), R-N^6-phenylisopropyl-adenosine (R-PIA), N-ethylcarboxamidoadenosine (NECA) and S-N^6-phenylisopropyladenosine (S-PIA) were 0.09, 0.13, 0.23 and 7.2 uM respectively. DPCPX shifted the R-PIA concentration response curve to the right in a dose-dependent manner. In isolated atrial myocytes adenosine derivatives caused an inhibition of isoprenaline-induced elevation of cAMP level and at a maximal effective concentration about 80% of the isoprenaline effect was abolished. CCPA, the adenosine analogue with the highest A_1 selectivity, was the most potent agonist with an IC_{50}-value of 85 nM. The radioligand [^{3}H]DPCPX showed a saturable binding to a crude atrial membrane preparation with a B_{max}-value of 200 fmol/mg protein and a K_D-value of 1.5 nM.

The order of potency in competition experiments with agonists was CCPA > R-PIA > NECA >> S-PIA. The stereoselectivity of the receptor to the two PIA-enantiomeres was clearly demonstrated in functional and in competition experiments. Our results show that the adenosine receptors in guinea pig atrial tissue are coupled to the K^+-channel as well as to the adenylate cyclase. The data from functional and binding studies are consistent with the A_1 receptor subtype.

B8. 2-CHLORO-N^6-[^{3}H]CYCLOPENTYLADENOSINE - A RADIOLIGAND WITH HIGH AFFINITY AND HIGH SELECTIVITY FOR A_1 ADENOSINE RECEPTORS.

K.-N. KLOTZ, M.J. LOHSE, U. SCHWABE, G. CRISTALLI, S. VITTORI, M. GRIFANTINI, Pharmakologisches Institut, Universität Heidelberg, Im Neuenheimer Feld 366, D-6900 Heidelberg, FRG; Dipartimento di Scienze Chimiche, Università di Camerino, Via S. Agostino 1, I-62032 Camerino, Italy

2-Chloro-N^6-cyclopentyladenosine (CCPA) has been shown to be an agonist with high affinity and an about 10,000 fold selectivity for A_1 adenosine receptors (Lohse et al., 1988). Competition of CCPA for the antagonist radioligand 8-cyclopentyl-1,3-[^{3}H]dipropylxanthine ([^{3}H]DPCPX) gave K_i-values of 0.2 and 19 nM for the high and low- affinity state, respectively. From the precursor 2-chloro-N^6-cyclopentenyladenosine the tritiated derivative [^{3}H]CCPA was prepared. [^{3}H]CCPA bound in a reversible manner to A_1 receptors in rat brain membranes. In saturation experiments a K_D-value of 0.2 nM was estimated for the high affinity state. The presence of GTP shifted the receptors to a low affinity state with a K_D-value of 13 nM. The high affinity of [^{3}H]CCPA enabled to detect A_1 receptors in tissues with very low receptor density. Saturation experiments were performed with membranes from rat ventricular myocytes and a K_D-value of 0.6 nM and a B_{max}-value of 16 fmol/mg protein was calculated. Binding to solubilized A_1 receptors from rat brain membranes was tested to further characterize [^{3}H]CCPA. The time course of association showed similar characteristics as has been shown for N^6-[^{3}H]phenylisopropyladenosine ([^{3}H]PIA). Binding equilibrium in the absence of divalent cations was reached within 3 h and Mg^{2+} markedly reduced association rate. [^{3}H]CCPA exhibited also high-affinity binding at solubilized receptors with a K_D-value of again 0.2 nM. The high A_1 selectivity of CCPA seemed to be preserved for the tritiated compound. A_2 receptors in human platelet membranes or solubilized receptors bound essentially no [^{3}H]CCPA in concentrations up to 400 nM. In conclusion, [^{3}H]CCPA is a new radioligand with high affinity for A_1 receptors and virtually no affinity for A_2 receptors. [^{3}H]CCPA is a suitable agonist to label A_1 receptors in tissues with very low receptor density owing to the subnanomolar affinity.

Lohse M.J., Klotz K.-N., Schwabe U., Cristalli G., Vittori S., and Grifantini M. (1988) 2-Chloro-N^6-cyclopentyladenosine: a highly selective agonist at A_1 adenosine receptors. Naunyn-Schmiedeberg's Arch. Pharmacol. 337, 687-689.

B10. RAPID SEPARATION OF PLATELET AND PLACENTAL ADENOSINE RECEPTORS FROM ADENOSINE A_2-LIKE BINDING PROTEIN. S. ZOLNIEROWICZ, C. WORK, K.A. HUTCHISON, I.H. FOX. The Human Purine Research Center, Departments of Internal Medicine and Biological Chemistry, University Hospitals, The University of Michigan, Ann Arbor, Michigan

The existence of the ubiquitous adenosine A_2-like binding protein makes it difficult to examine the binding properties of adenosine receptors using 5'-N-[^{3}H]-ethylcarboxamidoadenosine ([^{3}H]NECA). To avoid this problem, we have developed a rapid and simple method to separate adenosine receptors from the adenosine A_2-like protein. Human platelet and placental membranes were solubilized with the zwitterionic detergent 3-[3-(cholamidopropyl)-dimethylammonio]-1-propanesulfonate (CHAPS). The soluble platelet extract was precipitated with polyethylene glycol 8000 and the fraction enriched in adenosine receptors was isolated from precipitate by differential centrifugation. The adenosine A_2-like binding protein was removed from the soluble placental extract with hydroxylapatite and adenosine receptors were precipitated with polyethylene glycol. [^{3}H]NECA bound to adenosine receptor enriched fractions and was displaceable by both 100 uM NECA and 100 uM R-N^6-phenylisopropyladenosine (R-PIA). The specificity of the [^{3}H]NECA binding is typical for an adenosine A_2 receptor for platelets and an adenosine A_1 receptor for placenta.

This method leads to enrichment of adenosine A_2 receptors for platelets and adenosine A_1 receptors for placenta. This provides a useful preparation technique for pharmacological studies of adenosine receptors.

B11. PURIFICATION AND CHARACTERIZATION OF THE ADENOSINE A_2-LIKE BINDING SITE FROM HUMAN PLACENTAL MEMBRANES. KEVIN A. HUTCHISON AND IRVING H. FOX. The Human Purine Research Center, Departments of Internal Medicine and Biological Chemistry, University Hospitals, The University of Michigan, Ann Arbor, Michigan, 48109-0108

We have purified and characterized the adenosine A_2-like binding site from human placental membranes. 5'-N-Ethylcarboxamido[2,8-^{3}H]adenosine binds to this site with a Kd of 240 nM and a Bmax of 13.0 pmol/mg in human placental membranes. The adenosine A_2-like binding site was purified after extraction from placental membranes with 0.1% CHAPS. The purification included ammonium sulfate precipitation and concanavalin-A, DEAE-sephadex, and Sepharose 6B gel filtration chromatographies. The protein was purified 127 fold to homogeneity with a final specific activity of 1.5 to 1.9 nmol/mg protein and a 5.5 to 8.1% yield of binding activity from the membranes. The purified protein had similar binding properties and an identical potency order for displacement of [^{3}H]NECA by adenosine analogs as the initial membranes. Sodium dodecylsulfate polyacrylamide gel electrophoresis of purified protein revealed a single band at 98 kDa which coeluted with [^{3}H]NECA binding activity during sepharose 6B gel filtration chromatography. In 0.1% Triton X-100, the binding complex has a Stokes radius of 70 Å, a sedimentation coefficient of 6.9 S, and a partial specific volume of 0.698 ml/g. The detergent-protein complex has a calculated molecular mass of 230 kDa. The estimated frictional ratio, f/f_0, is 1.5. The native binding complex appears to consist of a dimer of identical subunits. The function of this ubiquitous protein remains unclear.

B12. COVALENT INCORPORATION OF 3'-O-(4-BENZOYL)BENZOYLATP INTO A P_2 PURINOCEPTOR IN TRANSFORMED MOUSE FIBROBLAST. L. ERB*, K. D. LUSTIG*, A. H. AHMED‡, F. A. GONZALEZ‡, G. A. WEISMAN*. *Departments of Biochemistry and Food Science & Nutrition, Univ. of Missouri-Columbia, Columbia, MO 65211 ‡Section of Biochemistry, Molecular and Cell Biology, Cornell University, Ithaca, NY 14853

Adenosine-5'-triphosphate (ATP) and 3'-O-(4-benzoyl)benzoylATP (BzATP), a photoaffinity analog of ATP, induced an increase in plasma membrane permeability to Na^+ and K^+ and, subsequently, to normally impermeant metabolites including nucleotides in transformed 3T6 mouse fibroblasts. Photoincorporation of ≥ 20 μM BzATP into 3T6 cells at 4 °C, a non-permeabilizing temperature, followed by removal of unbound BzATP, resulted in the efflux of $^{86}Rb^+$ and the release of a prelabeled pool of cytoplasmic nucleotides when the temperature was shifted to 37 °C. Consistent with the hypothesis that BzATP had been incorporated into a P_2 purinoceptor, analogs of ATP which caused an increase in nucleotide permeability in 3T6 cells incubated at 37 °C also inhibited the photoincorporation of BzATP, whereas ATP analogs which had no effect on photoincorporation also had no effect on nucleotide permeability. Photoincorporation of BzATP was inhibited by ATP ($IC_{50} = 50$ μM), adenosine-5'-O-(1-thiotriphosphate) (ATPaS, $IC_{50} = 10$ μM) and adenosine-5'-O-(3-thiotriphosphate) (ATPgS, $IC_{50} = 20$ μM). The inhibitory effect of ATP on BzATP photoincorporation could be overcome by increasing the length of the photoilluminiation period, a characteristic feature of photoaffinity labeling in the presence of nonreactive ligands. GTP, ITP, UTP and adenosine did not affect BzATP photoincorporation nor did the photoincorporation of BzGTP, BzITP or BzCTP at 4 °C have any effect on plasma membrane permeability after a temperature shift to 37 °C. In 3T6 cells incubated at 37 °C, an increase in plasma membrane permeability to nucleotides was induced by ATP analogs modified in the ribose or phosphate moiety but not by analogs modified in the adenine moiety. The rank order of agonist potency was BzATP ($EC_{50} = 15$ μM) > ATP ($EC_{50} = 50$ μM) $\approx$ ATPaS > 3'-amino,3'-deoxyATP ($EC_{50} = 150$ μM) > ATPgS ($EC_{50} = 175$ μM) > adenosine-5'-tetraphosphate (AP_4, $EC_{50} = 400$ μM) > AMP-CPP ($EC_{50} = 1$ mM) > ADPbS ($EC_{50} = 2$ mM). These findings suggest that BzATP can be covalently incorporated into a P_2 purinoceptor in 3T6 cells that is pharmacologically distinct from the previously described P_{2X} and P_{2Y} subclasses. (Supported by NIH Grants GM-36887 and AM-11789 and by NSF Grant PCM 81-21029.)

B14. AUTORADIOGRAPHIC ANALYSIS OF BRAIN ADENOSINE A_1 AND A_2 RECEPTORS FOLLOWING CHRONIC THEOPHYLLINE EXPOSURE TO RATS.

C.R. LUPICA[1], R. BERMAN[1], M. WILLIAMS[2] AND M.F. JARVIS[2]. [1]Department of Psychology, Wayne State University, Detroit, MI 48202. [2]Research Department, CIBA-GEIGY Corp. Summit, NJ 07901

Adenosine is involved in a wide variety of physiological processes; moreover, a specific contribution in neuronal communication has been indicated by the identification of discrete distributions of adenosine receptor subtypes in mammalian brain (Fastbom et al., 1987; Jarvis et al., 1989a). While the adenosine A_1 receptor has received extensive study, characterization of the adenosine A_2 receptor has been limited by the lack of available radioligands with high affinity and selectivity for this receptor subtype. Autoradiographic localization of A_2 receptors, which are specifically concentrated in the rat striatum, has been acomplished using the non-selective agonist, [^{3}H]NECA, in the presence of 50 nM CPA to block binding to the A_1 receptor (Bruns et al., 1986; Jarvis et al., 1989a). This highly selective distribution of A_2 receptors was recently confirmed using the 180-fold selective A_2 receptor agonist, [^{3}H](2-p-(2-carboxyethyl) phenethylamino)-5'-N-carboxamidoadenosine (CGS 21680) (Jarvis et al., 1989b; Jarvis and Williams, 1989).

Chronic exposure to adenosine receptor antagonists (i.e. methylxanthines such as caffeine and theophylline) either in the diet (Boulenger et al., 1983) or by systemic injection (Murray, 1982; Marangos et al., 1984; Szot et al., 1987) has been shown to produce small (5-20%), but reliable increases in adenosine A_1 receptor number without significantly affecting receptor affinity. However, observations of methylxanthine-induced increases in adenosine receptors have been limited to cortical and cerebellar regions of the rat brain and only with ligands which label A_1 receptors. In the present study, quantitative autoradiography was used to localize the binding of [^{3}H]CHA and [^{3}H]CGS 21680 to brain adenosine A_1 and A_2 receptors, respectively, in control and theophylline-treated rats. Theophylline treatment consisted of seven daily injections of 75 mg/kg i.p. theophylline followed by an additional seven daily injections of 100 mg/kg. Control animals received appropriate saline injections. Receptor binding and autoradiographic methodology were conducted as previously described (Jarvis, 1988; Jarvis et al., 1989a).

Chronic theophylline treatment resulted in moderate increases in [^{3}H]CHA binding to A$_1$ receptors in various regions of the rat brain with the greatest increases (25 - 40%) observed in the molecular layer of the cerebellum as well as in the hippocampus and superior colliculus. [^{3}H]CGS 21680 binding to A$_2$ receptors was highly localized in the striatal region of the rat brain (caudate nucleus, nucleus accumbens and olfactory tubercle), however, no significant differences in the density of [^{3}H]CGS 21680 binding were observed between theophylline and saline treated animals. The present results are consistent with and extend previous observations of regional variations in methylxanthine-induced increases in A$_1$ receptors. In contrast, since neither striatal A$_1$ and A$_2$ receptors were significantly altered by the theophylline treatment, it would appear the present results may also indicate possible regional variations in theophylline availability in the rat brain.

Boulenger, et al., 1983, Life Sci, 32, 1135.
Fastbom, et al., 1987, Neurosci. 27, 813.
Jarvis, M. 1988, In Receptor Localization, Ligand Autoradiography, Leslie and Altar eds, A.R. Liss, NY, p.95.
Jarvis et al., 1989a, Brain Res. 484, 111.
Jarivs et al., 1989b, J. Pharmacol. Exp. Ther., in press.
Marangos et al., 1984, Life Sci. 34, 899.
Murray, T. 1982, Eur. J. Pharmacol. 82, 113.
Szot et al., 1987, Neuropharmacol. 26, 1173.

B15. **CELLULAR LOCALIZATION OF ENDOGENOUS ADENOSINE IN PERIPHERAL TISSUES.** K. M. BRAAS, S. H. SNYDER, Department of Neuroscience, The Johns Hopkins University, School of Medicine, Baltimore, MD, USA 21205, Department of Anatomy and Neurobiology, University of Vermont, College of Medicine, Burlington, VT, USA 05405

The purine nucleoside adenosine influences numerous physiological processes throughout the body, including effects on neuronal communication within the CNS. Immunocytochemical localizations of endogenous adenosine in neuronal perikarya and fibers in discrete regions of the brain and retina (1,2), support a neurotransmitter or neuromodulator role for adenosine. Caffeine and other xanthines exert a variety of effects as antagonists at tissue adenosine receptors. We have examined tissues for endogenous adenosine immunoreactivity in order to facilitate an understanding of peripheral xanthine actions.

Adult male Sprague-Dawley rats under sodium pentobarbital anesthesia were perfused intracardially with phosphate buffer followed by 2.5% glutaraldehyde. Tissues were washed, dehydrated, and embedded in paraffin media. Paraffin-embedded tissue sections were immunocytochemically stained using the avidin-biotin-peroxidase complex (ABC) technique (1,2). Specific sensitive rabbit antisera directed against the adenosine derivative laevulinic acid (O^2',3'-adenosine acetal), which is capable of detecting as little as 1 pmol of adenosine by radioimmunoassay, have been characterized for immunocytochemistry previously (1). Adult male ACI rats were injected daily with 10 mg/kg deoxycoformycin for three days (3); control animals received vehicle only. Twenty-four hours following the final injection, the animals were perfused as described above, and the tissues processed for immunocytochemistry.

Xanthines exhibit diuretic effects, and tissues involved in control of water balance exhibit adenosine immunoreactivity. Intense adenosine immunoreactivity was localized to the collecting tubules of the kidney. In addition, lower staining was found in the kidney juxtaglomerular complex and brush border of the proximal convoluted tubules. Neuronal fibers in the neurointermediate lobe of the pituitary gland and cells of the adrenal cortex zona glomerulosa, which are involved in fluid and electrolyte balance also stain for adenosine. The testes exhibited differential staining across the tubules indicating an association with sperm cells; this localization parallels that observed for adenosine receptors and could account for the actions of xanthines on sperm motility. Increased catecholamine secretion evoked by caffeine may derive from endogenous adenosine identified in catecholamine containing cells of the adrenal medulla. In addition, adenosine containing myocytes were observed in the heart atrium and ventricle, and may play a role in the effects of xanthines on cardiac contractility. Adenosine immunoreactivity was also observed in the epithelium of the esophagus and stomach, and in serous cells of the submaxillary gland. These sites of endogenous adenosine may be involved in xanthine effects on gastric secretion.

Severe combined immunodeficiency has been associated with impaired adenosine deaminase resulting in increased tissue levels of adenosine. We treated animals with the adenosine deaminase inhibitor deoxycoformycin, which has been shown to cause morphological changes in immune tissues and altered adenosine deaminase expression (3). Like adenosine deaminase, adenosine immunoreactivity was observed in the reticuloendothelial Kupffer cells of the liver. In addition, prominent adenosine staining was observed in the medullary cords of lymph nodes, while little to no staining was observed in the thymus. Tissue morphology and adenosine staining were altered in the lymph nodes and thymus of deoxycoformycin treated rats, while liver and kidney staining remained the same. (1) Braas, et al. (1986) J. Neurosci. 6:1952; (2) Braas, et al. (1987) PNAS 84:3906; (3) Chechik, et al. (1983) JNCI 71:999.

C2. STIMULATION OF CYCLIC AMP PHOSPHODIESTERASE ACTIVITY IN HUMAN PLATELETS BY THROMBIN. ROBERT ALVAREZ, KURT LITTSCHWAGER, LINDA OSBURN AND DANA L. STAHL. Institute of Experimental Pharmacology, Syntex Research, Palo Alto, California 94304

Thrombin increases the specific activity of a membrane-associated, high-affinity form of cyclic AMP phosphodiesterase (Type IV) when incubated with intact human platelets. The apparent activation process occurs in the same concentration range of thrombin that is required to inhibit adenylate cyclase activity (0.032 to 1.0 units/ml). The capacity of thrombin to lower intracellular cyclic AMP in platelets, however, was not significantly diminished in the presence of an inhibitor of cyclic nucleotide phosphodiesterase, 1-methyl-3-isobutyl xanthine (MIX). Cyclic AMP accumulation, adenylate cyclase and cyclic AMP phosphodiesterase activity were also measured in platelets permeabilized with saponin. Following incubation with thrombin (1 unit/ml) cyclic AMP levels decreased 26 ±4% and this was accompanied by a decrease in the rate of synthesis of cyclic AMP from 240 ±8 to 150 ±15 pmoles/min/mg protein. Under these conditions, using a substrate concentration of 1 uM, there was no significant change in the rate of degradation of cyclic AMP. These observations suggest that inhibition of adenylate cyclase activity is the primary biochemical event responsible for the lowering of total intracellular cyclic AMP by saturating concentrations of thrombin. The possibility that activation of membrane-associated cyclic AMP phosphodiesterase contributes to the lowering of cyclic AMP in an intracellular pool or compartment has not been excluded. Thrombin also increases cyclic AMP phosphodiesterase activity in lysed human platelets and this effect is decreased in the presence of benzamidine, a serine protease inhibitor. In contrast to the stimulation observed with intact platelets, relatively high concentrations of thrombin (1-10 units/ml) are required for the apparent proteolytic activation of cyclic AMP phosphodiesterase in lysed cells.

C3. CELLULAR PATHOGENESIS OF NEPHROGENIC DIABETES INSIPIDUS (NDI) IN MICE: UNIQUE ROLE OF ABNORMAL cAMP-PHOSPHODIESTERASE. T.P. DOUSA, S. TAKEDA, P. MORGANO, J.-T. LIN, S. HOMMA, A.K. COFFEY, H. VALTIN, Nephrology Research Unit, Mayo Clinic, Rochester, MN and Department of Physiology, Dartmouth Medical School, Hanover, NH.

A strain of mice (DI +/+ severe) with hereditary nephrogenic diabetes insipidus (NDI-mice) fail to increase urine osmolality and to increase the number of intramembranous particle (IMP) clusters in the luminal membrane of papillary collecting ducts (PCD) in response to [Arg]-vasopressin (AVP). We now examined whether abnormal AVP-stimulated cAMP metabolism in PCD is a cause of this disorder. Studies were conducted on PCD microdissected from control mice (C-mice) and NDI-mice. Intact PCD from NDI-mice incubated in vitro failed to increase cAMP levels in response to AVP, but stimulation of adenylate cyclase by AVP in permeabilized PCD was preserved. However, the specific activity of cAMP phosphodiesterase (PDE) was markedly higher in PCD of NDI-mice compared to C-mice. Properties of PDE in PCD from NDI-mice suggested abnormally high activity of isozyme type PDE-III. Indeed, incubation of PCD form

NDI-mice with selective inhibitors of PDE-III, rolipram (RP) and cilostamide (CS), completely restored the AVP-dependent cAMP accumulation in PCD. The cAMP accumulation was restored by RP added alone, but not by CS alone. However, addition of CS with RP together markedly potentiated the effect of RP. Incubation of PCD of NDI-mice with RP + CS also restored the increase of luminal IMP clusters in response to AVP. Administration of RP + CS to NDI mice in vivo restored the antidiuretic response to AVP (decrease in urine flow and increase in urine osmolality). The extracts of homogenate from papilla of C-mice and from NDI-mice were separated by FPLC chromatography on a Mono-Q anionic column using linear Na^+Ac gradient. The cAMP-PDE activity eluted at 0.65 - 0.8 M Na^+Ac, sensitive to inhibition by RP, was markedly higher in extracts from kidneys of NDI mice than in kidneys from C-mice. We suggest that functional unresponsiveness in collecting ducts, namely PCD, of NDI-mice to AVP is due to failure to increase cellular content of cAMP, which is caused by pathologically high activity of isozyme PDE-III. We hypothesize that rapid breakdown of cAMP by abnormal rolipram-sensitive PDE-III activity in cells of PCD is the key pathogenic factor of hereditary NDI syndrome in mice. Thus, the hereditary NDI syndrome in mice is a unique disease in which a genetically determined refractoriness of cells to polypeptide hormone is due to pathologically rapid catabolism of the second messenger, in this instance cAMP.

C4. CAMP-SPECIFIC PARTICULATE PHOSPHODIESTERASE (CAP-PDE) IN CARDIAC VENTRICLE OF RAT AND GUINEA PIG. M. KLOCKOW, K. HERGESELL, R. JONAS, Pharmaceutical Research, E. Merck, Frankfurter Straße 250, D-6100 Darmstadt, FRG.

In the supernatant of canine heart ventricle a cAMP specific PDE has been found that is specifically inhibited by Rolipram or Ro 20-1724 but not by cGMP or typical inhibitors of CGI-PDE (PDE III) (R. E. Weishaar et al., Circ. Res. 61:539 (1987)). We have isolated high activity of this cAMP specific PDE (CAP-PDE) tightly bound to myofibrils of rat and guinea pig heart ventricle (N.-S-Arch. Pharmacol. 339, R 53 (1989)). The Km values for CAP-PDE (rat: 5.7 ± 0.16 μM, 1 prep.; g.-p.: 5.7 ± 1.2 μM, 5 prep.) are 10 times higher than those for CGI-PDE, and in contrast to CGI-PDE, CAP-PDE has no hydrolytic activity with respect to cGMP. The inhibitory activity (IC50) of various PDE inhibitors determined at 0.3 μM cAMP clearly differentiates CAP-PDE from CGI-PDE:

	CAP-PDE(rat) μM	CAP-PDE (g.-p.) μM	CGI-PDE (g.-p.) μM
Rolipram	0.3 ± 0.1	0.43 ± 0.18	> 30
Ro 20-1724	1.8 ± 0.4	3.6 ± 0.4	50 ± 10
Etazolate	0.8 ± 0.2	1.1 ± 0.2	12 ± 2
Isomazol	90 ± 20	90 ± 17	17.3 ± 2
EMD 55 123	1.6 ± 0.3	3.2 ± 0.3	6 ± 2
Milrinone	18 ± 1	25 ± 1	0.59 ± 0.23
cGMP	>> 60	>> 30	0.46 ± 0.26
Enoximone	>> 60	>> 30	3.1 ± 0.6
CI 930	> 60	> 30	0.16 ± 0.05
EMD 49 931	26 ± 3	34 ± 7	0.05 ± 0.03

EMD 49 931: 5-(5-methyl-3-oxo-4,5-dihydro-2H-pyridazinyl-(6))-2-(pyr-azolyl-(3)-benzimidazol.
EMD 55 123: 5-(1-Ethoxycarbonyl-4,4-dimethyl-1,2,3,4-tetrahydro-chinolin-7-yl)-6-methyl-3,6-dihydro-1,3,4-thiadiazin-2-on

Inhibitors of CAP-PDE increase the cAMP-level in isolated heart cells of rats more effectively that CGI-PDE inhibitors.

Isoprenaline :	0 μM	0.3 μM	1.0 μM	3.0 μM
Buffer	0.31 ± 0.01	0.78 ± 0.14	0.95 ± 0.05	1.03 ± 0.08
8 μM CI 930	0.59 ± 0.01	1.39 ± 0.06	1.59 ± 0.16	1.59 ± 0.10
8 μM EMD49931	0.64 ± 0.00	1.56 ± 0.03	1.98 ± 0.14	1.89 ± 0.01
1 μM Ro201724	0.60 ± 0.05	2.02 ± 0.12	2.52 ± 0.02	2.63 ± 0.11
1 μM Rolipram	0.45 ± 0.06	3.62 ± 0.22	4.47 ± 0.12	4.90 ± 0.40

C6. CYCLIC AMP PHOSPHODIESTERASES IN HUMAN T-LYMPHOCYTES.

JAMES B. POLSON, STEVEN A. ROBICSEK, JOSEPH J. KRZANOWSKI and ANDOR SZENTIVANYI, University of South Florida, College of Medicine, Tampa, Florida 33612-4799.

Several investigations of human lymphocyte cyclic nucleotide phosphodiesterase (PDE) have been published (reviewed by Epstein & Hachisu, Adv. Cyclic Nucleotide and Protein Phosphorylation Res. 16: 303-324, 1984). Since B-lymphocytes reportedly contain 62 times more cAMP-PDE than T-lymphocytes (Cooper et al., Acta Derm. Venereol. Suppl. 114: 41-47, 1985), it may be necessary to study a more homogenous population of T-lymphocytes than has been used in previous investigations in order to demonstrate the existence of different forms of T-lymphocyte PDE. The objective of the present investigation was to determine whether T-lymphocytes contain different cAMP-PDE isoforms by studying purified cell preparations from human peripheral blood.

The buffy coats from 450 ml units of human peripheral blood were purified by ficoll-hypaque density gradient centrifugation followed by adherence of cells to plastic culture flasks to remove monocytes. B-lymphocytes were removed by adherence to nylon wool columns, and the eluted T-lymphocytes were further purified by centrifugation through percoll density gradients. The resulting cell preparations contained at least 98% lymphocytes of which approximately 88% were T-lymphocytes and less than 2% were B-lymphocytes. These cells were sonicated and centrifuged at 23,000 x g for 30 min to separate soluble and particulate material. Cyclic AMP-PDE was assayed by a modification of the method described by Thompson and Appleman (Biochem. 10: 311-316, 1971).

Four times as much total PDE activity was detected in the soluble fraction compared to particulate using 1 μM cAMP as substrate. Phytohemagglutinin (PHA)-induced blastogenesis was accompanied by increases in both soluble (threefold) and particulate (twofold) PDE after 4 days of incubation.

Three compounds were found to produce different degrees of inhibition of the particulate and soluble fractions. Cyclic GMP (1 μM) produced 48% inhibition of the particulate enzyme and 2% inhibition of the soluble PDE (P < 0.01). The respective inhibitions by 50 μM CI 930 (3 (2 H)-pyridazinone, 4,5-dihydro-6-[4-(1H-imidazol-1-yl)phenyl]-5-methyl-monohydrochloride) were 98% and 17% (P < 0.01). By contrast, 10 μM Ro 20-1724 (d,l-1,4-[3-butoxy-4-methoxybenzyl]-2-imidazolidinone) produced 16% and 52% inhibition, respectively (P < 0.05).

The soluble PDE could be separated into 3 peaks of activity by ion exchange high pressure liquid chromatography, and only 2 of these peaks were inhibited by Ro 20-1724 in concentrations up to 20 μM.

The results lead to the conclusion that human T-lymphocytes contain different forms of cAMP-PDE that can be detected in purified cell populations. The data suggest possible similarity of the major soluble and particulate isozymes to forms of PDE that have been reported previously in a variety of cell types (see review by Beavo, Adv. Second Messenger and Phosphoprotein Res. 22: 1-38, 1988).

C7. USES OF THE YEAST cAMP-PHOSPHODIESTERASE GENE IN MAMMALIAN CELLS: EFFECTS AND REVERSAL BY cAMP ANALOGUES.

MICHIEL M. VAN LOOKEREN CAMPAGNE, ERICA WU*, ROBERT D. FLEISCHMANN*, MICHAEL M. GOTTESMAN*, FERNANDO VILLALBA, BERND JASTORFF# AND RICHARD H. KESSIN, Anatomy and Cell Biology, Columbia University, 630 W 168th Street, New York, NY 10032. * Laboratory of Molecular Biology, National Cancer Institute, Bethesda, MD 20892. # Organische Chemie, Bremen University, Loebener Straße, D-2800 Bremen, F.R.G.

A genomic DNA fragment from Saccharomyces cerevisiae which contains the SRA5(PDE2) gene, coding for the low Km cAMP-phosphodiesterase, was transfected into Chinese hamster ovary cells. Clones carrying the yeast cAMP-phosphodiesterase gene were capable of growth in the presence of cholera toxin, which slows the growth of untransfected cell lines by elevating their cAMP levels. The cholera toxin resistant transfected cell lines expressed high levels of yeast cAMP-phosphodiesterase mRNA and yeast cAMP-phosphodiesterase activity. Basal intracellular cAMP levels were not significantly affected by the presence of the yeast cAMP-phosphodiesterase gene, but elevation of cAMP levels in response to cholera toxin was suppressed. We have investigated the pharmacology of 25 cAMP analogues and 11 phosphodiesterase inhibitors. This study revealed a very different pharmacology than for other reported phosphodiesterases, and has allowed us to select cAMP analogues that are both potent activators of the cAMP-dependent protein kinase in vivo, and that are not hydrolyzed by the yeast cAMP-phosphodiesterase. These analogues can be used to remove the block imposed on the activation of the cAMP-dependent protein kinase by the presence of high amounts of yeast cAMP-phosphodiesterase activity in tissue culture cells. Transfection with vectors carrying the SRA5 gene provides a method to block the rise in cAMP levels associated with hormonal stimulation and to determine whether the increase is essential for the biological consequences of hormone or growth factor stimulation.

C8. CHARACTERIZATION OF cGMP-STIMULATED PHOSPHODIESTERASES FROM BOVINE BRAIN AND CALF LIVER BY PHOTOAFFINITY LABELLING AND PEPTIDE MAPPING. TAKAYUKI TANAKA, STEVEN HOCKMAN, AND VINCENT MANGANIELLO, Laboratory of Cellular Metabolism, NHLBI, NIH, Bethesda, MD 20892.

The purified particulate cGMP-stimulated PDE exhibits a slightly greater subunit Mr than soluble forms from either calf liver or bovine brain, based on Coomassie staining of PAGE-SDS gels or Western immunoblots. Characteristics of cGMP-stimulated phosphodiesterases from bovine brain and calf liver were further investigated by photoaffinity labelling and peptide mapping (Cleveland et al., J.B.C. 252, 1102, 1977). The purified particulate PDE from bovine was directly photolabelled with 32 nM [^{32}P]cGMP or 700 nM [^{32}P]cAMP in the presence of 200 μM IBMX, suggesting [^{32}P]cGMP and [^{32}P]cAMP were interacting with high affinity, presumably regulatory, sites. Photolabelling with [^{32}P]cGMP was inhibited by much lower concentrations of cGMP than of 8-Br-cGMP or cAMP; autoradiograms indicated a much higher affinity (more than ⌐100 fold) for cGMP than 8-Br-cGMP or cAMP. Exposure of bovine brain particulate and calf liver soluble PDEs to V8 protease according to the procedure of Cleveland et al., generated 5-6 major peptides with similar Mr values from both PDEs. Two unique peptides of Mr ⌐27,000 from bovine brain particulate PDE and Mr ⌐22,000 from calf liver soluble PDE were produced by treatment with V8 protease. This suggests at least some differences in the amino acid sequence of the both PDEs. Photolabelling of the brain particulate and liver soluble PDEs, followed by digestion with V8 protease, indicated that [^{32}P]cGMP was associated with the same two low molecular weight peptides of similar Mr values (⌐12,000 kDa). The [^{32}P]cGMP was not found in association with those peptides that differed in digests from brain particulate and liver soluble PDEs. Taken together, these findings suggest the existence of cGMP-stimulated PDE isoenzymes with similar (perhaps conserved) and different domains in brain particulate and liver soluble cGMP-stimulated PDEs. These results further imply that cGMP binding sites are apt to be located in conserved regions.

C9. THREE POTENTIAL cAMP AFFINITY LABELS; INACTIVATION OF HUMAN PLATELET LOW K$_m$ cAMP PHOSPHODIESTERASE BY 8-[(4^m-BROMO-2,3,-DIOXOBUTYL)THIO]-ADENOSINE 3',5'-CYCLIC MONOPHOSPHATE. P. G. GRANT, D. L. DECAMP, J. M. BAILEY, R. W. COLMAN, R. F. COLMAN, Thrombosis Research Center, Temple University School of Medicine, Philadelphia, PA 19140 and Department of Chemistry and Biochemistry, University of Delaware, Newark, DE 19716

Three new analogs of cAMP have been synthesized and characterized: 2-[(4-bromo-2,3,-dioxobutyl)thio]-adenosine 3',5'-cyclic monophosphate (2-BDB-TcAMP), 2-[(3-bromo-2-oxopropyl)thio]-adenosine 3',5'-cyclic monophosphate (2-BOP-TcAMP), 8-[(4-bromo-2,3-

dioxobutyl)thio]-adenosine 3',5'-cyclic monophosphate (8-BDB-TcAMP). The bromoketo moiety has the ability to react with the nucleophilic side chains of several amino acids while the dioxobutyl group can interact with arginine. These cAMP analogs were tested for their ability to inactivate the low K_m (high affinity) cAMP phosphodiesterase from human platelets. At short incubation times the 2-BDB-TcAMP, the 2-BOP-TcAMP and the 8-BDB-TcAMP were competitive inhibitors of cAMP hydrolysis by the phosphodiesterase with K_i values of 0.96 $\pm$ 0.12 mM, 0.70 $\pm$ 0.12 mM and 10.5 $\pm$ 0.9 mM respectively. However, 2-BDB-TcAMP and 2-BOP-TcAMP did not irreversibly inactivate the phosphodiesterase at pH values from 6.0 to 7.5 and at concentrations up to 10 mM. These results indicate that although the 2-substituted TcAMP analogs bind to the enzyme, there are no reactive amino acids in the vicinity of the 2-position of the cAMP binding site. In contrast, incubation of the platelet low K_m cAMP phosphodiesterase with 8-BDB-TcAMP resulted in a time dependent, irreversible inactivation of the enzyme with a second order rate constant of 0.031 $\pm$ 0.009 min^{-1} mM^{-1}. Addition of the substrates, cAMP and cGMP, and the product, AMP, to the reaction mixture resulted in marked decreases in the inactivation rate suggesting that the inactivation was due to reaction at the active site of the phosphodiesterase. The present results indicate the importance of the purine ring location of the alkylating group for inactivation of the phosphodiesterase. These compounds should prove to be useful in studying the active sites of cAMP phosphodiesterases and cAMP binding sites on other proteins.

D2. DOES ADENOSINE BEAR AN INHIBITORY SIGNAL ON INTERMEDIARY METABOLISM IN ISOLATED RAT HEPATOCYTES.

ALAIN LAVOINNE, Groupe Biochimie et Physiopathologie Digestive et Nutritionnelle, UFR Médecine de Rouen, BP97,76803 Saint Etienne Rouvray, FRANCE.

Pharmacological doses of adenosine are known to induce various modifications of gluconeogenesis depending on the substrates used (Lund et al., 1975, Biochem. J., 152:593; Lavoinne et al., 1987, Biochem. J., 246:449). Adenosine induced also various modifications of ketogenesis (Lavoinne, unpublished data). Up to now, however, the mechanism of its effect on intermediary metabolism is largely unknown. Therefore, we tried to elicit the mechanism by which adenosine affects these two pathways by using inhibitors of its metabolism, namely deoxycoformycin for adenosine deaminase (DCF) and iodotubercidin for adenosine kinase (iodoT).

Hepatocytes were obtained from female Wistar rats (200-300g) starved 24 h before use.

0.5 mM-adenosine induced no change in the gluconeogenic flux from 10 mM-dihydroxyacetone, a decrease from 10 mM-glutamine (- 16%; n=5; P<0.02) and an increase from 5 mM-asparagine (+29%; n=4; P<0.02) In the presence of 10 μM-DCF, adenosine induced a decrease in gluconeogenesis (-21% for dihydroxyacetone; -51% for glutamine; -18% for asparagine). With DCF, adenosine inhibited also gluconeogenesis from 10 mM-lactate, -pyruvate, - serine or -alanine. Whatever the gluconeogenic substrate used, this inhibitory effect was suppressed in the presence of 10 μM- iodoT or in the presence of both 10 μM-iodoT and 10 μM-DCF. Therefore, adenosine conversion through adenosine kinase is responsible for the inhibition of the gluconeogenic flux whatever the gluconeogenic substrate used. In the presence of DCF, this inhibitory effect appeared between 0.1 and 0.2 mM-adenosine and was maximum at 0.5 mM-adenosine.

Moreover, in the absence of added substrate, 0.5 mM-adenosine inhibited ketogenesis (-30% ; n=5; P<0.05). This inhibitory effect persisted in the presence of DCF, greatly decreased in the presence of iodoT (-13%; n=5; P<0.05), and was suppressed in the presence of both iodoT and DCF. Thus, adenosine conversion through adenosine kinase is also involved in the inhibition of ketogenesis.

Adenosine conversion through adenosine kinase was associated with an increase in the ATP content (+115%) and a large intracellular Pi consumption (-30%). Supplementation of Pi in the medium (4 or 8 mM) did not modify the inhibitory effect of adenosine on gluconeogenesis. Moreover, in the presence of 8 mM extra phosphate added, adenosine conversion into adenine nucleotides was not significantly increased, but the inhibition of gluconeogenesis was reinforced. These results allowed to conclude that neither the decrease in Pi concentration nor the increase in the ATP pool induced by adenosine may explain these effects.

These results suggested that adenosine conversion through adenosine kinase bears an unknown signal inducing inhibition of both gluconeogenesis and ketogenesis.

D4. EFFECTS OF R-PHENYLISOPROPYLADENOSINE ON ISOLATED CARDIAC MUSCLE OF ADULT AND SENESCENT F344 RATS.

S. MONTAMAT, R. OLSON, R. VESTAL, Clinical Pharmacology and Gerontology Research Unit, VA Medical Center, Boise, ID, and Departments of Medicine and Pharmacology, University of Washington, Seattle, WA

Adenosine is known to mediate negative inotropic and chronotropic effects in cardiac tissue. Using R-phenylisopropyladenosine (R-PIA), a metabolically stable adenosine analog selective for inhibitory (A1) adenosine receptors, effects of adenosine on adult and senescent rat heart were investigated. In separate experiments, adenosine deaminase (ADA) was administered to evaluate the effects of removing endogenous adenosine by conversion to the inactive metabolite inosine. Adult (6-8 months) and senescent (23-24 months) male F344 rats were sacrificed and the hearts rapidly removed. Right atria (RA), left atria (LA), and right papillary muscles (RPAP) were removed and placed in a 15 ml muscle bath (30°C) containing Krebs-bicarbonate buffer (pH 7.4). The buffer was continuously bubbled with a mixture of 95% O_2 and 5% CO_2. Each muscle was affixed to a force transducer. RA were allowed to contract spontaneously, and LA and RPAP were stimulated via punctate electrodes to contract isometrically at a rate of 30/min. Using high speed oscillographic recordings, the following parameters were obtained: resting force, peak developed force (DF), maximal rate of both rise in force and relaxation, time to peak force, and 90% relaxation time. Muscles were stabilized for 120 minutes before studies were conducted. For R-PIA and isoproterenol (ISO) dose-response studies, parameters were measured following a 15-minute exposure at each concentration, which was increased in a cumulative manner. ISO dose-response studies were performed in the absence and presence of R-PIA (10^{-5} M). Time-vehicle control studies were conducted, and the results have been corrected using these control data. Spontaneous heart rate in the RA of both adult and senescent animals decreased with increasing dose of R-PIA. A slight shift to the right occurred in the dose-response curve of the senescent animals, but there was no difference statistically in the IC_{50}, the concentration required to inhibit heart rate by 50% (5.0 x 10^{-9} M, adults vs. 1.9 x 10^{-8} M, senescent; p>0.05). Developed force in the RPAP was unchanged in both adult and senescent animals at concentrations up to 10^{-5} M. Developed force in the LA of adult and senescent animals was decreased with increasing concentrations of R-PIA, reaching maximal effect at 10^{-7} M. The dose-response curve for the senescent animals was shifted to the right with lesser maximal inhibition than the adults, but there was no significant difference in the IC_{50} (8.3 x 10^{-9} M, adults vs. 1.2 x 10^{-8} M, senescent; p>0.05) or maximal inhibition (adult vs. senescent at 10^{-7}, 10^{-6}, and 10^{-5} M; p>0.05). The EC_{50}, or concentration required for 50% of the maximal effect during isoproterenol dose-response, was significantly changed with the addition of R-PIA in the adult preparations (4.59 x 10^{-9} M, control vs. 1.02 x 10^{-7} M, R-PIA ; p<0.005), but not in the senescent preparations.

In summary, RPAP contractility was unchanged by R-PIA administration, suggesting that ventricular tissue is not responsive to direct stimulation of adenosine receptors in adult or senescent rats. RA chronotropy was similarly affected in both adult and senescent rats, as was LA contractility; however, a trend for decreased sensitivity was seen in the senescent rats. Significant differences between the adult and senescent rats were noted in LA responses to ISO in the presence of R-PIA. Possible explanations for the lesser effect in older rats are: 1) greater endogenous production of adenosine, or 2) decreased A1 receptor sensitivity. Endogenous adenosine may obscure age differences in inotropic and chronotropic responses to R-PIA.

D5. FUNCTIONAL CHARACTERIZATION OF ADENOSINE RECEPTORS ON VENTRICULAR MYOCYTES.

J. NEUMANN, N. BEHNKE, U. MENDE, W. MÜLLER, W. SCHMITZ, H. SCHOLZ, B. STEIN, Abteilung Allgemeine Pharmakologie, Universitäts-Krankenhaus Eppendorf, D-2OOO Hamburg 2O, Martinistraße 52, FRG.

In isolated electrically stimulated (1 Hz) guinea-pig ventricular myocytes the adenosine derivatives (-)-N^6-phenylisopropyladenosine (PIA, 0.001-100 µmol/l) and 5'-N-ethylcarboxamideadenosine (NECA, 0.001-100 µmol/l) reduced contractile response (IC_{25}:

400

0.08 µmol/l, 0.10 µmol/l, respectively) in the presence of isoprenaline (0.01 µmol/l). All experiments were performed in the presence of adenosine deaminase (1 µg/ml) in order to exclude interference from endogenous adenosine. The effect of PIA was accompanied by a reduction in cAMP content (IC_{25}: 0.25 µmol/l) while NECA did not change cAMP content. The effects of PIA and NECA on contractile response could be antagonized by the A_1-adenosine receptor antagonist 1,3-dipropyl-8-cyclopentylxanthine (DPCPX, 0.3 µmol/l, pA_2: 8.4 and 7.3, respectively). The reduction of cAMP content by PIA could also be antagonized by DPCPX (pA_2: 8.3), but NECA increased cAMP content in the additional presence of DPCPX to 120 % at most. The A_2-adenosine receptor antagonist 9-chloro-2-(2-furanyl)-5,6-dihydro-1,2,4-triazolo(1,5-c)quinazolin-5-imine (CGS 15943A, 0.01 µmol/l) did not alter the effects of PIA and NECA on contractile response. It did not antagonize the reduction of cAMP content by PIA. However, in the additional presence of CGS 15943A NECA at 0.1-10 µmol/l reduced cAMP content by 30 % at most. The reduction of cAMP content by PIA (1 µmol/l) was abolished by using myocytes from guinea pigs pretreated with pertussis toxin (18 µg/100 g, PTX-myocytes). NECA increased cAMP content in PTX-myocytes to 130 %. In order to estimate the efficacy of in vivo PTX pretreatment, in vitro ADP-ribosylation of membranes of PTX-myocytes and control myocytes with [^{32}P]-NAD in the presence of PTX was performed. Autoradiography of SDS-PAGE showed labelling in the 40 kDa region. Labelling at about 40 kDa in PTX-myocytes was greatly reduced compared with controls. In order to identify the GTP-binding proteins (G-proteins) involved, immunoblots of membranes from guinea-pig hearts with peptide-antibodies were performed.

An antiserum raised against a sequence common to all alpha-subunits of G-proteins, except G_z, detected an immunoreactive protein at about 40 kDa. This antigen probably was a G_i-like G-protein. The 40 kDa PTX-sensitive alpha$_{i2}$ is suggested to be involved in the inhibition of adenylate cyclase. Another PTX-sensitive G-protein, namely G_o, which is highly abundant in brain tissue, was not detectable by an antiserum directed against a peptide which is only found in the sequence of the alpha-subunit of G_o.

It is concluded that both A_1- and A_2-adenosine receptors are present on guinea-pig ventricular myocytes. The physiological importance of cardiac A_2-adenosine receptors which are coupled to adenylate cyclase in a stimulatory fashion needs to be elucidated. The contractile response is mediated by A_1-adenosine receptors. A_1-adenosine receptors are coupled via PTX-sensitive G-proteins in an inhibitory fashion to adenylate cyclase. (Supported by the Deutsche Forschungsgemeinschaft.)

D6. ONTOGENY OF PURINERGIC RESPONSES IN ISOLATED TISSUES FROM NEONATAL RATS. J. NICHOLLS, S.M.O. HOURANI & I. KITCHEN. Department of Biochemistry, Molecular Toxicology and Pharmacology Group, University of Surrey, Guildford, Surrey, GU2 5XH

Purines have pharmacological actions on many tissues, which are mediated by P_1- and P_2-purinoceptors recognising adenosine and ATP respectively, and P_2-purinoceptors have been subdivided into P_{2X} (contractile) and P_{2Y} (relaxant) (1). The ontogeny of peripheral purinoceptors in Wistar rats was investigated at 2, 5, 10, 15 days and in adults using the duodenum, which relaxes to adenosine and to ATP, and the urinary bladder in which ATP contracts. Concentration-response curves (10^{-7} - 10^{-4}M) were obtained non-cumulatively, and carbachol (10^{-7}M) was used to precontract the duodenum so that relaxations to the purines could be observed. Potency estimates were obtained from linear regression analysis of the concentration response curves.

In the duodenum, relaxations to adenosine did not change significantly with age, but the non-degradable ATP analogue, AMP-PCP, was significantly less potent at day 10 but equipotent to the adult at 5 and 15 days. ATP was always more potent than AMP-PCP at causing relaxations but was less potent in neonates than in the adult, and before 15 days the relaxations were followed by contractions, particularly at high concentrations. These contractions were not blocked by indomethacin (2.5 x 10^{-5}M).

401

In the bladder adenosine was less potent in the adult than in the neonate at inhibiting contraction. The potency of AMP-PCP in causing contractions varied with age, being similar to the adult at 2 and 5 days but much greater at 10 and 15 days. The potency of ATP was always lower than that of AMP-PCP, but varied with age in a similar manner.

These results show that purinoceptors are present in neonatal rats from as early as 2 days and that P_1 effects do not vary greatly with age, whereas P_2 effects vary in a complex manner. For AMP-PCP, contractions via P_{2X}-purinoceptors in the bladder are greatest at 10-15 days, whereas the relaxations of the duodenum via P_{2Y}-purinoceptors are lowest at 10 days. The contractions observed with ATP in the duodenum before 15 days may not be mediated by P_{2X}-purinoceptors, as they were never observed with AMP-PCP, which is more potent than ATP at this receptor subtype (1), but are unlikely to be the result of prostaglandin synthesis as has been reported for the guinea-pig taenia caeci (2), since indomethacin was without effect.

1. Burnstock, G. & Kennedy, C. (1985) Gen. Pharmac. 16, 433-440.
2. Brown, C.M. & Burnstock, G. (1981) Eur. J. Pharmacol., 69, 81-86.

D8. A NOVEL POPULATION OF PURINE RECEPTORS IN THE RAT COLON.

S.J. BAILEY, S.M.O. HOURANI. Department of Biochemistry, University of Surrey, Guildford, Surrey, GU2 5XH

The effects of the adenine derivatives adenosine (Ado), adenosine 5'-monophosphate (AMP), adenosine 5'-diphosphate (ADP), adenosine 5'-triphosphate (ATP), adenosine 5'(2-fluoro)-diphosphate (ADPbF), b,g-methylene-ATP (AMPPCP) and its enantiomer L-AMPPCP were examined on the rat colon muscularis mucosae preparation in the presence of atropine (1mM) (Bailey & Jordan, 1984). L-AMPPCP was inactive at concentrations up to 100mM, but each of the other purines contracted the colon in the concentration range 1 - 100mM. ATP and ADP were the most potent compounds tested, in that the maximum responses achieved at 100mM, expressed as a percentage of the response to 1mM substance P (SP), were 53.4 $\pm$ 2.7% and 41.2 $\pm$ 3.8%, respectively. Maximum responses to the other four purines ranged from 22.5 - 29.2%. Responses to ATP and ADP also differed from the other four purines in being slower in onset, requiring 70 - 90s to achieve peak response, as compared with 45 - 60s for the other purines.

The effects of each agonist were also investigated in the presence of the adenosine receptor antagonist 8-(p-sulphophenyl)theophylline (8SPT, 50mM). Dose response curves to ATP, ADP and SP were not significantly shifted, whilst those to Ado, AMP and AMPPCP were shifted approximately 50-fold to the right. The curve for ADPbF was also shifted to the right with a dose ratio of 5-10.

These results suggest that the rat colon muscularis mucosae preparation possesses a mixed population of purine receptors of the P_1 and P_2 types. The low potency of AMPPCP, and the lack of activity of L-AMPPCP suggest that the P_2 receptor is of the P_{2Y} subtype (Burnstock & Kennedy, 1985; Hourani et al., 1986). This is the first tissue reported to contain a P_{2Y} receptor mediating contractile effects. Furthermore, the susceptibility of responses to AMPPCP to blockade by 8SPT suggests that this agonist may in fact be less active at P_{2Y} receptors than previously thought, but possess some activity at a P_1 receptor. Similarly, ADPbF, previously suggested to be a selective agonist at P_{2Y} receptors (Hourani et al., 1988) may also have activity at P_1 receptors. Finally, the P_1 receptors involved in this tissue may differ from previously described A_1 and A_2 subtypes as the inhibitory effect of 8SPT was approximately ten times greater than that previously reported (Collis et al., 1987).

Bailey SJ & Jordan CC (1984) Br. J. Pharm 82: 441-4451.
Burnstock G & Kennedy C (1985) Gen. Pharm. 15: 433-440.
Collis MG, Jacobson KA & Tomkins DM (1987) Br. J. Pharm. 92; 69-75.
Hourani SMO, Loizou GD & Cusack NJ (1986) Eur. J. Pharm. 131; 99-103.
Hourani SMO, Welford LA, Loizou GD & Cusack NJ (1988) Eur. J. Pharm. 147: 131-136.

D9. THE ROLE OF ADP-RECEPTOR, AGGREGIN, IN THROMBIN- AND PLASMIN-INDUCED PLATELET AGGREGATION. R.N. PURI, H. BRADFORD, C.-J. HU, R. MATSUEDA, H. UMEYAMA, R.F. COLMAN, R.W. COLMAN.

Thrombosis Research Center, Temple University, 3400 North Broad Street, Philadelphia, PA 19140 USA; Sankyo Co., Ltd., Tokyo, Japan; Department of Biochemistry, University of Delaware, Newark, DE 19711

Covalent modification of the ADP-binding protein on the platelet surface by the nucleotide analog, 5'-fluorosulfonylbenzoyl adenosine (FSAB), leads to inhibition of ADP- induced platelet shape change and aggregation. ADP-binding protein has been shown to be different from glycoprotein-IIIa. These considerations led to the conclusion that aggregin ($M_r = 100$ kDa) is the putative ADP receptor on the platelet surface. We previously demonstrated that thrombin- and plasmin-induced platelet aggregation are accompanied by the cleavage of aggregin and these events are indirectly mediated by the intracellularly activated calpain expressed on the surface. Binding of thrombin and plasmin to their respective receptors on the platelet surface is a mandatory requirement for these proteases to induce the above platelet responses. High molecular weight kininogen (HK) is a potent inhibitor of calpain and binds reversibly to platelets, at the plasma concentration. HK inhibits thrombin- and plasmin-induced aggregation and cleavage of aggregin, supporting the cellular mechanism described above. 5,5'-Dithiodinitrobenzoic acid (DTNB), leupeptin, antipain, and EGTA also inhibit the thrombin- and plasmin-induced platelet responses. The concentrations of DTNB, leupeptin, antipain and EGTA used in such experiments had no effect on the amidolytic activity of thrombin and plasmin, nor did they inhibit the shape change induced by these proteases. In order to understand more about the nature of molecular mechanism(s) of inhibition by HK of the platelet responses induced by the two proteases, a number of peptide disulfides that combine the structural features of DTNB, leupeptin, or the inhibitory sequence of domain 2 of HK have been synthesized and evaluated for their ability to inhibit platelet responses induced by thrombin and plasmin, and other agonists. Only one peptide, Phe-Gln-Val-Val-Cys(NpyS)-Gly-NH_2[NpyS=2-thiopyridine] (P1), analogous to the peptide corresponding to inhibitory sequence of the domain 2 of HK, Gln-Val-Val-Ala-Gly(Pd2), inhibited thrombin- and plasmin-induced platelet aggregation and cleavage of aggregin. P1 also inhibited platelet calpain. However, P1 had no effect on the shape change induced by thrombin and plasmin. P1 did not inhibit platelet aggregation induced by ADP, collagen, A23187, PMA, A23187 + PMA, and U46619. P1 failed to inhibit the amidolytic activity of thrombin and plasmin or the fibrinogenolytic activity of thrombin. P1 did not raise the cAMP levels of intact washed platelets, nor did it affect the ability of iloprost to raise the platelet cAMP levels. Computer assisted three-dimensional analysis of P1 showed it to be isosteric with Pd2. Replacement of alanine in Pd2 by serine, cysteine, or the function Cys(NpyS) appear to have no effect on the three-dimensional disposition of the resulting peptides. This is consistent with the fact that the inhibitory sequence of egg white cystatin (Glyn-Leu-Val-Ser-Gly) contains serine instead of alanine. Such 3D analysis also showed that structural elements present in P1 may enable it to interact with the binding and catalytic domains of calpain simultaneously. In conclusion, these investigations show that cleavage of the platelet ADP receptor, aggregin, plays a major role in the mechanism of thrombin- and plasmin-induced platelet aggregation. The cellular and molecular mechanisms by which this cleavage occurs and can be inhibited have been delineated to a considerable extent.

D10. **RECEPTOR MEDIATED EFFECTS OF ADENOSINE ON THE AIRWAYS OF THE ALLERGIC RABBIT.** ALI, S., MUSTAFA, S.J., ATKINSON, L., METZGER, W. J., East Carolina University, School of Medicine, Greenville, NC 27858- 4354

Aerosolized adenosine causes a rapid and profound bronchoconstriction in asthmatic patients but not in normals. Further, theophylline, an adenosine receptor antagonist inhibits the bronchoconstriction provoked by adenosine. In the present study, we have employed an allergic rabbit model which is similar to the human model of allergic asthma with respect to late phase asthmatic responses. Neonates were immunized within 24 hrs of birth with ragweed in alum to preferentially produce ragweed specific IgE. These rabbits also have hyperreactive airways as determined by histamine challenges ($PC_{50} = 1.95 + 0.30$ mg/ml). Aerosolized adenosine caused bronchoconstriction in these allergic rabbits ($PC_{50} = 1.63 + 0.84$ mg/ml; however it did not cause bronchoconstriction in normal rabbits. CGS-15943, a non- xanthine adenosine receptor antagonist, inhibited the adenosine- induced bronchoconstriction.

Under in vitro conditions, adenosine produced dose-dependent contraction of isolated airway smooth muscle from the allergic rabbit but had no effect on the smooth muscle of normal rabbits. The peripheral airways were more responsive to the contracting effect of adenosine as compared to the central airways. Treatment with CGS-15943A (10^{-7}M) significantly antagonized the contraction produced by adenosine.

The existence of specific adenosine binding sites in the rabbit lung was evaluated using the radioligand 8-cyclopentyl 1-3- [^{3}H] dipropylxanthine ([^{3}H]DPCPX), an adenosine receptor antagonist. Saturation analysis revealed specific binding sites only in the lung of allergic rabbits. No specific binding could be detected in the lung of normal rabbits. The specific binding sites in allergic animals were saturable at ?1 nM. Scatchard analysis indicated a single class of binding sites with a K = the specific binding sites with specific agonists and antagonists suggested a primarily A_1 receptor binding site.

These results suggest that under both in vivo and in vitro conditions adenosine produced contraction of the bronchi of allergic rabbits but not of normal rabbits. The adenosine- induced effects in this model are likely mediated through specific adenosine receptors.

D11. ADENOSINE PRODUCES PULMONARY VASOCONSTRICTION THROUGH EICOSANOID GENERATION. ITALO BIAGGIONI, LANDON S. KING, JOHN H. NEWMAN, Departments of Medicine and Pharmacology, Vanderbilt University, Nashville, TN.

Adenosine is thought to produce vasodilatation in all vascular beds with the exception of the kidney. Due to its theoretical potential as a pulmonary vasodilator, we studied the hemodynamic effects of adenosine in the pulmonary vasculature of chronically instrumented awake sheep. Adenosine produced significant pulmonary vasoconstriction instead of the expected vasodilatation. This effect is produced via activation of specific cell surface adenosine receptors, since it was blocked by the adenosine-receptor antagonists theophylline and 1,3-dipropyl-8-(p-sulfophenyl)xanthine. The cell type involved in adenosine-induced pulmonary vasoconstriction appears to be located within the lung, since vasoconstriction was blunted when adenosine was infused into the left atrium, distal to the lung. However, adenosine does not directly vasoconstrict the pulmonary vasculature, because its effect could be completely abolished by cyclooxygenase inhibition with indomethacin or ibuprofen, by thromboxane receptor antagonism with SQ 29,548, and by inhibition of thromboxane synthesis. Thus, adenosine produced pulmonary vasoconstriction through generation of a thromboxane/endoperoxide product. Generation of thromboxane A2 within the lungs was demonstrated following adenosine administration. Whether endogenous adenosine is involved in the generation of pulmonary vasoconstriction seen in pathophysiological states remains to be determined.

D12. ACTIVITY OF XANTHINES AND ADENOSINE ANALOGS AS RELAXANTS OF GUINEA PIG TRACHEA. L.E. BRACKETT, M.T. SHAMIM, J.W. DALY, NIDDK, NIH, Bethesda, MD.

A variety of xanthines and adenosine analogs cause relaxation of carbachol-stimulated guinea pig trachea in vitro. Structural analogs of caffeine, theophylline, and enprofylline were examined for their potency as tracheal relaxants. All caffeine analogs tested are more potent than caffeine (EC50 = 551±81 uM) except the 8-p-sulfophenyl analog. 1,3,7-Tripropylxanthine and 1,3,7-tripropargylxanthine are among the more potent analogs with EC50 values of 12±1.3 and 65±11 uM, respectively. The 8-cyclohexyl-, 8-cyclopentyl- and 8-phenyl-derivatives of caffeine are relatively potent (EC50 = 75 uM). The theophylline analog, 1,3-di-n-propylxanthine is approximately two times more active than theophylline (EC50 = 162±17 uM). 3-Isobutyl-1-methylxanthine (EC50 = 7±1.8 uM) and 1-isoamyl-3-isobutylxanthine (EC50 = 37±4.2 uM) are among the most potent tracheal relaxants. The 8-substituted theophylline analogs are weak or inactive relaxants except 8-cyclopentyl- and 8-cyclohexyltheophylline, which are more potent or as potent as theophylline. In contrast to enprofylline (EC50 = 56+9 uM), 8-substituted enprofylline analogs are weak or inactive as relaxants with the exception of the 8-cyclohexyl analog. The potency of adenosine analogs as tracheal relaxants showed a rank order of: 5'-N-ethylcarboxamidoadenosine > 2-chloroadenosine > 2-phenylaminoadenosine = N6-cyclohexyladenosine. The rank order of potency suggests that adenosine analogs act at A2 adenosine receptors to cause tracheal relaxation. The xanthines do not appear to relax trachea through interaction with extracellular receptors, but instead the potency of xanthines as tracheal relaxants may reflect activity as phosphodiesterase inhibitors.

D13. INFLUENCE OF PURINOCEPTOR AGONISTS ON LUNG SURFACTANT SECRETION. SEAMUS A. ROONEY, ALASDAIR M. GILFILLAN, LAURICE I. GOBRAN AND MATTHIAS GRIESE, Department of Pediatrics, Yale University School of Medicine, New Haven, CT 06510, U.S.A.

Pulmonary surfactant is a phospholipid-rich material, composed largely of phosphatidylcholine (PC), which lines the alveolar surface of the lung. A deficiency in surfactant in premature infants leads to the respiratory distress syndrome which is a major cause of illness in this age group. Surfactant is synthesized in the type II alveolar epithelial cell, stored in lamellar inclusion bodies and secreted by exocytosis. Regulation of PC secretion has been extensively studied in isolated type II cells in primary culture and shown to be regulated by a variety of different signal transduction mechanisms. These include beta-adrenergic agonists, agents which activate protein kinase C, Ca++ ionophores and arachidonate metabolites. In addition, there is now functional evidence for at least three purinoceptors in the type II cell: A2 and P2 receptors which stimulate PC secretion and an A1 receptor which inhibits.

Type II cells were isolated from the lungs of adult rats, cultured overnight in medium containing [methyl-3H]choline chloride, then washed and incubated for a further 2 h in fresh medium without radioactivity but containing the purinoceptor agonists of interest. At the end of this period lipids were extracted from the cells and medium with a mixture of chloroform and methanol. Secretion is expressed as the percentage of total 3H-PC (cells + medium) in the medium after the 2 h incubation. The basal rate of PC secretion was approx. 1% and this was increased 2-fold by adenosine and up to 5-fold by ATP. Adenosine analogs stimulated to the same extent as adenosine, the rank order of potency being NECA > CADO = L-PIA $\geq$ adenosine > D-PIA. The EC50 for NECA was 8.9 x 10-8 M. NECA- stimulated secretion was accompanied by a 34-fold increase in cAMP level; both increases were antagonized by 5 x 10-5 M theophylline. These data are consistent with A2 receptor mediation of the stimulatory effect of adenosine on PC secretion. Radioligand binding studies have shown the existence of NECA specific binding in a type II cell membrane fraction. The stimulatory effect of adenosine was abolished by adenosine deaminase. In contrast, the effect of ATP was not antagonized by adenosine deaminase or by theophylline and it was not accompanied by increased cAMP concentration. It was, however, antagonized by alpha, beta- methylene ATP. Several ATP analogs also stimulated PC secretion but ATP itself was by far the most potent agonist with an EC50 of 4.1 x 10-6 M. The potency order was ATP > beta, gamma-methylene ATP = 2-methylthio ATP = 2-deoxy ATP $\geq$ 8- bromo ATP > alpha, beta-methylene ATP. The effect of ATP was not antagonized by Cibacron Blue. These data are consistent with P2 mediation of the stimulatory effect of ATP on PC secretion. The potency order, however, is not consistent with that reported for either the P2x or P2y receptor subtypes. Inclusion of adenosine deaminase (3 units/ml) in the culture medium markedly increased the stimulatory effect of NECA (as well as that of terbutaline and forskolin) on PC secretion but not that of ATP (or other agents not increasing cAMP levels). This stimulatory effect of adenosine deaminase was diminished by the A1 selective agonists N6-cyclopentyladenosine and 1-deaza-2-chloro-N6-cyclopentyladenosine. These data suggest that adenosine deaminase removes adenosine acting at an A1 receptor to inhibit PC secretion. In summary, these data provide functional evidence for the existence of three purinoceptors in type II cells. Their functional role, if any, in the physiological regulation of surfactant secretion remains to be established. (Supported by NIH grant HL-31175).

D15. NATURAL KILLER CELLS (NK) MEDIATE THE MODULATION OF ANTIBODY PRODUCTION IN MICE BY TUBERCIDIN (Tub) AND 2-FLUOROADENINE ARABINOSIDE MONOPHOSPHATE (FaraAMP). TERESA PRIEBE, LUIS RUIZ AND J. ARLY NELSON, Department of Experimental Pediatrics, Univ. Tex. M. D. Anderson Cancer Center, Houston, TX 77030.

Previous results from this laboratory demonstrated that treatment of mice with the adenosine analog Tub reduced NK cytotoxicity while stimulating antibody production (Cancer Res. 48:4799, 1988). A soluble precursor for a deoxyadenosine analog, FaraAMP, produced opposite effects, i.e., it stimulated NK cytotoxicity at doses that inhibited antibody formation.

Since NK cells have been reported to play a suppressor role in immunoglobulin induction, it was hypothesized that the actions of Tub and FaraAMP on antibody production occurred secondary to their opposing effects on NK cells. To test this hypothesis, effects of these nucleoside analogs on primary antibody response to sheep red blood cells were evaluated in a C57BL/6J mutant mouse lacking NK cell activity (the beige mutation, C57BL/6J-bg/bg). As previously found with C3H/He mice, NK cell activity was inhibited (Tub, doses 2-6mg/kg/day for 3 days) or stimulated (FaraAMP, doses 75-250mg/kg/day for 3 days) in heterozygous mice (C57BL/6J-bg/+). In support of the hypothesis, these nucleosides were without effect on primary antibody formation in the homozygous mutant mice at doses that clearly stimulated (Tub) or inhibited (FaraAMP) this immune response in the heterozygous animals. This result was corroborated in C57BL wild-type mice by abrogation of NK cell activity using monoclonal antibody to the surface glycosphingolipid, ganglio-n-tetraosylceramide (anti-asialoGM1). Similar experiments using the potent ADA inhibitor, deoxycoformycin, suggested that enhancement of primary antibody formation by the inhibitor was independent of the inhibitor's effect on NK cell activity. We conclude that under the conditions of drug treatment, Tub and FaraAMP modulate primary antibody formation in mice indirectly via NK cells. (Supported by NIH Grant HD-13951).

D16. **EFFECTS OF INTRATHECAL INJECTION OF R-PIA AND NECA ON NOCICEPTION AND MOTOR FUNCTION IN THE RAT.** R. KARLSTEN, T. GORDH JR, P. HARTVIG, C. POST, Department of Anesthesiology, University hospital, Uppsala, Sweden.

This study was designed to evaluate potential differences in antinociceptive effects and motor impairment in the rat following intrathecal administration of an A1-receptor agonist (R-PIA) and an A2-receptor agonist (NECA). We also intended to evaluate dose-response differences in regard to these parameters and if possible, find dose-intervals with antinociceptive effect without motor impairment.

Using implanted intrathecal catheters with the tip in the lumbar region, R-PIA was injected in doses 1, 3.3, 10 and 100 nmol and NECA in doses 1, 3.3, and 10 nmol.

Antinociception was measured using the tail immersion test and in the case of R-PIA also the hot plate test. Motor impairment was evaluated during spontaneous movement in a free space.

R-PIA induced a significant antinociceptive effect in the tail immersion and the hot plate test in doses from 3.3 nmol within 15 minutes following injection. Motor impairment was only seen at the highest dos (100 nmol) beginning with a partial paresis of the hind legs after 60 minutes and progressing to full paresis of the hind legs after 90 minutes.

NECA induced a significant antinociceptive response in tail immersion test at all doses but not of the same magnitude as was noted for R-PIA. Following the injection of NECA a clear dose-dependent impairment of motor function could be seen at all doses within 15 minutes and with a duration well exceeding that of the antinociceptive effect. Because of the severe motor impairment, the hot plate test could not be used in the studies of NECA.

These results suggests that the motor impairment is an A2-receptor mediated effect and that the receptor selectivity of R-PIA could be diminished at high doses. Therefore an A1-receptor agonist may be more suitable to produce an antinociceptive effect without motor disturbances. Since the results of the tail immersion test are unaffected by motor impairment, the drugs might inhibit activity in descending motor pathways without interfering with the spinal withdrawal reflex.

D17. **INTERACTIONS BETWEEN ADENOSINE AND THE AUTONOMIC NERVOUS SYSTEM IN CARDIOVASCULAR CONTROL.** ITALO BIAGGIONI, ROGELIO MOSQUEDA-GARCIA, ROSE MARIE ROBERTSON, DAVID ROBERTSON, Departments of Medicine and Pharmacology, Vanderbilt University, Nashville TN.

Adenosine is classically considered an inhibitory neuromodulator. It inhibits the release on many neurotransmitters and reduces neural activity. Adenosine also has neuromodulatory actions on autonomic mechanisms involved in cardiovascular control. These mechanisms are activated by stimulation of either mechano- or chemo-receptors (**afferent limb**). These signals are then integrated in the **brainstem**, where sympathetic discharge (**efferent limb**) is modulated. 1) In the **efferent limb**, adenosine has been extensively shown to inhibit norepinephrine release presynaptically and to oppose catecholamine actions postsynaptically. 2) We have shown that adenosine acts in discrete centers of the **brainstem** (Nucleus Tractus Solitarii and Area Postrema) to decrease blood pressure, heart rate and sympathetic activity. However, these actions are similar to those produced by excitatory amino acids or electrical stimulation of these nuclei and therefore are not those that would be expected from an inhibitory neuromodulator. 3) In man, we have shown that intravenous adenosine raises heart rate and blood pressure, stimulates ventilation, and increases muscle sympathetic nerve activity, probably through activation of **afferent** chemoreceptors. Other investigators have reported stimulation by adenosine of myocardial and renal afferents, suggesting that this is a generalized rather than an isolated phenomena. The cellular mechanisms of these apparent excitatory actions of adenosine are under investigation.

D18. **CO-TRANSMISSION IN THE RABBIT ILEOCOLIC ARTERY.** J.M. BULLOCH, Pharmakologisches Institut, University of Freiburg, Freiburg, West Germany.

Investigations have now revealed numerous neuroeffector systems where ATP and NA act as co-transmitters, including blood vessels like the rabbit saphenous artery (Burnstock & Warland, 1987) and the rabbit ileocolic artery (von Kugelgen & Starke, 1985; Bulloch & Starke, 1988). It was the aim of the study to investigate, in the rabbit ileocolic artery, whether differential release of the co-transmitters would occur by varying the conditions of stimulation.

Rabbits of either sex (2 - 3 kg) were decapitated, the ileocolic arteries were dissected out and were cannulated at both ends. They were perfused at a constant rate of flow (2.7 ml/min) in such a way that the perfusate did not mix with the bathing fluid, as detailed by von Kugelgen & Starke (1985). Vasoconstriction
produced by drugs or electrical field stimulation was observed as an increase in perfusion pressure.

Vasoconstrictions produced by nicotine (30 -300umol/l) or DMPP (1,1 - dimethyl- 4 phenyl-piperazinium iodide) (30-300umol/l) were transient and were largely blocked by prazosin. Electrical field stimulation produced frequency- and pulse number-dependent vasoconstrictor responses. At longer pulse trains two phases can be differentiated. The first one (and the only one detected at short pulse trains) greatly reduced by mATP. The second component was only sensitive to prazosin at higher frequencies (20Hz). At 5Hz 100 pulses neither component was significantly affected by prazosin and the blockade of alpha2-autoinhibition by yohimbine potentiated both phases of the response. However, the addition of
prazosin to arteries where yohimbine was already present attenuated both phases of the response to 5Hz 100 pulses. Therefore yohimbine not only changed the magnitude of the response to electrical field stimulation but also the extent of NA/ATP co-transmission, suggesting that yohimbine may have a preferential
effect on the release of NA.

In conclusion, differential release of transmitter may occur in the rabbit ileocolic artery and appears to be dependent on the method and conditions of stimulation used to produce the response.

Burnstock,G. & Warland,J.J.I. (1987). Br.J.Pharmacol.,90, 111-120.
Bulloch, J. & Starke, K. (1988). In: Biosignalling in Cardiac and Vascular Systems.pp122-128. Pergamon Press.
von Kugelgen, I. & Starke, K. (1985). J. Physiol., 367, 435-455.

D20. **ADENOSINE MEDIATED DILATION IS PARTIALLY DEPENDENT ON ENDOTHELIUM AND pO2 IN GUINEA PIG AORTA.** J. P. HEADRICK, R. M. BERNE. University of Virginia, Dept. of Physiology. Charlottesville, 22908, VA, USA

Responses to adenosine (Ado), 2-chloro-adenosine (Cl-Ado), n6-cyclohexyladenosine (CHA), n-ethylcarboxamidoadenosine (NECA), acetylcholine (Ach), and nitroprusside (Np) were examined in intact and luminally rubbed aortic rings isolated from adult guinea pigs. All drugs relaxed PGF2a (3μM) constricted rings in a dose-dependent manner. Luminal rubbing abolished responses to Ach, had no significant effect on responses to Np and CHA, and significantly reduced responses to Ado, Cl-Ado, and NECA by 24%, 28%, and 36% respectively (P<0.001). The order of potency of the adenosine analogues in rubbed and un-rubbed rings was NECA>Cl-Ado>Ado>CHA. The magnitude of the endothelial-dependent component of the relaxations displayed an identical order of potency. Oxyhemoglobin (5 µM) and metyrapone (0.1 mM) had no affect on the shift in the Ado response produced by luminal rubbing, but significantly reduced responses to Ach in intact rings (P<0.001). Indomethacin (10 µM) had no effect on the Ado response. Equilibration of isolated rings with hypoxic perfusate (95% N2-5% CO2) significantly potentiated relaxations to Ado and Np (max. responses increased by 24% and 12% respectively, P<0.05), but attenuated the responses to Ach (max. response reduced by 36%, P<0.01). The hypoxic potentiation of the Ado and Np responses was comparable in luminally rubbed and un-rubbed rings, and was unaffected by oxyhemoglobin and metyrapone treatment. The results indicate that Ado mediated dilation in guinea pig aorta is in part produced by an endothelial mechanism unrelated to EDRF or prostanoid release. Both the endothelial dependent and independent dilations are consistent with the stimulation of A2 Ado receptors. Although the predominant action of adenosine is mediated by direct smooth muscle receptor activation, the relative contribution of the endothelium to relaxation is more pronounced at low Ado doses (eg. 45% of the response to 3 µM Ado). Finally, the response of vascular smooth muscle to Ado may also be dependent upon the degree of tissue oxygenation, an important consideration in studies examining adenosine mediated relaxations both *in vitro* and *in vivo*.

D21. MODULATION OF CARDIAC EFFICIENCY: A ROLE FOR ADENOSINE IN METABOLICALLY STIMULATED RAT HEART. J. P. HEADRICK, R. J. WILLIS.
Griffith University, Division of Science and Technology. Nathan, 4111, QLD, Australia.

Adenosine inhibits the chronotropic and inotropic actions of exogenously applied catecholamines in heart, and may mediate changes in coronary resistance in response to increased cardiac activity. A novel role for endogenous adenosine is maintenance of cardiac efficiency during increased catecholamine stimulation. A 10 min infusion of isoproterenol (50 nM) into isovolumic rat hearts perfused at a constant pressure of 100 mmHg resulted in increases in MVO2 (from 99±8 to 254±24 µl O2/min/g), rate-pressure product (20±3 to 82±5 x103 mmHg/min), coronary flow (12.8±1 to 22.4±2 ml/min/g), and adenosine release (256±21 to 2125±201 pmol/min/g). Cardiac efficiency, calculated as: rate-pressure product+MVO2 (mmHg/µl O2/g), also increases upon stimulation with isoproterenol (198±16 to 311±15 mmHg/µl O2/g). Treatment of stimulated hearts with adenosine deaminase (2-4 IU/ml) further increased the rate-pressure product (94±5 x103 mmHg/min) and MVO2 (345±28 µl O2/min/g), and reduced coronary flow (19.1±1 ml/min/g) and cardiac efficiency (272±11 mmHg/µl O2/g). Similar results were obtained in stimulated hearts treated with the adenosine antagonist 8-phenyltheophylline (5 µM), infused alone or with adenosine deaminase. The effect of adenosine inhibition was evident over a range of isoproterenol doses (0.4, 3.0, 50, and 75 nM), and a similar pattern of results was obtained in norepinephrine stimulated hearts (100 µM), with and without adenosine deaminase treatment. The data indicate that at least 30-40% of the observed metabolic vasodilation is mediated by endogenous adenosine, and that endogenous adenosine inhibits the response of the myocardium to catecholamines. The action of endogenous adenosine appears to be directed towards maintaining optimal efficiency during metabolic stimulation. The mechanism for this action is unclear, however, it is possible thatadenosine preferentially reduces the positive chronotropic actions of the catecholamines, allowing the more efficient positive inotropic action to improve cardiac output at a lower O2 cost. The significance of this modulatory role of adenosine *in vivo* has yet to be determined.

D23. THE ROLE OF ADENOSINE IN AUTOREGULATION OF THE CEREBRAL BLOOD FLOW: EFFECTS OF CAFFEINE ON VASODILATORY RESPONSE OF THE RAT PIAL ARTERIOLES TO HYPOTENSION. SEIJI MORII, RENZHI WANG, HIROTOSHI OHTAKA, KENZOH YADA, MAKOTO KATORI*.

Departments of Neurosurgery and Pharmacology*, Kitasato University, School of Medicine, 1-15-1, Kitasato, Sagamihara, Kanagawa, Japan.

Adenosine is a potent dilator of the cerebral resistant vessels and its concentration in the brain tissue increases during hemorrhagic hypotension. Therefore, adenosine has been proposed as a metabolic regulator of the cerebrovascular resistance and considered to play a role in autoregulation of the cerebral blood flow. However, a causal relationship between adenosine and hypotensive dilatation of the resistant vessels has not yet been established. The effect of caffeine, an adenosine receptor blocker, on the acute dilatory response of the pial arterioles to hypotension was investigated in order to define a possible role of adenosine in autoregulation of the cerebral blood flow.

Methods: Male Sprague-Dawley rats weighing 280 g to 330 g were anesthetized with subcutaneous injection of a-chloralose (50 mg/kg) and urethane (500 mg/kg), tracheostomized and mechanically ventilated. The abdominal aorta was cannulated via the femoral artery to monitor mean arterial pressure. Using a closed cranial window technique and a video microscopy, the pial microvessels were directly observed in situ. The diameter of the arterioles (20-50 mm) was continuously measured by a video analyzer. Caffeine was injected intraperitoneally. Four sets of experiments were designed. [1] The effect of caffeine (0.05, 0.20 mmol/g) on the resting diameter of the arterioles was studied. [2] Adenosine (10-6M) dissolved in mock CSF was perfused over the pial surface under the cranial window before and after injection of caffeine (0.20 mmol/g). [3] Changes in the diameter of the pial arterioles following stepwise reduction of systemic arterial pressure induced by exsanguination were measured in the untreated and caffeine (0.05, 0.20 mmol/g) pretreated animals. [4] The bilateral common carotid arteries were occluded in the untreated and caffeine (0.40 mmol/g) pretreated rats while observing the pial microvessels.

Results: [1] With 0.05 mmol/g of caffeine, the diameter was decreased by 4.3% (p<0.05), while 0.20 mmol/g did not affect the resting diameter. Mean arterial pressure did not change before and after administration of caffeine. [2] Cortical perfusion of 10-6M adenosine dilated the pial arterioles by 18.1% (p<0.01). The vasodilatory effect of exogenous adenosine was completely antagonized by caffeine. [3] The pial arterioles continued to dilate while arterial pressure was decreased. During slight hemorrhagic hypotension (105-95 mmHg), the diameter of the arterioles in the control animals increased by 8.1% (p<0.01), whereas the arterioles in the treated groups did not respond significantly. During moderate hypotension (75-65 mmHg), the dilatory response of the pial arterioles was attenuated in a dose dependent manner. Vasodilatation induced by severe hypotension (40-30 mmHg), however, was not inhibited by caffeine. [4] The bilateral common carotid occlusion markedly reduced back pressure of the internal carotid artery. It produced 39.1% increase in the diameter of the arterioles in the untreated rats, but only 11.9% increase was noted in the caffeine pretreated animals. Caffeine significantly (p<0.05) inhibited the dilatation caused by occlusion of the major arteries.

Discussion and conclusion: It has been reported that adenosine concentration in the brain increases within the autoregulatory range of mild hypotension. In the present study, caffeine, one of adenosine receptor blockers, was used to define the role of adenosine. It was demonstrated that caffeine antagonized vasodilatory effect of adenosine and attenuated the vasodilatory response of the cerebral resistant vessels to hemorrhagic hypotension and bilateral carotid occlusion. Thus, it is concluded that adenosine plays a role in autoregulation of the cerebral blood flow.

D24. ATP PRODUCES VASOCONSTRICTION VIA A P2x RECEPTOR AND VASODILATION VIA A P2y RECEPTOR IN THE FELINE PULMONARY VASCULAR BED. C. NEELY, D. PELLACK, R. BEESBURG, D. HAILE, 3400 Spruce Street, Philadelphia, Pennsylvania 19104.

Evidence exists supporting the presence of ATP-sensitive (P2) purinoceptors in isolated systemic vessels. Moreover, we have previously demonstrated under conditions of controlled pulmonary blood flow and left atrial pressure in spontaneously breathing intact, chest cats that $\alpha\beta$, -methylene adenosine 5' triphosphate ($\alpha\beta_x$-meATP) inhibited ATP-induced vasoconstriction at low (resting) pulmonary vascular (PV) tone without effecting the PV vasoconstrictor response to angiotensin II (1). The order of potency of ATP analogs to induce vasoconstriction at low (resting) PV tone was $\alpha\beta_x$-meATP $>>\alpha\beta$ -meATP > ATP. We have also demonstrated a tone-dependent response of the feline pulmonary circulation to ATP. That is, when the tone of the PV bed is raised to 30-40 mmHg by infusing U44069, ATP produces a dose-dependent vasodilator response that is not blocked by atropine, propranolol, meclofenamate, cimetidine or BWA1433U an adenosine (P1) receptor antagonist (2). These data suggest that the vasoconstrictor response of ATP is mediated by a P2x receptor and that the vasodilator response of ATP is not medicated by adenosine or the release of a vasodilator prostanoid. With the use of a putative P2y receptor antagonist reactive blue 2 it was the purpose of these experiments to evaluate the hypothesis that ATP induces vasodilation at elevated PV tone by acting on a P2y receptor.

ATP-induced changes in PV tone were examined in the closed chest, spontaneously breathing cat. PV tone was monitored in the left lower lobe of the lung which was isolated from the systemic circulation by a triple lumen balloon perfusion catheter placed in the lobar artery under fluoroscopic guidance. The lobe was perfused at 40 ml/min with blood withdrawn from the femoral artery. Also, left atrial pressure was measured with a double lumen catheter placed transseptally in the left lower lobe vein. Because pulmonary blood flow and left atrial pressure are held constant, changes in mean lobar arterial pressure represent a change in PV resistance. After allowing the animal to stabilize following initial anesthesia and catheter placements steady baselines of the mean lobar arterial, femoral arterial, and left atrial pressures were obtained. At low (resting) PV tone (10-15 mmHg) bolus injections of $\beta\gamma$-meATP (3, 10 and 30 %g) were injected into the lobar perfusion circuit. Reactive blue 2 was infused into the lobar perfusion circuit at 35 mg/kg/hr for 1 hour or less, and responses to $\beta\gamma$-meATP were repeated. PV tone was then raised to an elevated tone of 30-40 mmHg by infusing U44069 into the lobar perfusion circuit and control responses to ATP (0.3 %g), isoproterenol (0.03 %g), adenosine (1.0 %g), acetylcholine (0.03 %g), nitroglycerin (0.3 %g), and $\beta\gamma$-meATP (3.0 and 10.0 %g) were obtained before and after reactive blue 2 (35 mg/kg/hr) at the same level of PV tone.

Reactive blue 2 significantly attenuated the ATP vasodilator response without effecting those of isoproterenol, adenosine, acetylcholine, or nitroglycerin. At elevated PV tone the small dilator response to $\beta\gamma$-meATP (3.0 %g) was converted to a vasoconstrictor response. Moreover, at low (resting) PV tone, following reactive blue 2, pressor responses to $\beta\gamma$-meATP (10 and 30 %g) were significantly enhanced. These data suggest that in the feline PV bed ATP produces vasoconstriction by acting on a P2x receptor and vasodilation by acting on a P2y receptor.

1. Cahill B, et al.: FASEB J 2:A952, March 1988.
2. Neely CF, et al.: JPET 250:170-176, 1989.

D27. **PROTECTIVE EFFECTS OF AN ADENOSINE RECEPTOR AGONIST IN ANIMAL MODELS OF PARKINSON'S DISEASE.** T.M. LOFTUS AND P.J. MARANGOS. Gensia Pharmaceuticals, 11075 Roselle st., San Diego, Ca. 92121

In brain, adenosine or adenosine agonists are protective in animal models of epilepsy and cerebral ischemia. A mechanism proposed for this protective effect involves the inhibition of excitatory amino acid (EEA) release and its resultant neurotoxicity. Recently, excitotoxicity has been implicated in the methamphetamine(METH) animal model of Parkinson's disease (Sonsalla et al. 1989 Science 243: 398-400).The METH model is thought to simulate a proposed neurodegenerative mechanism by inducing the release of newly synthesized dopamine and its oxidation products. Another model, involving Methylphenyltetrahydropyridine (MPTP), is thought to act by glial conversion of the MPTP to MPP+ which is selectively taken up by dopaminergic neurons and produces superoxide in mitochondria by redox-cycling with neuromelanin as an electron source . In light of the recent suggestion of an EAA component in Parkinson's disease models, we have employed both of these model systems to evaluate the efficacy of the adenosine agonist N6 cyclohexyladenosine (CHA) in ameliorating the development of the parkinsonian syndrome.

Neurodegeneration was induced in male Swiss Webster mice by 4 systemic injections of either 20mg/kg of MPTP or 5mg/kg of Methamphetamine, 2 hours between each. CHA was administered in three injections of 0.5 mg/kg at 0,3 and 6 hours relative to the first toxin injection. Ketamine was coadministered with the toxins at 100mg/kg per injection. Animals were permitted to survive for seven days before sacrifice upon which the brains were removed and the striata dissected out and analyzed for tyrosine hydroxylase activity.

Both Meth and MPTP induced a 55-60% reduction in striatal tyrosine hydroxylase activity. Treatment with either CHA(p<.01) or Ketamine (p<.05) was effective in completely reversing the Meth induced decrement. Ketamine, while not tried by us, was not reported to be effective by Sonsalla et al. against MPTP. CHA on the other hand completely prevented the decrement in this model as well (p<.01). In both models, it was possible to partially reverse the protection afforded by CHA by coadministration of the adenosine antagonist, caffeine (30-50mg/kg).

CHA is therefore protective in two animal models of the Parkinson syndrome, suggesting the potential value of an adenosinergic theraputic approach to this disorder.

D28. EFFECTS OF SYSTEMICALLY AND CENTRALLY ADMINISTERED ADENOSINE AGONISTS ON CHEMICALLY AND ELECTRICALLY INDUCED SEIZURES.

J. B. WIESNER, R.M. KRIEGER, E.P. ROSSI, Gensia Pharmaceuticals, Inc., San Diego, California, USA 92121-1207

Adenosine agonists have previously been shown to possess anticonvulsive efficacy in kindled, audiogenic, and chemically induced experimental seizures. We have begun to establish dose/time response relationships for adenosine agonists using standard paradigms of chemically and electrically induced seizures. In preliminary studies we are comparing these relationships in mice to those in rats, and central (intracerebroventricular, ICV) administration to systemic (intraperitoneal, IP) administration, for the agonists N-ethylcarboximidadenosine (NECA), cyclo- hexyladenosine (CHA), and adenosine (ADO). The seizure paradigms utilized seizures induced by (a) subcutaneously-injected pentylenetetrazol (70-75 mg/kg) or (b) one of three intensities of 60-Hz current delivered for 0.2 seconds via corneal electrodes.

Pentylenetetrazol-induced seizures in mice (male CFW) were blocked by NECA with approximate ED50 of 0.1 mg/kg at 30 minutes after IP administration and 40 ng at 20 minutes after ICV administration of the agonist. Rats (male Sprague Dawley) were less sensitive to the effects of NECA, as evidenced by the lack of protection up to a dose of 1 mg/kg IP (only latency to seizure was increased) and lower potency by ICV administration. CHA blocked seizures in mice with approximate ED50 of 2 mg/kg IP and 100 ng ICV. ADO exerted a weak anticonvulsive effect which decayed rapidly: a dose of 100 mg/kg elicited 50% protection at short time intervals ($\leq$ 2 minutes post injection) but was ineffective at intervals > 10 minutes; ICV ADO (up to 270 ug) was ineffective.

Electroshock seizures were induced according to three paradigms (Swinyard, 1972) utilizing minimal threshold (MINT), maximal threshold (MAXT), or supramaximal (MES) current intensities. High doses of NECA (1 mg/kg IP or 100 ng ICV) or CHA (30 mg/kg IP or 3 ug ICV) were ineffective in MES but were differentially effective in the threshold current tests: NECA protected more effectively in MINT (60-80%) than in MAXT (0-33%) while CHA protected more in MAXT (67-100%) than in MINT (33-67%). These results indicate that the anticonvulsive effect of adenosine agonists are (1) centrally mediated; (2) a result of increased seizure threshold rather than decreased seizure spread; (3) differentially expressed in various paradigms of electroshock seizures; and (4) in the case of NECA more potently expressed in mice than in rats.

D29. CONTRARY EFFECTS OF ACUTE VERSUS CHRONIC CAFFEINE ADMINISTRATION ON ISCHEMICALLY INDUCED HIPPOCAMPAL DAMAGE IN THE GERBIL. - INVOLVEMENT OF ADENOSINE RECEPTORS?

KARL A. RUDOLPHI1, MANFRED KEIL2, JOHAN FASTBOM3, JOHN J. GROME1, BERTIL B. FREDHOLM3, 1Hoechst AG Werk Kalle-Albert, Wiesbaden, FRG, 2Hoechst AG, Frankfurt/M, FRG, 3Dept. Pharmacology, Karolinska Inst., Stockholm, Sweden.

The endogenous neuromodulator adenosine may protect against ischemic brain dam-age (1,2). In order to further explore this hypothesis, we have firstly studied the effect of single dose pretreatment with the adenosine receptor antagonists caffeine and dipropylcyclopentylxanthine (DPCPX) on the ischemically induced hippocampal damage in Mongolian gerbils. Secondly we were interested if adeno-sine A1-receptor upregulation by long-term caffeine treatment may influence the ischemic damage since recently a rapid downregulation of these receptors was suggested to contribute to the development of selective ischemic hippocampal necrosis (3).

As we have previously published for theophylline (4), the acute pretreatment with caffeine and DPCPX increased the degree of necrosis of pyramidal neurons in the CA1 area of the gerbil hippocampus following 3 min of forebrain ischemia, induced by bilateral carotid occlusion, with DPCPX being the most potent drug. In particular, the 10-min oral pretreatment with a single dose of caffeine
(30 mg/kg) or DPCPX (10 mg/kg) significantly enhanced the degree of ischemic hippocampal necrosis by 53 % (caffeine) or 55 % (DPCPX).

Four-week oral treatment with caffeine (0.2 % in drinking water) caused a significant (10-17 %) increase in the binding of the adenosine A1-receptor ligand [3H]-cyclohexyladenosine (CHA) to several brain regions, including the hippocampal CA1 area, in Mongolian gerbils. Animals subjected to such treatment exhibited significantly less neuronal damage in the CA1 region following 5-min ischemia than did control animals (50 % of the caffeine-treated animals showed no damage at all compared to 11% in the control group).

Our findings provide further evidence for a protective role of endogenous adenosine during ischemia.

(1) Dragunow, M. and Faull, R.L.M. (1988) Trends Pharmacol. Sci., 9:193-194
(2) Phillis, J.W., (1989) Cerebrovasc. Brain Metab. Rev. 1:26-54
(3) Lee, K.S., Tetzlaff, W. and Kreutzberg, G.W. (1986) Brain Res. 380:155-158
(4) Rudolphi, K.A., Keil, M. and Hinze, H.-J. (1987) J. Cereb. Blood Flow Metabol., 7:74-81

E1.   5'-(ISOBUTYLTHIO)-ADENOSINE   (SIBA)   ANALOGS   AS ANTIMUSCARINIC PHARMACOPHORES WITH MUSCARINIC RECEPTOR SUBTYPE SPECIFICITY. P. K. CHIANG, M. C. PANKASKIE*, R. M. SMEJKAL, Walter Reed Army Institute of Research, Washington, DC 20307-5100; *Department of Pharmaceutical Sciences, University of Nebraska Medical Center, Omaha, Nebraska 68105-1065

5'-(Isobutylthio)-adenosine (SIBA) and several of its analogs were tested for muscarinic receptor specificity as measured by their effects on the binding of various ligands in different regions of rat brain and on acetylcholine-induced contraction of guinea pig ileum. At 100 uM, SIBA and seven analogs caused an 11-30% inhibition of [3H]N-methylscopolamine ([3H]NMS) binding to homogenates of both whole brain and cerebral cortex, an area rich in M1 subtype receptors. In comparison, a 20-39% inhibition was observed for SIBA and the analogs in cerebellum, an area rich in M2 subtype receptors. At 100 uM, SIBA and most of the analogs tested did not inhibit the binding of [3H]quinuclidinylbenzilate ([3H]QNB) to the total population of muscarinic receptors. However, at low concentrations, these drugs showed small stimulatory effects on [3H]QNB binding sites and on NMS-inaccessible [3H]QNB binding sites (a subpopulation of [3H]QNB sites). The maximal stimulation, induced by 3-deaza-SIBA, was about 137% of control at 1 uM, after which stimulation decreased. The binding of [3H]pirenzepine ([3H]PZ) to the whole brain and cerebral cortex, in contrast, was inhibited by SIBA and the other analogs tested in a dose-dependent manner with a Ki value within the uM range. The most potent of these, 3-deaza-SIBA, was found to act in the manner of a mixed type inhibitor and to significantly alter the rates of association and dissociation of [3H]PZ to cerebral cortex receptors. Some of the analogs were tested for their ability to antagonize the acetylcholine-induced contraction of guinea pig ileum, a tissue containing predominantly M2 subtype receptors. The potency of the analogs was slightly better than that of pirenzepine, but several orders of magnitude lower than atropine. The order of potency for these analogs was opposite that determined for the binding of [3H]PZ to cerebral cortex. In addition, several functional groups appear to be necessary for activity. The acyclo analog, 9-[[2-(isobutylthio)ethoxy]methyl]-adenine had no antimuscarinic activity, pointing to the importance of an intact ribofuranose ring. A basic nitrogen at position 7 of the purine ring is also required for activity, since 7-deaza-SIBA was not inhibitory. The amino analog of SIBA, 5'-(isobutylamino)adenosine, as well as other amino derivatives of SIBA compounds, were also inactive, indicating that a sulfur atom may be essential for activity. The results of these studies suggest that structural modifications at some positions on adenosine confer antimuscarinic activity on the analogs. Furthermore, SIBA and some of its analogs appear to have differential effects on muscarinic receptor subtypes both in rat brain and in guinea pig ileum. Their actions appear to affect the binding of pirenzepine more than the other ligands tested, implying some specificity for the M1 receptor subtype.

412

E2. MAPPING THE N6-REGION OF THE ADENOSINE A1 RECEPTOR WITH COMPUTER GRAPHICS.

A.P. IJZERMAN, P.J.M. VAN GALEN, F.J.J. LEUSENIS, AND W. SOUDIJN, Center for Bio-Pharmaceutical Sciences, Div. of Medicinal Chemistry, PO Box 9502, 2300 RA Leiden, The Netherlands.

N6-substituted adenosine derivatives can be very potent and selective agonists at adenosine A1 receptors. The N6-region is usually described by a model introduced by Kusachi et al (1). Detailed structure-activity relationships for this region have been reported by Daly et al (2).

In the present study, the N6-region was investigated in further detail with the aid of computer graphics methods.

Energetically favorable conformations of 26 representative compounds from Daly's study were computed with the molecular mechanics program MM2P and the semi-empirical molecular orbital program MOPAC. All structures were superimposed with the Chem-X package and, employing the flexible fit option, the relative positions of the various substituents were optimized.

The resulting map of the N6-region allows Kusachi's model to be extended with two new subregions: a C subregion, accommodating cycloalkyl substituents and a B subregion, which is filled with bulky substituents, such as a norbornanyl moiety. Furthermore, there are three forbidden areas: regions where occupation leads to diminished receptor affinity. The map can be used to determine the contribution of various subregions quantitatively, allowing accurate prediction of affinities of compounds that were not used to construct the map.

(1) Kusachi, S., et al (1985) J. Med. Chem. 28, 1636.
(2) Daly, J., et al (1986) Biochem. Pharmacol. 35, 2467.

E3. A SIMPLE LIPOPHILICITY-BASED MODEL FOR THE DISCRIMINATION OF A1 RECEPTOR AGONISTS AND ANTAGONISTS.

G. FOLKERS, C.E. MÜLLER*, K. EGER, Pharmazeutisches Institut der Universität Tübingen, Auf der Morgenstelle 8, 7400 Tübingen, West-Germany *present address: LBC, NIDDK, NIH, Bethesda, MD 20892, USA

Compounds which exhibit affinity to adenosine receptors show broad structural variations. In case of agonistic behaviour the compounds can be classified as adenosine derivatives whereas the antagonists do not seem to have larger structural motifs in common. The most striking difference between the agonists and the antagonists is the lacking of the ribose moiety of the latter whereas the conserved 2',3'-OH function is a feature of the agonists. This means that during the binding process both structural classes show different contributions of hydrophobic interaction. Since changes in lipophilicity seem to be at least partially correlated with entropy, from a thermodynamic viewpoint the ant/agonistic effects can be seen as a different enthalpic and entropic contribution to the interaction energy. Therefore we looked in a very simple approach for a correlation of lipophilicity and A1 receptor affinity. The plot of log P versus Ki-value of a set of 21 compounds exhibits two discrete fields one containing all antagonists, the other all agonists. We paid special attention to the criteria of structural heterogeneity and high receptor affinity of the selected compounds.

Acknowledgement: C.E.M was supported by the Deutsche Forschungsgemeinschaft

E4. SYNTHESIS OF CHIRAL ADENOSINE RECEPTOR RECOGNITION UNITS VIA A SHARPLESS ASYMMETRIC EPOXIDATION PROCEDURE.

NELSEN L. LENTZ AND NORTON P. PEET. Merrell Dow Research Inst., 2110 E. Galbraith Rd., Cincinnati, OH 45215.

(R)-N6-Phenylisopropyladenosine (R-PIA) is a potent, A1-selective adenosine agonist. In order to incorporate the chiral phenylisopropyl recognition unit into other potential receptor agents, we required chiral 2-benzylpropionaldehyde (1) and/or the corresponding acid and alcohol. Preparation of 1, via epoxide 5 obtained by a Sharpless asymmetric epoxidation procedure,1 led to an interesting skeletal rearrangement which also provided chiral 2-phenylbutyraldehyde (9).

Treatment of butadiene monoxide (3) with phenylmercuric chloride (2) in the presence of a palladium catalyst gave E-4-phenyl-2-buten-1-ol (4).1 Oxidation of 4 with 3-

chloroperoxybenzoic acid gave epoxide **5**, which was opened with trimethylaluminum[1,2] to afford diol **6**. Oxidative cleavage of **6** with sodium metaperiodate gave the desired aldehyde **1**.

Co-produced with **6** during the reaction of **5** with trimethylaluminum was the rearranged diol **8**. Both **6** and **8** could arise from phenonium ion[1] **7**, a proposed intermediate. Oxidation of **8** with sodium metaperiodate gave aldehyde **9**. The ratio of diols **6** and **8** was dependent on reaction conditions.

Allylic alcohol **4** was subjected to Sharpless epoxidation conditions, which allowed the preparation of the enantiomers of aldehydes **1** and **9**. Mosher's esters made from the alcohols of these aldehydes (by reduction with lithium aluminum hydride) were used to establish enantiomeric excesses.

[1]Katsuki, T.; Sharpless, K.B. *J. Am. Chem. Soc.* **1980**, *102*, 5974.

[2]Larock, R.C.; Ilkka, S.J. *Tetrahedron Lett.* **1986**, *27*, 2211.

[3]Roush, W.R.; Adam, M.A.; Peseckis, S.M. *Tetrahedron Lett.* **1983**, *24*, 1377.

[4]Oshima, K.; Suzuki, T.; Saimoto, H.; Tomioka, H.; Nozaki, H. *Tetrahedron Lett.* **1982**, *23*, 3597.

[5]Cram, D.J. *J. Am. Chem. Soc.* **1949**, *71*, 3863.

E6. NON-PURINE AGONIST/ANTAGONISTS FOR THE ADENOSINE RECEPTOR. R. J. QUINN, M. CHEBIB, J. F. DEVLIN, M. J. DOOLEY, A. ESCHER, F. HARDEN, H. JAYASURIYA, R. W. MONI, G. B. RINTOUL, P. J. SCAMMELLS. Division of Science and Technology, Griffith University, Brisbane, Queensland 4111, Australia

Two approaches to the development of non-purine compounds as potential agonists / antagonists of the adenosine receptor are being examined. The synthetic program, utilizing radioligand receptor binding and computer assisted molecular modelling, has concentrated on analogues with modified five-membered rings. Such compounds may selectively interact with the receptors and avoid the other actions of purines. An alternative approach is to search for non-purine receptor ligand leads using receptor-based technology for screening. Such a lead could serve as a starting point for design of an optimal structure. The application of both approaches will be described.

Considerable information on the N6-binding domain is available, with rather less information on substituents in other positions. There are few compounds other than modified xanthines that function as adenosine antagonists. Previously there have been reports of two classes of non-purines which bind to the adenosine receptor, the pyrazolo[3,4-d]pyrimidines and the pyrazolo[4,3-d] pyrimidines.

A series of over 70 non-purine compounds have been synthesized,to date, and the results of the structure/activity studies will be reported.

A structure-activity study of pyrazolo[3,4-d]pyrimidine analogs (1) of the pharmacologically active marine natural product, 1-methylisoguanosine (2), which is resistant to deamination by adenosine deaminase and thus displays activity after oral administration, will be reported.

(1)

(2)

(3)

Triazolo analogues of purines were synthesized in order to determine their potential as adenosine agonists and antagonists. A number of dithio-1,2,3-triazolo[4,5-d]pyrimidines (3) were developed with varying sulphur substituents as potential adenosine agonists and antagonists. Compounds with significant affinity for the A1 receptor.have been found. As opposed to a lead compound it proved possible to obtain only monoalkylation and the position of this alkylation was determined by chemical correlation. This ability to monoalkylate has allowed the synthesis of a number of monoalkylated analogues and analogues alkylated with different groups at the two positions, thus resolving which position is more important for binding.

The syntheses of the corresponding b-D-ribofuranosyl series has been initiated only recently.

Utilizing the [3H]-PIA radioligand binding assay for screening, a number of extracts from marine animals have been found to inhibit binding. The extracts are being examined using binding assay-directed fractionation.

E8. CHARACTERIZATION OF HUMAN, GUINEA PIG AND RABBIT LIVER PURINOCEPTORS. S. KEPPENS, A. VANDEKERCKHOVE and H. DE WULF. Dept. Biochemistry, Fac. Medicine, Catholic University of Louvain. Campus Gasthuisberg, B-3000 Leuven, Belgium.

Introduction. In rat liver, submicromolar concentrations of ATP and ADP induce a Ca^{++}-dependent glycogenolysis (Keppens & De Wulf, 1985; Charest et al., 1985; Sistare et al. 1985) by interaction with specific and biologically active P2Y purinoceptors on the plasma membrane (Keppens & De Wulf, 1986)

415

Objectives. Since hormonal sensitivity greatly differs from one animal species to an other (e.g. adrenergic responsiveness) we studied ATP binding (and where possible its effect) to human, rabbit and guinea pig liver.

Methods used. We have used isolated hepatocytes and purified liver plasma membranes to characterize the purinoceptors with ATPa[35S] as radioligand. Determination of the binding affinities of the different structural analogs was done by competition experiments (Keppens & De Wulf, 1986).

Results. In both rabbit and guinea pig hepatocytes, ATP acts as a Ca++-dependent glycogenolytic agonist. Binding of ATPa[35S] to these cells is rapid, saturable within one min and easily reversible upon addition of an excess of ATP. Equilibrium binding studies reveal a class of binding sites with Kd ≈ 0.2 μM and Bmax 3 - 6 pmoles/million of cells for both species. Binding of ATPa[35S] to purified liver plasma membranes yields Kd values of 0.19, 0.10 and 0.10 μM for human, guinea pig or rabbit liver. Bmax values range from 24 pmoles/mg protein for human liver plasma membranes to about 50 pmoles/mg protein for rabbit and guinea pig. For 11 structural analogs of ATP we determined their binding affinities to liver plasma membranes prepared from man and rabbit and compared them with their previously determined binding characteristics for rat hepatocytes. There is a strong correlation between the Kd values of the rat and human liver receptors (r = 0.933, P = 0.0005) and also between those of rat and rabbit liver receptors (r = 0.996, P = 0.0001)

Conclusions. We concluce from these data that liver from the species studied possess specific P2Y purinoceptors.

Charest, R., Blackmore, P.F.,& Exton, J.H. (1985) J. Biol. Chem. 260, 15789-15794

Keppens, S. & De Wulf, H., (1985) Biochem. J.231, 797-799

Keppens, S. & De Wulf, H., (1986) Biochem. J.240, 367-371

Sistare, F.D., Picking, R.A. & Haynes, R.C. (1985) J.Biol.Chem. 260, 12744-12747

E9. **THE SELECTIVITY OF αß METHYLENE ATP (αß MeATP) AND SURAMIN AS PURINERGIC ANTAGONISTS IN THE RABBIT SAPHENOUS ARTERY.** JANE E. NALLY & T.C. MUIR, Dept. of Pharmacology, University of Glasgow, Glasgow G12 8QQ, Scotland.

Where purinergic substances have been proposed as co-transmitters (Sneddon & Burnstock, 1984), antagonism of p2-purinoceptors by αß MeATP or suramin (Dunn & Blakeley, 1988) has been used to analyse responses. Because of doubts about the selectivity of αß MeATP (Byrne & Large, 1986; Komori et al., 1988), the suitability of proposed antagonists should be assessed in the tissue being used. Accordingly, the selectivity of αß MeATP and suramin has been examined in the rabbit saphenous artery where co-transmission may exist (Burnstock & Warland, 1987; Muir & Nally, 1989).

Contractions were recorded from ring preparations. αß MeATP (10-6 - 10-4 uM), dose-dependently antagonised the responses to further doses of drug, and to ATP (10-5 - 10-3 uM) following initial contraction. Contractions to nerve stimulation (0.5msec, 100v, 15sec) were reduced by some 50% and subsequently abolished by the ‡1-adrenoceptor antagonist prazosin (10-7M). αß MeATP was ineffective against KCl contractions (KCl 1-1.6 x 10-3 uM), where transmitter release had been prevented by pretreatment with 6-hydroxydopamine (6-OHDA, 5 x 10-4 uM). Contractions to added noradrenaline (NA, 10-7 - 10-5 uM) and histamine (10-7 - 10-5 uM) were each unaffected. Evoked e.j.p.s (0.5msec, 100v, 1-32Hz), and the dose dependent depolarisation to ATP (10-2 uM) were abolished by αß MeATP (10-6 uM). The recovery rate of e.j.p.s and that of the depolarisation to ATP, on washout of αß MeATP, were similar. Changes in membrane resistance produced by hyperpolarising current (Abe & Tomita, 1968) were unaffected by αß MeATP (4 x 10-6 uM), suggesting the absence of a non-selective membrane effect. In contrast to αß MeATP, suramin alone produced no initial contraction, but at 10-3 uM, antagonised contractions to ATP (10-6 - 10-4 uM), αß MeATP (10-6 - 10-4 uM) and field stimulation (<8Hz). Suramin, like αß MeATP, had no effect on KCl responses after pre-treatment with 6-OHDA (5 x 10-4 uM), nor on the contractions to NA (10-7 - 10-5 uM), but antagonised the contractions to histamine (10-7 - 10-5 uM) and 5-hydroxytryptamine (10-7 -

10-5 uM). Evoked e.j.p.s were abolished by suramin (10-3 uM). In the rabbit saphenous artery, αß MeATP appears selective for the non-adrenergic co-transmitter. Suramin however appears less selective and may be less useful in the study of co-transmission in this artery. We acknowledge the support of ICI and SERC.

Burnstock, G. & Warland, J.I. (1987) Br. J. Pharmacol. 90 : 111
Byrne, N.G. & Large, W.A. (1986) Br. J. Pharmacol. 88 : 6
Dunn, P.M. & Blakeley, A.G.H. (1988) Br. J. Pharmacol. 93 : 243
Komori, S., Kwon, S-C. & Ohashi, H. (1988) Br. J. Pharmacol. 94 :9
Muir, T.C. & Nally, Jane E. (1989) Br. J. Pharmacol. 96 : 188P
Sneddon, P. & Burnstock, G. (1984) Eur. J. Pharmacol. 100 : 85

E11. CHARACTERIZATION OF THE BINDING OF [3H]CV 1808 IN RAT STRIATUM. MATTHEW A. SILLS, MICHAEL WILLIAMS, STEPHEN D. HURT1 AND PATRICIA S. LOO. Research Department, Pharmaceuticals Division, CIBA-GEIGY Corp., Summit, NJ 07901 and 1NEN Products, 549 Albany St., Boston, MA 02118.

CV 1808 (2-phenylaminoadenosine) has diverse effects on purinergic systems in the mammalian brain. In competition binding assays, CV 1808 inhibited the binding of [3H]CHA to the adenosine A-1 receptor with a Ki value of 1090 nM. In contrast, CV 1808 inhibited the binding of [3H]CGS 21680 to the adenosine A-2 receptor with a Ki value of 100 nM (Williams et al., FASEB J. 3: A1047, 1989). In addition, CV 1808 was found to interact at adenosine uptake sites, inhibiting the uptake of [3H]uridine with an IC50 value of 140 nM (Krulan et al., FASEB J. 3: A1048, 1989). In the present study, CV 1808 was radiolabeled and its binding in rat striatum characterized.

Striatal tissue from male Sprague-Dawley rats was homogenized by Polytron in Tris-HCl buffer (pH 7.4) containing 10 mM MgCl2. After several washes, 400 µl of the ADA-treated tissue was added to polypropylene tubes containing 50 µl [3H]CV 1808 (final concentration of 5 nM) and 50 µl of either buffer or drug. Tubes were incubated for 90 min at 0-4oC and then filtered through a Brandel cell harvester.

Using a radioligand concentration of 5 nM, the amount of specific binding of [3H]CV 1808 was found to be 65-70% of the total radioligand bound. In saturation experiments, 14-19 different concentration of [3H]CV 1808 were examined ranging from 0.4 to 380 nM. Computer analysis of the binding data indicated that a one-component binding model adequately described the binding data. Bmax and Kd values were 2300 fmol/mg protein and 74 nM, respectively.

Results from competition experiments indicated that CV 1808 was the most active compound in competing for the binding of [3H]CV 1808, displaying an IC50 value of 42 nM. The non-selective adenosine agonists NECA and MECA displayed IC50 values of 15 and 31 µM, respectively, whereas the A-2 selective agonist CGS 21680 generated an IC50 value of 29 µM. The adenosine uptake inhibitors dipyridamole and NBI were also weak inhibitors of [3H]CV 1808 binding, producing IC50 values of 45 and 5700 µM, respectively.

Interestingly, the adenosine A-1 selective compound CPA generated a biphasic inhibition curve in competing for the binding of [3H]CV 1808. Computer analysis indicated that a two-component binding model described the binding data significantly better than a one-component model. Furthermore, 37% of the binding was inhibited with an IC50 value of 0.3 µM, whereas 62% of the binding was inhibited with an IC50 value of 530 µM.

Based on the present data, it appears that [3H]CV 1808 is not binding to the classical adenosine A-1 or A-2 receptors. The identity of the binding component(s) labeled by [3H]CV 1808 is currently unclear.

E13. 4-AMINO-[1,2,4]TRIAZOLO[4,3-a]QUINOXALINES - HIGHLY SELECTIVE ADENOSINE ANTAGONISTS AND POTENTIAL ANTIDEPRESSANTS. REINHARD SARGES, HARRY R. HOWARD, RONALD G. BROWNE, B. KENNETH KOE. Pfizer Central Res., Pfizer, Inc., Groton, CT 06340,

Many representatives from a series of 4-amino-[1,2,4]triazolo[4,3-a]quinoxalines reduce immobility in Porsolt's behavioral despair model in rats upon acute administration and may therefore have therapeutic potential as novel and rapid acting antidepressant agents. Optimal activity in this test is associated with hydrogen, CF3 or small alkyl groups in the 1-position, with NH2, NH-acetyl or amines substituted with small alkyl groups in the 4-position, and with hydrogen or 8-halogen substituents in the aromatic ring. Furthermore, many of these 4-amino-[1,2,4]triazolo[4,3-a]quinoxalines bind very avidly, and in some cases very selectively, to adenosine A1 and A2 receptors. A1 affinity of these compounds was measured by their inhibition of tritiated CHA (N6-cyclohexyl-adenosine) binding in rat cerebral cortex homogenates and A2 affinity by their inhibition of tritiated NECA (5'-(N-ethylcarboxamido)-adenosine) binding to rat striatal homogenate in the presence of excess cold N6-cyclopentyladenosine. SAR studies show that best A1 affinity is associated with ethyl, CF3 or C2F5 in the 1-position, NH-iPr or NH-cycloalkyl in the 4-position, and with an 8-chloro substituent. Affinity at the A2 receptor is mostly dependent on the presence of an NH2 group in the 4-position, and is enhanced by phenyl, CF3 or ethyl in the 1-position. The most potent A1 ligand has an IC50 of 5.5 nM, the most potent A2 ligand has an IC50 of 22 nM. The most selective A2 ligand has an A2/A1 potency ratio of greater than 450. The most selective A1 ligands have selectivity ratios of >3000. Representatives from this series appear to act as antagonists at both A1 and A2 receptors since they antagonize the inhibiting action of CHA on norepinephrine-stimulated cAMP formation in fat cells and they decrease cAMP accumulation induced by adenosine in limbic forebrain slices. Thus certain members of this 4-amino-[1,2,4]triazolo[4,3-a]quinoxaline series are the most A1 or A2 selective non-xanthine adenosine antagonists known.

F1. **EXTRACELLULAR ATP ACTING ON SKELETAL MUSCLE ACTIVATES PURINERGIC RECEPTORS, PRODUCES SECOND MESSENGERS AND MODIFIES FREE [Ca2+]i.** E. HEILBRONN, H. ERIKSSON & J. HÄGGBLAD, Unit of Neurochemistry and Neurotoxicology University of Stockholm, S-106 91 Stockholm, Sweden.

Evoked release of adenosine-5'-triphosphate (ATP) occurs together with that of acetylcholine (ACh) from the motor nerve ending . ATP is also released from the depolarized skeletal muscle itself. Its hydrolysis, by ectoenzymes, is much slower than that of ACh. Extracellular ATP has occasionally been observed to improve skeletal muscle contraction. The present investigation is an effort to explain these observations. Primary cultures of chick, rat or mouse skeletal muscle cells were used. ATP was found to activate a membrane P2y receptor-G protein-phospholipase C cascade, producing a time and dose dependent increase in inositoltriphophate (IP3) and diacylglycerol (DAG) from phosphoinositide. Slowly hydrolyzing ATP derivatives GTP and GTP S had similar effects, while ADP, AMP and adenosine were less effective. Formed IP3 was found to regulate [Ca2+]i in two ways: A strong, rapid and transient Ca2+-signal (Fura-2) was independent of external Ca2+, indicating Ca2+ release from an internal source. Work with myotubes from dysgenic mice suggested SR. A more sustained signal followed and depended on extracellular Ca2+, but not on depolarization. A channel, selective or nonselective for Ca2+, may exist and be opened by intracellular IP3 (as eg in the plasma membrane of human T-lymphocytes). Depolarisation of myotube by K+ or carbachol opens dihydropyridine- and verapamil sensitive Ca2+-channels which are closed by external ATP and by phorbol esters. A rapid decrease in the depolarised [Ca2+]i was seen. Observed events, including extracellular ATP metabolizing enzymes, suggest a fluctuating system regulating skeletal muscle activity. Effects including increases in DAG levels due to PI-turnover may result in increases in protein kinase C activity and thus phosphorylation of muscle cell receptors and/or channels. These possibilities are currently investigated.

F3. **GUANINE NUCLEOTIDE ANALOGUES AS ANTAGONIST FOR G-PROTEIN MEDIATED SIGNAL TRANSDUCTION.** YEE-KIN HO, Department of Biological Chemistry, University of Illinois at Chicago, Health Sciences Center, 1853 West Polk Street, Chicago, Illinois 60612.

The involvement of a family of GTP-binding proteins, the G-proteins, in cellular signalling has been well documented. Our goal is to design antagonist targeted for G-proteins with the aim of interrupting the flow of information in receptor and effector enzyme coupling. Latent G-proteins contains bound-GDP and has a high affinity for binding to the receptor, but a low affinity for the effector enzyme. Activated receptor catalyzes the exchange of GTP for the bound GDP, the G-GTP complex dissociates from the receptor and acquires a high affinity for the effector enzyme. Transducin (T), the G-protein in photoreceptor cells, which couples the visual excitation signal from rhodopsin to the cGMP phosphodiesterase (PDE) was used as a model system. Structural analyses of T have suggested that the communication between the functional domains of T could be mechanically linked through two flexible hinge regions of the protein structure via the interaction the bound Mg2+ and the phosphate groups of GTP. This information provides the basis for the design of effective uncouplers.

Cr(III) ,• bidentate GTP complexes which eliminate the interaction of the -, and •- phosphate with the bound Mg2+ leads to the activation of PDE but not the dissociation of transducin from rhodopsin membrane. On the other hand, GTP analogues with a more flexible ribose structure such as the dialdehyde-GTP or dideoxy-GTP are capable of dissociating of T from rhodopsin, but they only exhibit 60 percent efficiency in activating PDE, i.e. partially uncouple the effector activation. This preliminary study clearly demonstrates that GTP analogues can be used as uncoupler for G-proteins and different coupling events of G-protein can be selectively modified by specific GTP analogues. This result not only provides hints for elucidating the molecular mechanism of G-protein mediated signal transduction, but also clues for designing antigonist for selective intervention of signaling cascades.

F4. **MECHANISMS FOR IMMUNE MODULATION BY PURINE NUCLEOSIDE ANALOGS.** J. ARLY NELSON AND TERESA PRIEBE, Dept. Exptl. Pediatr., Univ. Tex. M. D. Anderson Cancer Cntr., Houston, TX 77030.

Severe combined immunodeficiency disease associated with adenosine deaminase (ADA) deficiency is thought to result from accumulation of the enzyme substrates, adenosine (Ado) and/or deoxyadenosine (dAdo). As an alternative to the use of inhibitors of ADA to model the disease in animals, we have employed Ado and dAdo analogs that are poor substrates for ADA. The Ado analog tubercidin (Tub) and the dAdo analog 2-fluoroadenine arabinoside (given as its soluble monophosphate, FaraAMP) produce strikingly-distinct effects on immune function in mice. Specifically, Tub inhibits natural killer (NK) cell activity and cell-mediated lymphocytolysis at doses that stimulate primary antibody response, whereas exactly the opposite effects are seen with FaraAMP. Using the NK cell-deficient C57BL/6J mouse (beige mutation), we have demonstrated that the effects on antibody formation are mediated via NK cells. Therefore, the mechanisms for modulating this activity are of interest. The inhibition (Tub) and stimulation (FaraAMP) of NK cell activity are readily demonstrated in lymphocytes exposed to the drugs in vitro. We tested a possible role for Ado receptors in mediating these effects in vitro using specific receptor agonists and the antagonist, XAC. Ado receptor A2 agonists (PAA; NECA) and Ado were more potent inhibitors of NK cell activity than were Ado receptor A1 agonists (PIA; CPA) and dAdo. The A1 agonists and dAdo were more effective stimulators of NK cell activity. These results suggest that NK cells, unlike T- and B- lymphocytes, might have surface Ado receptors of type A1 as well as A2. Although XAC prevented inhibition by Ado, dAdo and the A2 agonists, it did not prevent inhibition produced by Tub. The nucleoside transport inhibitor, nitrobenzylthioinosine monophosphate, did partially prevent Tub-induced inhibition or stimulation by FaraAMP; however, the transport inhibitor did not modify the effects of Ado or dAdo. We conclude that Tub inhibits NK cell activity by a mechanism not involving Ado receptors and that FaraAMP might stimulate NK cell activity via an intracellular "P-site". (Supported by NIH Grant, HD-13951).

F6. **RESOLUTION OF A1 ADENOSINE RECEPTORS FROM CO-PURIFIED POLYPEPTIDES.** A. PATEL, M.L. ARROYO, R. MUNSHI, J. LINDEN, Box 449 Health Sciences Center, University of Virginia, Charlottesville, VA 22908.

A1 adenosine receptors and G proteins were co-eluted from an agonist affinity column with GTP. High affinity agonist binding was demonstrated following removal of GTP. Based on the specific activity of [35S]GTPγS binding (4.6 nmol/mg protein), receptor-G protein complexes (R-G complexes) are > 50% pure. Affinity purified R-G complexes contain a 35 kDa A1 adenosine receptor binding subunit (photoaffinity labeled), at least 2 G proteins (Gi1, Go, and ß 35/36), and an unidentified 34 kDa polypeptide (p34). We now describe techniques for resolving receptors from co-purified polypeptides.

The eluate from affinity columns were applied to DEAE columns and eluted with ammonium formate (salt) gradients in the presence of 0.1% (low) or 0.5% (high) CHAPS. A1 receptors and G protein β-subunits were monitored with specific radioligands, 125I-BW-A844U and [35S]GTPγS, respectively. Monoclonal antibodies were raised against p34 and used to assay it by ELISA. The p34 peptide was eluted at low salt, < 0.1 M, and its elution was unaffected by detergent or GTP. These fractions were completely devoid of radioligand binding activity for receptors and G proteins. Both receptors and G proteins eluted as single peaks with > 0.25 M salt in low detergent in the absence or presence of GTPγS. In high detergent without GTPγS, multiple peaks of receptors were eluted, one of which also contained G proteins. Addition of GTPÁS caused most receptors and G proteins to elute as well resolved single sharp peaks.

As an additional means of resolving 35 kDa receptors from 35/36 kDa G protein ß-subunits we utilized low cross-linked SDS-polyacrylamide gels, on which ß-subunits migrate anomalously with an apparent molecular mass of 39 kDa. However, receptors and p34 are not resolved by this procedure.

These data are consistent with the notion that guanine nucleotides influence the interaction between purified A1 receptors and G proteins and together with detergent affect their mobilities on DEAE columns. A combination of DEAE chromatography and low cross-linked SDS-PAGE can be used to resolve receptors from co-purified polypeptides with similar molecular masses, i.e. G protein subunits and p34.

F7. A1 ADENOSINE RECEPTORS OF BOVINE BRAIN COUPLE TO Go AND Gi1 BUT NOT TO Gi2. R. MUNSHI, I-H. PANG*, P.C. STERNWEIS*, J. LINDEN, Box 449 University of Virginia Health Sciences Center, Charlottesville, VA 22908 and *The University of Texas Southwestern Medical Center, Dallas, TX 75235.

A1 adenosine receptors and associated guanine nucleotide binding proteins (G proteins) have been co-purified from bovine brain on an aminobenzyladenosine-Affi-Gel-202 affinity column [Munshi, R. and Linden, J. (1989) J. Biol. Chem. 264:14853-14859]. In purified receptor preparations, pertussis toxin catalyzes the ADP-ribosylation of two G protein α-subunits with molecular masses of 39 (Go) and 41 (Giα1 and/or Giα3) kDa, respectively. These data suggest that A1 adenosine receptors may not couple to the pertussis toxin-sensitive 40 kDa α-subunit corresponding to Giα2. With the use of specific antisera we now report that purified receptor preparations immunoreact with antiserum against Goα, Giα1, Gβ35 and Gβ36. Antiserum against Giβ3 reacts very weakly. The purified fractions do not cross-react with specific antiserum raised against Giα2. These data provide further evidence that A1 adenosine receptor-Giα2 complexes do not co-elute from the agonist affinity column.

To further investigate which G proteins can couple to purified A1 receptors, G proteins and receptors were reconstituted into phospholipid vesicles. G protein β-subunits present in purified receptor preparations were inactivated with N-ethylmaleimide (NEM). Reconstitution of these receptors (0.36 pmol) and added Gßγ (40 pmol) with 36 pmol of purified bovine brain Goα or Giα1 partially restored GTPγS-sensitive high affinity binding of the agonist 125I-aminobenzyladenosine. When the same amount of purified Giα2 was used, no high affinity agonist binding was detected.

In conclusion A1 receptors couple to Go, Giα1 and possibly to Giα3. Three lines of evidence suggest that bovine brain A1 adenosine receptors do not couple to Giα2: i) Pertussis toxin does not label a 40 kDa polypeptide in purified receptor-G protein preparations. ii) Antiserum against Giα2 does not cross-react with the purified fractions. iii) Purified Giα2 does not reconstitute high affinity agonist binding of an NEM-pretreated receptor preparation.

We are grateful to Drs. Susanne Mumby and Alfred G. Gilman for their gift of antisera.

F8.   A PERMISSIVE ROLE OF PERTUSSIS TOXIN-SENSITIVE G-PROTEIN IN ENHANCEMENT OF P2-PURINERGIC OR α1-ADRENERGIC RECEPTOR-LINKED PHOSPHOLIPASE C SYSTEM BY P1-PURINERGIC AGONISTS. F.OKAJIMA, K.SATO, M.NAZAREA, K.SHO, Y.KONDO.   Department of Physical Biochemistry, Institute of Endocrinology, Gunma University, Maebashi, 371 JJapan.

Extracellular ATP and other purinergic agonists were found to inhibit cAMP accumulation by depressing adenylate cyclase as an "inhibitory action" and/or to stimulate arachidonate release in association with phospholipase C or A2 activation and Ca2+ mobilization as "stimulatory actions" in FRTL-5 cells. These purinergic agonists could be divided into three goups based on a difference of sensitivities of their responses to islet-activating protein (IAP), pertussis toxin and adenosine (P1-receptor) antagonists.   The "inhibitory action" of group 1 agonists represented by ATP was completely inhibited by the pretreatment of the cells with IAP, whereas the "stimulatory actions" were partially inhibited by the toxin, even when an about 41 kDa membrane protein(s) was completely ADP-ribosylated.   Only the IAP-sensitive part of the "stimulatory actions" was antagonized by 1,3-diethyl-8-phenylxanthine (DPX), an adenosine antagonists.   Group 2 agonists represented by GTP and 8-bromoadenosine 5'-triphosphate (Br-ATP)also exerted the "stimulatory actions" at 2 to 3 orders higher concentrations than ATP, although they were entirely insensitive to both IAP and DPX.   Group 3 agonists including adenosine 5'-O-(1-thiotriphosphate)(ATPα S), adenosine 5'-O-(2-thiodiphosphate) (ADPßS) and P1-agonists such as AMP and N6-(L-2-phenylisopropyl-adenosine (PIA) induced "inhibitory action", but did not show any "stimulatory actions".   Nevertheless, these group 3 agonists remarkably enhanded the "stimulatory actions" of group 2 agonists such as GTP and Br-ATP. Such permissive actions of group 3 agonists were sensitive to both IAP and DPX, as were shown for a part of the "stimulatory actions" of group 1 agonists as well as the "inhibitory actions" of both group 1 and 3 agonists.   FRTL-5 cells also responded to norepinephrine, an α1-adrenergic agonist resulted in the induction of "stimulatory actions" but not "inhibitory action" just like group 2 agonists. The "stimulatory actions" as induced by norepinephrine were also enhanced by PIA, a group 3 agonist in an IAP-sensitive manner. We conclude that an IAP substrate G-protein(s) which mediates the "inhibitory action" of purinergic agonists via a DPX-sensitive P1-type of purinergic receptor(s) may not directly link to the phospholipase C or A2 system, but enhance the system which links to a DPX-insensitive P2-receptor or α1-receptor, in an indirect or permissive manner.

F13.   INVOLVEMENT OF ADENOSINE IN LOW MAGNESIUM POTENTIATION OF HIPPOCAMPAL POTENTIALS. J. H. CONNICK, J. T. BARTRUP, J. M. BULLOCH & T. W. STONE, Dept. of Pharmacology, University of Glasgow, Scotland, UK.

Recently it has been reported that an increased excitability of hippocampal neurones produced by perfusion with low magnesium media is independent of the activation of NMDA receptors. Since we have described the magnesium dependency of the presynaptic inhibitory effects of adenosine on synaptic activity in the hippocampal slice, the present project was designed to determine whether there was a relationship between these two phenomena.
Slices were cut 500 m thick from the dorsal hippocampus of male rats and preincubated in an artificial csf at room temperature 20-24oC for at least one hour before use. Individual slices were transferred to a 0.5 ml perfusion chamber, through which medium was pumped at a rate of 3 ml per minute at 30oC. Recordings from the CA1 pyramidal cell layer were made with single glass microelectrodes. Population potentials were amplified and displayed on storage oscilloscopes, and were then plotted immediately onto a chart recorder.
50 μM adenosine reduced the orthodromic CA1 population spike to 15 ± 3.8% of control (n = 8). The concentration of adenosine producing a 50% reduction of population spike size was 34.5 ± 6.2   μM (n = 10). When the concentration of magnesium ions in the perfusing medium was reduced from the initial 2 mM to 1 mM or 0.6 mM an increase in the potential size was observed of 12.6 ± 2.8% of control (n = 8) or 15.3 ± 3.8% control (n = 10) respectively. The inclusion of AP5 had no effect on the increase of potential size.

Perfusing the slices with 8-phenyltheophylline at a concentration of 100 nM blocked the effects of added adenosine on the population spikes. The addition of xanthine resulted in an increase of potential size of approximately the same extent as that seen in 0.5 mM magnesium. When the bathing solution was changed from one containing 8-phenyltheophylline and 2 mM magnesium, to one containing xanthine and 0.5 mM magnesium, no further increment of potential size was observed. The same result was obtained by using the destructive enzyme adenosine deaminase.

The finding that 8-phenyltheophylline or adenosine deaminase, both of which should effectively neutralise the effects of endogenous adenosine on the evoked potentials, cause an increase of potential size comparable with that seen on lowering magnesium concentrations, argues strongly that the low magnesium effect is related to the presence of adenosine. This conclusion is supported by the fact that combinations of low magnesium and xanthine or deaminase solutions produced no further increments of potential size.

F16. **45Ca UPTAKE BY SYNAPTOSOMES STIMULATED ELECTRICALLY IN THE PRESENCE OF ADENOSINE.** M.L. GONCALVES, F. PINTO, J. A. RIBEIRO, Lab. Pharmacology, Gulbenkian Institute of Science, Oeiras, Portugal.

Adenosine partially inhibits 45Ca uptake by rat brain synaptosomes stimulated by potassium (e.g. Ribeiro et al., 1979; Wu et al.,1982),and totally blocks Ca uptake by guinea-pig myenteric plexus synaptosomes stimulated electrically (Shinozuka et al., 1985).

The present work was undertaken to compare the effect of adenosine on Ca uptake induced by both potassium and electrical stimulation in rat brain synaptosomes.

The experiments were carried out on synaptosomes prepared from rat brain homogenates incubated at 37 C in a medium (pH 7.4) containing (mM): NaCl 132; KCl 5; MgCl2 1.3; NaH2PO4 1.2; CaCl2 1.2; glucose 10. Calcium uptake was induced by means of electrical stimulation applied through platinum electrodes. Rectangular pulses of 20 V amplitude and 400 us duration with a frequency of 10 pulses/s were applied during 2 min. In some experiments the uptake of Calcium was induced by high potassium (60 mM). The uptake was terminated by adding a stop buffer containing 5 mM EGTA. This was followed by filtration through GF/C glass fiber filters. The radioactivity associated with the synaptosomes was determined in a scintillation spectrometer.

Electrical stimulation increased 45Calcium uptake by synaptosomes, an effect that reached a maximum 2 min after initiation of the stimulation. Adenosine tested in concentrations between 1 nM and 100 uM decreased the uptake of Calcium induced by electrical stimulation. This effect of adenosine was concentration-dependent for concentrations up to 1 uM. For higher adenosine concentrations (10 uM and 100 uM) the reduction in Calcium uptake was smaller. At a concentration of 1 uM adenosine almost abolished the uptake of Calcium induced by electrical stimulation (91.6 $\pm$ 3.3% of reduction, n=9). In the presence of high potassium the uptake of 45Calcium by synaptosomes was greatly increased and adenosine at a concentration of 1 uM reduced the uptake by 43.9 $\pm$ 6.1% (n=4). The inhibitory effect of adenosine (1 uM) on Calcium uptake by synaptosomes electrically stimulated was antagonized by 5 uM of 1,3-dipropyl-8-p-sulfophenylxanthine.

These results suggest that the calcium channels opened by depolarization of the presynaptic membrane during electrical stimulation are more sensitive to adenosine than those opened by potassium depolarization. The effect of adenosine results from activation of a xanthine-sensitive adenosine receptor.

The authors acknowledge the technical assistance of Mr. Artur Costa.

RIBEIRO, J.A., SA-ALMEIDA, A.M. and NAMORADO, J.M. (1979) Biochem. Pharmacol. 28, 1297-1300.
SHINOZUKA, K., MAEDA, T. and HAYASHI (1985) Eur. J. Pharmacol. 113, 417-424.
WU,P.H.,PHILLIS,J.W.,THIERRY,D.L.(1982) J.Neurochem. 39, 700-708.

F17. **THE EFFECTS OF ADENOSINE ON INTRACELLULAR CALCIUM IN PLATELETS ARE LINKED TO ADENYLATE CYCLASE ACTIVATION.** SUBIR PAUL, IGOR FEOKTISTOV, ALAN S. HOLLISTER, DAVID ROBERTSON AND ITALO

BIAGGIONI, Departments of Medicine and Pharmacology, Vanderbilt University, Nashville, TN and Cardiology Institute, Tomsk, USSR.

Platelet aggregation and secretion are associated with a rise in intracellular calcium concentration ([Ca2+]i). Adenosine has been postulated as an endogenous inhibitor of platelet aggregation. The antiaggregatory effects of adenosine are related to activation of adenylate cyclase. We now studied the effect of adenosine on the rise in [Ca2+]i and platelet aggregation produced by thrombin in human platelets. [Ca2+]i was determined by Fura-2. Adenosine inhibited the slope of the first phase of aggregation and the rise in [Ca2+]i produced by thrombin in a dose-dependent manner. The dose that produced 50 % inhibition of both aggregation and the rise in [Ca2+]i was approximately 100 nM. The inhibitory effects of adenosine on [Ca2+]i were shared by its stable analogs, NECA being approximately 10-fold more potent that L-PIA, suggesting that these effects were mediated through adenosine Ra receptors. Furthermore, caffeine antagonized the effects of adenosine on platelet aggregation and [Ca2+]i. The effects of adenosine on [Ca2+]i appear to be mediated through a rise in intracellular cAMP, since they were prevented by the adenylate cyclase inhibitor 2',5'-dideoxyadenosine 1 mM, and were potentiated by phosphodiesterase inhibition with papaverine 1 μM. Adenosine also inhibits the rise in [Ca2+]i produced by thrombin in a calcium-free medium, suggesting that adenosine inhibits both calcium influx as well as the release of calcium from intracellular stores.

F19. **ADENOSINE STIMULATES THE SODIUM-POTASSIUM (Na-Pi) SYMPORTER AND LOWERS CYCLIC AMP IN CULTURED EPITHELIAL CELLS FROM PROXIMAL TUBULE OF OPOSSUM KIDNEY (OK).** R. COULSON,1 S.J. SCHEINMAN.2 Dept. of Veterana Affairs, Medical Center, Bay Pines FL 33504 1 and Departments of Internal Medicinee, USF, Tampa 1 and SUNY-HSC, Syracuse, NY.2.

We have shown tthat DPX, an adenosinen receptor antagonist, inhibits the Na-Pi symporter and elevates cAMP on OK cells (Coulson and Scheinman, JPET 248:589, 1989). We hypothesize that endogenous adenosine stimulates this symporter via an A1 receptor coupled to inhibition of cyclase. To test this hypothesis we have treated OK cells with PIA, an A1 agonist, and measured: Na-dependent uptake of 32-Pi , adenylate cyclase and cAMP [basal and in the presence of bovine 1-34 parathyroid hormone (PTH)]. Results: PIA (1 uM) significantly stimulated basal (+31%) and PTH (10 pM)-inhibited (+26%) uptake of Pi into OK cells. Feeding the cells 1 mM caffeine for 5-7d significantly elevated basal (+57%) and PTH-inhibited (+40%) Pi uptake. Caffeine-feeding did not augment PIA's effect on Pi uptake. PIA at 1 uM did not lower basla or PTH-elevated cAMP unless the cells were fed caffeine and cAMP-phosphodiesterasee was inhibited with Ro 20-1724. Under these conditions, PIA lowered basal and PTH-elevated cAMP to the same extent (-35%). 0.1 mM DDA, as P site agonist, was more potent than 1 uM PIA as an inhibitor of adenylate cyclase but had no effect on the uptake of Pi. Conclusions: These data support a role for adenosine as a stimulator of Na-dependent Pi tptake into proximal epithelial cells. Caffeine feeding may stimulate Pi uptake and lower cAMP level by up-regulating an A1 receptor coupled to adenylate cyclase. Adenosine's stimulation of Pi uptake may involve second messengers in addition to cAMP since inhibition of adenylate cyclase via the P-site did not affect Pi uptake.

F20. **EFFECTS OF INTRACAUDATE PERFUSIONS OF R(-)N6-(2-PHENYL-ISOPROPYL)ADENOSINE (R-PIA) AND 2-PHENYLAMINOADENOSINE (CV-1808) ON EXTRACELLULAR CAUDATE DOPAMINE (DA) AS MEASURED BY MICRODIALYSIS.** M. E. MORGAN1,2, R. E. VESTAL1,2,3, 1Clinical Pharmacology and Gerontology Research Units, VA Medical Center, Boise, ID; 2Dept. of Pharmaceutical Sciences, Idaho State Univ., College of Pharmacy, Pocatello, ID; 3Depts. of Medicine and Pharmacology, Univ. of Washington, School of Medicine, Seattle, WA.

Adenosine (ADO) modulates central nervous system function. Present information suggests that ADO effects are due to an interaction with specific ADO receptors. It is assumed that ADO A1 receptors are inhibitory and ADO A2 receptors are excitatory. In general, studies have shown that synthesis and release of caudate DA are attenuated by ADO or metabolically stable

ADO analogues. These experiments describe the effects of intracaudate perfusion and systemic administration of R-PIA, an ADO A1 specific analogue, and CV-1808, an ADO A2 specific analogue, on caudate dopaminergic function in awake behaving animals.

Male Sprague-Dawley rats weighing 250-350 gm were anesthetized with chloral hydrate (150 mg/kg, i.p.) and ketamine (50 mg/kg, i.p.), a guide cannula was implanted into left caudate 4.0 mm below the dural surface and a probe holder was set in place. The apparatus was held in place with cranioplastic cement. The guide cannula was covered and plugged with a stylus. Animals were allowed to recover from surgery for 3-5 days. On the day of the experiment, the stylus was removed and replaced with a microdialysis probe (2 mm dialysis tip). Microdialysis was used to deliver simultaneously intracaudate doses of ADO analogues and collect extracellular fluid (ECF) samples of DA, 3,4-dihydroxyphenylacetic acid (DOPAC) and homovanillic acid (HVA). R-PIA was dissolved in acidified Dulbecco's phosphate buffer containing 1.2 mM CaCl2. CV-1808 was dissolved in 0.05% DMSO in acidified Dulbecco's phosphate buffer containing 1.2 mM CaCl2. Drug vehicle for baseline and control perfusions were the appropriate Dulbecco's solutions for each analogue. The final pH of all solutions was 6.00 and was stable throughout the experiment. ADO analogue concentrations were determined by HPLC-UV. Actual intracaudate ADO analogue dose, which varies from probe to probe, was determined by subtracting the input concentration from the output concentration and is reported as mean ± S.E.M. Probes were perfused at a rate of 2.5 %l/min with drug vehicle or ADO analogue. Dialysates were collected every 20 min. The perfusion schedule was as follows: drug vehicle perfused until baseline ECF DA levels stabilized, ADO analogue or drug vehicle perfused for 20 min, then drug vehicle perfused for an additional 120 min. Preliminary experiments indicated that a 40 min drug washout yielded a drug free dialysate. Caudate ECF concentrations of DA, DOPAC and HVA were determined by HPLC-EC.

Dialysate percent recoveries determined in vitro at 37°C were as follows: DA, 12.1 ± 1.3; DOPAC, 16.1 ± 2.2; and HVA, 14.7 ± 1.5 (mean ± S.D., n = 9). ADO analogue in dialysate did not interfere with the detection of DA or its metabolites. A 20 min intracaudate perfusion of R-PIA (5.0 ± 0.7 nmoles, n = 6) increased caudate DA ECF levels to 230% of control. During the drug washout period, DA levels decreased to 60 and 70% of control (40 and 120 min, respectively). In comparison, a 20 min intracaudate perfusion of CV-1808 (8.6 ± 0.7 nmoles, n = 5) decreased caudate DA ECF levels to 53% of control. During drug washout, DA levels decreased to 10 and 61% of control (20 and 120 min, respectively). DOPAC and HVA levels were not significantly altered. In a different group of rats, R-PIA (3 mg/kg, i.p.) did not significantly effect caudate DA, DOPAC or HVA ECF levels.

These are the first data to show reciprocal purinergic regulation of caudate dopaminergic function. ADO receptors have been found on corticostriatal and intrinsic caudate neurons. They have not been localized on dopaminergic terminals. For this reason, the effects observed maybe mediated by caudate interneurons and/or a striatonigral negative neuronal feedback pathway.

F22. ACTIONS OF ADENOSINE ON THE DISPOSITION OF [3H](-)-NORADRENALINE IN ISOLATED RAT ATRIA. K. KURAHASHI*, D.M. PATON**, *Department of Pharmacology, University of Kyoto, Kyoto, Japan and **Department of Oral Biology, University of Alberta, Edmonton, Alberta, Canada.

Adenosine produces inhibition of neurotransmission in peripheral adrenergic neurons through the activation of presynaptic A1 receptors. The objectives of this research were to determine the actions of adenosine on the disposition of noradrenaline in peripheral adrenergic neurons. For this purpose isolated rat atria were used and labelled with 0.3 x 10-6M [3H]noradrenaline ([3H]NA) for 60 min unless otherwise indicated. [3H]NA was separated from its [3H] metabolites by alumina and Dowex column chromatography.

Adenosine (1 x 10-5M) had no effect on the neuronal transport of [3H]NA as it did not impair its accumulation over 1 or 10 min, nor did it alter the ability of tyramine (3 x 10-6M) to release [3H]NA. It also did not alter the permeability of the neuronal plasma membrane as it did not modify the accumulation of [3H] metabolites in the tissue, or the passive efflux of [3H]NA from neurons when the neuronal transporter was inhibited by cocaine and the metabolism of [3H]NA was inhibited by pargyline.

Adenosine (1 x 10-5M) did not alter the release of [3H]NA by tyramine (3 x 10-6M), an indirect-acting sympathomimetic amine causing the displacement of NA from intraneuronal stores. It did, however, significantly reduce the release of [3H]NA by potassium chloride (30 or 70 mM). Potassium-induced release of NA is, under these conditions, calcium-dependent and exocytotic in nature.

The deamination of [3H]NA was not altered by adenosine (1 x 10-5M) during the uptake of [3H]NA or during release induced by tyramine. Exposure of atria to potassium (70 mM) not only caused a ten-fold increase in the efflux of [3H]NA but also caused a 40% reduction in the efflux of [3H]DOPEG, the principal deaminated metabolite of [3H]NA. Adenosine (1 x 10-5M) caused an approximate 50% reduction in the release of [3H]NA by potassium but did not alter the efflux of [3H]DOPEG, but the addition of adenosine caused a significant increase in efflux of the deaminated metabolite.

It was concluded that adenosine does not significantly modify the passive permeability of the plasma membrane of adrenergic neurons, the neuronal transport of NA or the displacement of NA from intraneuronal stores. It does, however, significantly impair the calcium-dependent exocytotic release of the transmitter. While adenosine does not inhibit the deamination of NA by MAO, it does under certain circumstances modify the extent of deamination of NA. This may, in turn, reflect the effect of potassium (30 mM) on the neuronal transport of [3H]NA and hence its susceptibility to deamination by MAO.

Supported by grants from the Medical Research Council of New Zealand and the Central Research Fund of the University of Alberta.

F24. ADENOSINE RECEPTOR COUPLING TO ADENYLYL CYCLASE OF VENTRICULAR MYOCYTE MEMBRANES. DOBSON, J.G., JR., ROMANO, F.D., FENTON, R.A., RODIN, S.M., GEORGE, E.E, Department of Physiology, University of Massachusetts Medical School, Worcester, MA 01655, U.S.A.

Adenosine is thought to serve as a negative feedback modulator of -adrenergic catecholamine induced metabolic and mechanical responses in the ventricular myocardium. This antiadrenergic effect of adenosine is due to an attenuation of β-adrenergic stimulated myocardial membrane adenylyl cyclase activity. The action is mediated by adenosine A1 receptors on the postsynaptic membrane. The present study was undertaken to investigate adenosine receptor coupling to adenylyl cyclase of ventricular myocyte membranes. Phenylisopropyladenosine, an A1 receptor agonist, at 1 μM inhibited by 50% isoproterenol-stimulated adenylyl cyclase activity. This inhibition by phenylisopropyladenosine attenuated equipotent forskolin-stimulated adenylyl cyclase activity by only 25%. Phenylisopropyladenosine had no effect on adenylyl cyclase stimulation by 5'-guanylyl imidodiphosphate. Phenylaminoadenosine, an A2 agonist, at >10 μM stimulated adenylyl cyclase activity. This effect was not significantly inhibited by theophylline or 0.1 μM 1,3-dipropyl-8-cyclopentylxanthine (DPCPX). The latter is more A1 specific and antagonized the phenylisopropyladenosine inhibition of isoproterenol-stimulated adenylyl cyclase activity. N-ethylcarboxamidoadenosine, a non-selective adenosine receptor agonist at 10 μM, had no effect on adenylyl cyclase activity in the absence of DPCPX, but markedly stimulated adenylyl cyclase in the presence of DPCPX. These results indicate that both A1 and A2 receptors exist on the ventricular myocyte sarcolemma. More importantly these findings suggest that adenosine inhibition of catecholamine-stimulated adenylyl cyclase activity is due primarily to an alteration in -adrenergic receptor-mediated transduction, although a direct inhibition of the catalytic component may also be involved.

F25. ADENOSINE RECEPTORS IN CULTURED ATRIAL MYOCYTES: COUPLING TO MODULATION OF CONTRACTILITY AND ADENYLATE CYCLASE ACTIVITY AND IDENTIFICATION BY RADIOLIGAND BINDING. BRUCE T. LIANG. Cardiovascular Div., Brigham and Women's Hospital and Harvard Med. School. 75 Francis St., Boston, MA 02115.

Adenosine receptors in a spontaneously contracting atrial myocyte culture from 14 day chick embryos were characterized by radioligand binding studies and by examining the involvement of G-protein in coupling these receptors to a high-affinity state and to the adenylate cyclase and the myocyte contractility. Binding of the antagonist radioligand [3H] CPX was rapid, reversible, and saturable and was to a homogeneous population of sites with a Kd value of 2.1 ±

0.2 nM and an apparent Bmax of 26.2 $\pm$ 3 f mole/mg protein (n=10, $\pm$ S.E.). Guanyl-5-yl(b,g-imido) diphosphate, Gpp (NH)p, had no effect on either the Kd or the Bmax and CPX reversed the N6-R-phenyl-2-propyladenosine (R-PIA)-induced inhibition of adenylate cyclase activity and contractility, indicating that [3H] CPX is an antagonist radioligand. Competition curves for [3H] CPX binding by a series of reference adenosine agonists were consistent with labeling of an A1 adenosine receptor and were better fit by a two-site model than by a one-site model. ADP-ribosylation of the G-protein by the endogeneous NAD+ in the presence of pertussis toxin shifted the competition curves from bi to monophasic with Ki values similar to those of the KL observed in the absence of prior pertussis intoxication. The adenosine agonists were capable of inhibiting both the adenylate cyclase activity and myocyte contractility in either the absence or the present of isoproterenol. The A1 adenosine receptor-selective antagonist CPX reversed these agonist effects. The order of ability of the reference adenosine receptor agonists in causing these inhibitory effects was similar to the order of potency of the same agonists in inhibiting the specific [3H] CPX binding (R-PIA > S-PIA or NECA). These data indicate that the adenosine receptor coupled to inhibition of adenylate cyclase activity and to the negative inotropic effect is the A1 subtype. Pertussis treatment, which prevented formation of the high-affinity state of the receptor, also uncoupled the adenosine receptor from inhibition of the adenylate cyclase activity and the negative inotropic effect. Taken together, the present study indicates that adenosine receptors of the A1 subtype are present on the spontaneously contracting atrial myoctyes and are negatively coupled to adenylate cyclase and to the contractile state. The cultured embryonic chick atrial myocyte preparation represents a useful model system for characterizing the cardiac A1 adenosine receptor.

F26. INTERACTION OF THE CARDIAC INOTROPIC DRUGS AMRINONE, MILRINONE AND PIMOBENDAN WITH ENDOGENOUS ADENOSINE. JANINE M. MANUEL*, DAVID M. PATON**, *Department of Pharmacology and Clinical Pharmacology, University of Auckland, New Zealand and **Department of Oral Biology, University of Alberta, Edmonton, Alberta, Canada.

Endogenous adenosine has been shown to modulate the positiveinotropic action of isoprenaline, and there are also reports of endogenous adenosine being involved in the cardiac actions of adenosine. The purpose of the present study was to examine the possible role of endogenous adenosine in the positive inotropic actions of pimobendan and the bipyridine derivatives, amrinone and milrinone, and the site of origin of adenosine so involved.

In these studies contractile responses of isolated guinea-pig atria were recorded isometrically at 30oC. Atria were driven at 5 Hz (pulse duration 2 msec, supramaximal voltage) and allowed to equilibrate for 60 min. Pimobendan, amrinone and milrinone were dissolved in DMSO, and responses to these drugs were studied non-cumulatively.

Pimobendan (0.3-1.2 x 10-5M), amrinone (10-6-10-4M) and milrinone (10-6-10-4M) all produced concentration-dependent positive inotropic responses. These responses were significantly inhibited by the adenosine receptor antagonist, 8-phenyltheopylline (3x10-5M), the response to 6 x 10-6M pimobendan being reduced from 38+/-7% (mean+/-SEM) to 17+/-3%. The addition of exogenous adenosine deaminase (1 unit/ml) also significantly reduced their positive inotropic actions, the response to milrinone (10-5M) being reduced from 44+/-22% to 22+/-3%. The inhibitor of adenosine transport, dipyridamole (10-6M) significantly impaired their actions, the response to milrinone (10-5M) being reduced from 42+/-4% to 19+/-5%.

The finding that responses to the inotropic drugs were reduced by 8-phenyltheophylline and by adenosine deaminase suggests that a portion of their positive inotropic actions results from antagonism of the actions of endogenous adenosine. Adenosine itself has negative inotropic and chronotropic actions on atria through activation of A1 receptors.

The finding that dipyridamole reduced their actions suggests that the endogenous adenosine interacting with the inotropic drugs originated intracellularly and utilized the adenosine transporter to reach the extracellular space.

Supported by the Medical Research Council of New Zealand and the Central Research Fund of the University of Alberta.

F27. IRREVERSIBLE INACTIVATION OF ADENYLATE CYCLASE BY THE "P"-SITE AGONIST 2',5'DIDEOXY-,3'-p-FLUOROSULFONYLBENZOYL-ADENOSINE.

SIU-MEI HELENA YEUNG AND ROGER A. JOHNSON, Department of Physiology and Biophysics, School of Medicine, Health Sciences Center, State University of New York at Stony Brook, Stony Brook, New York 11794-8661, U.S.A.

2',5'-Dideoxy,3'-p-fluorosulfonylbenzoyl-adenosine (2'5'dd3'FSBA) was synthesized and found to be an affinity label for the "P"-site of adenylate cyclase. This compound inactivated a crude detergent-dispersed adenylate cyclase from rat brain and the partially purified enzyme from bovine brain with apparent IC50s of ~10-30 uM and ~100-150 uM, respectively. Both crude and purified preparations were inhibited irreversibly; only ten percent of the initial activity of the crude adenylate cyclase and 50% of the activity of the purified enzyme were recovered after extensive dialysis of enzyme incubated with 0.1 mM 2'5'dd3'FSBA. The irreversible inactivation of adenylate cyclase by 2'5'dd3'FSBA was completely blocked by 0.1 mM 2'5'dideoxyadenosine, a potent "P"-site agonist. Since the incubation of adenylate cyclase with 2'5'dd3'FSBA was done in the presence of 0.1 mM ATP, it is unlikely that inactivation by 2'5'dd3'FSBA was due to an effect on the catalytic site of the enzyme. Dithiothreitol protected the enzyme from irreversible inactivation by 2'5'dd3'FSBA, but reversible inhibition of the enzyme was still observed, though with reduced potency. When 2 mM dithiothreitol was added after 30 min preincubation with 2'5'dd3'FSBA, the rat brain enzyme was partially (~80%) reactivated. The data suggest that 2'5'dd3'FSBA may irreversibly inactivate adenylate cyclase by reacting with a cysteinyl-moiety in proximity to the "P"-site domain of the enzyme. Thus, 2'5'dd3'FSBA, and especially its radiolabeled analog, should prove to be a useful probe for structural studies of adenylate cyclase, particularly with regard to the "P"-site.

G2. INHIBITION OF INTERLEUKIN-2 AND PHORBOL ESTER DEPENDENT PROLIFERATION OF A MURINE T-CELL LINE BY cAMP ANALOGUES AND ISOBUTYLMETHYLXANTHINE.

Y. S. YIM, R. HURWITZ, Division of Hematology-Oncology, Dept. of Pediatrics and Cell Biology, Baylor College of Medicine, One Baylor Plaza, Houston, TX, 77030.

Investigators have previously demonstrated that cAMP analogs and phosphodiesterase inhibitors prevent lectin-stimulated proliferation of peripheral blood mononuclear cells. Both cAMP and cGMP phosphodiesterase activities increase in these mitogen-stimulated cells. We have demonstrated that a 63 kDa CaM-dependent phosphodiesterase is induced during T-lymphocyte mitogenesis. This enzyme has immunologic and chemical properties similar to those of the 63 kDa CaM-dependent phosphodiesterase isolated from brain. We therefore decided to explore the effects of cyclic nucleotides and of a phosphodiesterase inhibitor on the proliferation of an IL-2-dependent murine T-cell line (HT-2). This cell line was grown in RPMI 1640 containing IL-2 (10 units/100 ml) and fetal calf serum (10%). PMA (100 nM) in place of IL-2 also supports cell growth. Cells were washed and incubated in media without IL-2 for 4 hours prior to addition of mitogens. Twenty two hours after the initiation of the culture, IL-2-dependent proliferation was measured by pulsing each culture (5 x 103 cells in 200 %l media) with 1 %Ci [3H]-thymidine for 4 hours. IL-2-dependent proliferation of HT-2 cells could be inhibited by dibutyryl-cAMP (IC50 ~25 %M), 8-Br-cAMP (IC50 ~25 %M) or IBMX (IC50 ~10 %M) when the inhibitor was added at the same time as the mitogen. PMA-dependent proliferation could be inhibited by dibutyryl-cAMP (IC50 ~35 %M), or IBMX (IC50 ~10 %M) under similar incubation conditions. cGMP analogs had no effect on proliferation induced by either IL-2 or PMA. Cells remained >90% viable (as determined by trypan blue exclusion) throughout all experiments. In order to determine if the inhibition of cell proliferation can be correlated with the time of addition of the inhibitor, dibutyryl cAMP (250 %M) was added to the cultures at various time points following the addition of either IL-2 or PMA. The ability of this analog to inhibit cell proliferation decreased as the time following the initiation of the culture increased. When the analog was added with the mitogen, the proliferation of the cells was inhibited by ~90%. When the analog was added 16 hours after mitogen, the proliferation of the cells was inhibited by only 50% and when added at 22 hours, cell proliferation was unaffected when the cells were harvested at 24 hours. However, cAMP analogs continued to inhibit proliferation when the cells were monitored at 48 hours. These data suggest that cAMP may affect more than one point of the cell cycle during proliferation and that appropriately designed phosphodiesterase inhibitors may have therapeutic potential for mediating T-cell immune responses.

G4. CYCLIC NUCLEOTIDE PHOSPHODIESTERASES FROM BOVINE AORTIC ENDOTHELIAL CELLS AND SMOOTH MUSCLE: CHARACTERIZATION AND DRUG SENSITIVITIES.

C. LUGNIER V. SCHINI* Laboratoire de Pharmacologie Cellulaire et Moléculaire, CNRS URA 600, Université Louis Pasteur de Strasbourg, B.P. 24, 67401 Illkirch Cedex, France.*Department of Physiology and Biophysics, Mayo Clinic and Mayo Foundation,Rochester,MN 55905, USA.

Vascular endothelial cells play an important role in the regulation of smooth muscle tone and might be implicated in cardiovascular diseases. Stimulation of endothelial cells in culture by various vasoactive agonists (prostacyclin, bradykinin, atriopeptins and acetylcholine) increases either cyclic AMP (cAMP) or their cyclic GMP (cGMP) contents, and is associated with a release of cGMP in the extracellular medium. Cyclic nucleotide phosphodiesterase (PDE) which play a major role in controlling cAMP and cGMP levels have been isolated from various tissues, including smooth muscle cells, but the PDEs of the vascular endothelium had not been studied until now. In the present work, the PDEs were isolated from cultured bovine aortic endothelial cells and from bovine aortic smooth muscle, using identical techniques (HPLC) and their drug sensitivities were determined.

In smooth muscle, cGMP hydrolytic activity (determined in presence of calmodulin) represented more than 70% of the total cyclic nucleotide hydrolytic activity. In opposite, cAMP hydrolytic activity in endothelial cells represented more than 70% of the total hydrolytic activity. Only two PDE forms were found in endothelial cells: one form, cGS-PDE, hydrolyzed both cAMP and cGMP and was activated by cGMP , the other, cAMP-PDE, hydrolyzed specifically cAMP and was not sensitive to cGMP. The latter form was also present in vascular smooth muscle which otherwise contained , a cGMP-specific PDE, (cGMP-PDE) and a calmodulin-activated PDE which preferentially hydrolyzed cGMP (CaM-PDE). Endothelial cAMP-PDE was inhibited similarly to smooth muscle cAMP-PDE with decreasing potencies by trequinsin > rolipram > dipyridamole > papaverine > IBMX > AAL 05 > zaprinast. Endothelial cGS-PDE was, similarly to cardiac cGS-PDE, relatively insensitive to zaprinast which specifically inhibited smooth muscle cGMP-PDE and was inhibited by dipyridamole. The ineffectiveness of zaprinast to inhibit the cGMP hydrolytic activity of cGS-PDE is in accordance the reported lack effect of zaprinast to modify cGMP level in endothelial cells.

The low cGMP hydrolytic activity observed in endothelial cells could be paralleled with the extracellular release from endothelial cells of cGMP induced by atriopeptin-II. The differences in PDE forms contained in endothelial and smooth muscle cells point out differences in the respective role of each cyclic nucleotide in the vascular tone and suggest further investigations.

Author Index

Subject Index